Foundations in Signal Processing, Communications and Networking

Volume 13

More information about this series at http://www.springer.com/series/7603

Rudolf Ahlswede

Combinatorial Methods and Models

Rudolf Ahlswede's Lectures on Information Theory 4

Edited by

Alexander Ahlswede
Ingo Althöfer
Christian Deppe
Ulrich Tamm

Springer

Author
Rudolf Ahlswede (1938–2010)
Department of Mathematics
University of Bielefeld
Bielefeld
Germany

Editors
Alexander Ahlswede
Bielefeld
Germany

Ingo Althöfer
Faculty of Mathematics and Computer Science
Friedrich-Schiller-University Jena
Jena
Germany

Christian Deppe
Department of Mathematics
University of Bielefeld
Bielefeld
Germany

Ulrich Tamm
Faculty of Business and Health
Bielefeld University of Applied Sciences
Bielefeld
Germany

ISSN 1863-8538 ISSN 1863-8546 (electronic)
Foundations in Signal Processing, Communications and Networking
ISBN 978-3-319-53137-3 ISBN 978-3-319-53139-7 (eBook)
DOI 10.1007/978-3-319-53139-7

Library of Congress Control Number: 2017936898

Mathematics Subject Classification (2010): 94-XX, 94BXX

Printed on acid-free paper

This Springer imprint is published by Springer Nature
The registered company is Springer International Publishing AG
The registered company address is: Gewerbestrasse 11, 6330 Cham, Switzerland

As long as algebra and geometry proceed along separate paths, their advance was slow and their applications limited. But when these sciences joined company, they drew from each other fresh vitality and hence forward marched on at a rapid pace towards perfection.

Joseph Louis Langrange

Preface[1]

After an introduction to classical information theory, we present now primarily own methods and models, which go considerably beyond it. They were also sketched in our Shannon Lecture 2006. There are two main components: our combinatorial approach to information theory in the late 1970s, where probabilistic source and channel models enter via the skeleton, a hypergraph based on typical sequences, and our theory of identification, which is now generalized to a general theory of information transfer (GTIT) incorporating also as ingredient a theory of common randomness, the main issue in cryptology. We begin with methods, at first with collections of basic covering, colouring, and packing lemmas with their proofs, which are based on counting or the probabilistic method of random choice.

Of course, these two methods are also closely related: the counting method can be viewed as the method of random choice for uniform probability distributions. It must be emphasized that there are cases where the probabilistic method fails, but the greedy algorithm (maximal coding) does not or both methods have to be used in combination. A striking example, Gallager's source coding problem, is discussed. Particularly useful is a special case of the Covering Lemma, called the link. It was used by Körner for zero-error problems, which are packing problems, in his solution of Rényi's problem. Very useful are also two methods, the elimination technique and the robustification technique, with applications for AV-theory and unidirectional memories.

Colouring and covering lemmas find also applications in many lectures on combinatorial models of information processing:

- Communication complexity,
- Interactive communication,
- Write-efficient Memories and
- ALOHA.

[1]This is the original preface written by Rudolf Ahlswede for the second 1000 pages of his lectures. This volume consists of the first third of these pages.

They are central in the theory of identification, especially in the quantum setting, in the theory of common randomness, and in the analysis of a complexity measure by Ahlswede, Khachatrian, Mauduit, and Sárkozy for number theoretical crypto-systems.

Bielefeld, Germany Rudolf Ahlswede

Words and Introduction of the Editors

Rudolf Ahlswede was one of the worldwide-accepted experts on information theory. Many key developments in this area are due to him. In particular, he made big progress in multi-user theory. Furthermore, with identification theory, he introduced a new research direction. Rudolf Ahlswede died in December 2010.

The fourth volume of Rudolf Ahlswede's lectures on information theory is focused on combinatorics. Rudolf Ahlswede's original motivation to study combinatorial aspects of information theory problems were zero-error codes: in this case, the structure of the coding problems usually drastically changes from probabilistic to combinatorial. The best example is Shannon's zero-error capacity where independent sets in graphs have to be examined. The extension to multiple access channels leads to the Zarankiewicz problem.

On his initiative, professorships for combinatorics and complexity theory were established in Bielefeld University. He made contacts to the leading institutes worldwide. In his own research, combinatorics became more and more important such that in the big special research unit "Discrete Structures in Mathematics" at Bielefeld University, Rudolf Ahlswede was the head of two projects on "Models with Information Exchange" and "Combinatorics of Sequence Spaces", respectively. Rudolf Ahlswede also became very renowned for his research on combinatorics: let us only mention that with Levon Khachtrian, he settled the famous $4m$ conjecture of Paul Erdös and that the well-known Ahlswede—Daykin inequality (also called Four Function Theorem (FFT)) even takes his name. Bollobas wrote in his book "Combinatorics" about that result:

> *At the first glance the FFT looks too general to be true and, if true, it seems too vague to be of much use. In fact, exactly the opposite is true: the Four Functions Theorem (FFT) of Ahlswede and Daykin is a theorem from "the book". It is beautifully simple and goes to the heart of the matter. Having proved it, we can sit back and enjoy its power enabling us to deduce a wealth of interesting results. This is precisely the reason why this section is rather long: it would be foolish not to present a good selection of the results one can obtain with minimal effort from the FFT.*

The history of the idea of the AD-inequality is very interesting. As Daykin came to a visit to Bielefeld, Ahlswede was just wallpapering. He stood on the ladder, and Daykin wanted to tell him from a newly proven inequality. The declaration was complicated, and Ahlswede said that probably a more general (and easier) theorem should hold. He made directly—on the ladder—a proposal which already was the AD-inequality.

The lecture notes he selected for this volume concentrate on the deep interplay between coding theory and combinatorics. The lectures in Part I (Basic Combinatorial Methods for Information Theory) are based on Rudolf Ahlswede's own research and the methods and techniques he introduced.

A code can combinatorially be regarded as a hypergraph, and many coding theorems can be obtained by appropriate colourings or coverings of the underlying hypergraphs. Several such colouring and covering techniques and their applications are introduced in Chap. 1.

Chapter 2 deals with codes produced by permutations. Finally, in Chap. 3, applications of one of Rudolf Ahlswede's favourite research fields—extremal problems in combinatorics—are presented. In particular, he analysed Kraft's inequality for prefix codes as the LYM property in the poset imposed by a tree. This led to a generalization to arbitrary posets.

Rudolf Ahlswedes' results on diametric and intersection theorems were already included in the book Lectures on Advances in Combinatorics (with V. Blinovsky).

Whereas the first part concentrates on combinatorial methods in order to analyse classical codes as prefix codes or codes in the Hamming metric, the second part of this book is devoted to combinatorial models in information theory. Here, the code concept already relies on a rather combinatorial structure, as in several concrete models of multiple access channels (Chap. 4) or more refined distortions (Chap. 5). An analytical tool coming into play, especially during the analysis of perfect codes, are orthogonal polynomials (Chap. 6).

Finally, the editors would like to tell a little bit about the state of the art at this point. Rudolf Ahlswede's original plan was to publish his lecture notes containing in total a number of about 4000 pages in three very big volumes. With the publisher, he finally agreed to subdivide each volume in 3–4 smaller books. The first three books which appeared so far, indeed, were the first "big" volume on which Rudolf Ahlswede had concentrated most of his attention, so far, and which was almost completely prepared for publication by himself. Our editorial work with the first three volumes, hence, was mainly to take care of the labels and enumeration of the formulae, theorems, etc., and to correct some minor mistakes. Starting with this volume, the situation is a little different. Because of Rudolf Ahlswede's sudden death, his work here was not yet finished and some chapters were not completed. We decided to delete some sections with which we did not feel comfortable or which were just fragmentary.

Our thanks go to Regine Hollmann, Carsten Petersen, and Christian Wischmann for helping us typing, typesetting, and proofreading. Furthermore, our thanks go to

Bernhard Balkenhol who combined the first approx. 2000 pages of lecture scripts in different styles (AMS-TeX, LaTeX, etc.) to one big lecture script. Bernhard can be seen as one of the pioneers of Ahlswede's lecture notes.

Alexander Ahlswede
Ingo Althöfer
Christian Deppe
Ulrich Tamm

Contents

Part I
Combinatorial Methods for Information Theory

Chapter 1
Covering, Coloring, and Packing Hypergraphs

Definition 1.1 A hypergraph $\mathcal{H} = (\mathcal{V}, \mathcal{E})$ consists of a (finite) vertex set $\mathcal{V}$ and a set of hyper-edges $\mathcal{E}$, where each edge $E \in \mathcal{E}$ is a subset of $E \subset \mathcal{V}$.

The vertices will usually be labelled by $\mathcal{V} = (v_1, \dots, v_I)$, the edges by $\mathcal{E} = (E_1, \dots, E_J)$, where $I, J \in \mathbb{N}$ with $I = |\mathcal{V}|$ and $1 \leq J \leq 2^{|\mathcal{E}|}$.

The concept was introduced by Claude Berge with the additional assumption $\bigcup\limits_{E \in \mathcal{E}} E = \mathcal{V}$, which we dropped in [3, 4] for convenience. At that time many mathematicians saw no reason to have a new name for what was called a "set system", in particular in Combinatorics, and they felt that this fancy name smelled like standing for a general concept of little substance. They missed that by viewing the structure as generalizations of graphs many extensions of concepts, ideas, etc. from Graph Theory were suggested.

Also a useful property is the duality when looking at the incidence properties $v \in E$ or $v \notin E$. Keeping this structure one can interpret $\mathcal{E}$ as vertex set and $\mathcal{V}$ as edge set, where v equals the set $\{E \in \mathcal{E} : v \in E\}$, and thus get the dual hypergraph $\mathcal{H}^* = (\mathcal{E}, \mathcal{V})$.

One can also describe the hypergraph isomorphically as a bipartite graph with two vertex sets $\mathcal{V}$ and $\mathcal{E}$ and the given incidence structure as vertex-vertex adjacency. Another equivalent description is in terms of a 0-1-matrix with $|\mathcal{V}|$ rows and $|\mathcal{E}|$ columns.

We consider only finite $\mathcal{H}$, that is, $|\mathcal{V}| < \infty$.

Basic parameters of hypergraphs are

- $\deg(v) = |\{E \in \mathcal{E} : v \in E\}|$, the degree of v;
- $\deg(E) = |\{v \in \mathcal{V} : v \in E\}| = |E|$, the degree of E;
- $d_{\mathcal{V}} = \min\limits_{v \in \mathcal{V}} \deg(v)$, $D_{\mathcal{V}} = \max\limits_{v \in \mathcal{V}} \deg(v)$, $\overline{d}_{\mathcal{V}} = \sum_{v \in \mathcal{V}} \frac{|v|}{|\mathcal{V}|}$;
- $d_{\mathcal{E}} = \min\limits_{E \in \mathcal{E}} \deg(E)$, $D_{\mathcal{E}} = \max\limits_{E \in \mathcal{E}} \deg(E)$, $\overline{d}_{\mathcal{E}} = \sum_{E \in \mathcal{E}} \frac{|E|}{|\mathcal{E}|}$;

A. Ahlswede et al. (eds.), *Combinatorial Methods and Models*,
Foundations in Signal Processing, Communications and Networking 13,
DOI 10.1007/978-3-319-53139-7_1

For the analysis of complex source coding problems a new concept is important, which we introduce.

Definition 1.2 We call $\mathcal{H}_2 = (\mathcal{V}, \mathcal{E}, (\mathcal{E}_j)_{j=1}^J)$ with $\mathcal{V} = \{v_1, \dots, V_I\}$ and $\mathcal{E} = \{E_1, \dots, E_j\}$ a 2-hypergraph if for every j $\mathcal{E}_j = \{E_j^m : 1 \le m \le M_j\}$ is a family of subsets, called subedges, of E_j.

The study of coding problems for correlated sources motivated the following concept.

Definition 1.3 As usual, let $\mathcal{H} = (\mathcal{V}, \mathcal{E})$, $\mathcal{V} = \{v_1, \dots, v_I\}$, $\mathcal{E} = \{E_1, \dots, E_J\}$ be a hypergraph.

Additionally, we are given sub–probability distributions Q on the set of edges $\mathcal{E}$ and Q_E on every edge E, i.e., mappings $Q : \mathcal{E} \to \mathbb{R}_+$ and $Q_E : E \to \mathbb{R}_+$ such that

$$\sum_{j=1}^{J} Q(E_j) \le 1, \sum_{v \in E} Q_E(v) \le 1.$$

The quadruple $\big(\mathcal{V}, \mathcal{E}, Q, (Q_E)_{E \in \mathcal{E}}\big)$ is denoted as **weighted hypergraph**.

1.1 Covering Hypergraphs

Definition 1.4 A *covering* of a hypergraph: $\mathcal{H} = (\mathcal{V}, \mathcal{E})$ is a subset $\mathcal{C} \subset \mathcal{E}$ of the edges such that every vertex $v \in \mathcal{V}$ is contained in some edge $E \in \mathcal{C}$:

$$\bigcup_{E \in \mathcal{C}} = \mathcal{V}.$$

Lemma 1.1 (Covering) *For any hypergraph* $(\mathcal{V}, \mathcal{E})$ *with*

$$\min_{v \in \mathcal{V}} deg(v) \ge d$$

there exists a covering $\mathcal{C} \subset \mathcal{E}$ *with*

$$|\mathcal{C}| \le \left\lceil \frac{|\mathcal{E}|}{d} \log |\mathcal{V}| \right\rceil.$$

where $d = \min_{v \in \mathcal{V}} |\{E \in \mathcal{E} : v \in E\}|$.

It is a consequence of the next lemma.

As usual $P(\mathcal{E})$ is the set of all PD's on $\mathcal{E}$ and the indicator function $1_E : \mathcal{V} \to \{0, 1\}$ is defined by

$$1_E(v) = \begin{cases} 1 & v \in E \\ 0 & v \notin E \end{cases} \quad \text{for every} v \in \mathcal{V} (E \in \mathcal{E}) .$$

Lemma 1.2 (Covering) *There exists a covering* $\mathcal{C} \subset \mathcal{E}$ *of* $\mathcal{H}$ *with*

$$|\mathcal{C}| \leq \left\lceil \min_{P \in \mathcal{P}(\mathcal{E})} \left(\min_{v \in \mathcal{V}} \sum_{E \in \mathcal{E}} 1_E(v) P(E) \right)^{-1} \log |\mathcal{V}| \right\rceil .$$

Proof Select edges $E^{(1)}, \ldots, E^{(k)}$ independently according to some PD $P \in \mathcal{P}(\mathcal{E})$.

The probability that some vertex $v \in \mathcal{V}$ is not contained in one of these edges is

$$\Pr\left(v \notin \bigcup_{i=1}^{k} E^{(i)} \right) = \Pr(v \notin E^{(1)}) \cdot \cdots \cdot \Pr(v \notin E^{(k)})$$

$$= \left(1 - \sum_{E \in \mathcal{E}} 1_E(v) P(E) \right)^k \qquad (1.1.1)$$

$$\Pr\left(\exists v \in \mathcal{V} : v \notin \bigcup_{i=1}^{k} E^{(i)} \right) \leq |\mathcal{V}| \max_{v \in \mathcal{V}} \left(1 - \sum_{E \in \mathcal{E}} 1_E(v) \mathcal{P}(E) \right)^k$$

If the RHS < 1, the *probability* for the existence of a covering is *positive* and therefore a covering exists.

Taking the logarithm on both sides we obtain as condition for the size of the covering

$$\max_{v \in \mathcal{V}} k \log\left(1 - \sum_{E \in \mathcal{E}} 1_E(v) P(E) \right) + \log |\mathcal{V}| < 0.$$

Since $x - 1 \geq \log x$ for all $x \in \mathbb{R}_+$ we get $-k \min_{v \in \mathcal{V}} \sum_{E \in \mathcal{E}} 1_E(v) P(E) + \log |\mathcal{V}| < 0$ and as condition for the existence of a covering

$$|\mathcal{C}| \leq \left\lceil \left(\min_{v \in \mathcal{V}} \sum_{E \in \mathcal{E}} 1_E(v) P(E) \right)^{-1} \log |\mathcal{V}| \right\rceil .$$

Since we are free to choose a PD $P \in \mathcal{P}(\mathcal{E})$, the result follows. □

Remark Choose P as *uniform* PD: $\mathcal{P}(E) = \frac{1}{|\mathcal{E}|}$ for all $E \in \mathcal{E}$. Since $\sum\limits_{E \in \mathcal{E}} 1_E(v) = \deg(v) \geq d$ we obtain $|\mathcal{C}| \leq \left(\frac{d}{|\mathcal{E}|} \right)^{-1} \log |\mathcal{V}|$, that is, Lemma 1.1.

Definition 1.5 A covering $\mathcal{C} \triangleq \{E_1, \ldots, E_k\}$ of a hypergraph $\mathcal{H} = (\mathcal{V}, \mathcal{E})$ is called *c-balanced* for some constant $c \in \mathbb{N}$, if no vertex occurs in more than c edges of $\mathcal{C}$.

Lemma 1.3 (Covering) *A hypergraph* $\mathcal{H} = (\mathcal{V}, \mathcal{E})$ *with* $d_{\mathcal{V}} > 0$ *has a c-balanced covering* $\mathcal{C} = \{E_1, \ldots, E_k\}$, *if*

(i) $k \geq \lceil |\mathcal{E}| d_{\mathcal{V}}^{-1} \cdot \log |\mathcal{V}| \rceil + 1$ *(covered prob.* $> \frac{1}{2}$)
(ii) $c \leq k \leq c|\mathcal{E}| D_{\mathcal{V}}^{-1}$
(iii) $\exp\left\{ -D\left(\lambda || \frac{D_{\mathcal{V}}}{|\mathcal{E}|}\right) k + \log |\mathcal{V}| \right\} < \frac{1}{2}$ *for* $\lambda \triangleq \frac{c}{k}$ *(not balanced prob.* $< \frac{1}{2}$)

Remark There is also some confusion concerning the scopes of analytical and combinatorial methods in probabilistic coding theory, particularly in the theory of identification. We present a covering (or approximation) lemma for hypergraphs, which especially makes strong converse proofs in this area transparent and dramatically simplifies them.

Lemma 1.4 (Covering) *Let* $\mathcal{H} = (\mathcal{V}, \mathcal{E})$ *be an e-uniform hypergraph (all edges have cardinality e) and P a PD on* $\mathcal{E}$.

Consider a PD Q on $\mathcal{V}$:

$$Q(v) \triangleq \sum_{E \in \mathcal{E}} P(E) \frac{1}{e} 1_E(v).$$

The following holds:

Fix $\varepsilon, \tau > 0$, *then there exists a set of vertices* $\mathcal{V}_0 \subset \mathcal{V}$ *and edges* $E^{(1)}, \dots, E^{(2)} \in \mathcal{E}$ *such that for*

$$\bar{Q}(v) \triangleq \frac{1}{L} \sum_{i=1}^{L} \frac{1}{e} 1_E^{(i)}(v)$$

(i) $Q(\mathcal{V}_0) \leq \tau$
(ii) $(1-\varepsilon)Q(v) \leq \bar{Q}(v) \leq (1+\varepsilon)Q(v)$ *for all* $v \in \mathcal{V} \setminus \mathcal{V}_0$
(iii) $L \leq \left\lceil \frac{|\mathcal{V}|}{e} \frac{2 \ln 2 \log(2|\mathcal{V}|)}{\varepsilon^2 \tau} \right\rceil$

For ease of application we formulate and prove a slightly more general version of this:

Lemma 1.5 *Let* $\mathcal{H} = (\mathcal{V}, \mathcal{E})$ *be a hypergraph, with a measure* Q_E *on each edge E, such that* $Q_E(v) \leq \eta$ *for all* E, $v \in E$. *For a probability distribution P on* $\mathcal{E}$ *define*

$$Q = \sum_{E \in \mathcal{E}} P(E) Q_E,$$

and fix $\varepsilon, \tau > 0$. *Then there exist vertices* $\mathcal{V}_0 \subset \mathcal{V}$ *and edges* $E_1, \dots, E_L \in \mathcal{E}$ *such that with*

$$\bar{Q} = \frac{1}{L} \sum_{i=1}^{L} Q_{E_i}$$

the following holds:

$$Q(\mathcal{V}_0) \le \tau;$$
$$\forall v \in \mathcal{V} \setminus \mathcal{V}_0 \quad (1-\varepsilon)Q(v) \le \bar{Q}(v) \le (1+\varepsilon)Q(v);$$
$$L \le \left\lceil \eta|\mathcal{V}|\frac{2\ln 2\log(2|\mathcal{V}|)}{\varepsilon^2\tau} \right\rceil.$$

Proof Define i.i.d. random variables $Y_1, \ldots, Y_L$ with

$$\Pr\{Y_i = E\} = P(E) \quad \text{for } E \in \mathcal{E}.$$

For $v \in \mathcal{V}$ define $X_i = Q_{Y_i}(v)$. Clearly $\mathbb{E}X_i = Q(v)$, hence it is natural to use a large deviation estimate to prove the bounds on $\bar{Q}$. We find

$$\Pr\left\{\frac{1}{L}\sum_{i=1}^{L} X_i \notin [(1-\varepsilon)Q(v), (1+\varepsilon)Q(v)]\right\} \le 2\exp\left(-L\cdot\frac{\varepsilon^2 Q(v)}{2\eta\ln 2}\right).$$

Now we define

$$\mathcal{V}_0 = \left\{v \in \mathcal{V} : Q(v) < \frac{1}{|\mathcal{V}|}\tau\right\},$$

and observe that $Q(\mathcal{V}_0) \le \tau$. Hence,

$$\Pr\left\{\exists v \in \mathcal{V}\setminus\mathcal{V}_0 : \frac{1}{L}\sum_{i=1}^{L} Q_{Y_i}(v) \notin [(1-\varepsilon)Q(v), (1+\varepsilon)Q(v)]\right\}$$
$$\le 2|\mathcal{V}|\exp\left(-L\cdot\frac{\varepsilon^2\tau}{2\eta|\mathcal{V}|\ln 2}\right).$$

The RHS becomes less than 1, if

$$L > \eta|\mathcal{V}|\frac{2\ln 2\log(2|\mathcal{V}|)}{\varepsilon^2\tau},$$

hence there exist instances E_i of the Y_i with the desired properties.

The interpretation of this result is as follows: Q is the expectation measure of the measures Q_E, which are sampled by the Q_E. The lemma says how close the sampling average $\bar{Q}$ can be to Q. In fact, assuming $Q_E(E) = q \le 1$ for all $E \in \mathcal{E}$, one easily sees that

$$||Q - \bar{Q}||_1 \le 2\varepsilon + 2\tau.$$

□

1.1.1 Multiple Coverings for Hypergraphs and Approximation of Output Statistics

We assume that in $\mathcal{H} = (\mathcal{V}, \mathcal{E})$ all edges $E \in \mathcal{E}$ have the same cardinality D. For a uniform distribution P on $\mathcal{E}$ we can define the associated (output) distribution Q,

$$Q(v) = \sum_{E \in \mathcal{E}} D(E) 1_E(v) = \sum_{E \in \mathcal{E}} \frac{1}{|\mathcal{E}|} 1_E(v). \tag{1.1.2}$$

Our goal is to find an $\mathcal{E}^* \subset \mathcal{E}$ as small as possible such that the distribution Q^*,

$$Q^*(v) \triangleq \sum_{E \in \mathcal{E}^*} \frac{1}{|\mathcal{E}^*|} 1_E(v) \quad \text{for } v \in \mathcal{V} \tag{1.1.3}$$

is a good approximation of Q in the following sense. For some $\mathcal{V}^* \subset \mathcal{V}$

$$\sum_{u \in \mathcal{V}^*} Q(v) \leq \delta \tag{1.1.4}$$

and

$$(1 - \varepsilon) Q(v) \leq Q^*(v) \leq (1 + \varepsilon) Q(v) \quad \text{for } v \in \mathcal{V} \setminus \mathcal{V}^* \tag{1.1.5}$$

Lemma 1.6 (Multiple Covering) *For the uniform hypergraph* $\mathcal{H} = (\mathcal{V}, \mathcal{E})$, *and any* $\varepsilon, \delta > 0$ *there is a* $\mathcal{E}^* \subset \mathcal{E}$ *and a* $\mathcal{V}^* \subset \mathcal{V}$ *such that for* Q^* *defined in (1.1.2), (1.1.3) hold and*

$$|\mathcal{E}^*| \leq \frac{\delta |\mathcal{V}|}{\varepsilon^2 |E|} \log |\mathcal{V}|.$$

Remark The result also holds for multiple edges.

Proof By standard random choice of $E^{(1)}, E^{(2)}, \ldots, E^{(l)}$ we know by Lemma... that for $v \in \mathcal{V}$

$$\mathcal{P}_{v,\varepsilon} \triangleq \Pr\left(\frac{1}{\ell} \left| \sum_{i=1}^{\ell} 1_{E^{(i)}}(v) - \frac{\deg(v)}{|\mathcal{E}|} \right| \geq \frac{\deg(v)}{|\mathcal{E}|} \varepsilon \right)$$

$$\leq \exp\left\{ -\ell D \left(\frac{\deg(v)}{|\mathcal{E}|} (1 + \varepsilon) \middle\| \frac{\deg(v)}{|\mathcal{E}|} \right) \right\} \leq \exp\left\{ -\ell \frac{\varepsilon^2 \deg(v)}{|\mathcal{E}| \ln 2} \right\}.$$

Define with the previously defined average degree $\overline{d}_{\mathcal{V}}$ (which here is $\overline{d}_{\mathcal{V}} = \frac{|\mathcal{E}||E|}{|\mathcal{V}|}$)

$$\mathcal{V}^* = \{ v \in \mathcal{V} : \deg(v) < \delta \overline{d}_{\mathcal{V}} \}$$

and notice that (1.1.2) holds:

$$Q(\mathcal{V}^*) = \frac{1}{|\mathcal{E}|} \cdot \frac{1}{|E|} \sum_{v \in \mathcal{V}^*} \deg(v) \le \frac{|\mathcal{V}|\delta \overline{d}_{\mathcal{V}}}{|\mathcal{E}||E|} = \delta. \tag{1.1.6}$$

Further, for $v \in \mathcal{V} \setminus \mathcal{V}^*$ we have $\deg(v) \ge \delta \overline{d}_{\mathcal{V}}$ and by (1.1.4)

$$P_{v,\varepsilon} \le \exp\left\{-\ell \frac{\varepsilon^2 \delta \overline{d}_{\mathcal{V}}}{|\mathcal{E}| \ln 2}\right\} \tag{1.1.7}$$

$$\sum_{v \in \mathcal{V}^*} P_{v,\varepsilon} \le \|\mathcal{V}| \exp\left\{-\ell \frac{\varepsilon \delta \overline{d}_{\mathcal{V}}}{|\mathcal{E}| \ln 2}\right\} < 1 \tag{1.1.8}$$

if

$$\ell > \frac{\delta |\mathcal{V}| \ln 2}{\varepsilon^2 |E|} \log |\mathcal{V}| \tag{1.1.9}$$

and the desired $\mathcal{E}^*$ exists. □

By the same arguments applied to a class of uniform hypergraphs $\{\mathcal{H}_s = (\mathcal{V}_s, \mathcal{E}_s) : s = 1, \dots, S\}$ we get the

Lemma 1.7 (Simultaneous Multiple Covering) *With P_s, Q_s, Q_s^* and $\mathcal{V}_s^*$ defined analogously as for a single hypergraph and any $\varepsilon, \delta > 0$, for $s = 1, \dots, S$*

(i) $\sum_{v \in \mathcal{V}_s^*} Q_s(v) \le \delta$;

(ii) $(1 - \varepsilon) Q_s(v) \le Q_s^*(v) \le (1 + \varepsilon) Q(v)$ *for* $v \in \mathcal{V}_s^*$ *for suitable* $\mathcal{E}_s^*$ $(1 \le s \le S)$ *with*

$$|\mathcal{E}_s^*| \le \max_s \frac{\delta |\mathcal{V}_s|}{\varepsilon^2 |E_s|} \log(|\mathcal{V}| \cdot S).$$

1.2 Coverings, Packings, and Algorithms

1.2.1 Fractional Packings and Coverings

For a hypergraph $\mathcal{H} = (\mathcal{V}, \mathcal{E})$ recall the packing problem to find the maximum number of disjoint edges $\nu(\mathcal{H})$.

Let $\tau(\mathcal{H})$ denote the minimum number of vertices in a set τ to represent the edges. Here representation means that every edge contains a vertex of τ. The problem is known as transversal problem and also as covering problem in the dual hypergraph, where $\mathcal{E}$ takes the role of the vertex set and $\mathcal{V}$ takes the role of the edge set. Many cases have been studied. $\mathcal{E}$ may consist of graphic objects like edges, paths, circuits, cliques or objects in linear spaces like bases, flats, etc.

Perhaps the simplest way to look at a hypergraph and its dual is in terms of the associated bipartite graph describing the vertex-edge incidence structure.

The estimations of $\nu(\mathcal{H})$ and $\tau(\mathcal{H})$ are in general very difficult problems.

We follow here a nice partial theory due to Lovász, which starting from the obvious inequality

$$\nu \leq \tau$$

aims at characterizing classes with equality.

A guiding idea is to approximate τ in terms of *fractional matching* (which in a way occurred already in covering product spaces). A weight function $w : \mathcal{E} \to \mathbb{R}_+$ is called a fractional matching of $\mathcal{H}$, if

$$\sum_{E \ni v} w(E) \leq 1 \qquad \text{for all } v \in \mathcal{V},$$

and

$$\nu^*(\mathcal{H}) = \max_{w \text{ fractional matching}} \sum_E w(E)$$

is the *fractional matching or packing number.*

Analogously, a weight function $\varphi : \mathcal{V} \to \mathbb{R}_+$ is called a *fractional (vertex edge) cover*, if

$$\sum_{v \in E} \varphi(v) \geq 1 \qquad \text{for all } E \in \mathcal{E},$$

and

$$\tau^*(\mathcal{H}) = \max_{\varphi \text{ fractional cover}} \sum_{v \in \mathcal{V}} \varphi(v)$$

is the *fractional cover number.*

Notice that by the Duality Theorem of linear programming

$$\nu^*(\mathcal{H}) = \tau^*(\mathcal{H})$$

and thus we have

$$\nu(\mathcal{H} \leq \nu^*(\mathcal{H}) = \tau^*(\mathcal{H}) \leq \tau(\mathcal{H}). \tag{1.2.1}$$

A remarkable fact is the following: if equality holds for all partial hypergraphs in one of the two inequalities then it holds also in the other [8].

Whereas ν and τ are optima of discrete linear programs, which are generally hard to calculate, the value $\nu^* = \tau^*$ is the optimum of an ordinary linear program and easy to compute especially if $\mathcal{H}$ has nice symmetry properties.

Recall our conventions about the notation for degrees $d_\mathcal{E}$, $D_\mathcal{E}$, and $D_\mathcal{V}$. Using weight functions with

$$w(E) = D_\mathcal{V}^{-1}, \qquad \varphi(v) = d_\mathcal{E}^{-1}$$

we get the

Lemma 1.8

$$|\mathcal{E}|D_{\mathcal{V}}^{-1} \le \tau^* \le |\mathcal{V}|d_{\mathcal{E}}^{-1}.$$

Now, if $d_{\mathcal{V}} = D_{\mathcal{V}}$ *and* $d_{\mathcal{E}} = D_{\mathcal{E}}$, *that is, the hypergraph is* $r = d_{\mathcal{E}}$*-uniform and* $d = d_{\mathcal{V}}$*-regular, then*

$$\tau^* = \frac{|\mathcal{E}|}{d} = \frac{|\mathcal{V}|}{r}.$$

Examples

1. $\mathcal{H} = \{[n], \binom{[n]}{k}\}$, $\binom{n}{k}\binom{n-1}{k-1}^{-1} = \frac{n}{k} \le \tau^* \le \frac{n}{k}$
2. $\mathcal{H} = \{\mathbb{F}_2^k, \mathcal{P}_{k,s}\}$, where $\mathcal{P}_{k,s}$ is the set of all $(k-s)$-dimensional planes in $\mathbb{F}_2^k$, here
$$\begin{bmatrix} k \\ s \end{bmatrix} 2^s \begin{bmatrix} k \\ s \end{bmatrix}^{-1} = 2^s \le \tau^* \le 2^k 2^{-(k-s)} = 2^s.$$

1.2.2 A Greedy Algorithm to Estimate $\tau(\mathcal{H})$, $\tau^*(\mathcal{H})$ from Above

For $v \in \mathcal{V}$ let $\mathcal{E}(v)$ be the edges from $\mathcal{E}$ containing v.

Select vertices $v_1, v_2, \ldots, v_i$ successively, v_{i+1} being a point with

$$\deg_{\mathcal{V}_{i+1}}(v_{i+1}) = D_{\mathcal{V}_{i+1}} \tag{1.2.2}$$

where $\mathcal{H}_{i+1} = (\mathcal{V}_{i+1}, \mathcal{E}_{i+1})$ with $\mathcal{V}_{i+1} = \mathcal{V} \setminus \{v_1, v_2, \ldots, v_i\}$ and $\mathcal{E}_{i+1} = \mathcal{E} \setminus \bigcup_{j=1}^{i} \mathcal{E}(v_j)$.

The procedure stops at t if $v_1, v_2, \ldots, v_t$ represent all edges.

Lemma 1.9 (Covering, Lovász 1975, [9]) *For any hypergraph* $\mathcal{H} = (\mathcal{V}, \mathcal{E})$

$$\tau(\mathcal{H}) \le (1 + \frac{1}{2} + \cdots + \frac{1}{D_{\mathcal{V}}})\tau^*(\mathcal{H}) < (1 + \log D_{\mathcal{V}})\tau^*(\mathcal{H})$$

For the proof we need the following tool.

A k-matching (or packing) of $\mathcal{H}$ is a family $\mathcal{M}$ of edges from $\mathcal{E}$, where an edge can occur several times, such that

$$\sum_{E \in \mathcal{M}} 1_E(v) \le k \quad \text{for all } v \in \mathcal{V}. \tag{1.2.3}$$

Let $\nu_k(\mathcal{H})$ denote the maximum number of edges in a k-matching. Then $\nu_1(\mathcal{H}) = \nu(\mathcal{H})$ is our familiar quantity, the maximum number of disjoint edges.

A k-matching is simple if no edge occurs in it more than once. Let $\tilde{\nu}_k \le \nu_k$ be the maximum number of edges in simple k-matchings.

(There are analogous concepts for covers, but they are not used here.)

We prove now the essential auxiliary result.

Lemma 1.10 *If for any hypergraph $\mathcal{H}$ any greedy cover algorithm produces t covering vertices, then*

$$t \leq \frac{\tilde{\nu}_1}{1 \cdot 2} + \frac{\tilde{\nu}_2}{2 \cdot 3} + \cdot + \frac{\tilde{\nu}_{D_\mathcal{V}-1}}{(D_\mathcal{V}-1) \cdot D_\mathcal{V}} + \frac{\tilde{\nu}_{D_\mathcal{V}}}{D_\mathcal{V}} \tag{1.2.4}$$

(Clearly $\tilde{\nu}_{D_\mathcal{V}} = |\mathcal{E}|$).

Since $\tau \leq t, \tilde{\nu}_i \leq \nu_i \leq i\nu \leq i\nu^* = i\tau^*$, insertion of these inequalities in (1.2.4) gives the first inequality in the Covering Lemma. For the second we use $\frac{1}{2} + \frac{1}{3} + \frac{1}{n} + \cdots + \frac{1}{d} \leq \int_1^d \frac{1}{x} dx = \log |_1^d = \log d$.

Proof Let t_j denote the number of steps in the greedy cover algorithm in which the chosen vertex covers j new edges. After $t_{D_\mathcal{V}} + t_{D_\mathcal{V}-1} + \cdots + t_{j+1}$ steps the hypergraph $\mathcal{H}_i'$ formed by the uncovered edges has degree $D_{\mathcal{V}_i'} \leq i$ and hence $|\mathcal{E}_i'| \leq \tilde{\nu}_i$. Also $|\mathcal{E}_i'| = it_i + \cdots + 2t_2 + t_1$, and therefore

$$it_i + \cdots + 2t_2 + t_1 \leq \tilde{\nu}_i \qquad \text{for } i = 1, 2, \ldots, D_\mathcal{V}. \tag{1.2.5}$$

Now for $d = D_\mathcal{V}$ by the following multiplication of the inequalities and additions

$$\sum_{i=1}^{d-1} \frac{1}{i(i+1)} (it_i + \cdots + 2t_2 + t_1) + \frac{1}{d} (dt_d + \cdots + t_1) \leq \sum_{i=1}^{d-1} \frac{\tilde{\nu}_i}{i(i+1)} + \frac{\tilde{\nu}}{d} \tag{1.2.6}$$

and therefore

$$\begin{aligned} &\sum_{i=1}^{d} \left(\frac{1}{i(i+1)} + \cdots + \frac{1}{d(d-1)} + \frac{1}{d} \right) it_i \\ &= \sum_{i=1}^{d} t_i \qquad \left(\text{because } \frac{1}{d(d-1)} + \frac{1}{d} = \frac{1}{d-1} \text{ etc.}\right) \\ &= t, \end{aligned}$$

which together with (1.2.6) gives the claimed inequality (1.2.4) □

Remark Comparison with Covering Lemma 1.9 shows that for the dual hypergraph $\mathcal{H}^*$ the factor $\log |\mathcal{E}|$ is replaced by $(1 + \log D_\mathcal{V})$, which is smaller, if

$$D_\mathcal{V} < \frac{1}{2} |\mathcal{E}|.$$

But this is not always the case!

1.2.3 Applications

Lemma 1.11 (Covering (the link)) *Let G be a group and $A \subset G$. Then there exists a set $B \subset G$ such that $AB = G$ and*

$$|B| \leq \frac{|G|}{|A|}(1 + \log |A|).$$

It is instructive to use for G the cyclic group $\mathbb{Z}_n$. Then the last lemma implies Theorem 2 of [7]. For comparison of our bound and the bound of Lóvasz the ratio $r = \frac{|A|}{|\mathbb{Z}_n}$ is relevant. Indeed

$$A(r) = \frac{1}{r} \log n \quad \text{and} \quad L(r) = \frac{1}{r}(1 + \log n)$$

have as maximal difference

$$\max_{0 \leq r \leq 1} (L(r) - A(r)) = \max_{0 \leq r \leq 1} \frac{1}{r}(1 + \log r)$$

Now

$$\begin{aligned} \left(\frac{1 + \log r}{r}\right)' &= \frac{\frac{1}{r} \cdot r - (1 + \log r)}{r^2} \\ &= -\frac{\log r}{r^2} = 0, \end{aligned}$$

implies $r = 1$

$$\left(-\frac{\log r}{r^2}\right)' = \frac{-r + 2r \log r}{r^4}$$

for $r = 1 := -1$ maximum.

So we are at most better by 1!

Problem Try to improve the estimate by Lóvasz in order to get rid of this 1.

Actually for $r = 1$ both bounds seem bad! For $A = \mathbb{Z}_n$ this is obvious, however, it needs analysis for $\frac{|A_n|}{|\mathbb{Z}_n|} \to 1$ as $n \to \infty$.

Let $l < k$ be constant and let n go to infinity, then for the hypergraph $\mathcal{H}_{l,k,n} = \left(\binom{[n]}{l}, \binom{[n]}{k}\right)$ Rödl [10] proved that

$$\tau(\mathcal{H}_{l,k,n}) = (1 + o(1)) \frac{\binom{n}{k}}{\binom{k}{l}} \qquad \text{as } n \to \infty.$$

Notice that $\tau^*(\mathcal{H}_{l,k,n}) = \frac{\binom{n}{k}}{\binom{k}{l}}$ and therefore the factor $1 + \log\binom{k}{l}$ is far and the factor $\log\binom{n}{k}$ is very far from optimal.

Problem Show that $\min(1 + \log D_{\mathcal{V}}, \log|\mathcal{E}|)$ is in general the best possible factor.

1.3 Application to the k-Tuple Chromatic Number χ_k

For a graph G $\chi_k(G)$, called k-tuple chromatic number [13] of G, is the least number l for which it is possible to assign a k-subset $C(v)$ of $\{1, \dots, l\}$ to every vertex x of G such that $C(v) \cap C(v') = \emptyset$ for $(v, v') \in \mathcal{E}$. Of couse $\chi(G) = \chi_1(G)$ is the ordinary chromatic number.

It is shown in [9] with the help of Lemma 1.9 that

Theorem 1.1 (Lovász)

$$\chi_k(G) \geq \frac{k}{1 + \log\alpha(G)}\chi(G)$$

and with the help of Coloring Lemma 1.2 that

Theorem 1.2 (Lovász) *For any graph G*

$$\chi(G) \leq (1 + \log\alpha(G)) \max_{G'} \frac{|\mathcal{V}(G')|}{\alpha(G')},$$

where G' ranges over all induced subgraphs of G.

1.4 On a Problem of Shannon in Graph Theory

1.4.1 Introduction

While investigating his zero-error capacity of a DMC which is equivalent to the optimal rate of the stability number or maximal number of independent vertices $\alpha(G^n)$, where the graph $G(\mathcal{V}, \mathcal{E})$ is associated with the transmission matrix W of the DMC by defining $\mathcal{V} = \mathcal{X}$ and $\mathcal{E} = \{(x, x')) :$ where $W(\cdot, x)$ and $W(\cdot|x')$ have common support$\} \cup \{(x, x) : x \in \mathcal{X}\}$, the loops, and the cartesian product of two graphs G and H, denoted by $G \times H$ is defined as follows: $\mathcal{V}(G \times H) = \mathcal{V}(G) \times \mathcal{V}(H)$ and $\mathcal{E}(G \times H) = \{\big((g, h), (g', h')\big) :$ iff $(g, g') \in \mathcal{E}(G)$ and $(h, h') \in \mathcal{E}(H)\}$.

Shannon [12] investigated when

$$\alpha(G \times H) = \alpha(G) \cdot \alpha(H) \tag{1.4.1}$$

holds for graphs and found a partial answer in terms of *preserving functions*.

$\sigma : \mathcal{V}(G) \to \mathcal{V}(G)$ is called preserving if $(v, v') \in \mathcal{E}(G)$ implies that

$$(\sigma(v), \sigma(v')) \notin \mathcal{E}(G). \tag{1.4.2}$$

Theorem 1.3 (Shannon 1956, [12]) *If there exists a preserving function $\sigma : \mathcal{V}(G) \to \mathcal{V}(G)$ such that $\sigma(\mathcal{V})$ is an independent set of vertices in G, then (1.4.1) holds for every finite graph H. In this case G is called universal.*

1.5 A Necessary and Sufficient Condition in Terms of Linear Programming for G to be Universal

Let G be a finite graph with $\mathcal{V}(G) = \{g_1, \dots, g_n\}$ and let $\{C_1, \dots, C_s\}$ be a fixed ordering of all the different cliques of G. Define

$$\alpha_i^{(j)} = \begin{cases} 1, & \text{if } g_i \in C_j \\ 0, & \text{if } g_i \notin C_j \end{cases}$$

and the following polytope in the n-dimensional Euclidean space

$$P_G = \left\{ x = (x_1, x_2, \dots, x_n) : \sum_{i=1}^{n} \alpha_i^{(j)} x_i \leq 1,\ x_i \geq 0,\ 1 \leq j \leq s \right\}.$$

Theorem 1.4 (Rosenfeld 1967, [11]) *A finite graph G is universal if and only iff*

$$\max_{x \in P_G} \sum_{i=1}^{n} x_i = \alpha(G). \tag{1.5.1}$$

Proof Necessity of (1.5.1)
(i) W.l.o.g. we may assume that $\{g_1, \dots, g_\alpha(G)\} = A$ is an independent set of vertices in G. Choose

$$x_i = \begin{cases} 1 & \text{for } 1 \leq i \leq \alpha(G) \\ 0 & \text{for } i > \alpha(G). \end{cases}$$

Since no vertices in A are contained in the same clique it is obvious that for every j

$$\sum_{\alpha=1} \alpha_i^{(j)} x_i \leq 1 \qquad \text{while} \sum_{i=1}^{n} x_i = \alpha(G).$$

Therefore we always have

$$\max_{x \in P_G} \sum_{i=1}^{n} x_i \geq \alpha(G).$$

(ii) Suppose G is not universal, i.e., there exists a graph H with

$$\alpha(G \times H) > \alpha(G)\alpha(H)$$

(the inequality $\alpha(G \times H) \geq \alpha(G)\alpha(H)$ is obvious). Let $A \subset G \times H$ be a maximum independent set of vertices in $G \times H$. (i.e., $|A| = \alpha(G \times H)$). Define $A_i = \{h : (g_i, h) \in A\} \subset \mathcal{V}(H)$. Since $(g_i, h) \to (g_i, h')$ if $h \to h'$ and A is independent it follows that A_i is an independent set of vertices in $\mathcal{V}(H)$ and therefore $|A_i| \leq \alpha(H)$. Furthermore, if $A'_i = \{(g_i, h) : h \in A_i\}$ then $A = \bigcup_{i=1}^{n} A'_i$ and the union is disjoint. Now choose $x_i = \frac{1}{\alpha(H)}|A_i|$ and verify that

$$\sum_{i=1}^{n} = \frac{1}{\alpha(H)} \sum_{i=1}^{n} |A'_i| = \frac{|A|}{\alpha(H)} = \frac{\alpha(G \times H)}{\alpha(H)} > \alpha(G). \tag{1.5.2}$$

Let us show that for every j

$$\sum_{i=1}^{n} \alpha_i^{(j)} x_i \leq 1.$$

If

$$C_j = \{g_{i_1}, \ldots, g_{i_k}\}, \qquad \text{then } \sum_{i=1}^{n} \alpha_i^{(j)} x_i = \sum_{l=1}^{n} x_{i_l}. \tag{1.5.3}$$

Since $g_{i_r} \to g_{i_t}$ with $1 \leq r, t \leq k$ it follows that $\bigcup_{l=1}^{k} A_{i_l}$ is an independent set of vertices in $\mathcal{V}(H)$ and the union is disjoint. Hence we get

$$\alpha(H) \sum_{l=1}^{k} x_{i_l} = \sum_{l=1}^{k} |A_{i_l}| = \left| \bigcup_{l=1}^{k} A_{i_l} \right| \leq \alpha(H)$$

and $\sum_{l=1}^{k} x_{i_l} = \sum_{i=1}^{k} \alpha_i^{(l)} x_i \leq 1$. Thus (1.5.2) and (1.5.3) prove the necessity of condition (1.5.1). □

1.5.1 Shannon's Condition Is Not Necessary

We show now that the existence of a preserving function for G, while being sufficient for G to be universal, is not necessary for G to be universal. For this first notice that for a preserving function σ and for an independent set of vertices $A \subset \mathcal{V}(G)$ $\sigma(A)$ is independent and $|\sigma(A)| = |A|$, because otherwise two vertices in A have the same image, which violates (1.4.2). Therefore we also have $\alpha(\sigma(G)) = \alpha(G)$. Since $\sigma^{-1}(v)$ is a complete subgraph of G it follows that $\mathcal{V}(G)$ is covered by $|\mathcal{V}(\sigma(G))|$ complete subgraphs.

Therefore a necessary condition for the existence of a preserving function σ such that $|\sigma(\mathcal{V}(G))| = \alpha(\sigma(G))$ is that $\mathcal{V}$ is covered in G by $\alpha(G)$ complete subgraphs.

Let G_1 and G_2 be two disjoint pentagons and G_3 a set of 5 vertices such that $\mathcal{V}_3 \cap \mathcal{V}_1 \cup \mathcal{V}_2 = \emptyset$. Adjoin by an edge each vertex of $\mathcal{V}_3$ to all the vertices of $\mathcal{V}_1$ and $\mathcal{V}_2$. Let H be the graph defined by these relations, then we have

$$|\mathcal{V}(H)| = 15, \quad \alpha(H) = 5 \qquad (\mathcal{V}_3 \text{ is independent})$$

Since a pentagon cannot be covered by less than 3 complete subgraphs (which are of cardinality 2), it is obvious that H cannot be covered by less than 6 complete subgraphs. Thus Shannon's condition of the existence of a preserving function cannot hold for H.

On the other hand to show that H is universal observe that all the cliques of H are triangles, every vertex of H is contained in exactly 10 different cliques is 50, therefore we have $1 \le j \le 50$, $\sum_{i=1}^{15} \alpha_i^{(j)} x_i \le 1$ implies $\sum_{j=1}^{50} \sum_{i=1}^{15} \alpha_i^{(j)} x_i \le 50$, however $\sum_{j=1}^{50} \sum_{i=1}^{15} \alpha_i^{(j)} x_i = 10 \sum_{i=1}^{15} x_i \le 50$ implies $\sum_{i=1}^{15} x_i \le 5 = \alpha(H)$, and by Theorem 1.4 H is universal.

(iii) To prove the "sufficiency" of condition (1.5.1) suppose that

$$\max_{x \in P_G} \sum_{i=1}^{n} x_i > \alpha(G)$$

Since the coefficients of the linear inequalities determining P_G are non-negative integers we may assume that all the components $(x_1, \dots, x_n)$ of a maximizing point are *rational*. Let β be the least common multiple of all the denominators of the x_i's. Let $y_i = \beta \cdot x_i$, $1 \le i \le n$ be non-negative integers satisfying

$$\sum_{i=1}^{n} y_i > \alpha(G) \cdot \beta \tag{1.5.4}$$

$$\sum_{i=1}^{n} \alpha_i^{(j)} y_i \le \beta, \qquad 1 \le j \le s. \tag{1.5.5}$$

Using these inequalities we shall construct a graph H for which (1.4.1) does not hold. This will complete the proof.

Let A_i, $1 \le i \le n$, be a family of disjoint sets with $|A_i| = y_i$ and define $\mathcal{V}(H) = \bigcup_{i=1}^{n} A_i$.

Two vertices $u, v \in \mathcal{V}(H)$ are joined by an edge if $u = v$ or for $i \ne j$ $u \in A_i$ and $v \in A_j$ and $g_i \ne g_j$. Thus any set A_i is independent. Let $\mathcal{U} = \{u_1, \dots, u_i\}$ be an independent set of $\mathcal{V}(H)$, We may assume that for some t $\mathcal{U} \cap A_i \ne \emptyset$, $1 \le i \le t$, and $\mathcal{U} \cap A_i = \emptyset$ for $i > t$. Since $\mathcal{U}$ is independent so is $\bigcup_{i=1}^t A_i$. It follows from the definition of H that the set $\{g_1, \ldots, g_t\}$ is a complete subgraph of G and therefore it is contained in a clique of G. Here we have $\sum_{i=1}^t x_i \leq 1$ and this implies $\sum_{i=1}^t y_i \leq \beta$ and $|\bigcup_{i=1}^t A_i| \leq \beta$. This means that

$$\alpha(H) \leq \beta. \tag{1.5.6}$$

Consider now $D = \{(g_i, h) : h \in A_i\} \subset \mathcal{V}(G \times H)$. If $(g, h), (g', h') \in D$, then $(g, g') \in \mathcal{E}(G)$ and $h, h' \in \mathcal{E}(H)$ and therefore $((g, h), (g', h')) \notin \mathcal{E}(G \times H)$. If $(g, g') \notin \mathcal{E}(G)$, then obviously $((g, h), (g', h')) \notin \mathcal{E}(G \times H)$ and D is an independent set of vertices in $\mathcal{V}(G \times H)$. Using (1.5.4), (1.5.5), and (1.5.6) we obtain

$$\alpha(G \times H) \geq |D| = \left| \bigcup_{i=1}^n A_i \right| = \sum_{i=1}^n y_i > \alpha(G) \cdot \beta \geq \alpha(G) \cdot \alpha(H).$$

□

1.5.2 Characterizing Universality in Terms of Integer Linear Programming

The condition (1.4.1) can be described as follows: G is β-universal for any integer β if and only if for any set of non-negative integers x_i satisfying $\sum_{i=1}^n \alpha_i^{(j)} x_i \leq \beta$, $1 \leq j \leq s$, one has

$$\sum_{i=1}^n x_i \leq \alpha(G)\beta. \tag{1.5.7}$$

Indeed, suppose G is not β-universal, i.e., there exists a β for which $\max_{x \in P_G} \sum_{i=1}^n x_i > \alpha(G)\beta$. If $\{g_1, \ldots, g_{\alpha(G)}\}$ is an independent set of vertices in G, choose $y_i = x_i + 1$, $1 \leq i \leq \alpha(G)$, $y = x_i$ for $i > \alpha(G)$. It is obvious that $\sum_{i=1}^n \alpha_i^{(j)} y_i \leq \beta + 1$ while $\sum_{i=1}^n y_i > \alpha(G)(\beta + 1)$.

This shows that if G is not β-universal it is also not $(\beta + 1)$-universal. Since the number of non-isomorphic graphs with n vertices is finite, it follows that there exists an integer $\beta(n)$ such that G is universal if and only if it is $\beta(n)$-universal. The function $\beta(n)$ non-decreasing in n. The values for $n \leq 5$ can be computed using Shannon's observation that all graphs with at most 5 vertices are universal except for the pentagon which is not 2-universal. Hence $\beta(n) = 0$ for $n \leq 4$ and $\beta(5) = 2$. Using that the β in part (iii) is the determinant of a matrix of order n with 0's and 1's only and therefore $\beta < n!$, we get $\beta(n) < n!$

Finally, one can use Theorem 1.4 to estimate $\alpha(G \times H)$. Given G and H one can calculate $a = \max \sum_{i=1}^n x_i$ subject to $\sum_{i=1}^n \alpha_i^{(j)} x_i \leq \alpha(H)$, $1 \leq j \leq s_G$, where x_i

is a non-negative integer and $b = \max \sum_{i=1}^{m} y_i$, $\sum_{i=1}^{m} \beta_i^{(j)} y_i \leq \alpha(G)$, $1 \leq j \leq s_H$, where $\beta_i^{(j)}$ has the same meaning for H as $\alpha_i^{(j)}$ for G.

Obviously $\alpha(G \times H) \leq \min\{a, b\}$.

1.6 The Basic Coloring Lemmas

Definition 1.6 A coloring ϕ of a hypergraph $(\mathcal{V}, \mathcal{E})$ is a mapping of the vertex set $\mathcal{V}$ into some finite set. If actually $\phi : \mathcal{V} \to \{1, \dots, L\}$, then we speak of an L-coloring.

Several types of coloring will be introduced. Colorings of hypergraphs turn out to be a very useful tool in Multi-User Source Coding Theory. Thus, most of the coloring lemmas presented can be applied for instance to prove the Slepian–Wolf Theorem, either if the maximum error is considered or for the average-error concept.

Also a powerful application of the hypergraph approach will be demonstrated for arbitrarily varying sources, where the achievable rates for arbitrarily varying sources with and without side information at the decoder are characterized.

1.6.1 Colorings Which Are Good in All Edges

The first coloring lemma presented was motivated by list reduction. Remember that the list reduction lemma was central in the derivation of the capacity formula for the DMC with feedback. The question now is: do we really need the feedback in order to apply the list reduction lemma? In other words, the list size should be reduced to 1, since then the sender needs no information about the received word and feedback is not essential. It turns out that this is not possible. However the following lemma shows that a reduction on "small" list size is possible.

Lemma 1.12 (Coloring) *Let $\mathcal{H} = (\mathcal{V}, \mathcal{E})$ be a hyper-graph with $\max_{E \in \mathcal{E}} |E| \leq L$. Further let $|\mathcal{E}| \cdot L < t!$ for some $t \in \mathbb{N}$. Then there exists a coloring (of the vertices) $\phi : \mathcal{V} \to \{1, \dots, L\}$ with*

$$|\phi^{-1}(i) \cap E| < t \text{ for all } i = 1, \dots, L \text{ and all edges } E \in \mathcal{E}.$$

(in all edges every color occurs at most t times)

Proof For every subset $A \subset \mathcal{V}$ define

$$\mathcal{F}(A) \triangleq \left\{\phi : \mathcal{V} \to \{1, \dots, L\} : \phi \text{ is constant on } A\right\}$$

So the set $\mathcal{F}_E^t$ of colorings that are bad (with more than t colors) for an edge E is given by

$$\mathcal{F}_E^t = \bigcup_{A \subset E, |A| \geq t} \mathcal{F}(A)$$

and the set $\mathcal{F}^t$ of all bad colorings is just $\mathcal{F}^t = \bigcup_{E \in \mathcal{E}} \mathcal{F}_E^t$. Denoting by $\mathcal{F} \triangleq \big\{\phi : \mathcal{V} \to \{1, \ldots, L\}\big\}$ the set of all colorings with at most L colors, we shall show that $\frac{|\mathcal{F}^t|}{|\mathcal{F}|} < 1$. Then of course $|\mathcal{F}^t| < |\mathcal{F}|$ and the existence of at least one "good" coloring as required in the lemma is immediate. Therefore observe that $|\mathcal{F}| = L^{|\mathcal{V}|}$ and that

$$|\mathcal{F}^t| \leq |\mathcal{E}| \cdot \binom{L}{t} \cdot L \cdot L^{|\mathcal{V}|-t},$$

since for every edge $E \in \mathcal{E}$, $|E| \leq L$ by the assumptions and since one of L possible colors is needed to color the vertices in A, and there are $L^{|\mathcal{V}|-t}$ possible colorings of the vertices outside A.

From this follows that

$$\frac{|\mathcal{F}^t|}{|\mathcal{F}|} \leq |\mathcal{E}| \cdot \binom{L}{t} \cdot L^{1-t} \leq |\mathcal{E}| \frac{L}{t!} < 1$$

by the assumption. □

Definition 1.7 A coloring as in the previous lemma is called an (L, t)-coloring. We call an $(L, 1)$-coloring a strict coloring.

Strict colorings usually require an enormous number of colors. The next lemma concerns strict colorings.

To a hyper-graph $(\mathcal{V}, \mathcal{E})$ we can assign a graph $(\mathcal{V}, \mathcal{E}^*)$, where the vertex set is the same as before and two vertices are connected if they are both contained in some edge $E \in \mathcal{E}$. A graph is a special hyper-graph. A strict vertex coloring of $(\mathcal{V}, \mathcal{E}^*)$ is also a strict vertex coloring of $(\mathcal{V}, \mathcal{E})$, and vice versa.

Lemma 1.13 (Coloring) *Let $(\mathcal{V}, \mathcal{E}^*)$ be a graph with maximum degree $D_{\mathcal{V}} \leq D$. Then there exists a strict coloring with L colors, if*

$$L \geq D + 1.$$

Proof We proceed by a greedy construction. Color the vertices iteratively in any way such that no two adjacent vertices get the same color. If the procedure stops before all vertices are colored, then necessarily one vertex $v \in \mathcal{V}$ must have $\deg(v) \geq D + 1$, contradicting the hypothesis. □

Definition 1.8 A coloring of $\mathcal{H} = (\mathcal{V}, \mathcal{E})$ is denoted by ϕ_λ, $0 < \lambda < 1$, if for every edge $E \in \mathcal{E}$ at least $(1-\lambda)|E|$ colors occur only once.

Lemma 1.14 (Coloring) *A hyper-graph $\mathcal{H} = (\mathcal{V}, \mathcal{E})$, $\mathcal{E} = \{E_1, \dots, E_J\}$ has an L-coloring ϕ_λ with $\lambda \in (0,1)$, $L \in \mathbb{N}$ and $L \geq D_\mathcal{V}$ if*

$$D_\mathcal{V} < \lambda L \quad \textit{and}$$

$$\sum_{j=1}^{J} \exp\left\{-|E_j| \cdot D\left(\lambda || \frac{|E_j|}{L}\right)\right\} < 1.$$

Proof The vertices $v_1, \dots, v_I$ in the hypergraph $\mathcal{H}$ are colored independently at random according to the uniform distribution. Hence the color of vertex v_i can be regarded as the realization of a random variable X_i taking values in $\{1, \dots, L\}$ with $\Pr(X_i = \ell) = \frac{1}{L}$ for $\ell = 1, \dots, L$. Furthermore $X_1, \dots, X_I$ are required to be independent. This coloring procedure will be denoted as *L-standard random coloring* from now on. It will be used many times.

Define now for $i = 1, \dots, I$; $j = 1, \dots, J$ random variables

$$f_i^j(X_1, \dots, X_I) \triangleq \begin{cases} 1, & \text{if } X_i \neq X_{i'} \text{ for all } i' \text{ with } v_{i'} \in E_j \cap \{v_1, \dots, v_{i-1}\} \\ 0, & \text{else.} \end{cases}$$

We can view the coloring procedure as an iterative coloring of the vertices in E_j. Then f_i^j takes the value 1 (the coloring is "good" for E_j in step i), if v_i gets a color, which has not occurred in E_j until step i. For instance, the coloring of the edge $E_j \triangleq (v_3, v_7, v_8, v_{11})$ is "good" in step 8 ($f_8^j(X_1, \dots, X_I) = 1$) if v_8 has another color than v_3 and v_7.

Clearly, if $\sum_{v_i \in E_j} f_i^j(X_1, \dots, X_I) \geq (1 - \frac{\lambda}{2})|E_j|$, then at most $\frac{\lambda}{2}|E_j|$ colors occur more than once in E_j, and therefore $(1-\lambda)|E_j|$ vertices are colored correctly.

We upperbound now

$$\Pr\left(\sum_{v_i \in E_j} f_i^j(X_1, \dots, X_I) < (1 - \frac{\lambda}{2})|E_j|\right)$$

using the Chernoff Bound.

It is clear from the definition of the f_i^j's that this expression depends only on those random variables X_i with $v_i \in E_j$. So for each edge E_j we can concentrate on the random variables X_i corresponding to vertices $v_i \in E_j$. Let us denote these random variables as $X_1, \dots, X_t$, $t \triangleq |E_j|$. Accordingly, the f_i^j's are relabelled.

Obviously (with $\epsilon_1, \dots, \epsilon_{i-1} \in \{0, 1\}$) for $i = 1, \dots, |E_j|$

$$\Pr\big(f_i^j = 1 \mid f_{i-1}^j = \epsilon_{i-1}, \dots, f_1^j = \epsilon_1\big) \geq \frac{L - (i-1)}{L} \geq L - \frac{|E_j|}{L} \tag{1.6.1}$$

(at most $i - 1$ colors have been used before vertex v_i is colored).

In order to apply Bernstein's trick we now consider the random variables $\overline{f}_i^j = 1 - f_i^j$. Obviously

$$\Pr\left(\textstyle\sum_{v_i \in E_j} f_i^j < (1 - \frac{\lambda}{2})|E_j|\right) = \Pr\left(\textstyle\sum_{v_i \in E_j} \overline{f}_i^j > \frac{\lambda}{2} \cdot |E_j|\right)$$
$$\leq \exp\big\{-|E_j| \cdot D\big(\tfrac{\lambda}{2} || \tfrac{|E_j|}{L}\big)\big\},$$

since with 1.6.1 the expected value

$$\mathbb{E}\big(\overline{f}_i^j \mid \overline{f}_{i-1}^j, \dots, \overline{f}_1^j\big) = \Pr\big(\overline{f}_i^j = 1 \mid \overline{f}_{i-1}^j, \dots, \overline{f}_1^j\big) \leq \frac{|E_j|}{L}.$$

Notice that the f_i^j's are not independent. However, since

$$\Pr\big(f_t^j = \epsilon_t, \dots, f_1^j = \epsilon_1\big)$$
$$= \textstyle\prod_{s=2}^t \Pr\big(f_s^j = \epsilon_s \mid f_{s-1}^j = \epsilon_{s-1}, \dots, f_1^j = \epsilon_1\big) \cdot \Pr\big(f_1^j = \epsilon_1\big)$$

□

In the following we shall introduce some refinements of Coloring Lemma 1.14 which are suitable for problems in Coding Theory discussed later on.

Definition 1.9 We denote by ϕ_λ^2, $0 < \lambda < 1$, a vertex coloring of $\mathcal{H}_2$ for which in every subedge E_j^m ($m = 1, \dots, M_j$, $j = 1, \dots, J$) at least $(1 - \lambda)|E_j^m|$ colors occur which occur only once in $E_j \in \mathcal{E}$

Lemma 1.15 (Coloring) *A 2-hyper-graph* $\mathcal{H}_2 = \big(\mathcal{V}, \mathcal{E}, (\mathcal{E}_j)_{j=1}^J\big)$ *has a* ϕ_λ^2 *coloring with L colors, if*

$$D_{\mathcal{V}} < \frac{\lambda}{2} L \quad \textit{and}$$

$$\sum_{j=1}^J \sum_{m=1}^{M_j} \exp\{-|E_j^m| \cdot D(\frac{\lambda}{2} || \frac{|E_j|}{L}\} < 1.$$

Proof We use again the standard random L-coloring $(X_1, \dots, X_I)$, thus the color of the vertex v_i, $i = 1, \dots, I$, is regarded as the realization of the random variable X_i taking values in $\{1, \dots, L\}$.

For $i = 1, \dots, I, m = 1, \dots, M_j, j = 1, \dots, J$ random variables $f_i^{j,m}$ are defined by

$$f_i^{j,m}(X_1, \dots, X_I) \triangleq \begin{cases} 1, & \text{if } X_i \neq X_i' \\ & \text{for all } v_{i'} \in \left(E_j^m \cap \{v_1, \dots, v_{i-1}\}\right) \cup (E_j \setminus E_j^m) \\ 0, & \text{else} \end{cases}$$

Hence $f_i^{j,m}$ takes the value 1 (the coloring of vertex v_i is good in subedge E_j^m), if the color of v_i is different from all the colors of its predecessors in E_j^m and all the colors that occurred "outside" E_j^m. We upperbound now $\Pr\left(\sum_{i \in E_j^m} f_i^{j,m} < (1 - \frac{\lambda}{2})|E_j^m|\right)$ by application of Bernstein's trick as in Lemma 1.14.

As above for $E_j^m \triangleq \{v_{i_s} : s = 1, \dots, |E_j^m|, i_1 < i_2 < \dots < i_{|E_j^m|}\}$ we can estimate

$$\Pr\left(f_{i_s}^{j,m} = 1 \mid f_{i_{s-1}}^{j,m} = \epsilon_{s-1}, \dots, f_{i_1}^{j,m} = \epsilon_1\right)$$

$$\geq \frac{L - (s-1) - (|E_j| - |E_j^m|)}{L} \geq \frac{L - |E_j|}{L} \geq 1 - \frac{|E_j|}{L}$$

The same reasoning as in the proof of Coloring Lemma 1.14 yields now

$$\Pr\left(\min_{j=1,\dots,J}\left(\min_{m=1,\dots,M_j} \frac{1}{|E_j^m|}\sum_{i \in E_j} f_i^{j,m}\right) < 1 - \frac{\lambda}{2}\right)$$

$$\leq \sum_{j=1}^{J}\sum_{m=1}^{M_j} \exp\left\{-|E_j^m| \cdot D\left(\frac{\lambda}{2}\|\frac{|E_j|}{L}\right)\right\}.$$

□

Let $(\mathcal{V}, \mathcal{A}, (\mathcal{F}_E)_{E \in \mathcal{A}})$ and $(\mathcal{V}, \mathcal{B}, (\mathcal{F}_E)_{E \in \mathcal{B}})$ be two 2-hypergraphs with the same vertex set $\mathcal{V}$ and $\mathcal{A} \cap \mathcal{B} = \varnothing$. Define $\mathcal{H}_2 \triangleq (\mathcal{V}, \mathcal{A} \cup \mathcal{B}, (\mathcal{F}_E)_{E \in \mathcal{A} \cup \mathcal{B}})$. We are interested in colorings ϕ_λ^2 of $\mathcal{H}_2$ which, in addition, are strict on $(\mathcal{V}, \mathcal{A})$.

Those colorings automatically color all subedges out of $\bigcup_{E \in \mathcal{A}} \mathcal{F}_E$ strictly and we need not be concerned with them. Write $\mathcal{B}$ as $\mathcal{B} = \{E_1, \dots, E_J\}$ and denote the subedges by E_j^m, $1 \leq m \leq M_j$, $1 \leq j \leq J$. Let $(\mathcal{V}, \mathcal{A}^*)$ be the graph associated with $(\mathcal{V}, \mathcal{A})$ as in Coloring Lemma 1.13 and let $D_\mathcal{V}$ denote the maximal degree of the vertices in this graph. We are now prepared to state

Lemma 1.16 (Coloring) *Let $\mathcal{H}_2 = (\mathcal{V}, \mathcal{A} \cup \mathcal{B}, (\mathcal{F}_E)_{E \in \mathcal{A} \cup \mathcal{B}})$ be a 2-hypergraph with*

$$\mathcal{A}, \mathcal{B}, E_j (1 \leq j \leq J), E_j^m \quad (1 \leq m \leq M_j, 1 \leq j \leq J),$$

and $D_\mathcal{V}$ as just described. For $L \geq D^* + 1 + d$, $\mathcal{H}_2$ has a coloring ϕ_λ^2 which is strict on $(\mathcal{V}, \mathcal{A})$ if

$$\frac{|E_j|}{d} < \frac{\lambda}{2} \text{ for all } j = 1, \ldots, J \text{ and}$$

$$\sum_{j=1}^{J} \sum_{m=1}^{M_j} 2 \exp\{-|E_j^m| \cdot D(\frac{\lambda}{2} || \frac{|E_j|}{d})) < 1$$

Proof The idea of the proof consists of a combination of the ideas for the proofs of Coloring Lemmas 1.13 and 1.15 as follows. We color the vertices $v_1, v_2, \ldots$ iteratively as in the proof of Lemma 1.13 except that now we have, since $L \geq D + 1 + d$, in each step at least d colors available one of which we choose at random according to the uniform distribution on any d available colors (those with smallest values in $\{1, \ldots, L\}$ for instance). Thus we get a strict L-coloring of $(\mathcal{V}, \mathcal{A})$ as before. What do we get for $\big(\mathcal{V}, \mathcal{B}, (\mathcal{F}_E)_{E \in \mathcal{B}}\big)$? This random coloring procedure can be described by a sequence of RV's $X_1, \ldots, X_I$.

Those RV's are not independent or identically distributed. We overcome this additional difficulty by substituting the functions $f_i^{j,m}$ defined in the proof of Lemma 1.14 by the following two types of functions corresponding to the events: "the color of v_i is different from all the colors of its predecessors in E_j^m" and "the color of v_i is different from all the colors in $E_j \setminus E_j^m$".

For $m = 1, \ldots, M_j$, $j = 1, \ldots, J$ and $i = 1, \ldots, I$ define RV's

$$g_i^{j,m}(X_1, \ldots, X_I) \triangleq \begin{cases} 1 & \text{if } X_i \neq X_{i'}, \text{ for all } v_{i'} \in E_j \text{ with } i' < i \\ 0 & \text{otherwise} \end{cases}$$

and

$$G_i^{j,m}(X_1, \ldots, X_I) \triangleq \begin{cases} -1 & \text{if } X_i = X_{i'}, \text{ for some } v_{i'} \in E_j \setminus E_j^m \\ 0 & \text{otherwise} \end{cases}$$

Clearly, if

$$\sum_{i \in E_j^m} (g_i^{j,m} + G_i^{j,m}) \geq (1 - \lambda)|E_j^m|$$

then at least a fraction of $(1 - \lambda)$ vertices in E_j^m is colored correctly within E_j.

We can use

$$\Pr\left\{\sum_{i\in E_j^m}(g_i^{j,m}+G_i^{j,m})<(1-\lambda)|E_j^m|\right\}$$

$$\leq \Pr\left\{\sum_{i\in E_j^m} g_i^{j,m} < (1-\frac{\lambda}{2})|E_j^m|\right\} + \Pr\left\{\sum_{i\in E_j^m} G_i^{j,m} < -\frac{\lambda}{2}|E_j^m|\right\}.$$

As in the previous proof one shows that

$$\Pr\left\{\sum_{i\in E_j^m} g_i^{j,m} < (1-\frac{\lambda}{2})|E_j^m|\right\} \leq \exp\{-|E_j^m|\cdot D(\frac{\lambda}{2}||\frac{|E_j|}{d})\}.$$

By symmetry the same bound holds for the second term. □

Remarks *It must be emphasized, that this seems to be the first proof combining the greedy and the random choices*

In this context we should also mention the work of Beck et al. in "Combinatorial Discrepancy Theory" (cf. J. Beck: Irregularities of distribution, in Surveys in Combinatorics 1985). Here we are given a hypergraph $\mathcal{H}=(\mathcal{V},\mathcal{E})$, the vertices of which have to be colored with two colors as uniform as possible with respect to the hyper-edges. As in Lemma 1.12 we want to achieve that each color meets each subset considered in approximately the same number of elements. Thereto we choose as range of the coloring ϕ the set $\{+1,-1\}$, hence $\phi:\mathcal{V}\to\{1,-1\}$ and for each edge $E\in\mathcal{E}$ we define

$$d(\phi,E)\triangleq\left|\sum_{v\in E}\phi(v)\right|.$$

The *combinatorial discrepancy* now is defined as follows

$$D(\mathcal{E})\triangleq\min_{f:\mathcal{X}\to\{1,-1\}}\max_{E\in\mathcal{E}} d(\phi,E)$$

1.7 Colorings Which Are Good in Average

1.7.1 Weighted Hypergraphs

The motivation of this definition is as follows. We want to give a proof for the Slepian–Wolf Theorem, when the average error is considered. In this case Q

corresponds to a probability distribution P^n_X on $\mathcal{X}^n$ and the Q_E's correspond to conditional probability distributions $P^n_{Y|X}(\cdot|x^n)$, $x^n \in \mathcal{X}^n$.

For a coloring $\phi : \mathcal{V} \to \{1, \dots, L\}$ we define g^E_v for all vertices $v \in \mathcal{V}$ and edges $E \in \mathcal{E}$ by

$$g^E_v = \begin{cases} 1, & \text{if } \phi(v) = \phi(v') \text{ for some } v' \in E \setminus \{v\} \\ 0, & \text{else} \end{cases}$$

Definition 1.10 The coloring $\phi : \mathcal{V} \to \{1, \dots, L\}$ has average error $\overline{\lambda}$, if

$$\sum_{E\in\mathcal{E}} \sum_{v\in E} g^E_v \cdot Q_E(v) \cdot Q(E) = \overline{\lambda}.$$

Lemma 1.17 (Coloring) *The weighted hypergraph $\mathcal{H} = \big(\mathcal{V}, \mathcal{E}, Q, (Q_E)_{E\in\mathcal{E}}\big)$ defined as above can be colored with L colors and average error $\overline{\lambda}$, $0 < \overline{\lambda} < 1$, if*

$$D_{\max} \cdot L^{-1} \leq \overline{\lambda}.$$

Proof Again the standard random L-coloring is used. Hence the color of each vertex v_i, $i = 1, \dots, I$ is a random variable X_i. Then for all $E \in \mathcal{E}$

$$\mathcal{E} g^E_v (X_1, \dots, X_I) \leq \frac{|E| - 1}{L} < \frac{|E|}{L} \leq D_{\max} \cdot L^{-1},$$

since at most $|E| - 1$ colors are used to color the vertices in $E \setminus \{v\}$. Therefore

$$\mathcal{E} \sum_{E\in\mathcal{E}} \sum_{v\in E} Q_E(v) Q(E) g^E_v (X_1, \dots, X_I) \leq D_{\max} \cdot L^{-1},$$

since Q and Q_E, $E \in \mathcal{E}$ are sub-probability distributions. □

Remark Coloring Lemma 1.4 can be applied to prove the average-error version of the Slepian-Wolf Theorem (see Chap. 2). To see this, let $\mathcal{V} \triangleq \mathcal{Y}^n$, $\mathcal{E} \triangleq \big(\mathcal{T}^n_{Y|X,\delta'}(x^n)\big)_{x^n \in \mathcal{T}^n_{X,\delta}}$, $Q(x^n) \triangleq P^n_X(x^n)$ and $Q_E(y^n) \triangleq P_{Y|X}(y^n|x^n)$. Then the conditions of Lemma 1.17 are fulfilled, if we color $\mathcal{Y}^n$ with $L \geq \max_{E\in\mathcal{E}} |E| \cdot \frac{1}{\overline{\lambda}}$ colors.

This is an abstract version of Cover's argument [5]. Notice that not the AEP-property, but only the value of $D_{\max}$ is important. Another proof of the average-error version of the Slepian–Wolf Theorem, in which the AEP-property is not important, will be done by "orthogonal colorings" presented next.

1.7.2 Orthogonal Colorings

Let $\mathcal{V}$ and $\mathcal{W}$ be finite sets and let C be a subset of $\mathcal{V} \times \mathcal{W}$. Observe that $(\mathcal{V}, \mathcal{W}, \mathcal{C})$ is a bipartite graph.

If P is a probability distribution on $\mathcal{V} \times \mathcal{W}$ concentrated on C (its *carrier*), i.e., $P(v, w) \neq 0$ implies $(v, w) \in C$, then we call $(\mathcal{V}, \mathcal{W}, \mathcal{C}, \mathcal{P})$ a *stochastic bipartite graph.*

Definition 1.11 Let φ and Ψ be colorings of $\mathcal{V}$ and $\mathcal{W}$, respectively. Then $\rho \triangleq (\varphi, \Psi)$ is denoted as *orthogonal coloring* of $\mathcal{V} \times \mathcal{W}$ and colors in particular all edges in C. Clearly, if for $(v, w) \in C$

$$\rho(v, w) \neq \rho(v', w') \text{ for all } v', w' \in C \setminus \{v, w\}$$

then knowing $\rho(v, w)$ the pair (v, w) can be identified. Conversely, if this is not the case, then (v, w) cannot be identified or decoded correctly. Suppose, (v, w) occurs with probability $P(v, w)$, then the average error probability $\overline{\lambda}(\rho)$ is given by

$$\overline{\lambda}(\rho) = P\big(\{(v, w) : |\rho^{-1}(\rho(v, w) \cap C| > 1\}\big).$$

We call $\rho = (\varphi, \Psi)$ an $\mathbf{L_1 \times L_2}$**-coloring**, if

$$||\varphi|| \leq L_1 \text{ and } ||\Psi|| \leq L_2.$$

Finally, $C_{|w}$ (resp. $C_{|v}$) denotes the **cross section**

$$\{v \in \mathcal{V} : (v, w) \in C\} \text{ (resp. } \{w \in \mathcal{W} : (v, w) \in C\}).$$

Lemma 1.18 (Coloring) *The standard random orthogonal $L_1 \times L_2$-coloring of the bipartite stochastic graph $(\mathcal{V}, \mathcal{W}, C, P)$ has an expected average error probability less than*

$$\frac{N_2}{L_2} + \frac{N_1}{L_1} + \frac{N}{L_1 \cdot L_2},$$

where $N \triangleq |C|$, $N_1 \triangleq \max_{w \in \mathcal{W}} |C_{|w}|$, and $N_2 \triangleq \max_{v \in \mathcal{V}} |C_{|v}|$.

Proof Given L_1 and L_2 let us color $\mathcal{V}$ and $\mathcal{W}$ independently at random in the standard way. Hence, the colorings are regarded as random variables

$$X^{|\mathcal{V}|} \triangleq X_1, \dots, X_{|\mathcal{V}|}, Y^{|\mathcal{W}|} \triangleq Y_1, \dots, Y_{|\mathcal{W}|},$$

where the X_i's and Y_i's are independent and identically distributed according to the uniform distribution with $\text{Prob}(X_i = \ell_1) = \frac{1}{L_1}$ for $\ell_1 = 1, \dots, L_1; i = 1, \dots, |\mathcal{V}|$ and $\text{Prob}(Y_j = \ell_2) = \frac{1}{L_2}$ for $\ell_2 = 1, \dots, L_2; j = 1, \dots, |\mathcal{W}|$.

In order to upperbound the average error probability $\mathcal{E}\overline{\lambda}\big(X^{|\mathcal{V}|}, Y^{|\mathcal{W}|}\big)$, we break up the probability of incorrectly coloring some $(v, w) \in \mathcal{V} \times \mathcal{W}$ in three partial events. These events depend on the location of the same color (in one of the cross-sections or outside). In each of these partial events we can make use of the independence in the standard random coloring and simply count all possibilities.

$$
\begin{aligned}
\text{Prob}\big((X_v, Y_w) = (X_{v'}, Y_{w'}) \text{ for some } (v', w') \in C \smallsetminus \{v, w\}\big)
&\le \text{Prob}\big((X_v, Y_w) = (X_v, Y_{w'}) \text{ for some } w' \in C_{|v|} \smallsetminus \{w\}\big)\\
&+ \text{Prob}\big((X_v, Y_w) = (X_{v'}, Y_w) \text{ for some } v' \in C_{|w|} \smallsetminus \{v\}\big)\\
&+ \text{Prob}\big((X_v, Y_w) = (X_{v'}, Y_{w'})\\
&\quad \text{for some } v', w' \in C, v' \neq v, w' \neq w\big)\\
&\le \tfrac{1}{L_2}\big(|C_{|v|}| - 1\big) + \tfrac{1}{L_1}\big(|C_{|w|}| - 1\big)\\
&\quad + \tfrac{1}{L_1 \cdot L_2}\big(|C| - |C_{|v|}| - |C_{|w|}|\big)\\
&\le \tfrac{1}{L_2}|C_{|v|}| + \tfrac{1}{L_1}|C_{|w|}| + \tfrac{1}{L_1 \cdot L_2}|C|.
\end{aligned}
$$

Therefore,

$$
\begin{aligned}
\mathcal{E}\overline{\lambda}\big(X^{|\mathcal{V}|}, Y^{|\mathcal{W}|}\big) &\le \textstyle\sum_{(v,w)\in\mathcal{V}\times\mathcal{W}} P(v, w) \left(\frac{|C_{|v|}|}{L_2} + \frac{|C_{|w|}|}{L_1} + \frac{|C|}{L_1 \cdot L_2}\right)\\
&\le \textstyle\sum_{(v,w)\in\mathcal{V}\times\mathcal{W}} P(v, w) \left(\frac{N_2}{L_2} + \frac{N_1}{L_1} + \frac{N}{L_1 \cdot L_2}\right)\\
&\le \tfrac{N_2}{L_2} + \tfrac{N_1}{L_1} + \tfrac{N}{L_1 \cdot L_2}
\end{aligned}
$$

for $N_2 = \max_{v\in\mathcal{V}} |C_{|v|}|$, $N_1 = \max_{w\in\mathcal{W}} |C_{|w|}|$ and $N = |C|$, since
$\sum_{(v,w)\in\mathcal{V}\times\mathcal{W}} P(v, w) = 1.$ □

Notice that only the parameters of the *carrier* C are important and no AEP-property is used.

1.7.3 Universal Colorings of Internally Weighted Hypergraphs

The last coloring lemma we present is a generalization of Coloring Lemma 1.14 to *internally weighted hypergraphs* $\big(\mathcal{V}, \mathcal{E}, (Q_j)_{j=1}^{J}\big)$, where as usual $\mathcal{V} = \{v_1, \ldots, v_I\}$ and $\mathcal{E} = \{E_1, \ldots, E_J\}$ denote the sets of vertices and edges, respectively.

Additionally, for every hyper-edge $E_j, 1 \le j \le J$, there is an additive measure $Q_j : E_j \to \mathbb{R}_+$. Especially, Q_j might be a probability distribution on the vertices in E_j. If, e.g., Q_j is chosen as the uniform distribution $\left(Q_j(v) = \frac{1}{|E_j|}\text{ for all } v \in E_j\right)$, then, in the following discussion, we are in the situation considered in Lemma 1.14. For a coloring ϕ of $\mathcal{V}$ we define g_i^j for $i = 1, \dots, I$; $j = 1, \dots, J$ by

$$g_i^j \triangleq \begin{cases} 1 & \text{if } \mathcal{C}(i) = \mathcal{C}(i') \text{ for some } i' \in E_j \setminus \{i\} \\ 0 & \text{otherwise} \end{cases}$$

Hence, $g_i^j = 1$, exactly if in E_j there is another vertex $v_{i'}$ which has the same color as v_i.

Definition 1.12 We say that ϕ has **goodness** λ^* for the internally weighted hypergraph, if

$$\sum_{v_i \in E_j} g_i^j \cdot Q_j(v_i) \le \lambda^* Q_j(E_j) \text{ for all } j = 1, \dots, J.$$

So it is required that in each edge at most a fraction of $\lambda^* Q_j(E_j)$ (notice that $Q_j(E_j) = \sum_{v_i \in E_j} Q_j(v_i)$) is "badly colored". In the paragraph on "Colorings which are good in average" we introduced the average error for weighted hypergraphs. In addition to internally weighted hypergraphs those hypergraphs are equipped with a probability distribution on the hyper-edges; and the concept of average error (averaged over all edges) corresponds to the concept of goodness (for all edges) for internally weighted hypergraphs. In order to extend Coloring Lemma 1.14 to internally weighted hypergraphs we still have to require a uniformity condition, namely

$$Q_j(v_i) \le \frac{1}{\sqrt{|E_j|}} Q_j(E_j) \text{ for all } i \in E_j, j = 1, \dots, J \qquad (1.7.1)$$

Hence, no vertex should be too important in any edge of the hypergraph.

Lemma 1.19 (Coloring) *Assume that the internally weighted hypergraph* $\left(\mathcal{V}, \mathcal{E}, (Q_j)_{j=1}^J\right)$ *satisfies the uniformity condition (1.7.1). Then it has for* $L \ge d_{\max}$ *a coloring with* L *colors and goodness* λ^*, $0 < \lambda^* < 1$, *if for some* $\alpha < 0$

$$\sum_{j=1}^{J} \exp\left\{\alpha\left(\frac{\lambda^*}{2} - \frac{|E_j|}{L}\right) \cdot Q_j(E_j) + \frac{\alpha^2}{2} Q_j(E_j)^2\right\} < \frac{1}{2}.$$

Proof We use again the standard random coloring with L colors of $\mathcal{V}$ and define for an edge $E \in \mathcal{E}$ the random variables

$$f_i(X_1,\dots,X_I) \triangleq \begin{cases} 1 & \text{if } X_i \neq X_{i'} \text{ for all } i' < i, v_{i'} \in E \\ 0 & \text{otherwise} \end{cases},$$

$$F_i(X_1,\dots,X_I) \triangleq \begin{cases} 1 & \text{if } X_i \neq X_{i'} \text{ for all } i' > i, v_{i'} \in E \\ 0 & \text{otherwise} \end{cases}.$$

If Q is the weight on the edge E, then

$$\sum_{v_i \in E} Q(v_i) f_i \geq (1 - \frac{\lambda^*}{2}) \cdot Q(E) \text{ and } \sum_{v_i \in E} Q(v_i) F_i \geq (1 - \frac{\lambda^*}{2}) \cdot Q(E)$$

implies that the weight of the correctly colored vertices in E is greater than $(1 - \lambda^*)Q(E)$.

In the previous coloring lemmas we could apply Bernstein's trick, since the f_i's were identically distributed. Here we have the (weighted) random variables $Q(v_i) \cdot f_i$ which are no longer identically distributed. However, with the same argumentation as in the proof of Lemma 1.14, we can apply the more general Chernoff bound, in which the exponential function is estimated by the first three terms of its Taylor series.

For $\alpha < 0$ (and hence $\alpha' \triangleq \alpha \cdot \ell n 2 < 0$. We will use α' in the expansion of the exponential function, which is to the base 2)

$$\Pr\left(\sum_{v_i \in E} Q(v_i) \cdot f_i < (1 - \frac{\lambda^*}{2}) Q(E)\right)$$

$$\leq \exp\{-\alpha(1 - \frac{\lambda^*}{2}) Q(E)\} \cdot \prod_{v_i \in E} \mathcal{E} \exp\{\alpha Q(v_i) f_i\}$$

Further, as pointed out above,

$$\prod_{v_i \in E} \mathcal{E} \exp\{\alpha Q(v_i) f_i\} = \prod_{v_i \in E} \left(\Pr(f_i = 0) + \Pr(f_i = 1) \cdot \exp\{\alpha Q(v_i)\}\right)$$

$$\leq \prod_{v_i \in E} \left(\frac{|E|}{L} + \frac{L - |E|}{L} \exp\{\alpha Q(v_i)\}\right)$$

$$\leq \prod_{v_i \in E} \left(\frac{|E|}{L} + \frac{L - |E|}{L} (1 + \alpha' Q(v_i) + \frac{[\alpha' Q(v_i)]^2}{2})\right)$$

$$= \prod_{v_i \in E} \left(1 + \frac{L - |E|}{L} (\alpha' Q(v_i) + \frac{[\alpha' Q(v_i)]^2}{2}) \right)$$

$$= \exp \left\{ \sum_{v_i \in E} \log \left(1 + \frac{L - |E|}{L} \left(\alpha' Q(v_i) + \frac{[\alpha' Q(v_i)]^2}{2} \right) \right) \right\}$$

$$\leq \exp \left\{ \alpha' Q(E) \frac{L - |E|}{L} + \sum_{v_i \in E} \frac{[\alpha' Q(v_i)]^2}{2} \right\},$$

since $\log(1 + x) \leq x$ for all $x > -1$.

Since by the assumption $\sum_{i \in E} Q(v_i)^2 \leq Q(E)^2$ we have

$$\Pr \left(\sum_{v_i \in E} Q(v_i) f_i < (1 - \frac{\lambda^*}{2}) Q(E) \right)$$

$$\leq \exp \left\{ -\alpha' (1 - \frac{\lambda^*}{2}) Q(E) + \alpha' Q(E) \frac{L - |E|}{L} + \sum_{v_i \in E} \frac{[\alpha' Q(v_i)]^2}{2} \right\}$$

$$\leq \exp \{ \alpha' \frac{\lambda^*}{2} Q(E) - \alpha' \frac{|E|}{L} Q(E) + \frac{\alpha'^2}{2} Q(E)^2 \}.$$

Since the same estimation holds for $\Pr\left(\sum_{v_i \in E} Q(v_i) F_i) < (1 - \frac{\lambda^*}{2}) Q(E) \right)$, summation over all edges yields the statement of the theorem as a sufficient condition for the existence of a coloring as required. □

1.8 Orthogonal Coloring of Rectangular Hypergraphs $(\mathcal{V} \times \mathcal{W}, \mathcal{E})$

We recall that the pair $(\mathcal{V} \times \mathcal{W}, \mathcal{E})$ is a rectangular hypergraph, if $\mathcal{V}$, $\mathcal{W}$ are finite sets and $\mathcal{E}$ is a family of subsets of $\mathcal{V} \times \mathcal{W}$. $C = \bigcup_{E \in \mathcal{E}} E$ is the carrier of the hypergraph.

If φ (resp. ψ) is a coloring of $\mathcal{V}$ (resp. $\mathcal{W}$), then $\rho = (\varphi, \psi)$ is an orthogonal coloring of $\mathcal{V} \times \mathcal{W}$. The following two types of colorings are needed in coding an arbitrarily varying correlated source (AVCS) with and without side information at the decoder.

We denote by $\rho_\lambda = (\varphi_\lambda, \psi_\lambda)$ an orthogonal coloring of $(\mathcal{V} \times \mathcal{W})$ for which in every edge E, $E \in \mathcal{E}$, at least $(1 - \lambda)|E|$ colors occur.

A stronger notion of orthogonal coloring, denoted by ρ_λ^2 is defined by the requirement, that in every edge E, $E \in \mathcal{E}$, at least $(1-\lambda)|E|$ colors occur, which *occur only once in* C.

More generally one can consider an orthogonal 2-hypergraph $(\mathcal{V} \times \mathcal{W}, \mathcal{E}, (\mathcal{E}_j)_{j=1}^J)$, that is a 2-hypergraph with rectangular vertex set. We denote again by $\rho_\lambda^2 = (\varphi_\lambda, \psi_\lambda)$ an orthogonal coloring of $(\mathcal{V} \times \mathcal{W})$ for which in every subedge E_j^m, $m = 1, \dots, M_j$, $j = 1, \dots, J$ at least $(1-\lambda)|E_j^m|$ colors occur, which only occur once in E_j.

We study hypergraphs with one edge E only. Using standard random coloring $(V^{|\mathcal{V}|}; W^{|\mathcal{W}|}) = (V_1, \dots, V_{|\mathcal{V}|}; W_1, \dots, W_{|\mathcal{W}|})$ with L_1, L_2 colors we are interested in estimating the probability $P_{\lambda,E}$ of not obtaining a coloring ρ_λ.

$F \subset \mathcal{V} \times \mathcal{W}$ is called a *diagonal*, if no two elements of F have the same first or second component. *We analyze now random coloring for certain types of edges by decomposing them into diagonals*. This leads us to Lemmas 1.6, 1.7, 1.8 and 1.9.

Other approaches are conceivable and we propose the

Problem What can be said about the RV's $\{(V_i, W_j) : (i,j) \in E\}$ for a general set $E \subset \mathcal{V} \times \mathcal{E}$? Which probabilistic inequalities or laws can be established? In particular, given N, for which sets with $|E| = N$ does random coloring perform most poorly?

1.8.1 Types of Edges and Partitioning into Diagonals

We define now four types of edges which occur in the coding problems for AVCS's. Let ζ, ζ_1, ζ_2 be reals with $0 < \zeta_1, \zeta_2; \zeta_1, \zeta_2 \leq \zeta$.

An edge $E \subset \mathcal{V} \times \mathcal{W} = \mathcal{X}^n \times \mathcal{Y}^n$ is said to be of *ζ-point type*, if for $D = |E|$

$$D \leq n^\zeta. \tag{1.8.1}$$

Define now $d_1 = \min\{|E_{|w}| : E_{|w} \neq \phi, w \in \mathcal{W}\}$, $d_2 = \min\{|E_{|w}| : E_{|w} \neq \phi, v \in \mathcal{V}\}$, $D_1 = \max\{|E_{|w}| : E_{|w}, w \in \mathcal{W}\}$, $D_2 = \max\{|E_{|w}| : E_{|w}, v \in \mathcal{V}\}$, that is, the minimal and maximal sizes of cross-sections of E.

E is said to be of *$(\zeta, \zeta_1, \zeta_2)$-diagonal type*, if

$$D_1 \leq n^{\zeta_1}, D_2 \leq n^{\zeta_2}, D \geq n^\zeta, \tag{1.8.2}$$

and of *$(\zeta, \zeta_1, \zeta_2)$-rectangle type*, if

$$d_1 \geq n^{\zeta_1}, d_2 \geq n^{\zeta_2}, D \geq n^\zeta. \tag{1.8.3}$$

Finally, E is said to be of *(ζ_1, ζ_2)-column type*, if for $\mathcal{V}(E, v) = \{v' : \exists w \text{ with } (v, w), (v', w) \in E\}$

$$|\mathcal{V}(E, v)| \leq n^{\zeta_1} \quad \text{for all } v \in \mathcal{V} \text{ and } d_2 \geq n^{\zeta_2}. \tag{1.8.4}$$

The (ζ_1, ζ_2)-*row type* is defined analogously. Those two types may be called *line type*.

Our first result concerns partitions of an arbitrary edge E into diagonals. With E we associate a graph $G(E)$ with vertex set E: the vertices (v, w) and (v', w') are connected, iff $v = v'$ or $w = w'$. $\deg(v, w)$ counts the number of vertices connected with (v, w).

Proposition 1.1 *Let $E \subset \mathcal{V} \times \mathcal{W}$ satisfy $\max_{(v,w)\in E} deg(v, w) \leq T - 1$, then there exists a partition $\{F_1, \dots, F_t\}$ of E into diagonals, such that*

(i) $t \leq T$ and
(ii) $|F_i| \leq \frac{|E|}{2T}$, $1 \leq i \leq t$.

Proof Clearly, by Lemma 1.13 one can color the vertices with T colors such that adjacent vertices have different colors. A set of vertices with the same color forms a diagonal and we have a partition of E into $t \leq T$ diagonals.

To show (ii), let us choose among the partitions into T or fewer diagonals one, say $\{F_1, \dots, F_t\}$, with a minimal number of diagonals having a cardinality $< \frac{|E|}{2T}$. Suppose now that for instance $|F_1| = \alpha |E| T^{-1}$ for $0 < \alpha < \frac{1}{2}$. From $\sum_{i=1}^{t} |F_i| = |E|$ we conclude that for some $i \neq 1$, $|F_i| \geq |E| T^{-1}$. Let A_i be the set of vertices from F_i, which are connected with a vertex from F_1. The structure of $G(E)$ is such that $|A_i| \leq 2|F_1| = 2\alpha |E| T^{-1}$. Choose a subset $B_1 \subset F_1 \setminus A_i$ with $|B_1| = \lceil (1 - 2\alpha)|E|(2T)^{-1} \rceil$ and define two new diagonals $F_1' = F - 1 \cup B_i$, $F_i' = F_i \setminus B_i$. Then $|F_1'| \geq |E|(2T)^{-1}$ and $|F_1'| \geq \lceil |E|(2T)^{-1} \rceil - \lceil |E|(2T)^{-1} \rceil + \alpha |E| T^{-1} \geq |E|(2T)^{-1}$. This contradicts the definition of the partition $\{F_1, \dots, F_t\}$ and (ii) is proved. □

Our next result is for edges of rectangle or diagonal type.

Proposition 1.2 *$E \subset \mathcal{V} \times \mathcal{W} = \mathcal{X}^n \times \mathcal{Y}^n$ can be partitioned into diagonals $\{F_1, \dots, F_t\}$ such that*

(i) $t \leq |E| n^{-\min(\zeta_1, \zeta_2)}$, $|F_i| \geq 4^{-1} n^{\min(\zeta_1, \zeta_2)}$ for $1 \leq i \leq t$, if E is of $(\zeta, \zeta_1, \zeta_2)$-rectangle type;
(ii) $t \leq n^{\max(\zeta_1, \zeta_2)}$, $|F_i| \geq 4^{-1} n^{\zeta - \max(\zeta_1, \zeta_2)}$ for $1 \leq i \leq t$, if E is of $(\zeta, \zeta_1, \zeta_2)$-diagonal type.

Proof Apply Proposition 1.1 with:

(i) $T = D_1 + D_2$. Since $|E| \geq \max(D_2 n^{\zeta_1}, D_1 n^{\zeta_2})$ we have $D_1 \leq |E| n^{-\zeta_2}$, $D_2 \leq |E| n^{-\zeta_1}$ and therefore the bound on t. Also $|F_i| \geq E|(2T)^{-1} \geq (2(n^{-\zeta_1} + n^{-\zeta_2}))^{-1} \geq 4^{-1} n^{\min(\zeta_1, \zeta_2)}$.
(ii) $T = n^{\zeta_1} + n^{\zeta_2}$. Obviously, $t \leq T$ implies $t \leq 2n^{\max(\zeta_1, \zeta_2)}$ and $|F_i| \geq 2^{-1} n^{\zeta} (n^{\zeta_1} + n^{\zeta_2})^{-1} \geq 4^{-1} n^{\zeta - \max(\zeta_1, \zeta_2)}$. □

1.8.2 Coloring Most Points Correctly in Their Neighborhood

If in addition to a hypergraph $(\mathcal{V}, \mathcal{E})$, $\mathcal{V} = \{v_1, \dots, v_{|\mathcal{V}|}\}$, $\mathcal{E} = \{E_1, \dots, E_J\}$ we are given $u_1 \in E_i$, $1 \leq i \leq J$, $u_i < u_j$ for $i < j$, then we speak of a neighborhood system or matching system (NS). Here E_i is the neighborhood of u_i.

We are interested in colorings of the vertices $\mathcal{V}$, denoted by μ_λ, such that for at least $(1-\lambda)J$ u_i's

$$\mu_\lambda(u_i) \neq \mu_\lambda(v) \quad \text{for all } v \in E_i, v \neq u_i. \tag{1.8.5}$$

Let $V^{|\mathcal{V}|} = V_1, \dots, V_{|\mathcal{V}|}$ be a standard random L-coloring of $\mathcal{V}$. For $1 \leq i \leq J$ define

$$g_i(V_i, \dots, V_{u_i}) = \begin{cases} 1 \text{ if } V_{u_i} \neq V_j \text{ for all } j < u_i \text{ and } j \in E_i \\ 0 \text{ otherwise} \end{cases} \tag{1.8.6}$$

and

$$G_i(V_{u_i}, V_{u_i+1}, \dots, V_{|\mathcal{V}|}) = \begin{cases} 1 \text{ if } V_{u_i} \neq V_j \text{ for all } j > u_i \text{ and } j \in E_i \\ 0 \text{ otherwise.} \end{cases} \tag{1.8.7}$$

Clearly, if $\sum_{i=1}^{J} g_i \geq (1-\lambda)J$ and $\sum_{i=1}^{J} G_i \geq (1-\lambda)J$, then we have a μ_λ coloring.

Now observe that for $1 \leq i \leq J$

$$\Pr(g_i = 1 | g_{i-1} = \varepsilon_{i-1}, \dots, g_1 = \varepsilon_1) \geq \frac{L - |E_i|}{L} \geq \frac{L - D_{\max}}{L}, \tag{1.8.8}$$

and by the usual arguments

$$\Pr\left(\sum_{i=1}^{J} g_i < (1-\lambda)J\right) \leq \exp\left\{\left[h(\lambda) + \lambda \log \frac{D_{\max}}{L}\right] J\right\},$$

if $D_{\max}(L - \mathcal{D}_{\max})^{-1}(1-\lambda) \leq 1$ or, equivalently, if $L \geq D_{\max}\lambda^{-1}$.

Since the same inequality holds, if g_i is replaced by G_i, we have proved

Lemma 1.20 (Coloring) *For an NS $(\mathcal{V}, (E_i, u_i)_{1 \leq i \leq J})$ standard random L-coloring leads to a coloring μ_λ, $0 < \lambda < 1$, with a probability greater than*

$$1 - 2\exp\left\{\left[h(\frac{\lambda}{2}) + \frac{\lambda}{2} \log \frac{D_{\max}}{L}\right] J\right\}, \ \textit{if } \frac{L}{\lambda} \geq D_{\max}.$$

1.8.3 One-Sided Balanced Colorings of Rectangular Hypergraphs

Let $E \subset \mathcal{V} \times \mathcal{W}$ be arbitrary, let L be any positive integer let $D_2 = \max_{v \in \mathcal{V}} |E_{|v}|$. For an L-coloring φ of $\mathcal{V}$ we consider

$$b_\varphi(l) = \sum_{v \in \varphi^{-1}(l)} |E_{|v}| \text{ and } b_\varphi = \max_{1 \leq l \leq L} b_\varphi(l).$$

Lemma 1.21 (Coloring) *If $V^{|\mathcal{V}|}$ denotes the standard random L-coloring on $\mathcal{V}$, then for any $\alpha > 0$*

$$Pr(b_{V^{|\mathcal{V}|}} > \alpha \max(|E|L^{-1}, D_2)) \leq L \cdot \exp\left\{\frac{\alpha}{2} \max(|E|L^{-1} D_2^{-1}, 1) + |E|L^{-1} D_2^{-1}\right\}.$$

Proof Define for $v \in \mathcal{V}$, $1 \leq l \leq L$,

$$f_v^l = \begin{cases} 1 \text{ if } V_v = l \\ 0 \text{ otherwise.} \end{cases}$$

Then for $\gamma > 0$

$$\begin{aligned} &\Pr(\sum_{v \in \mathcal{V}} |E_{|v}| f_v^l > \alpha \max(|E|L^{-1}, D_2)) \\ &\leq \exp\{-\alpha\gamma \max(|E|L^{-1}, D_2)\} \prod_{v \in \mathcal{V}} \mathbb{E} \exp\{\gamma |E_{|v}| f_v^l\}. \end{aligned}$$

Now

$$\begin{aligned} \prod_v \mathbb{E} \exp\{\gamma |E_{|v}| f_v^l\} &= \prod_v \left(\frac{1}{L} \exp(\gamma |E_{|v}| + \frac{L+1}{L}\right) \\ &= \prod_v \left(1 + \frac{1}{L}\left(\gamma |E_{|v}| + \frac{\gamma^2}{2!}|E_{|v}|^2 + \dots\right)\right) \\ &\leq \exp\left\{\sum_v \frac{1}{L}\left(\gamma |E_{|v}| + \frac{\gamma^2}{2!}|E_{|v}|^2 + \dots\right)\right\} \\ &\leq \exp\left\{\sum_v \frac{1}{L}\gamma |E_{|v}|(1 + \gamma D_2 + \gamma^2 D_2^2 + \dots)\right\}. \end{aligned} \tag{1.8.9}$$

For $\gamma = \frac{1}{2} D_2^{-1}$ this is equal to $\exp\{|E|L^{-2} D_2^{-1}\}$ and the probability in question is smaller than

$$\exp\{-\alpha \frac{1}{2} \max(|E|L^{-1} D_2^{-1}, 1) + |E|L^{-1} D_2^{-1}\}.$$

Since this holds for all $1 \leq l \leq L$, the result follows. □

Remark In applications we choose $\alpha = 4 + 2e^{\varepsilon n}$ and thus get the double exponential bound $L \exp\{-e^{\varepsilon n}\}$.

1.8.4 Orthogonal Coloring of a Long Diagonal Within an Edge

We consider the situation $F \subset E \subset \mathcal{V} \times \mathcal{W}$, where F is a diagonal, $|F| = d$.

We use the standard orthogonal random $L_1 \times L_2$-coloring $(V^{|\mathcal{V}|}; W^{|\mathcal{W}|})$:

$$L_1 \geq \gamma\lambda^{-1} D_1; \;\; L_2 \geq \gamma\lambda^{-1}\alpha \max(|E|L_1^{-1}, D_2). \tag{1.8.10}$$

Here $0 < \lambda < \frac{1}{4}$, $\alpha > 2$, and $\gamma > 1$. D_1, D_2 are the maximal sizes of cross-sections of E.

We estimate now

$$\Pr((V^{|\mathcal{V}|}, W^{|\mathcal{W}|}) \text{ is not a } \rho_{4\lambda}^2 \text{ coloring of } F \text{ within } E)$$

from above by a sum $p_1 + p_2 + p_3$ of three probabilities.

Step 1. Denote the elements of F by (a_i, b_i), $1 \leq i \leq d$, and consider the NS system $(\mathcal{V}, E_{|b_i}, a_i)_{1 \leq i \leq d})$.

Lemma 1.20 gives a bound on the probability p_i that $V^{|\mathcal{W}|}$ is not a $\mu_{2\lambda}$ of this NS system.

Step 2. Lemma 1.21 gives a bound on

$$p_2 = \Pr(b_{V^{|\mathcal{V}|}} > \alpha \max(|E|L_1^{-1}, D_2)).$$

The property

$$b_{V^{|\mathcal{V}|}} \leq \alpha \max(|E|L_1^{-1}, D_2)$$

implies that for all l, $1 \leq l \leq L_1$,

$$H_l := \bigcup_{V_v = l} E_{|v} \quad \text{satisfies}$$

$$|H_l| \leq \alpha \max(|E|L_1^{-1}, D_2).$$

Define the NS system $(\mathcal{W}, (G_i, b_i)_{1 \le i \le d})$, where $G_i = H_l$ iff $b_i \in H_l$.

Step 3. Apply now Lemma 1.20 with $L = L_2$ to the NS system $(\mathcal{W}, (G_i, b_i)_{1 \le i \le d})$ in order to obtain a bound on p_3, the probability that $W^{|\mathcal{W}|}$ is not a $\mu_{2\lambda}$ coloring of $\mathcal{W}, (G_i, b_i)_{1 \le i \le d})$.

L_1 and L_2 are chosen in (1.8.10) such that

$$\begin{aligned} p_1 + p_2 + p_3 \le 2\exp&\left\{\left[h(\lambda) + \lambda \log \frac{D_1}{L_1}\right] d\right\} \\ &+ L_1 \exp\left\{-\frac{\alpha}{2} \max(|E| L_1^{-1} D_2^{-1}, 1) + |E| L_1^{-1} D_2^{-1}\right\} \\ &+ 2\exp\left\{\left[h(\lambda) + \lambda \log \frac{D_2'}{L_2}\right] d\right\}, \end{aligned}$$

where $D_2' - \alpha \max(|E| L_1^{-1}, D_2)$, and such that the RHS is smaller than

$$4\exp\left\{\left[h(\lambda) + \lambda \log \frac{\lambda}{\gamma}\right]\right\} + L_1 \exp\left\{-\left(\frac{\alpha}{2} - 1\right)\right\}.$$

We have thus proved

Lemma 1.22 (Coloring) *Let $E \subset \mathcal{V} \times \mathcal{W}$ be an edge with D_1, D_2 as maximal sizes of cross-sections and let L_1, L_2 be integers with $L_1 \ge \gamma \lambda^{-1} D_1$,*

$$L_2 \ge \gamma \lambda^{-1} \alpha \max(|E| L_1^{-1}, D_2), \quad \textit{where } 0 < \lambda < \frac{1}{4}, \alpha > 2, \textit{ and } \gamma > 1.$$

Then the orthogonal random $L_1 \times L_2$-coloring $(\mathcal{V}^{|\mathcal{V}|}, W^{|\mathcal{W}|})$ is a $\rho_{4\lambda}^2$ of a diagonal $F \subset E$ with a probability greater than

$$1 - 4\exp\left\{\left[h(\lambda) + \lambda \log \frac{\lambda}{\gamma}\right] |F|\right\} - L_1 \exp\left\{-\left(\frac{\alpha}{2} - 1\right)\right\}.$$

As an immediate consequence of this lemma we get

Lemma 1.23 (Coloring) *Let $(\mathcal{V} \times \mathcal{W}, \mathcal{E}, (\mathcal{E})_{j=1}^J)$ be a 2-hypergraph and let $D_1^* = \max_{1 \le j \le J} D_{1j}$, $D_2^* = \max_{1 \le j \le J} D_{2j}$, where D_{1j}, D_{2j} are the maximal sizes of cross-sections of $E_j \in \mathcal{E}$.*

For integers L_1, L_2 with $L_1 \ge \gamma \lambda^{-1} D_1^*$, $L_2 \ge \gamma \lambda^{-1} \alpha \max(\max_{1 \le j \le J} |E_j| L_1^{-1}, D_2^*)$, $0 < \lambda < \frac{1}{4}, \alpha > 2, \gamma > 1$ the orthogonal $L_1 \times L_2$-coloring $(V^{|\mathcal{V}|}, W^{|\mathcal{W}|})$ is a $\rho_{4\lambda}^2$ of the 2-hypergraph with a probability greater than

$$1 - N\left(4\exp\left\{\left[h(\lambda) + \lambda \log \frac{\lambda}{\gamma}\right] d^*\right\} - L_1 \exp\left\{-\left(\frac{\alpha}{2} - 1\right)\right\}\right),$$

if every subedge E^i_j *can be partitioned into diagonals of length* $\geq d^*$ *and if* $N \geq d^{*-1} \sum_{i,j} |E^i_j|$.

1.9 Balanced Colorings

We present now an easy consequence of Lemma 1.21, which we state for the ease of reference as

Lemma 1.24 (Balanced Coloring) *Let* $(\mathcal{V}, \mathcal{E})$ *be a hypergraph and let* L *be an arbitrary positive integer. If for* $\alpha > 0$

$$L \sum_{i=1}^{M} \exp\left\{-\frac{\alpha}{2} \max(|E_j|L^{-1}, 1) + |E_j|L^{-1}\right\} < 1, \tag{1.9.1}$$

then there exists an L*-coloring* Φ *of* $\mathcal{V}$ *with*

$$|E_j \cap \Phi^{-1}(l)| \leq \alpha \max(|E_j|L^{-1}, 1) \textit{ for } 1 \leq j \leq M, 1 \leq l \leq L.$$

Proof Apply Lemma 1.21 with $D_2 = 1$, $E = E_j$ and sum over j. □

Remark Using in (1.8.9) (Sect. 1.8) and (1.9.1) the exp-function to the basis 2 and $\gamma = 1$ one easily verifies that in (1.9.1) $\frac{\alpha}{2}$ can be replaced by α. Moreover, then (1.9.1) can be replaced by

$$LM\, 2^{1-\alpha} < 1 \quad \text{for } \alpha > 1. \tag{1.9.2}$$

The idea of balance came up also in connection with coverings and also partitions. We address now ε-balanced vertex colorings with L colors, i.e., a function $\phi : \mathcal{V} \to \{1, 2, \dots, L\}$ such that

$$\left| \frac{|\phi^{-1}(l) \cap E|}{|E|} - \frac{1}{L} \right| < \frac{\varepsilon}{L} \quad \text{for every } 1 \leq l \leq L \text{ and } E \in \mathcal{E}. \tag{1.9.3}$$

It is instructive to notice that this is equivalent to

$$\frac{1-\varepsilon}{L} < \frac{|\phi^{-1}(l) \cap E|}{|E|} < \frac{1+\varepsilon}{L} \tag{1.9.4}$$

for every $1 \leq l \leq L$ and $E \in \mathcal{E}$.

Throughout, logarithms and exponents are to the base 2. Natural logarithms are denoted by ln.

Lemma 1.25 (Two-sided balanced colorings of hypergraphs) *Let $\mathcal{H} = (\mathcal{V}, \mathcal{E})$ be a hypergraph with $d_{\mathcal{E}} > 0$. Then for any $0 < \varepsilon < \frac{1}{2}$ and $L < d_{\mathcal{E}}\varepsilon^2/\ln(2|\mathcal{E}|)$ there exists an ε-balanced vertex coloring with L colors.*

Proof Use the standard random coloring of the vertices, that is, the i.i. uniformly distributed RV's $X_1, X_2, \ldots, X_{|\mathcal{V}|}$ with L values, and let $Y_{vl} = \begin{cases} 1 \text{ if } X_v = l \\ 0 \text{ otherwise} \end{cases}$. Thus for the random coloring ϕ we have $|\phi^{-1}(l) \cap E| = \sum_{v \in E} Y_{vl}$ and the standard large deviation bound for the binomial distribution [1] gives for every l, $1 \le l \le L$ and $E \in \mathcal{E}$ that

$$\Pr\left\{|\phi^{-1}(l) \cap E| < \frac{1-\varepsilon}{L}|E|\right\} \le \exp\left\{-|E| D\left(\frac{1-\varepsilon}{L}\bigg\|\frac{1}{L}\right)\right\}$$

$$\Pr\left\{|\phi^{-1}(l) \cap E| > \frac{1+\varepsilon}{L}|E|\right\} \le \exp\left\{-|E| D\left(\frac{1+\varepsilon}{L}\bigg\|\frac{1}{L}\right)\right\}$$

This gives

$$D\left(\frac{1+\varepsilon}{L}\bigg\|\frac{1}{L}\right) - \frac{\varepsilon^2}{L\ln 2},$$

and Calculus shows that this is a convex function of ε in the interval $-\frac{1}{2} \le \varepsilon \le \frac{1}{2}$, with minimum equal to 0 attained at $\varepsilon = 0$. It follows that the probability that (1.9.3) does not hold for the random coloring is upper bounded by $|\mathcal{V}| \cdot 2\exp\{-d_{\mathcal{E}}\varepsilon^2/L\ln 2\}$ under the hypothesis of the lemma this bound is less than 1, and the assertion follows. □

Instead of hypergraphs with edgewise balancedness measured cardinality wise, that is, in terms of uniform distributions on the edges, we consider now more general pairs $(\mathcal{V}, \mathcal{P})$ with vertex set $\mathcal{V}$ and a set of PD's $\mathcal{P} \subset \mathcal{P}(\mathcal{V})$ and look for colorings which are balanced for every $P \in \mathcal{P}$.

Lemma 1.26 (Balanced coloring for PD's) *For $(\mathcal{V}, \mathcal{P})$ let $0 < \varepsilon \le \frac{1}{9}$ and $d > 0$ such that for*

$$E(P, d) = \left\{v : P(v) \le \frac{1}{d}\right\} \tag{1.9.5}$$

$$P(E(P(d))) \ge 1 - \varepsilon \quad \textit{for all } P \in \mathcal{P}. \tag{1.9.6}$$

Thus the probability that the standard random coloring ϕ of $\mathcal{V}$ fails to satisfy

$$\left|\frac{P(\phi^{-1}(l) \cap E(P, d))}{P(E(P, d))} - \frac{1}{L}\right| < \frac{\varepsilon}{L} \tag{1.9.7}$$

simultaneously for $1 \le l \le L$, $P \in \mathcal{P}$ is less than

$$2|\mathcal{P}|L\exp(-(\varepsilon^2/3L)d) \tag{1.9.8}$$

and less than 1*, if*

$$L < \frac{\varepsilon^2 d}{3\log 2|\mathcal{P}|}. \tag{1.9.9}$$

Corollary 1.1 *Under the assumption of the previous lemma, in particular the variational distance of the distribution of ϕ from the uniform distribution on $\{1, 2, \ldots, L\}$ is less than 3ε, i.e.,*

$$\sum_{l=1}^{L} |P(\phi^{-1}(l)) - \frac{1}{L}| < 3\varepsilon \quad \textit{for all } P \in \mathcal{P}$$

with probability specified in (1.9.8) and number of colors L specified in (1.9.9).

Proof of Lemma 1.26. We have

$$P(\phi^{-1}(l) \cap E(P,d)) = \sum_{v\in E(P,d)} P(v)Y_{vl}. \tag{1.9.10}$$

Chernoff bounding gives that for any $A \subset \mathcal{V}$

$$\begin{aligned}
&\Pr\left\{\sum_{v\in A} P(v)Y_{vl} > \frac{1+\varepsilon}{L}P(A)\right\} \\
&= \Pr\left\{\exp\left[\beta\sum_{v\in A} P(v)Y_{vl}\right] > \exp\left(\beta\frac{1+\varepsilon}{L}P(A)\right)\right\} \\
&\leq E\left(\exp\left[\beta\sum_{v\in A} P(v)Y_{vl}\right]\right)\exp\left(-\beta\frac{1+\varepsilon}{L}P(A)\right) \\
&= \left(-\beta\frac{1+\varepsilon}{L}P(A)\right)\prod_{v\in A}\left[1+\frac{1}{L}(\exp(\beta P(v))-1)\right]
\end{aligned} \tag{1.9.11}$$

where $\beta > 0$ is arbitrary, and similarly

$$\begin{aligned}
&\Pr\left\{\sum_{v\in A} P(v)Y_{vl} < \frac{1-\varepsilon}{L}P(A)\right\} \\
&\leq \exp\left(\beta\frac{1-\varepsilon}{L}P(A)\right)\cdot\prod_{v\in A}\left[1+\frac{1}{k}(\exp(-\beta P(v))-1)\right].
\end{aligned} \tag{1.9.12}$$

Apply (1.9.11) to $A = E(P,d)$ with $\beta = \varepsilon d$. Then for $v \in A = E(P,d)$ we have $\beta P(v) \leq \varepsilon$, by (1.9.5), and, therefore,

$$\exp(\beta P(v)) - 1 = \sum_{j=1}^{\infty} \frac{(\beta P(v) \ln 2)^j}{j!}$$

$$< \beta P(v) \left[1 + \frac{1}{2} \sum_{j=1}^{\infty} (\varepsilon \ln 2)^j \right] \ln 2 = \beta P(v)(1 + \varepsilon^*) \ln 2$$

where

$$\varepsilon^* = \frac{\varepsilon \ln 2}{2(1 - \varepsilon \ln 2)}.$$

Using the inequality $1 + t \ln 2 \leq \exp t$, it follows that the last product in (1.9.11) is upper bounded by

$$\exp \left[\sum_{v \in E(P,d)} \frac{1}{L} \beta P(v)(1 + \varepsilon^*) \right] = \exp \left[\frac{\varepsilon}{L} (1 + \varepsilon^*) P(E(P, d)) \right].$$

Thus (1.9.11) gives, using the assumption (1.9.6) and recalling that $\beta = \varepsilon d$,

$$\Pr \left\{ \sum_{v \in E(P,d)} P(v) Y_{vl} > \frac{1 + \varepsilon}{L} P(E(P, d)) \right\} < \exp \left[\frac{-\beta}{L} (\varepsilon - \varepsilon^*) P(E(P, d)) \right]$$
$$< \exp \left(-\frac{\varepsilon d (\varepsilon - \varepsilon^*)(1 - \varepsilon)}{L} \right) < \exp \left(-\frac{\varepsilon^2}{3L} d \right). \tag{1.9.13}$$

Here, in the last step, we used that

$$(\varepsilon - \varepsilon^*)(1 - \varepsilon) = \varepsilon \left(1 - \frac{\ln 2}{2(1 - \ln 2)} \right) (1 - \varepsilon) > \frac{\varepsilon}{3}$$

if $\varepsilon < 3 - 2 \log e$, and that condition does hold by the assumption $\varepsilon \leq \frac{1}{9}$. It follows from (1.9.12) in a similar but even simpler way (as $\exp(-\beta P(v))$ can be bounded by $\beta P(v)(-1 + \frac{1}{2} \varepsilon \ln 2) \ln 2$) that the LHS of (1.9.12) is also bounded by $\exp(-(\varepsilon^2/3L)d)$.

Recalling (1.9.10), we have thereby shown that the probability that (1.9.7) does not hold for a randomly chosen ϕ is $< 2|\mathcal{P}| \exp(-(\varepsilon^2/3L)d)$. Hence this probability is less than 1 if $L \leq (\varepsilon^2/3 \log(2|\mathcal{P}|))d$. This completes the proof of Lemma 1.26, because (1.9.8) is an immediate consequence of (1.9.7). □

In the theory of AVC with feedback the following generalization was needed.

Lemma 1.27 *Let $\mathcal{V}, \mathcal{P}(\mathcal{E}(P))_{P \in \mathcal{P}})$ be given, where $\mathcal{V}$ is a finite set of vertices, $\mathcal{P} \subset \mathcal{P}(\mathcal{V})$, and for every $P \in \mathcal{P}$, $\mathcal{E}(P)$ is a set of edges in $\mathcal{V}$.*

Assume that for all $P \in \mathcal{P}$

$$\alpha(P) = \max\{P(v) : v \in \bigcup_{E \in \mathcal{P}(P)} E\} < 1. \tag{1.9.14}$$

Now, if there are positive numbers $\eta(P)$ for all $P \in \mathcal{P}$ such that for $L \geq 2$ and $\delta \in (0, 1)$

$$\left(\frac{1}{\alpha(P)}\right)^{1-\delta} \left[\eta(P) - \frac{e}{2L}\alpha^{\delta}(P)P(E)\right] > \ln\left\{2L \sum_{P \in \mathcal{P}} |\mathcal{E}(P)|\right\}, \tag{1.9.15}$$

then there is an L-coloring ϕ of $\mathcal{V}$, which satisfies for all $P \in \mathcal{P}$, $E \in \mathcal{E}(P)$, and $l \in \{1, 2, \dots, L\}$

$$\left| P\left(\phi^{-1}(l) \cap E\right) - \frac{1}{L}P(E) \right| < \eta(P). \tag{1.9.16}$$

Furthermore, for $\delta = \frac{1}{4}$, $\eta(P) = 2\alpha^{\frac{1}{4}}(P)$, and $\alpha = \max_{P \in \mathcal{P}} \alpha(P)$

$$\alpha^{-\frac{1}{2}} > \ln\left[2L \sum_{P \in \mathcal{P}} |\mathcal{E}(P)|\right] \tag{1.9.17}$$

implies (1.9.15) and thus (1.9.16) holds.

Proof We use the standard random L-coloring ϕ. Next we introduce the RV's

$$Y_{vl} = \begin{cases} 1, & \text{if } v \text{ gets color } l \\ 0 & \text{otherwise} \end{cases}$$

and

$$Z_l^P(E) = \sum\nolimits_{v \in E} P(v) Y_{vl} \text{ for } P \in \mathcal{P}.$$

With Bernstein's version of Chebyshev's inequality

$$\begin{aligned}
&\Pr\left(Z_l^P(E) > \tfrac{1}{L}P(E) + \eta(P)\right\} \\
&\leq \exp_e\left\{-\alpha(P)^{-(1-\delta)}\left[\tfrac{1}{L}P(E) + \eta(P)\right]\right\} \cdot \mathbb{E}\exp_e\left\{\alpha(P)^{-(1-\delta)} \sum\nolimits_{v \in E} P(v)Y_{vl}\right\} \\
&= \exp_e\left\{-\alpha(P)^{-(1-\delta)}\left[\tfrac{1}{L}P(E) + \eta(P)\right]\right\} \cdot \prod\nolimits_{v \in E} \mathbb{E}\exp_e\left\{\alpha(P)^{-(1-\delta)} P(v)Y_{vl}\right\} \\
&= \exp_e\left\{-\alpha(P)^{-(1-\delta)}\left[\tfrac{1}{L}P(E) + \eta(P)\right]\right\} \cdot \prod\nolimits_{v \in E}\left(\tfrac{L-1}{L} + \tfrac{1}{L}\exp_e\left\{\alpha(P)^{-(1-\delta)}P(v)\right\}\right).
\end{aligned}$$

Using Lagrange's remainder formula for the Taylor series of the exponential function we continue with the upper bound

$$\exp_e\left\{-\alpha(P)^{-(1-\delta)}\left[\frac{1}{L}P(E)+\eta(P)\right]\right\}\cdot$$

$$\prod_{v\in E}\left\{1+\frac{1}{L}\left[\alpha(P)^{-(1-\delta)}P(v)+\frac{[\alpha(P)^{-(1-\delta)}P(v)]^2\cdot e}{2}\right]\right\}$$

and since $\ln(1+x)<x$ for $x>0$ with the upper bound

$$\begin{aligned}
&\exp_e\left\{-\alpha(P)^{-(1-\delta)}\left[\frac{1}{L}P(E)+\eta(P)-\frac{1}{L}\sum_{v\in E}P(v)-\frac{e}{2L}\alpha(p)^{-(1-\delta)}\sum_{v\in E}P^2(v)\right]\right\}\\
=\;&\exp_2\left\{-\alpha(P)^{-(1-\delta)}\left[\eta(P)-\frac{e}{2L}\alpha(P)^{-(1-\delta)}\sum_{v\in E}P^2(v)\right]\right\}\\
\leq\;&\exp_2\left\{-\alpha(P)^{-(1-\delta)}\left[\eta(P)-\frac{e}{2L}\alpha(P)^{-(1-\delta)}\cdot\sum_{v\in E}\alpha(P)P(v)\right]\right\},
\end{aligned}$$

because $P(v)\leq\alpha(P)$ for $v\in E$.

The last upper bound equals

$$\exp_2\left\{-\alpha(P)^{-(1-\delta)}\left[\eta(P)-\frac{e}{2L}\alpha(P)^{\delta}P(E)\right]\right\}.$$

Analogously,

$$\Pr\left\{Z_l^P(E)<\frac{1}{L}P(E)-\eta(P)\right\}\leq\exp_e\left\{-\alpha(P)^{-(1-\delta)}\left[\eta(P)-\frac{e}{2L}\alpha(P)^{\delta}P(E)\right]\right\}$$

for all $P\in\mathcal{P}$, $E\in\mathcal{E}(P)$ and $l\in\{1,2,\ldots,L\}$.

Finally, since $\left(\frac{1}{\alpha(P)}\right)^{\frac{3}{4}}\left[2\alpha(P)^{\frac{1}{4}}-\frac{e}{2L}\alpha(P)^{\frac{1}{4}}P(E)\right]>\left(\frac{1}{\alpha(P)}\right)^{\frac{1}{2}}\geq\alpha^{-\frac{1}{2}}$, (1.9.17) implies (1.9.15). □

1.10 Color Carrying Lemma and Other Concepts and Results

1.10.1 Color Carrying Lemma

Definition 1.13 We say that a hypergraph $\mathcal{H}=(\mathcal{V},\mathcal{E})$ carries M colors, if there is a vertex coloring with M colors such that all these colors occur in every edge. Let $M(\mathcal{V},\mathcal{E})$ be the maximal number of colors carried by $\mathcal{H}$.

Lemma 1.28 (Color carrying lemma) *For every hypergraph* $\mathcal{H}=(\mathcal{V},\mathcal{E})$

$$M(\mathcal{V},\mathcal{E})\geq D_{\min}\cdot\log^{-1}(|\mathcal{E}|\cdot D_{\min})$$

Proof Use a standard random coloring of $\mathcal{V}$ with M colors. The probability that not every edge carries M colors is not larger than

$$\sum_{E \in \mathcal{E}} \sum_{i=1}^{M} (1 - \frac{1}{n})^{|\mathcal{E}|}$$

and for

$$|\mathcal{E}| M (1 - \frac{1}{n})^{D_{\min}} < 1$$

or even for $\log |\mathcal{E}| M < \frac{1}{n} \log D_{\min}$ and for $2 \leq M < \frac{D_{\min}}{\log |\mathcal{E}| M}$ a proper coloring exists, which implies

$$M(\mathcal{V}, \mathcal{E}) \geq (\log |\mathcal{E}| D_{\min})^{-1} D_{\min},$$

because $M \leq D_{\min}$. □

We turn to another fundamental problem in Graph Theory. Let $\mathcal{H} = \binom{[n]}{k}$ be a complete k-uniform hypergraph $\mathcal{H} = (\mathcal{V}, \mathcal{E})$, $|\mathcal{V}| = n$, $|E| = k$, $E \in \mathcal{E}$. Let L positive numbers $r_1, \dots, r_L$ be given. Consider a coloring of the edges of $\mathcal{H}$ by the "colors" $1, \dots, L$, that is, each edge has its own number from $[L]$. The question is, what is the minimal n_0 such that when $n > n_0$ for an arbitrary coloring of $\mathcal{H}$ there exists at least one color i such that the edges having this color generate a complete subgraph (clique) $\mathcal{H}_i = (\mathcal{V}_i, \mathcal{E}_i)$ with the number of vertices $|\mathcal{V}_i| > r_i$.

Theorem 1.5 (Ramsey 1930) *Let $r_1, \dots, r_L$, L be positive numbers. There exists n_0 such that for $n > n_0$ a k-uniform hypergraph with n vertices whose edges are colored by L numbers contains a monochromatic complete subhypergraph with $n' \geq r_i$ vertices.*

1.10.2 Other Basic Concepts

Notice that we called an L-coloring *strict* if in every edge $E \in \mathcal{E}$ all vertices have different colors. The minimum L such that a strict L-coloring of $\mathcal{H}$ exists we denote the (strong) *chromatic number* $\chi(\mathcal{H})$.

By Lemma 1.12 $\chi(\mathcal{H}) \leq d_{\max} + 1$. A much deeper result classifies the cases where $d_{\max} + 1$ colors are needed.

Theorem 1.6 (Brooks) *Let $\mathcal{G}$ be a connected graph. If $\mathcal{G}$ is neither a complete graph nor a cycle of odd length, then $\chi(\mathcal{G}) \leq d_{\max}$ holds. Otherwise it equals $d_{\max} + 1$*

In general $\chi(\mathcal{G})$ can be much smaller than $d_{\max}$ (for instance for a star).

Another concept of strict coloring due to Erdös and Hajnal requires that every edge (of cardinality at least 2) has at least 2 different colors. For graphs both concepts coincide. In coding correlated sources our concept finds more applications.

Dual to our concept of a chromatic number of a hypergraph there is the following concept.

Definition 1.14 The chromatic index ind($\mathcal{H}$) of a hypergraph is the minimal number of colors needed to color all edges in $\mathcal{E}$ such that no vertex occurs in two edges of the same color.

While investigating switching circuits Shannon proved that the chromatic index ind($\mathcal{G}$) of a multigraph $\mathcal{G}$ is less than $\frac{3}{2}d_{\max}$. A deep improvement, due to Vizing [14], is

Theorem 1.7 (Vizing)

$$d_{\max} \leq ind(\mathcal{G}) \leq d_{\max} + 1.$$

Note that in contrast to Brooks' result, here we have a strong lower bound!

Unfortunately we missed the following coloring concept. Strict colorings often require a large number of colors. In many cases it suffices to work with several coloring $f_1, \dots, f_k : \mathcal{V} \to \mathbb{N}$.

Definition 1.15 A collection of functions $\mathcal{F} = (f_1, \dots, f_k)$ is a **perfect hashing** of $\mathcal{H} = (\mathcal{V}, \mathcal{E})$, if for every $E \in \mathcal{E}$ there is an f_i which is injective on the edge E (**perfectly hashes** the edge E).

This means, that $\mathcal{H}$ can be decomposed into hypergraphs $\mathcal{H}_i = (\mathcal{V}, \mathcal{E}_i)$, $\bigcup_i \mathcal{E}_i = \mathcal{E}$ such that f_i is a strict coloring of $\mathcal{H}$.

In applications it is relevant that $\sum_{i=1}^{k} ||f_i||$ (where $||\ ||$ denotes the cardinality of the range of such a function) is usually much smaller than $\chi(\mathcal{H})$. *The idea of hashing can be combined with any of the coloring concepts introduced and not just with strict colorings.*

We close this chapter with a few historical remarks. Whereas our "Theory of hypergraph coloring" is essentially of an asymptotic type, but often good enough for an asymptotic theory of coding, the traditional work on graph coloring is primarily concerned with exact bounds. The book [2] mostly shaped in this spirit.

The most famous result on coloring is the Four Color Theorem, which states that every planar graph has chromatic number $\chi \leq 4$, and was proved by Appel and Haken with extensive help by a computer.

There is a related open conjecture concerning the structure of L-colorable graphs. Hadwiger's conjecture [6] says that a graph $\mathcal{G}$ with $\chi(\mathcal{G}) \geq L$ contains a subgraph contractible to the complete graph $\mathcal{K}_L$.

According to Wagner the case $n = 5$ is equivalent to the Four Color Conjecture and therefore proved. Robertson, Sanders Seymour, and Thomas settled Hadwiger's conjecture for $L \leq 6$ and thereby also gave a new proof of the Four Color Theorem.

References

1. A. Ahlswede, I. Althöfer, C. Deppe, U. Tamm (eds.), *Storing and Transmitting Data, Rudolf Ahlswede's Lectures on Information Theory 1*, Foundations in Signal Processing, Communications and Networking, vol. 10, 1st edn. (Springer, 2014)
2. R. Ahlswede, V. Blinovsky, *Lectures on Advances in Combinatorics* (Springer, Berlin, 2008)
3. G. Birkhoff, Three observations on linear algebra. Univ. Nac. Tucumán. Revista A. **5**, 147–151 (1946)
4. Z. Blázsik, M. Hujter, A. Pluhaár, Z. Tuza, Graphs with no induced C4 and 2K2. Discret. Math. **115**, 51–55 (1993)
5. T.M. Cover, J.A. Thomas, *Elements of Information Theory*, 2nd edn. (Wiley, New York, 2006)
6. H. Hadwiger, Über eine Klassifikation der Streckenkomplexe. Vierteljschr. Naturforsch. Ges. Zürich **88**, 133–143 (1943)
7. G.G. Lorentz, On a problem of additive number theory. Proc. Am. Math. Soc. **5**(5), 838–841 (1954)
8. L. Lóvasz, Minimax theorems for hypergraphs, in *Hypergraph Seminar*. Lecture Notes in Mathematics, vol. 441 (Springer, Berlin, 1974), pp. 111–126
9. L. Lóvasz, On the ratio of optimal integral and fractional covers. Discret. Math. **13**, 383–390 (1975)
10. V. Rödl, On a packing and covering problem. Eur. J. Comb. **5**, 69–78 (1985)
11. M. Rosenfeld, On a problem of C.E. Shannon in graph theory. Proc. Am. Math. Soc. **18**, 315–319 (1967)
12. C.E. Shannon, The zero error capacity of a noisy channel. I.R.E. Trans. Inf. Theory **IT–2**, 8–19 (1956)
13. S. Stahl, n-tuple colorings and associated graphs. J. Comb. Theory (B) **29**, 185–203 (1976)
14. V.G. Vizing, A bound on the external stability number of a graph. Dokl. Akad. Nauk SSSR **164**, 729–731 (1965)

Further Readings

15. M.O. Albertson, J.P. Hutchinson, On six-chromatic toroidal graphs. Proc. Lond. Math. Soc. **3**(41), 533–556 (1980)
16. R. Aharoni, I. Ben-Arroyo, A.J.Hoffman Hartman, Path-partitions and packs of acyclic digraphs. Pacific J. Math. **118**, 249–259 (1985)
17. S. Benzer, On the topology of the genetic fine structure. Proc. National Acad. Sci. U. S. A. **45**(11), 1607–1620 (1959)
18. C. Berge, *Théorie des graphes et ses applications* (Dunod, Paris, 1958)
19. C. Berge, Les problémes de coloration en Théorie des Graphes. Publ. Inst. Statist. Univ. Paris **9**, 123–160 (1960)
20. C. Berge, Färbung von Graphen, deren sämtliche bzw. deren ungerade Kreise starr sind, Wiss. Zeitschrift der Martin-Luther-Universität Halle-Wittenberg, 114–115 (1961)
21. C. Berge, *The Theory of Graphs and its Applications* (Methuen, London, 1961), p. 95
22. C. Berge, Sur un conjecture relative au probleme des codes optimaux, Comm. 13ieme Assemblee Gen. URSI, Tokyo (1962)
23. C. Berge, *Perfect graphs, in Six Papers on Graph Theory* (Indian Statistical Institute, Calcutta, Research and Training School, 1963), pp. 1–21
24. C. Berge, Une Application de la Théorie des Graphes à un problème de Codage, in *Automata Theory*, ed. by E.R. Caianiello (Academic Press, New York, 1966), pp. 25–34
25. C. Berge, *Some classes of perfect graphs, in Graph Theory and Theoretical Physics* (Academic Press, New York, 1967), pp. 155–165

26. C. Berge, The rank of a family of sets and some applications to graph theory, in *Recent Progress in Combinatorics (Proceedings of the Third Waterloo Conference on Combinatorics, 1968)* (Academic Press, New York, 196), pp. 49–57
27. C. Berge, Some classes of perfect graphs, in *Combinatorial Mathematics and its Applications, Proceedings of the Conference Held at the University of North Carolina, Chapel Hill, 1967* (University of North Carolina Press, 539–552, 1969)
28. C. Berge, Graphes et Hypergraphes, Monographies Universitaires de Mathématiques, No. 37. Dunod, Paris (1970)
29. C. Berge, Balanced matrices. Math. Program. **2**(1), 19–31 (1972)
30. C. Berge, Graphs and Hypergraphs, Translated from the French by Edward Minieka, North-Holland Mathematical Library, vol. 6. North-Holland Publishing Co., Amsterdam-London; American Elsevier Publishing Co., Inc., New York (1973)
31. C. Berge, A theorem related to the Chvátal conjecture, in *Proceedings of the Fifth British Combinatorial Conference (University of Aberdeen, Aberdeen, 1975)*, Congressus Numerantium, No. XV, Utilitas Math., Winnipeg, Man. (1976), pp. 35–40
32. C. Berge, k-optimal partitions of a directed graph. Eur. J. Comb. **3**, 97–101 (1982)
33. C. Berge, Path-partitions in directed graphs, in *Combinatorial Mathematics*, ed. by C. Berge, D. Bresson, P. Camion, J.F. Maurras, F. Sterboul (North-Holland, Amsterdam, 1983), pp. 32–44
34. C. Berge, A property of k-optimal path-partitions, in *Progress in Graph Theory*, ed. by J.A. Bondy, U.S.R. Murty (Academic Press, New York, 1984), pp. 105–108
35. C. Berge, On the chromatic index of a linear hypergraph and the Chvátal conjecture, in *Annals of the New York Academy of Sciences*, vol. 555, ed. by G.S. Bloom, R.L. Graham, J. Malkevitch, C. Berge (1989), pp. 40–44
36. C. Berge, *Hypergraphs, Combinatorics of Finite Sets, Chapter 1, Section 4* (North-Holland, New York, 1989)
37. C. Berge, On two conjectures to generalize Vizing's Theorem. Le Matematiche **45**, 15–24 (1990)
38. C. Berge, The q-perfect graphs I: the case $q = 2$, in *Sets, Graphs and Numbers*, ed. by L. Lovász, D. Miklós, T. Szönyi. Colloq. Math. Soc. Janós Bolyai, vol. 60 (1992), pp. 67–76
39. C. Berge, The q-perfect graphs II, in *Graph Theory, Combinatorics and Applications*, ed. by Y. Alavi, A. Schwenk (Wiley Interscience, New York, 1995), pp. 47–62
40. C. Berge, The history of the perfect graphs. Southeast Asian Bull. Math. **20**(1), 5–10 (1996)
41. C. Berge, Motivations and history of some of my conjectures, in *Graphs and combinatorics (Marseille, 1995)* (1997), pp. 61–70 (Discrete Math. 165–166)
42. C. Berge, V. Chvátal (eds.), *Topics on Perfect Graphs*. Annals of Discrete Mathematics, vol. 21 (North Holland, Amsterdam, 1984)
43. C. Berge, P. Duchet, Strongly perfect graphs, in *Topics on Perfect Graphs*, ed. by C. Berge, V. Chvátal. North-Holland Mathematics Studies, vol. 88 (North-Holland, Amsterdam, 1984), pp. 57–61 (Annals of Disc. Math. 21)
44. C. Berge, A.J.W. Hilton, On two conjectures about edge colouring for hypergraphs. Congr. Numer. **70**, 99–104 (1990)
45. C. Berge, M. Las, Vergnas, Sur un théorème du type König pour hypergraphes. Ann. New York Acad. Sci. **175**, 32–40 (1970)
46. I. Ben-Arroyo Hartman, F. Sale, D. Hershkowitz, On Greene's Theorem for digraphs. J. Graph Theory **18**, 169–175 (1994)
47. A. Beutelspacher, P.-R. Hering, Minimal graphs for which the chromatic number equals the maximal degree. Ars Combinatoria **18**, 201–216 (1983)
48. O.V. Borodin, A.V. Kostochka, An upper bound of the graph's chromatic number, depending on the graph's degree and density. J. Comb. Theory B **23**, 247–250 (1977)
49. A. Brandstädt, V.B. Le, J.P. Spinrad, *Graph classes: a survey* (SIAM Monographs on Discrete Mathematics and Applications (SIAM, Philadelphia, 1999)
50. R.C. Brigham, R.D. Dutton, A compilation of relations between graph invariants. Networks **15**(1), 73–107 (1985)

51. R.C. Brigham, R.D. Dutton, A compilation of relations between graph invariants: supplement I. Networks **21**, 412–455 (1991)
52. R.L. Brooks, On colouring the nodes of a network. Proc. Camb. Philos. Soc. **37**, 194–197 (1941)
53. M. Burlet, J. Fonlupt, Polynomial algorithm to recognize a Meyniel graph, in *Topics on Perfect Graphs*, ed. by C. Berge, V. Chvátal. North-Holland Mathematics Studies, vol. 88 (North-Holland, Amsterdam, 1984), pp. 225–252 (Annals of Discrete Math. 21)
54. K. Cameron, On k-optimum dipath partitions and partial k-colourings of acyclic digraphs. Eur. J. Comb. **7**, 115–118 (1986)
55. P.J. Cameron, A.G. Chetwynd, J.J. Watkins, Decomposition of snarks. J. Graph Theory **11**, 13–19 (1987)
56. P. Camion, *Matrices totalement unimodulaires et problèmes combinatoires* (Université Libre de Bruxelles, Thèse, 1963)
57. P.A. Catlin, Another bound on the chromatic number of a graph. Discret. Math. **24**, 1–6 (1978)
58. W.I. Chang, E. Lawler, Edge coloring of hypergraphs and a conjecture of Erdös-Faber-Lovász. Combinatorica **8**, 293–295 (1988)
59. C.-Y. Chao, On a problem of C. Berge. Proc. Am. Math. Soc. **14**, 80 (1963)
60. M. Chudnovsky, G. Cornuejols, X. Liu, P. Seymour, K. Vuskovic, Recognizing Berge graphs. Combinatorica **25**, 143–186 (2005)
61. M. Chudnovsky, N. Robertson, P. Seymour, R. Thomas, The strong perfect graph theorem. Ann. Math. **164**, 51–229 (2006)
62. V. Chvátal, Unsolved problem no. 7, in *Hypergraph Seminar*, ed. by C. Berge, D.K. Ray-Chaudhuri. Lecture Notes in Mathematics, vol. 411 (Springer, Berlin, 1974)
63. V. Chvátal, Intersecting families of edges in hypergraphs having the hereditary property, in *Hypergraph Seminar (Proceedings of the First Working Seminar, Ohio State University, Columbus, Ohio, 1972; dedicated to Arnold Ross)*. Lecture Notes in Mathematics, vol. 411 (Springer, Berlin, 1974), pp. 61–66
64. V. Chvátal, On certain polytopes associated with graphs. J. Comb. Theory Ser. B **18**, 138–154 (1975)
65. V. Chvátal, On the strong perfect graph conjecture. J. Comb. Theory Ser. B **20**, 139–141 (1976)
66. V. Chvátal, Perfectly ordered graphs, in *Topics on Perfect Graphs*, ed. by C. Berge, V. Chvátal. North-Holland Mathematics Studies, vol. 88 (North-Holland, Amsterdam, New York, 1984), pp. 63–65 (Annals of Disc. Math. 21)
67. V. Chvátal, Star-cutsets and perfect graphs. J. Comb. Theory Ser. B **39**(3), 189–199 (1985)
68. V. Chvátal, J. Fonlupt, L. Sun, A. Zemirline, Recognizing dart-free perfect graphs. SIAM J. Comput. **31**(5), 1315–1338 (2002)
69. V. Chvátal, D.A. Klarner, D.E. Knuth, Selected combinatorial research problems, Technical report STAN-CS, 72-292 (1972)
70. J. Colbourn, M. Colbourn, The chromatic index of cyclic Steiner 2-design. Int. J. Math. Sci. **5**, 823–825 (1982)
71. M. Conforti, G. Cornuéjols, Graphs without odd holes, parachutes or proper wheels: a generalization of Meyniel graphs and of line graphs of bipartite graphs. J. Comb. Theory Ser. B **87**, 300–330 (2003)
72. M. Conforti, M.R. Rao, Structural properties and decomposition of linear balanced matrices. Math. Program. Ser. A B **55**(2), 129–168 (1992)
73. M. Conforti, M.R. Rao, Articulation sets in linear perfect matrices I: forbidden configurations and star cutsets. Discret. Math. **104**(1), 23–47 (1992)
74. M. Conforti, M.R. Rao, Articulation sets in linear perfect matrices II: the wheel theorem and clique articulations. Discret. Math. **110**(1–3), 81–118 (1992)
75. M. Conforti, M.R. Rao, Testing balancedness and perfection of linear matrices. Math. Program. Ser. A **61**(1), 1–18 (1993)
76. M. Conforti, G. Cornuéjols, A. Kapoor, K. Vuskovic, A mickey-mouse decomposition theorem, in *Integer Programming and Combinatorial Optimization (Copenhagen, 1995)*. Lecture Notes in Computer Science, vol. 920 (Springer, Berlin, 1995), pp. 321–328

77. M. Conforti, G. Cornuéjols, A. Kapoor, K. Vuskovic, Even and odd holes in cap-free graphs. J. Graph Theory **30**(4), 289–308 (1999)
78. M. Conforti, G. Cornuéjols, M.R. Rao, Decomposition of balanced matrices. J. Comb. Theory Ser. B **77**(2), 292–406 (1999)
79. M. Conforti, B. Gerards, A. Kapoor, A theorem of Truemper. Combinatorica **20**(1), 15–26 (2000)
80. M. Conforti, G. Cornuéjols, A. Kapoor, K. Vuskovic, Balanced 0, $\pm$1 matrices I, decomposition. J. Comb. Theory Ser. B **81**(2), 243–274 (2001)
81. M. Conforti, G. Cornuéjols, A. Kapoor, K. Vuskovic, Balanced 0, $\pm$1 matrices II, recognition algorithm. J. Comb. Theory Ser. B **81**(2), 275–306 (2001)
82. M. Conforti, G. Cornuéjols, G. Gasparyan, K. Vuskovic, Perfect graphs, partitionable graphs and cutsets. Combinatorica **22**(1), 19–33 (2002)
83. M. Conforti, G. Cornuéjols, A. Kapoor, K. Vuskovic, Even-hole free graphs, part I: decomposition theorem. J. Graph Theory **39**(1), 6–49, vol. 40 (2002)
84. M. Conforti, G. Cornuéjols, A. Kapoor, K. Vuskovic, Even-hole free graphs, part II: recognition algorithm. J. Graph Theory **40**(4), 238–266 (2002)
85. M. Conforti, G. Cornuéjols, K. Vuskovic, Decomposition of odd-hole-free graphs by double star cutsets and 2-joins. Discret. Appl. Math. **141**(1–3), 41–91 (2004)
86. M. Conforti, G. Cornuéjols, K. Vuskovic, Square-free perfect graphs. J. Comb. Theory B 257–307 (2004)
87. G. Cornuéjols, *Combinatorial optimization: packing and covering, in CBMS-NSF Regional Conference Series in Applied Mathematics*, vol. 74 (SIAM, Philadelphia, 2001)
88. G. Cornuéjols, The strong perfect graph conjecture, in *Proceedings of the International Congress of Mathematicians III: Invited Lectures Beijing* (2002), pp. 547–559
89. G. Cornuéjols, W.H. Cunningham, Compositions for perfect graphs. Discret. Math. **55**(3), 245–254 (1985)
90. G. Cornuéjols, B. Reed, Complete multi-partite cutsets in minimal imperfect graphs. J. Comb. Theory Ser. B **59**(2), 191–198 (1993)
91. I. Csiszár, J. Körner, *Information Theory. Coding Theorems for Discrete Memoryless Systems*. Probability and Mathematical Statistics (Academic Press Inc., New York, 1981)
92. I. Csiszár, J. Körner, L. Lovász, K. Marton, G. Simonyi, Entropy splitting for antiblocking corners and perfect graphs. Combinatorica **10**(1), 27–40 (1990)
93. W.H. Cunningham, J.A. Edmonds, A combinatorial decomposition theory. Can. J. Math. **32**(3), 734–765 (1980)
94. C.M.H. de Figueiredo, S. Klein, Y. Kohayakawa, B. Reed, Finding skew partitions efficiently. J. Algorithms **37**, 505–521 (2000)
95. B. Descartes, A three colour problem, Eureka (April 1947; solution March 1948) and Solution to Advanced Problem No. 4526, Amer. Math. Monthy, vol. 61 (1954), p. 352
96. R.P. Dilworth, A decomposition theorem for partially ordered sets. Ann. Math. **2**, 161–166 (1950)
97. G.A. Dirac, Map-colour theorems. Can. J. Math. **4**, 480–490 (1952)
98. G.A. Dirac, On rigid circuit graphs. Abh. Math. Sem. Univ. Hamburg **25**, 71–76 (1961)
99. R.J. Duffin, The extremal length of a network. J. Math. Anal. Appl. **5**, 200–215 (1962)
100. R.D. Dutton, R.C. Brigham, INGRID: a software tool for extremal graph theory research. Congr. Numerantium **39**, 337–352 (1983)
101. R.D. Dutton, R.C. Brigham, F. Gomez, INGRID: a graph invariant manipulator. J. Symb. Comput. **7**, 163–177 (1989)
102. J. Edmonds, Minimum partition of a matroid into independent subsets. J. Res. Nat. Bur. Stand. Sect. B **69B**, 67–72 (1965)
103. J. Edmonds, Maximum matching and a polyhedron with 0, 1-vertices. J. Res. Nat. Bur. Stand. Sect. B **69B**, 125–130 (1965)
104. J. Edmonds, Paths, trees, and flowers. Can. J. Math. **17**, 449–467 (1965)
105. J. Edmonds, Lehman's switching game and a theorem of Tutte and Nash-Williams. J. Res. Nat. Bur. Stand. Sect. B **69B**, 73–77 (1965)

106. J. Edmonds, Optimum branchings. J. Res. Nat. Bur. Stand. Sect. B **71B**, 233–240 (1967)
107. J. Edmonds, Submodular functions, matroids, and certain polyhedra, in *Combinatorial Structures and their Applications (Proceedings of the Calgary International Conference, Calgary, Alberta, 1969)* (Gordon and Breach, New York, 1970), pp. 69–87
108. J. Edmonds, Matroids and the greedy algorithm, (Lecture, Princeton, 1967). Math. Programming **1**, 127–136 (1971)
109. J. Edmonds, *Edge-disjoint branchings, in Combinatorial Algorithms (Courant Computer Science Symposium 9, New York University, New York, 1972)* (Algorithmics Press, New York, 1973), pp. 91–96
110. J. Edmonds, Submodular functions, matroids, and certain polyhedra, in *Combinatorial optimization-Eureka, you shrink!*. Lecture Notes in Computer Science, vol. 2570 (Springer, Berlin, 2003), pp. 11–26
111. J. Edmonds, D.R. Fulkerson, Bottleneck extrema. J. Comb. Theory **8**, 299–306 (1970)
112. P. Erdös, Graph theory and probability. Canad. J. Math. **11**, 34–38 (1959)
113. P. Erdös, Problems and results in Graph Theory, in *Proceedings of the 5th British Combinatorial Conference*, ed. by C.St.J.A. Nash-Williams, J. Sheehan. Utilitas Math., vol. 15 (1976)
114. P. Erdös, A. Hajnal, On chromatic number of graphs and set-systems. Acta Math. Acad. Sci. Hung. **17**, 61–99 (1966)
115. P. Erdös, V. Faber, L. Lovász, Open problem, in *Hypergraph Seminar*, ed. by C. Berge, D. Ray Chaudhuri. Lecture Notes in Mathematics, vol. 411 (Springer, Berlin, 1974)
116. P. Erdös, C. Ko, R. Rado, Intersection theorems for systems of finite sets. Quart. J. Math. Oxford Ser. **2**(12), 313–320 (1961)
117. J. Fonlupt, J.P. Uhry, Transformations which preserve perfectness and H-perfectness of graphs. Ann. Discret. Math. **16**, 83–95 (1982)
118. J. Fonlupt, A. Zemirline, *A polynomial recognition algorithm for perfect $K_4 - \{e\}$-free graphs, Rapport Technique RT-16* (Artemis, IMAG, Grenoble, France, 1987)
119. J.-L. Fouquet, Perfect Graphs with no $2K_2$ and no K_6, Technical report, Universite du Maine, Le Mans, France (1999)
120. J.-L. Fouquet, F. Maire, I. Rusu, H. Thuillier, *Unpublished internal report* (Univ, Orléans, LIFO, 1996)
121. L.R. Ford, D.R. Fulkerson, *Flows in Networks* (Princeton University Press, Princeton, 1962)
122. D.R. Fulkerson, The maximum number of disjoint permutations contained in a matrix of zeros and ones. Can. J. Math. **16**, 729–735 (1964)
123. D.R. Fulkerson, *Networks, frames, blocking systems, in Mathematics of the Decision Sciences, Part 1, (Seminar, Stanford, California, 1967)* (American Mathematical Society, Providence, 1968), pp. 303–334
124. D.R. Fulkerson, The perfect graph conjecture and pluperfect graph theorem, in *2nd Chapel Hill Conference on Combinatorial Mathematics and its Applications, Chapel Hill, N.C.* (1969), pp. 171–175
125. D.R. Fulkerson, Notes on combinatorial mathematics: anti-blocking polyhedra, Rand corporation, Memorandum RM-6201/1-PR (1970)
126. D.R. Fulkerson, *Blocking polyhedra, in Graph theory and its Applications* (Academic, New York, 1970), pp. 93–111
127. D.R. Fulkerson, Blocking and anti-blocking pairs of polyhedra. Math. Program. **1**, 168–194 (1971)
128. D.R. Fulkerson, *Disjoint common partial transversals of two families of sets, in Studies in Pure Mathematics (Presented to Richard Rado)* (Academic Press, London, 1971), pp. 107–112
129. D.R. Fulkerson, Anti-blocking polyhedra. J. Comb. Theory Ser. B **12**, 50–71 (1972)
130. D.R. Fulkerson, On the perfect graph theorem, in *Mathematical Progamming (Proceedings of the Advanced Seminar, University of Wisconsin, Madison, Wisconsin, 1972)*, ed. by T.C. Hu, S.M. Robinson, Mathematical Research Center Publications, vol. 30, (Academic Press, New York, 1973), pp. 69–76
131. D.R. Fulkerson (ed.), *Studies in Graph Theory*. Studies in Mathematics, vol. 12 (The Mathematical Association of America, Providence, 1975)

132. Z. Füredi, The chromatic index of simple hypergraphs. Res. Problem Graphs Comb. **2**, 89–92 (1986)
133. T. Gallai, Maximum-minimum Sätze über Graphen. Acta Math. Acad. Sci. Hungar. **9**, 395–434 (1958)
134. T. Gallai, Über extreme Punkt- und Kantenmengen. Ann. Univ. Sci. Budapest. Eötvös Sect. Math. **2**, 133–138 (1959)
135. T. Gallai, Graphen mit triangulierbaren ungeraden Vielecken, Magyar Tud. Akad. Mat. Kutató Int. Közl. **7**, 3–36 (1962)
136. T. Gallai, On directed paths and circuits, in *Theory of Graphs*, ed. by P. Erdös, G. Katona (Academic Press, New York, 1968), pp. 115–118
137. T. Gallai, A.N. Milgram, Verallgemeinerung eines graphentheoretischen Satzes von Rédei. Acta Sci. Math. **21**, 181–186 (1960)
138. F. Gavril, Algorithms on circular-arc graphs. Networks **4**, 357–369 (1974)
139. J.F. Geelen, Matchings, Matroids and unimodular Matrices, Ph.D. thesis, University of Waterloo, 1995
140. D. Gernert, A knowledge-based system for graph theory. Methods Oper. Res. **63**, 457–464 (1989)
141. D. Gernert, Experimental results on the efficiency of rule-based systems, in *Operations Research '92*, ed. by A. Karmann et al. (1993), pp. 262–264
142. D. Gernert, Cognitive aspects of very large knowledge-based systems. Cogn. Syst. **5**, 113–122 (1999)
143. D. Gernert, L. Rabern, A knowledge-based system for graph theory, demonstrated by partial proofs for graph-colouring problems. MATCH Commun. Math. Comput. Chem. **58**(2), 445–460 (2007)
144. A. Ghouila-Houri, Sur une conjecture de Berge (mimeo.), Institut Henri Poincaré (1960)
145. A. Ghouila-Houri, Caractérisation des matrices totalement unimodulaires. C. R. Acad. Sci. Paris **254**, 1192–1194 (1962)
146. A. Ghouila-Houri, Caractérisation des graphes non orientés dont on peut orienter les arêtes de maniére à obtenir le graphe d'une relation d'ordre. C. R. Acad. Sci. Paris **254**, 1370–1371 (1962)
147. P.C. Gilmore, A.J. Hoffman, A characterization of comparability graphs and of interval graphs. Canad. J. Math. **16**, 539–548 (1964)
148. M. Gionfriddo, Zs. Tuza, On conjectures of Berge and Chvátal. Discret. Math. **124**, 76–86 (1994)
149. M.K. Goldberg, Construction of class 2 graphs with maximum vertex degree 3. J. Comb. Theory Ser. B **31**, 282–291 (1981)
150. M.C. Golumbic, *Algorithmic Graph Theory and Perfect Graphs*. Computer Science and Applied Mathematics (Academic Press, New York, 1980). Second edition, Annals of Discrete Mathematics 57, Elsevier, 2004
151. R. Gould, *Graph Theory* (Benjamin Publishing Company, Menlo Park, 1988)
152. C. Greene, D.J. Kleitman, The structure of Sperner k-families. J. Comb. Theory Ser. A **34**, 41–68 (1976)
153. M. Grötschel, L. Lovász, A. Schrijver, *Geometric Algorithms and Combinatorial Optimization* (Springer, Berlin, 1988)
154. A. Hajnal, J. Surányi, Über die Auflösung von Graphen in vollständige Teilgraphen. Ann. Univ. Sci. Budapest. Eötvös Sect. Math. **1**, 53–57 (1958)
155. G. Hajos, Über eine Art von Graphen, Int. Math. Nachr. **11** (1957)
156. F. Harary, C. Holtzmann, Line graphs of bipartite graphs. Rev. Soc. Mat. Chile **1**, 19–22 (1974)
157. M. Henke, A. Wagler, Auf dem Weg von der Vermutung zum Theorem: Die Starke-Perfekte-Graphen-Vermutung. DMV-Mitteilungen **3**, 22–25 (2002)
158. N. Hindman, On a conjecture of Erdös, Farber. Lovász about n-colorings. Canad. J. Math. **33**, 563–570 (1981)

159. C.T. Hoàng, Some properties of minimal imperfect graphs. Discret. Math. **160**(1–3), 165–175 (1996)
160. A.J. Hoffman, Some recent applications of the theory of linear inequalities to extremal combinatorial analysis. Proc. Sympos. Appl. Math. **10**, 113–127 (1960)
161. A.J. Hoffman, Extending Greene's Theorem to directed graphs. J. Comb. Theory Ser. A **34**, 102–107 (1983)
162. A.J. Hoffman, J.B. Kruskal, Integral boundary points of convex polyhedra, in *Linear Inequalities and Related Systems*. Annals of Mathematics Studies, vol. 38 (Princeton University Press, Princeton, 1956), 223–246
163. P. Horák, A coloring problem related to the Erdös-Faber-Lovász Conjecture. J. Comb. Theory Ser. B **50**, 321–322 (1990)
164. S. Hougard, A. Wagler, Perfectness is an elusive graph property, Preprint ZR 02–11, ZIB, 2002. SIAM J. Comput. **34**(1), 109–117 (2005)
165. T.C. Hu, Multi-commodity network flows. Oper. Res. **11**(3), 344–360 (1963)
166. R. Isaacs, Infinite families of non-trivial trivalent graphs which are not Tait colorable. Am. Math. Mon. **82**, 221–239 (1975)
167. T. Jensen, G.F. Royle, Small graphs with chromatic number 5: a computer search. J. Graph Theory **19**, 107–116 (1995)
168. D.A. Kappos, *Strukturtheorie der Wahrscheinlichkeitsfelder und -Räume, Ergebnisse der Mathematik und ihrer Grenzgebiete, Neue Folge, Heft 24* (Springer, Berlin, 1960)
169. H.A. Kierstead, J.H. Schmerl, The chromatic number of graphs which neither induce $K_{1,3}$ nor $K_5 - e$. Discret. Math. **58**, 253–262 (1986)
170. A.D. King, B.A. Reed, A. Vetta, An upper bound for the chromatic number of line graphs, Lecture given at EuroComb 2005, DMTCS Proc, AE, 151–156 (2005)
171. D. König, Über Graphen und ihre Anwendung auf Determinantentheorie und Mengenlehre. Math. Ann. **77**, 453–465 (1916)
172. D. König, Graphen und Matrizen. Math. Fiz. Lapok **38**, 116–119 (1931)
173. J. Körner, A property of conditional entropy. Studia Sci. Math. Hungar. **6**, 355–359 (1971)
174. J. Körner, An extension of the class of perfect graphs. Studia Sci. Math. Hungar. **8**, 405–409 (1973)
175. J.Körner, Coding of an information source having ambiguous alphabet and the entropy of graphs, in *Transactions of the 6th Prague Conference on Information Theory, etc., 1971, Academia, Prague* (1973), pp. 411–425
176. J. Körner, Fredman-Komlos bounds and information theory. SIAM J. Alg. Disc. Math. **7**, 560–570 (1986)
177. J. Körner, G. Longo, Two-step encoding for finite sources. IEEE Trans. Inf. Theory **19**, 778–782 (1973)
178. J. Körner, K. Marton, New bounds for perfect hashing via information theory. Eur. J. Comb. **9**(6), 523–530 (1988)
179. J. Körner, K. Marton, Graphs that split entropies. SIAM J. Discret. Math. **1**(1), 71–79 (1988)
180. J. Körner, A. Sgarro, A new approach to rate-distortion theory. Rend. Istit. Mat. Univ. di Trieste **18**(2), 177–187 (1986)
181. A.V. Kostochka, M. Stiebitz, Excess in colour-critical graphs, in Graph theory and combinatorial biology (Proceedings of Balatonlelle). Bolyai Society Mathematical Studies **7**, 87–99 (1996)
182. H.V. Kronk, The chromatic number of triangle-free graphs. Lecture Notes in Mathematics **303**, 179–181 (1972)
183. H.W. Kuhn, Variants of the Hungarian method for assignment problems. Naval Res. Logist. Q. **3**, 253–258 (1956)
184. E.L. Lawler, Optimal matroid intersections, in *Combinatorial Structures and Their Applications*, ed. by R. Guy, H. Hanani, N. Sauer, J. Schonheim (Gordon and Breach, 1970), p. 233
185. A. Lehman, On the width-length inequality, (Mimeo. 1965). Math. Program. **17**, 403–413 (1979)

186. P.G.H. Lehot, An optimal algorithm to detect a line graph and output its root graph. J. Assoc. Comput. Mach. **21**, 569–575 (1974)
187. C.G. Lekkerkerker, C.J. Boland, Representation of a finite graph by a set of intervals on the real line. Fund. Math. **51**, 45–64 (1962)
188. C. Linhares-Sales, F. Maffray, Even pairs in square-free Berge graphs, Laboratoire Leibniz Res. Rep. 51-2002 (2002)
189. N. Linial, Extending the Greene-Kleitman theorem to directed graphs. J. Comb. Theory Ser. A **30**, 331–334 (1981)
190. L. Lovász, On chromatic number of finite set-systems. Acta Math. Acad. Sci. Hungar. **19**, 59–67 (1968)
191. L. Lovász, Normal hypergraphs and the perfect graph conjecture. Discret. Math. **2**(3), 253–267 (1972)
192. L. Lovász, A characterization of perfect graphs. J. Comb. Theory Ser. B **13**, 95–98 (1972)
193. L. Lovász, On the Shannon capacity of a noisy channel. I.R.E. Trans. Inf. Theory **25**, 1–7 (1979)
194. L. Lovász, Perfect graphs, in *Selected Topics in Graph Theory*, ed. by L.W. Beineke, R.J. Wilson, vol. 2, (Academic Press, New York, 1983), pp. 55–87
195. L. Lovász, Normal hypergraphs and the weak perfect graph conjecture, in *Topics on Perfect Graphs*, ed. by C. Berge, V. Chvátal. North-Holland Mathematics Studies, vol. 88 (North-Holland, Amsterdam, 1984), pp. 29–42 (Ann. Disc. Math. 21)
196. F. Maffray, B.A. Reed, A description of claw-free perfect graphs. J. Comb. Theory Ser. B **75**(1), 134–156 (1999)
197. S.E. Markosjan, I.A. Karapetjan, Perfect graphs. Akad. Nauk Armjan. SSR Dokl. **63**(5), 292–296 (1976)
198. K. Marton, On the Shannon capacity of probabilistic graphs. J. Comb. Theory Ser. B **57**(2), 183–195 (1993)
199. R.J. McEliece, *The Theory of Information and Coding*, 2nd edn. Encyclopedia of Mathematics and its Applications, vol. 86 (Cambridge University Press, Cambridge, 2002)
200. R. Merris, *Graph Theory* (Wiley, New York, 2001)
201. H. Meyniel, On the perfect graph conjecture. Discret. Math. **16**(4), 339–342 (1976)
202. H. Meyniel, The graphs whose odd cycles have at least two chords, in *Topics on Perfect Graphs*, ed. by C. Berge, V. Chvátal (North-Holland, Amsterdam, 1984), pp. 115–120
203. H. Meyniel, private communication with C. Berge, 1985 (or 1986?)
204. H. Meyniel, A new property of critical imperfect graphs and some consequences. Eur. J. Comb. **8**, 313–316 (1987)
205. N.D. Nenov, On the small graphs with chromatic number 5 without 4-cliques. Discret. Math. **188**, 297–298 (1998)
206. J. Nesetril, K-chromatic graphs without cycles of length ≤ 7. Comment. Math. Univ. Carolina **7**, 373–376 (1966)
207. S. Olariu, Paw-free graphs. Inf. Process. Lett. **28**, 53–54 (1988)
208. O. Ore, *Theory of Graphs*. American Mathematical Society Colloquium publications, vol. 38 (American Mathematical Society, Providence, 1962)
209. M.W. Padberg, Perfect zero-one matrices. Math. Program. **6**, 180–196 (1974)
210. K.R. Parthasarathy, G. Ravindra, The strong perfect graph conjecture is true for $K_{1,3}$-free graphs. J. Comb. Theory B **21**, 212–223 (1976)
211. C. Payan, private communication with C. Berge (1981)
212. G. Polya, Aufgabe 424. Arch. Math. Phys. **20**, 271 (1913)
213. M. Preissmann, C-minimal snarks. Ann. Discret. Math. **17**, 559–565 (1983)
214. H.J. Prömel, A. Steger, Almost all Berge graphs are perfect. Comb. Probab. Comput. **1**(1), 53–79 (1992)
215. L. Rabern, On graph associations. SIAM J. Discret. Math. **20**(2), 529–535 (2006)
216. L. Rabern, A note on Reed's Conjecture, arXiv:math.CO/0604499 (2006)
217. J. Ramirez-Alfonsin, B. Reed (eds.), *Perfect Graphs* (Springer, Berlin, 2001)
218. F.P. Ramsey, On a problem of formal logic. Proc. Lond. Math. Soc. **2**(30), 264–286 (1930)

219. G. Ravindra, Strongly perfect line graphs and total graphs, in *Finite and infinite sets, Vol. I, II*, ed. by Eger, 1981; A. Hajnal, L. Lovász, V.T. Sós. Colloq. Math. Soc. János Bolyai, vol. 37, (North-Holland, Amsterdam, 1984), pp. 621–633
220. G. Ravindra, Research problems. Discret. Math. **80**, 105–107 (1990)
221. G. Ravindra, D. Basavayya, Co-strongly perfect bipartite graphs. J. Math. Phys. Sci. **26**, 321–327 (1992)
222. G. Ravindra, D. Basavayya, Co-strongly perfect line graphs, in Combinatorial Mathematics and Applications (Calcutta, Sankhya Ser. A, vol. 54. Special Issue **1988**, 375–381 (1988)
223. G. Ravindra, D. Basavayya, A characterization of nearly bipartite graphs with strongly perfect complements. J. Ramanujan Math. Soc. **9**, 79–87 (1994)
224. G. Ravindra, D. Basavayya, Strongly and costrongly perfect product graphs. J. Math. Phys. Sci. **29**(2), 71–80 (1995)
225. G. Ravindra, K.R. Parthasarathy, Perfect product graphs. Discret. Math. **20**, 177–186 (1977)
226. B. Reed, ω, Δ, and χ. J. Graph Theory **27**(4), 177–212 (1998)
227. B. Reed, A strengthening of Brooks' Theorem. J. Comb. Theory Ser. B **76**(2), 136–149 (1999)
228. J.T. Robacker, Min-Max theorems on shortest chains and disjoint cuts of a network. Research Memorandum RM-1660, The RAND Corporation, Santa Monica, California (1956)
229. N. Robertson, P. Seymour, R. Thomas, Hadwiger's conjecture for K_6-free graphs. Combinatorica **13**, 279–361 (1993)
230. N. Robertson, P. Seymour, R. Thomas, Excluded minors in cubic graphs. manuscript (1996)
231. N. Robertson, P. Seymour, R. Thomas, Tutte's edge-colouring conjecture. J. Comb. Theory Ser. B **70**, 166–183 (1997)
232. N. Robertson, P. Seymour, R. Thomas, Permanents, Pfaffian orientations, and even directed circuits. Ann. Math. **150**, 929–975 (1999)
233. F. Roussel, P. Rubio, About skew partitions in minimal imperfect graphs. J. Comb. Theory, Ser. B **83**, 171–190 (2001)
234. N.D. Roussopoulos, A max $\{m, n\}$ algorithm for determining the graph H from its line graph G. Inf. Process. Lett. **2**, 108–112 (1973)
235. B. Roy, Nombre chromatique et plus longs chemins. Rev. Fr. Automat. Inform. **1**, 127–132 (1967)
236. H. Sachs, On the Berge conjecture concerning perfect graphs, in *Combinatorial Structures and their Applications (Proceedings of the Calgary International Conference, Calgary, Alberta)* (Gordon and Breach, New York, 1969), pp. 377–384
237. M. Saks, A short proof of the the k-saturated partitions. Adv. Math. **33**, 207–211 (1979)
238. J. Schönheim, Hereditary systems and Chvátal's conjecture, in *Proceedings of the Fifth British Combinatorial Conference (University of Aberdeen, Aberdeen, 1975)*, Congressus Numerantium, No. XV, Utilitas Math., Winnipeg, Man. (1976), pp. 537–539
239. D. Seinsche, On a property of the class of n-colorable graphs. J. Comb. Theory B **16**, 191–193 (1974)
240. P. Seymour, Decomposition of regular matroids. J. Comb. Theory Ser. B **28**, 305–359 (1980)
241. P. Seymour, Disjoint paths in graphs. Discret. Math. **29**, 293–309 (1980)
242. P. Seymour, How the proof of the strong perfect graph conjecture was found. Gazette des Mathematiciens **109**, 69–83 (2006)
243. P. Seymour, K. Truemper, A Petersen on a pentagon. J. Comb. Theory Ser. B **72**(1), 63–79 (1998)
244. S. Sridharan, On the Berge's strong path-partition conjecture. Discret. Math. **112**, 289–293 (1993)
245. L. Stacho, New upper bounds for the chromatic number of a graph. J. Graph Theory **36**(2), 117–120 (2001)
246. M. Stehlík, Critical graphs with connected complements. J. Comb. Theory Ser. B **89**(2), 189–194 (2003)
247. P. Stein, Chvátal's conjecture and point intersections. Discret. Math. **43**(2–3), 321–323 (1983)
248. P. Stein, J. Schönheim, On Chvátal's conjecture related to hereditary systems. Ars Comb. **5**, 275–291 (1978)

249. F. Sterboul, *Les parametres des hypergraphes et les problemes extremaux associes* (Thèse, Paris, 1974), pp. 33–50
250. F. Sterboul, Sur une conjecture de V. Chvátal, in *Hypergraph Seminar*, ed. by C. Berge, D. Ray-Chaudhuri, Lecture Notes, in Mathematics, vol. 411, (Springer, Berlin, 1974), pp. 152–164
251. L. Surányi, The covering of graphs by cliques. Studia Sci. Math. Hungar. **3**, 345–349 (1968)
252. P.G. Tait, Note on a theorem in geometry of position. Trans. R. Soc. Edinb. **29**, 657–660 (1880)
253. C. Thomassen, Five-coloring graphs on the torus. J. Comb. Theory B **62**, 11–33 (1994)
254. K. Truemper, Alpha-balanced graphs and matrices and GF(3)-representability of matroids. J. Comb. Theory B **32**, 112–139 (1982)
255. A. Tucker, Matrix characterizations of circular-arc graphs. Pac. J. Math. **39**, 535–545 (1971)
256. A. Tucker, The strong perfect graph conjecture for planar graphs. Can. J. Math. **25**, 103–114 (1973)
257. A. Tucker, Perfect graphs and an application to refuse collection. SIAM Rev. **15**, 585–590 (1973)
258. A. Tucker, Structure theorems for some circular-arc graphs. Discret. Math. **7**, 167–195 (1974)
259. A. Tucker, Coloring a family of circular arcs. SIAM J. Appl. Math. **29**(3), 493–502 (1975)
260. A. Tucker, Critical perfect graphs and perfect 3-chromatic graphs. J. Comb. Theory Ser. B **23**(1), 143–149 (1977)
261. A. Tucker, The validity of the strong perfect graph conjecture for K_4-free graphs, in Topics on Perfect Graphs, ed. by C. Berge, V. Chvátal (1984), pp. 149–158 (Ann. Discret. Math. 21)
262. A. Tucker, Coloring perfect $(K_4 - e)$-free graphs. J. Comb. Theory Ser. B **42**(3), 313–318 (1987)
263. W.T. Tutte, A short proof of the factor theorem for finite graphs. Can. J. Math. **6**, 347–352 (1954)
264. W.T. Tutte, On the problem of decomposing a graph into n connected factors. J. Lond. Math. Soc. **36**, 221–230 (1961)
265. W.T. Tutte, Lectures on matroids. J. Res. Nat. Bur. Stand. Sect. B **69B**, 1–47 (1965)
266. W.T. Tutte, On the algebraic theory of graph colorings. J. Comb. Theory **1**, 15–50 (1966)
267. P. Ungar, B. Descartes, Advanced problems and solutions: solutions: 4526. Am. Math. Mon. **61**(5), 352–353 (1954)
268. J. von Neumann, A certain zero-sum two-person game equivalent to the optimal assignment problem, in *Contributions to the Theory of Games*. Annals of Mathematics Studies, No. 28, vol. 2 (Princeton University Press, Princeton, 1953), pp. 5–12
269. K. Wagner, Über eine Eigenschaft der ebenen Komplexe. Math. Ann. **114**, 570–590 (1937)
270. D.L. Wang, P. Wang, Some results about the Chvátal conjecture. Discrete Math. **24**(1), 95–101 (1978)
271. D.B. West, *Introduction to Graph Theory* (Prentice-Hall, Englewood Cliffs, 1996)
272. C. Witzgall, C.T. Zahn Jr., Modification of Edmonds' maximum matching algorithm. J. Res. Nat. Bur. Stand. Sect. B **69B**, 91–98 (1965)
273. Q. Xue, (C_4, Lotus)-free Berge graphs are perfect. An. Stiint. Univ. Al. I. Cuza Iasi Inform. (N.S.) **4**, 65–71 (1995)
274. Q. Xue, On a class of square-free graphs. Inf. Process. Lett. **57**(1), 47–48 (1996)
275. A.A. Zykov, On some properties of linear complexes. Russian Math. Sbornik N. S. **24**(66), 163–188 (1949)

Chapter 2
Codes Produced by Permutations: The Link Between Source and Channel Coding

2.1 Introduction

In [4] (Sects. 5 and 6) we suggested as a *program in coding theory to systematically investigate the symmetric group* $\mathcal{S}_n$ (the group of permutations) acting on the components $1, \dots, n$. The immediate use of this group is due to the fact that it leaves probability distributions specifying stationary memoryless multi-user sources and channels invariant. As a justification for his belief in this program he presented a general robustification technique, and he derived Slepian and Wolf's [25] source coding theorem for the DMC via a *covering lemma* (see Sect. 2.4). By this method source codes are built from channel codes.

Here we show that channel codes, which achieve capacity (and even the random coding bound), can also be built up iteratively by producing bigger codes from good smaller codes with suitable permutations $\pi_1, \dots, \pi_t$, say, which we call *code producers*. In particular, this is possible for subcodes consisting just of one codeword.

To fix ideas, we describe the production first in this case. Suppose we are given a single codeword $x^n = (x_1, \dots, x_n)$ of length n and the permutations operate on $\{1, \dots, n\}$. By $\pi_1 x^n$ we mean the n-sequence obtained from x^n by permuting the components of x^n according to π_1, i.e.,

$$\pi_1 x^n = \left(x_{\pi_1 1}, \dots, x_{\pi_1 n}\right).$$

Now we have two codewords x^n and $\pi_1 x^n$. Form now $\pi_2 x^n$ and $\pi_2 \circ \pi_1 x^n$, then $\pi_3 x^n$, $\pi_3 \circ \pi_1 x^n$, $\pi_3 \circ \pi_2 x^n$, $\pi_3 \circ \pi_2 \circ \pi_1 x^n$, etc. In each step we double the cardinality of our codeword set, if repetitions are counted with multiplicity. In this manner it is possible to construct simply structured codes. Note that in order to give a code book for such a code we have to list t permutations, say, instead of $\exp\{t\}$ codewords.

Finally, we prove right away a somewhat stronger result than just achievability of the random coding bound: the same set of permutations can serve for every positive rate below capacity as follows. If the rate is R, then use first t' permutations, where

A. Ahlswede et al. (eds.), *Combinatorial Methods and Models*,
Foundations in Signal Processing, Communications and Networking 13,
DOI 10.1007/978-3-319-53139-7_2

t' is minimal with the property $\exp\{t'\} \geq \exp\{nR\}$. Moreover, we also establish universality in the sense of Goppa [18], that is, the same set of permutations can be used for all channels of bounded alphabet sizes. The exact statements are given in Theorem 2.3. For ordinary codes Goppa proved universality with respect to the capacities and this result was sharpened by Csiszár, Körner, and Marton [10] to the universal achievability of the random coding bound. Those authors also proved that the expurgated bound can be achieved using a universal set of codewords, and Csiszár and Körner established in [8] the (universal) achievability of both bounds simultaneously. We do not know yet whether those results can be proved for our simply structured codes for we do not even know whether the expurgated bound can be achieved at all. The immediate reason is that expurgation destroys the algebraic structure.

We would like to draw attention to another *problem* of some interest. Generally speaking the idea of building bigger structures from smaller structures is very common in human life (also the reverse process, which is often an unfortunate fact), in science, and, especially, in engineering. It is often wasteful to build a new machine from scratch, if functioning parts are available and could be used. Code producers perform this task for all discrete memoryless channels with properly bounded alphabet sizes and all rates. However, they do so only for fixed block length n. Hence, it may be interesting to try now to build producers from smaller ones, that is to introduce "producers of producers".

Our *main tool* for proving Theorem 2.3 is a kind of maximal code method for abstract bipartite graphs, which was given in [4]. The method uses average errors. Other differences from Feinstein's maximal code method [14], which is for maximal errors, are explained in [4]. An important feature of the method is that while finding codewords iteratively, the error probability of any initial code can be linked to the error probability of the extended code.

Moreover, the selection of a codeword at each step can be done at random and the probability of finding good code extensions can be estimated rather precisely. These estimates are used in Sects. 2.5 and 2.6 to derive bounds on the probability that a randomly chosen (nonexpurgated or suitably expurgated) code achieves the best known error bounds. They are also used for showing the existence of universal code producers.

In applying the abstract maximal method to "channel graphs" the actual calculations of graphic parameters such as degrees, etc., involve information quantities.

We give applications of the abstract maximal coding method and of other methods of [4] to other graphs and hypergraphs of genuine information theoretical interest. There the graphic parameters *cannot* be described by *information quantities* and this will, as we hope, convince more people of the use of the abstract approach to Information Theory developed in [4].

2.2 Notation and Known Facts

Script capitals $\mathcal{X}, \mathcal{Y}, \dots$ will denote finite sets. The cardinality of a set $\mathcal{A}$ and of the range of a function f will be denoted by $|\mathcal{A}|$ and $||f||$, respectively. The letters P, Q will always stand for probability distributions (PD's) on finite sets, and $X, Y, \dots$ denote random variables (RV's).

Channels, Empirical Distributions, Generated Sequences

A stochastic matrix $W = \{W(y|x) : y \in \mathcal{Y}, x \in \mathcal{X}\}$ uniquely defines a DMC with input alphabet $\mathcal{X}$, output alphabet $\mathcal{Y}$, and transmission probabilities

$$W^n(y^n|x^n) = \prod_{t=1}^{n} W(Y_t|x_t)$$

for n-sequences $x^n = (x_1, \dots, x_n) \in \mathcal{X}^n$, $y^n = (y_1, \dots, y_n) \in \mathcal{Y}^n$, $n = 1, 2, 3, \dots$.

We denote by $\mathcal{P}$ the sets of all PD's on $\mathcal{X}$ and by $\mathcal{W}$ (resp. $\mathcal{V}$) the set of all channels with alphabets $\mathcal{X}, \mathcal{Y}$.

For positive integers n we set

$$\mathcal{P}_n = \{P \in \mathcal{P} : P(x) \in \{0, 1/n, 2/n, \dots, 1\} \text{ for all } x \in \mathcal{X}\}.$$

For any $P \in \mathcal{P}_n$, called *empirical distribution* (ED), we define the set $\mathcal{W}_n(P) = \{\tilde{W} \in \mathcal{W} : \tilde{W}(y|x) \in \{0, 1/(nP(x)), 2/(nP(x)), \dots, 1\} \text{ for all } x \in \mathcal{X}, y \in \mathcal{Y}\}$. $\mathcal{V}_n(P)$ is defined similarly.

The *ED* of a sequence $x^n \in \mathcal{X}^n$ is the distribution $P_{x^n} \in \mathcal{P}_n$ defined by letting $P_{x^n}(x)$ count the relative frequency of the letter x in the n-sequence x^n. The *joint ED* of a pair $(x^n, y^n) \in \mathcal{X}^n \times Y^n$ is the distribution P_{x^n,y^n} on $\mathcal{X} \times \mathcal{Y}$ defined analogously. For $P \in \mathcal{P}$, the set $\mathcal{T}_P^n$ of all P-typical sequences in $\mathcal{X}^n$ is given by

$$\mathcal{T}_P^n = \{x^n : P_{x^n} = P\}.$$

For $\tilde{W} \in \mathcal{W}$ a sequence $y^n \in \mathcal{Y}^n$ is said to be $\tilde{W}$*-generated by* x^n, if for all $(x, y) \in \mathcal{X} \times \mathcal{Y}$

$$P_{x^n,y^n}(x, y) = P_{x^n}(x) \cdot \tilde{W}(y|x).$$

The set of those sequences is denoted by $\mathcal{T}_{\tilde{W}}^n(x^n)$. Observe that $\mathcal{T}_P^n \neq \emptyset$ if and only if $P \in \mathcal{P}_n$ and $\mathcal{T}_{\tilde{W}}^n(x^n) \neq \emptyset$ if and only if $\tilde{W} \in \mathcal{W}_n(P_{x^n})$.

Entropy and Information Quantities

Let X be a RV with values in $\mathcal{X}$ and distribution $P \in \mathcal{P}$, and let Y be a RV with values in $\mathcal{Y}$ such that the joint distribution of (X, Y) on $\mathcal{X} \times \mathcal{Y}$ is given by

$$\Pr\{X = x, Y = y\} = P(x)\tilde{W}(y|x), \quad \tilde{W} \in \mathcal{W}.$$

Then for the entropy $H(X)$, conditional entropy $H(Y|X)$, and mutual information $I(X \wedge X)$ we shall also write $H(P)$, $H(\tilde{W}|P)$, and $I(P, \tilde{W})$, respectively. For $P, \tilde{P} \in \mathcal{P}$

$$D(\tilde{P}||P) = \sum_{x \in \mathcal{X}} \tilde{P}(x) \log \frac{\tilde{P}(x)}{P(x)}$$

denotes the relative entropy and for $\tilde{W}, \tilde{\tilde{W}} \in \mathcal{W}$ the quantity

$$D\left(\tilde{W}||\tilde{\tilde{W}}|P\right) = \sum_{x} P(x) D\left(\tilde{W}(\cdot|x)||\tilde{\tilde{W}}(\cdot|x)\right)$$

stands for the conditional relative entropy. Finally, for $x^n \in \mathcal{X}^n$, $y^n \in \mathcal{Y}^n$

$$I(x^n \wedge y^n) = \sum_{x} \sum_{y} P_{x^n, y^n}(x, y) \log \frac{P_{x^n, y^n}(x, y)}{P_{x^n}(x) \cdot P_{y^n}(y)}.$$

Elementary Properties of Typical Sequences and Generated Sequences

$$|\mathcal{P}_n| \leq (n+1)^{|\mathcal{X}|}, \tag{2.2.1}$$

$$|\mathcal{W}_n(P)| \leq (n+1)^{|\mathcal{X}| \cdot |\mathcal{Y}|}, \quad \text{for } P \in \mathcal{P}_n, \tag{2.2.2}$$

$$|\mathcal{V}_n(P)| \leq (n+1)^{|\mathcal{X}| \cdot |\mathcal{X}|}, \quad \text{for } P \in \mathcal{P}_n, \tag{2.2.3}$$

$$|\mathcal{T}_P^n| = \frac{n!}{\prod_{x \in \mathcal{X}} (nP(x))!}, \quad \text{for } P \in \mathcal{P}_n, \tag{2.2.4}$$

$$(n+1)^{-|\mathcal{X}|} \exp\{nH(P)\} \leq |\mathcal{T}_P^n| \leq \exp\{nH(P)\}, \quad \text{for } P \in \mathcal{P}_n. \tag{2.2.5}$$

For $P \in \mathcal{P}_n$, $\tilde{W} \in \mathcal{W}_n(P)$, $x^n \in \mathcal{T}_P^n$:

$$(n+1)^{-|\mathcal{X}| \cdot |\mathcal{Y}|} \exp\{nH(\tilde{W}|P)\} \leq |\mathcal{T}_{\tilde{W}}^n(x^n)| \leq \exp\{nH(\tilde{W}|P)\}. \tag{2.2.6}$$

For $P \in \mathcal{P}_n$, $\tilde{P} \in \mathcal{P}$, $x^n \in \mathcal{T}_P^n$:

$$\tilde{P}^n(x^n) = \exp\left\{-n\left(D(P||\tilde{P}) + H(P)\right)\right\}, \tag{2.2.7}$$

where $\tilde{P}^n$ is the n-fold extension of $\tilde{P}$. For $P \in \mathcal{P}$; $\tilde{W}, \tilde{\tilde{W}} \in \mathcal{W}$; $x^n \in \mathcal{T}_P^n$, $y^n \in \mathcal{T}_{\tilde{W}}^n(x^n)$:

$$\tilde{\tilde{W}}(y^n|x^n) = \exp\left\{-n\left(D\left(\tilde{W}||\tilde{\tilde{W}}|P\right) + H\left(\tilde{W}|P\right)\right)\right\}. \tag{2.2.8}$$

For $P \in \mathcal{P}_n$, $\tilde{W} \in \mathcal{W}_n(P)$, $y^n \in \mathcal{T}^n_{P\tilde{W}}$:

$$\begin{aligned}(n+1)^{-|\mathcal{X}|\cdot|\mathcal{Y}|}&\exp\left\{n\left(H(P) - I(P,\tilde{W})\right)\right\}\\ &\leq \left|\left\{x^n \in \mathcal{T}^n_P : y^n \in \mathcal{T}^n_{\tilde{W}}(x^n)\right\}\right| \\ &\leq \exp\left\{n\left(H(P) - I(P,\tilde{W})\right)\right\},\end{aligned} \tag{2.2.9}$$

where $P\tilde{W}$ denotes the PD on $\mathcal{Y}$ given by $P\tilde{W}(y) = \sum_x P(x)\tilde{W}(y|x)$ for $y \in \mathcal{Y}$.

Historical Sketch of the Bounds on the Reliability Function

An (n, N) code $\mathcal{C}$ for the DMC is a system of pairs $\{(u_i, \mathcal{D}_i) : i = 1, \dots, N\}$ with $u_i \in \mathcal{X}^n$ and pairwise disjoint subsets $\mathcal{D}_i \subset \mathcal{Y}^n$ $(i = 1, \dots, N)$. $\lambda(\mathcal{C}, W)$ denotes the average error probability of $\mathcal{C}$, i.e.,

$$\bar{\lambda}(\mathcal{C}, W) = \frac{1}{N}\sum_{i=1}^{N} W^n(\mathcal{D}_i^c|u_i),$$

where $\mathcal{D}_i^c = \mathcal{X}^n - \mathcal{D}_i$. $\lambda_{\max}(\mathcal{C}, W) = \max_i W^n(\mathcal{D}_i^c|u_i)$ denotes the maximal error of $\mathcal{C}$. $\mathcal{C}$ is called an ML code (maximum likelihood code), if for $i = 1, \dots, N$ the sets $\mathcal{D}_i$ consist of those n-words $y^n \in \mathcal{Y}^n$ such that

$$W^n(y^n|u_i) \geq W^n(y^n|u_j), \quad \text{for all } j \neq i,$$

$$W^n(y^n|u_i) > W^n(y^n|u_j), \quad \text{for all } j < i.$$

If we define for any rate R

$$\bar{\lambda}(n, R, W) = \min\left\{\bar{\lambda}(\mathcal{C}, W) : \mathcal{C} \text{ is an } (n, N) \text{ code with } N \geq \exp\{nR\}\right\},$$

then

$$E(R, W) = \limsup_{n\to\infty} -\frac{1}{n}\log\bar{\lambda}(n, R, W)$$

is the familiar reliability function for the DMC W.

Since Shannon discovered the coding theorem for the DMC in his famous paper [22] there has been considerable effort in improving bounds on the error probability for codes of a given rate or, equivalently, on the reliability function $E(R, W)$. Well-known upper bounds on $E(R, W)$ are the sphere packing bound $E_{sp}(R, W)$ and the straight line bound $E_{sl}(R, W)$. These bounds were derived by Shannon, Gallager, and Berlekamp [24]. $E_{sp}(R, W)$ was first established (with an incomplete proof) by Fano [13]. For rates $R > C$ Wolfowitz's strong converse [26] implies

$$\liminf_{n\to\infty} \bar{\lambda}(n, R, W) = 1.$$

For $R > C$ the problem is to evaluate

$$\liminf_{n\to\infty} -\frac{1}{n} \log\left(1 - \bar{\lambda}(n, R, W)\right).$$

Arimoto [6] extended the sphere packing exponent for rates above capacity, and finally Dueck and Körner [12] showed that this exponent is optimal. A partial result in this direction was obtained earlier by Omura [21].

The best known lower bounds for $R < C$ are the random coding bound $E_r(R, W)$, which was derived by Fano and given a simpler proof by Gallager [15], and the expurgated bound $E_{ex}(P, W)$, which is due to Gallager [15].

Our results here mainly concern those lower bounds. Csiszár, Körner, and Marton [10] have rederived those bounds via typical sequences incorporating earlier ideas of Haroutunian [19], Blahut [7], and Goppa [18]. Their approach leads to universal codes. The function $E_r(R, W)$ and to a certain extent also the function $E_{ex}(R, W)$ appear in the new derivations in a form somewhat more linked to information quantities than the familiar analytic expression [16]. The results of [10] are

Theorem 2.1 (Theorem R, Csiszár, Körner, and Marton [10]) *For every $R > 0$, $\delta > 0$, $n \geq n_0(|\mathcal{X}|, |\mathcal{Y}|, \delta)$, and every ED $P \in \mathcal{P}_n$ there exists an (n, N) code*

$$\mathcal{C} = \{(u_i, \mathcal{D}_i) : i = 1, \dots, N\} \quad \text{with } u_i \in \mathcal{T}_P^n \text{ and } \frac{1}{n}\log N \geq R - \delta$$

such that

$$\bar{\lambda}(\mathcal{C}, W) \leq exp\{-n(E_r(R, P, W) - \delta)\} \tag{2.2.10}$$

for any $W \in \mathcal{W}$, where

$$E_r(R, P, W) = \min_{\tilde{W}\in\mathcal{W}} D(\tilde{W}||W|P) + [I(P, \tilde{W}) - R]^+,$$

and $[t]^+ = \max\{0, t\}$.

Theorem 2.2 (Theorem EX, Csiszár, Körner, and Marton [10]) *For every $R > 0$, $\delta > 0$, $n \geq n_0(|\mathcal{X}|, |\mathcal{Y}|, \delta)$, and every ED $P \in \mathcal{P}_n$ there exist codewords*

$$u_1, \dots, u_N \in \mathcal{T}_P^n \quad \text{with } \frac{1}{n}\log N \geq R - \delta$$

such that for every $W \in \mathcal{W}$ the corresponding ML code

$$\mathcal{C}^W = \left\{\left(u_i, \mathcal{D}_i^W\right) : i = 1, \dots, N\right\}$$

(i.e., the $\mathcal{D}_i^W$ denote the maximum likelihood decoding sets with respect to W) satisfies

$$\bar{\lambda}(\mathcal{C}^W, W) \leq exp\{-n(E_{ex}(R, P, W) - \delta)\}$$

where

$$E_{ex}(R, P, W) = \min_{\substack{X,\tilde{X}P-distributed \\ I(X\wedge\tilde{X})\leq R}} Ed(X, \tilde{X}) + I(X \wedge \tilde{X}) - R$$

and $d(x, \tilde{x}) = -\log \sum_{y\in\mathcal{Y}} \sqrt{W(y|x) \cdot W(y|\tilde{x})}$ *for* $x, \tilde{x} \in \mathcal{X}$. $Ed(\cdot)$ *means the expectation of* $d(\cdot)$.

Actually, in [8] a unified description of the random coding and expurgated bound was given, but this description will not be used here.

2.3 The Main Result: Channel Codes Produced by Permutations

Let k, n be positive integers with $k \cdot n \leq n!$. We call any subset $\{\pi_1, \dots, \pi_{n\cdot k}\} \subset \mathcal{S}_n$ a *code producer*. Such a code producer works as follows. Assume we are given a DMC W with alphabet $\mathcal{X}, \mathcal{Y}$, where $|\mathcal{X}|, |\mathcal{Y}| \leq 2^k$ and we want to transmit one of 2^m messages ($m \leq n \cdot k$) over this channel using an n-length block code. First we identify the messages with m-sequences in $\{0, 1\}^m$, then we choose a proper ED $P \in \mathcal{P}_n$ and build the "canonical P-sequence" u_P defined by

$$u_P = (x_1, \dots, x_1, x_2, \dots, x_2, \dots, x_{|\mathcal{X}|}, \dots, x_{|\mathcal{X}|}) \in \mathcal{T}_P^n,$$

where $\mathcal{X} = \{x_1, \dots, x_{|\mathcal{X}|}\}$.

Suppose that message $z^m = (z_1, \dots, z_m) \in \{0, 1\}^m$ is to be sent over the channel W. Then the encoder puts

$$\pi_M^{z_m} \circ \pi_{m-1}^{z_{m-1}} \circ \cdots \circ \pi_z^{z_1} \circ id(u_P)$$

into the channel, where $id \in \mathcal{S}_n$ is the identity mapping and $\pi_i^0 = id$, $\pi_i^1 = \pi_i$ for $i = 1, \dots, n \cdot k$. Thus, the codeword set produced for the given parameters $\mathcal{X}$, $P \in \mathcal{P}_n$, and m is

$$\left\{\pi_m^{z_m} \circ \cdots \circ \pi_z^{z_1} \circ id(u_P) : z^m = (z_1, \dots, z_m) \in \{0, 1\}^m\right\}.$$

We denote the ML code with respect to the channel W for this codeword set by

$$\mathcal{C}(\pi_1, \dots, \pi_{n-k}, P, \mathcal{X}, \mathcal{Y}, W, R),$$

where

$$R = \frac{1}{n} \log 2^m.$$

Two sequences $\pi_M^{z_m} \circ \pi_1^{z_1} \circ id(u_P), \pi_M^{z'_m} \circ \cdots \circ \pi_1^{z'_1} \circ id(u_P)$ are considered as different if $z^m \neq z'^m$, even though they may represent the same element of $\mathcal{T}_P^n$. Therefore the cardinalities of the produced codeword sets are always powers of two.

If N is given and we want to produce an n-length block code with N messages ($R = (1/n)\log N$), then by $\mathcal{C}(\pi_1, \ldots, \pi_{nk}, P, \mathcal{X}, \mathcal{Y}, W, R)$ we mean always the code having 2^m codewords, where 2^m is the smallest power of 2 with $2^m \geq N$.

Theorem 2.3 (Ahlswede and Dueck [5]) *Fix a positive integer k and $\delta > 0$. Then for any $n \geq n_0(k, \delta)$ there exists a producer $\{\pi_1, \ldots, \pi_{n \cdot k}\} \subset \mathcal{S}_n$ with the properties*

$$\bar{\lambda}(\mathcal{C}(\pi_1, \ldots, \pi_{nk}, P, \mathcal{X}, \mathcal{Y}, W, R), W) \leq exp\{-n(E_r(R, P, W) - \delta)\},$$

for every $\mathcal{X}, \mathcal{Y}$ with $|\mathcal{X}|, |\mathcal{Y}| \leq 2^k$, for every $P \in \mathcal{P}_n$, for every channel W with alphabets $\mathcal{X}$ and $\mathcal{Y}$, and for every rate $R > 0$.

The theorem is an immediate consequence of the following basic result.

Lemma 2.1 *Fix alphabets $\mathcal{X}, \mathcal{Y}$ and $\delta > 0$. Then for every $n \geq n_0(|\mathcal{X}|, |\mathcal{Y}|, \delta)$, every ED $P \in \mathcal{P}_n$, and every code $\mathcal{C} = \{(u_i, \mathcal{D}_i) : i = 1, \ldots, N\}$; $u_i \in \mathcal{T}_P^n$ for $i = 1, \ldots, N$; there exists a permutation $\pi \in \mathcal{S}_n$ and suitable decoding sets $\mathcal{E}_1, \ldots, \mathcal{E}_N, \mathcal{E}_{1,\pi}, \ldots, \mathcal{E}_{N,\pi}$ such that the enlarged code*

$$\mathcal{C}'_\pi = \{(u_1, \mathcal{E}_1), \ldots, (u_N, \mathcal{E}_N), (\pi u_1, \mathcal{E}_{1,\pi}), \ldots, (\pi u_N, \mathcal{E}_{N,\pi})\}$$

satisfies for every $W \in \mathcal{W}$

$$\bar{\lambda}(\mathcal{C}'_\pi, W) \leq \bar{\lambda}(\mathcal{C}, W) + exp\{-n(E_r(R, P, W) - \delta)\}, \tag{2.3.1}$$

where $R = (1/n)\log N$.

Moreover, for a randomly (according to the uniform distribution on $\mathcal{S}_n$) chosen π (2.3.1) holds for all $W \in \mathcal{W}$ with a probability larger than $1 - exp\{-(\delta/2) \cdot n\}$.

The proof is based on the maximal coding idea of [4]. In its original form codewords are added iteratively to a given code. Here we add permutations iteratively and thus keep doubling the lengths of codes. The reader may find it easier to study first Theorems 2.5 and 2.6 in Sect. 2.5, whose proofs use the original form. These theorems are needed for the derivation of double exponential bounds on the probability that a randomly chosen code fails to meet the random coding or expurgated bound for the DMC. They also imply Theorems 2.1 and 2.2 and thus give an alternative proof of those theorems by maximal coding.

For the proof Lemma 2.1 we need Lemmas 2.2 and 2.3 below. They involve quantities which we now define.

Fix $R > 0$, $\delta > 0$, $P \in \mathcal{P}_n$, and let $\{u_1, \ldots, u_N\} \subset \mathcal{T}_P^n$ and $N \leq \exp\{nR\}$ be given.

For any pair $\tilde{W}, \tilde{\tilde{W}} \in \mathcal{W}$ we define the function $g_{\tilde{W},\tilde{\tilde{W}}}$ on $\mathcal{X}^n$ by

$$g_{\tilde{W},\tilde{\tilde{W}}}(u) = \left| \mathcal{T}^n_{\tilde{W}}(u) \cap \bigcup_{i=1}^{N} \mathcal{T}^n_{\tilde{\tilde{W}}}(u_i) \right|, \quad \text{for } u \in \mathcal{X}^n. \tag{2.3.2}$$

$g_{\tilde{W},\tilde{\tilde{W}}}(u)$ measures the size of intersections of sets generated by n and of sets generated by the given system of codewords.

Furthermore, for permutations $\pi \in \mathcal{S}_n$ we define the function $g^*_{\tilde{W},\tilde{\tilde{W}}}$ by

$$g^*_{\tilde{W},\tilde{\tilde{W}}}(\pi) = \sum_{i=1}^{N} g_{\tilde{W},\tilde{\tilde{W}}}(\pi u_i).$$

Let U be a RV equidistributed on $\mathcal{T}^n_P$ and let Π be a RV equidistributed on $\mathcal{S}_n$.

Lemma 2.2 *For every pair* $\tilde{W}, \tilde{\tilde{W}} \in \mathcal{W}$

(i) $Eg_{\tilde{W},\tilde{\tilde{W}}}(U) \leq (n+1)^{|\mathcal{X}|} exp\{n(H(\tilde{W}|P) - [I(P,\tilde{\tilde{W}}) - R]^+\}$, *where* $[t]^+ = \max\{0, t\}$.
Furthermore, for any $\delta > 0$, $\xi \geq 0$ *and* $n \geq n_0(\delta, |\mathcal{X}|, |\mathcal{Y}|)$;

(ii) $Pr\{g_{\tilde{W},\tilde{\tilde{W}}}(U) \geq (n+1)^{|\mathcal{X}|} exp\{n(H(\tilde{W}|P) - [I(P,\tilde{\tilde{W}}) - R - \xi]^+ + (3/4)\delta)\}$ *for some* $\tilde{W}, \tilde{\tilde{W}} \in \mathcal{W}\} \leq exp\{-n((\delta/2) + \xi)\}$.

Lemma 2.3 *For every pair* $\tilde{W}, \tilde{\tilde{W}} \in \mathcal{W}$

(i) $Eg_{\tilde{W},\tilde{\tilde{W}}}(\Pi) = N \cdot Eg_{\tilde{W},\tilde{\tilde{W}}}(U)$.
For any $\delta > 0$ *and* $n \geq n_0(\delta, |\mathcal{X}|, |\mathcal{Y}|)$;

(ii) $Pr\{g_{\tilde{W},\tilde{\tilde{W}}}(\Pi) \geq N \cdot exp\{n(H(\tilde{W}|P) - [I(P,\tilde{\tilde{W}}) - R]^+ + (3/4)\delta)\}$ *for some* $\tilde{W}, \tilde{\tilde{W}} \in \mathcal{W}\} \leq exp\{-n(\delta/2)\}$.

Proof of Lemma 2.2. Choose any $\tilde{W}, \tilde{\tilde{W}} \in \mathcal{W}$ and note that $g_{\tilde{W},\tilde{\tilde{W}}}$ is zero for sequences in $\mathcal{T}^n_P$ if $\tilde{W} \notin \mathcal{W}_n(P)$ or $\tilde{\tilde{W}} \notin \mathcal{W}_n(P)$. Let $P\tilde{W}$ denote the distribution on $\mathcal{Y}$ given by

$$P\tilde{W}(y) = \sum_x P(x)\tilde{W}(y|x), \quad \text{for } y \in \mathcal{Y}.$$

Note again that $g_{\tilde{W},\tilde{\tilde{W}}}$ is zero for sequences in $\mathcal{T}^n_P$ if $P\tilde{W} \neq P\tilde{\tilde{W}}$. Hence we assume that $\tilde{W}, \tilde{\tilde{W}} \in \mathcal{W}_n(P)$ and $P\tilde{W} = P\tilde{\tilde{W}}$.

$$
\begin{aligned}
Eg_{\tilde{W},\tilde{\tilde{W}}}(U) &= E\left|\mathcal{T}^n_{\tilde{W}}(U)\cap\bigcup_{i=1}^{N}\mathcal{T}^n_{\tilde{\tilde{W}}}(u_i)\right| \\
&\le \sum_{i=1}^{N} E\left|\mathcal{T}^n_{\tilde{W}}(U)\cap\mathcal{T}^n_{\tilde{\tilde{W}}}(u_i)\right| \\
&= N\cdot E\left|\mathcal{T}^n_{\tilde{W}}(U)\cap\mathcal{T}^n_{\tilde{\tilde{W}}}(u_i)\right| \quad \text{(by symmetry)} \\
&= N\cdot \sum_{y^n\in\mathcal{T}^n_{\tilde{\tilde{W}}}(u_i)} \Pr\left(y^n\in\mathcal{T}^n_{\tilde{W}}(U)\right).
\end{aligned}
$$

Since U is equidistributed over $\mathcal{T}^n_P$, we have for every $y^n\in\mathcal{Y}^n$ that

$$
\Pr(y^n\in\mathcal{T}^n_{\tilde{W}}(U)) = \frac{\left|\left\{x^n : x^n\in\mathcal{T}^n_P,\, y^n\in\mathcal{T}^n_{\tilde{W}}(x^n)\right\}\right|}{|\mathcal{T}^n_P|};
$$

therefore (2.2.5) and (2.2.9) yield.

$$
\begin{aligned}
Eg_{\tilde{W},\tilde{\tilde{W}}}(U) &\le N\cdot|\mathcal{T}^n_{\tilde{\tilde{W}}}(u_i)|\cdot\exp\left\{n(H(P)-I(P,\tilde{W})-H(P))\right\}\cdot(n+1)^{|\mathcal{X}|} \\
&\le N\cdot\exp\left\{n\left(H\left(\tilde{\tilde{W}}|P\right)-I\left(P,\tilde{W}\right)\right)\right\}\cdot(n+1)^{|\mathcal{X}|}. \qquad (2.3.3)
\end{aligned}
$$

By assumption, $P\tilde{W}=P\tilde{\tilde{W}}$ and thus, $I(P,\tilde{W})=H(P\tilde{W})-H(\tilde{W}|P)$.

We therefore get from (2.3.3)

$$
Eg_{\tilde{W},\tilde{\tilde{W}}}(U)\le N\cdot\exp\left\{n\left(H\left(\tilde{W}|P\right)-I\left(P,\tilde{\tilde{W}}\right)\right)\right\}\cdot(n+1)^{|\mathcal{X}|}. \qquad (2.3.4)
$$

On the other hand, it is obvious from the definition of $g_{\tilde{W},\tilde{\tilde{W}}}$ and from (2.2.6) that

$$
Eg_{\tilde{W},\tilde{\tilde{W}}}(U)\le E|\mathcal{T}^n_{\tilde{W}}(U)|\le\exp\left\{nH\left(\tilde{W}|P\right)\right\}. \qquad (2.3.5)
$$

Since $N\le\exp\{nR\}$, (2.3.4) and (2.3.5) imply (i). (ii) follows from (i) by applying Chebyshev's inequality. □

Proof of Lemma 2.3. Let $\tilde{W}, \tilde{\tilde{W}} \in \mathcal{W}$. Then

$$\begin{aligned}
Eg^*_{\tilde{W},\tilde{\tilde{W}}}(\Pi) &= \frac{1}{n!}\sum_{i=1}^{N}\sum_{\pi\in\mathcal{S}_n} g_{\tilde{W},\tilde{\tilde{W}}}(\pi u_i) \\
&= \frac{1}{n!}\sum_{i=1}^{N}\sum_{v\in\mathcal{T}_P^n} |\{\pi\in\mathcal{S}_n : \pi u_i = v\}| \cdot g_{\tilde{W},\tilde{\tilde{W}}}(v) \\
&= \frac{1}{n!}\prod_{x\in\mathcal{X}}(nP(x))!\sum_{i=1}^{N}\sum_{v\in\mathcal{T}_P^n} g_{\tilde{W},\tilde{\tilde{W}}}(v) \qquad (2.3.6)\\
&= N\cdot Eg_{\tilde{W},\tilde{\tilde{W}}}(U). \qquad (2.3.7)
\end{aligned}$$

Equation (2.3.7) follows from (2.2.4). Thus part (i) of the lemma is proved and part (ii) follows with Chebyshev's inequality. □

Proof of Lemma 2.1. Lemma 2.3 guarantees the existence of a permutation $\pi \in \mathcal{S}_n$ with

$$g^*_{\tilde{W},\tilde{\tilde{W}}}(\pi), g^*_{\tilde{W},\tilde{\tilde{W}}}(\pi^{-1}) \le N\cdot\exp\left\{n\left(H\left(\tilde{W}|P\right) - \left[I\left(P,\tilde{\tilde{W}}\right) - R\right]^+ + \frac{3}{4}\delta\right)\right\} \qquad (2.3.8)$$

for any pair $\tilde{W}, \tilde{\tilde{W}} \in \mathcal{W}$. π^{-1} denotes the inverse permutation of π.

Let $\mathcal{C} = \{(u_i, \mathcal{D}_i) : i = 1, \dots, N\}$ be a code for the given codeword set $\{u_1, \dots, u_N\} \subset \mathcal{T}_P^n$. Define now decoding sets

$$\mathcal{E}_i = \mathcal{D}_i - \left\{y^n : I(\pi u_j \wedge y^n) \ge I(u_i \wedge y^n) \text{ for some } j\right\}$$

for $i = 1, \dots, N$ and

$$\mathcal{E}_{i,\pi} = \pi\mathcal{D}_i - \left\{y^n : I(u_j \wedge y^n) \ge I(\pi u_i \wedge y^n) \text{ for some } j\right\}$$

for $i = 1, \dots, N$.

Notice that the sets $\mathcal{E}_1, \dots, \mathcal{E}_N, \mathcal{E}_{1,\pi}, \dots, \mathcal{E}_{N,\pi}$ are disjoint and set

$$\mathcal{C}' = \{(u_1, \mathcal{E}_1), \dots, (u_N, \mathcal{E}_N), (\pi u_1, \mathcal{E}_{1,\pi}), \dots, (\pi u_N, \mathcal{E}_{N,\pi})\}.$$

Now we have for every $W \in \mathcal{W}$

$$\begin{aligned}
\bar{\lambda}(\mathcal{C}', W) = \frac{1}{2N}\Bigg(&\sum_{i=1}^{N}(W^n(\mathcal{D}_i - \mathcal{E}_i|u_1) \\
&+W^n(\pi\mathcal{D}_i - \mathcal{E}_{i,\pi}|\pi u_1)) + 2N\bar{\lambda}(\mathcal{C}, W)\Bigg). \qquad (2.3.9)
\end{aligned}$$

First we estimate

$$\sum_{i=1}^{N} W^n(\mathcal{D}_i - \mathcal{E}_i|u_i) = \sum_{i=1}^{N} W^n\left(\left\{y^n | I(\pi u_j \wedge y^n) \geq I(u_i \wedge y^n) \text{ for some } j\right\} |u_i\right)$$

$$= \sum_{\substack{\tilde{W},\tilde{\tilde{W}} \in \mathcal{W}_n(P) \\ I(P,\tilde{W}) \leq I(P,\tilde{\tilde{W}})}} \sum_{i=1}^{N} W^n \left(\mathcal{T}^n_{\tilde{W}}(u_i) \cap \bigcup_{j=1}^{N} \mathcal{T}^n_{\tilde{\tilde{W}}}(\pi u_j) | u_i \right)$$

$$= \sum_{\substack{\tilde{W},\tilde{\tilde{W}} \in \mathcal{W}_n(P) \\ I(P,\tilde{W}) \leq I(P,\tilde{\tilde{W}})}} \sum_{i=1}^{N} W^n \left(\mathcal{T}^n_{\tilde{W}}(\pi^{-1} u_i) \cap \bigcup_{j=1}^{N} \mathcal{T}^n_{\tilde{\tilde{W}}}(u_j) | \pi^{-1} u_i \right)$$

$$= \sum_{\substack{\tilde{W},\tilde{\tilde{W}} \in \mathcal{W}_n(P) \\ I(P,\tilde{W}) \leq I(P,\tilde{\tilde{W}})}} \exp\left\{ -n \left(D\left(\tilde{W}||W|P\right) + H\left(\tilde{W}|P\right)\right)\right\} g^*_{\tilde{W},\tilde{\tilde{W}}}(\pi^{-1})$$

by (2.2.8). Now apply (2.3.8) to get

$$\sum_{i=1}^{N} W^n\left(\mathcal{D}_i - \mathcal{E}_i|u_i\right)$$
$$\leq \sum_{\substack{\tilde{W},\tilde{\tilde{W}} \in \mathcal{W}_n(P) \\ I(P,\tilde{W}) \leq I(P,\tilde{\tilde{W}})}} N \cdot \exp\left\{ -n \left(D\left(\tilde{W}||W|P\right) \right.\right.$$
$$\left.\left. + \left[I\left(P,\tilde{W}\right) - R\right]^+ - \frac{3}{4}\delta\right)\right\} \tag{2.3.10}$$
$$\leq N \cdot \exp\left\{-n\left(E_r(R,P,W) - \delta\right)\right\}, \text{ for } n \geq n_0\left(|\mathcal{X}|, |\mathcal{Y}|, \delta\right). \tag{2.3.11}$$

In (2.3.10) we have used $[I(P,\tilde{W}) - R]^+ \leq [I(P,\tilde{\tilde{W}}) - R]^+$. In the same way

$$\sum_{i=1}^{N} W^n\left(\pi\mathcal{D}_i - \mathcal{E}_{i,\pi}|\pi u_i\right)$$
$$\leq \sum_{\substack{\tilde{W},\tilde{\tilde{W}} \in \mathcal{W}_n(P) \\ I(P,\tilde{W}) \leq I(P,\tilde{\tilde{W}})}} \sum_{i=1}^{N} W^n \left(\mathcal{T}^n_{\tilde{W}}(\pi u_i) \cap \bigcup_{j=1}^{N} \mathcal{T}^n_{\tilde{\tilde{W}}}(u_j) | \pi u_i \right)$$
$$= \sum_{\substack{\tilde{W},\tilde{\tilde{W}} \in \mathcal{W}_n(P) \\ I(P,\tilde{W}) \leq I(P,\tilde{\tilde{W}})}} \exp\left\{ -n \left(D\left(\tilde{W}||W|P\right) + H\left(\tilde{W}|(P)\right)\right)\right\} g^*_{\tilde{W},\tilde{\tilde{W}}}(\pi)$$
$$\leq N \cdot \exp\left\{-n\left(E_r(R,P,W) - \delta\right)\right\}, \quad \text{for } n \geq n_0\left(|\mathcal{X}|, |\mathcal{Y}|, \delta\right). \tag{2.3.12}$$

The result now follows from (2.3.11), (2.3.12), and (2.3.9).

The second part of the claim of Lemma 2.1 follows directly from Lemma 2.3 (ii) and the argument given in this proof. □

Proof of Theorem 2.3. Lemma 2.1 states that if one chooses the permutation π randomly according to the equidistribution on $\mathcal{S}_n$, then the probability is at most $\exp\{-(\delta/2)n\}$ that (2.3.1) cannot be fulfilled. Since the number of ED's in $\mathcal{P}_n$ the number of different alphabets $\mathcal{X}, \mathcal{Y}$ with $|\mathcal{X}|, |\mathcal{Y}| \leq 2^k$ is "exponentially small" it is clear that one can obtain the theorem immediately from Lemma 2.1. □

2.4 Correlated Source Codes Produced by Permutations from Ordinary Channel Codes

Gallager [17] and Koselev [20] have derived a "random coding" error exponent for discrete memoryless correlated sources (DMCS's) $(X_t, Y_t)_{t=1}^{\infty}$ in case the decoder is informed about the outputs of one of the sources. Csiszár and Körner [8] improved those results by establishing what they considered to be the counterpart of the expurgated bound in source coding. Our results below confirm this view. In [4], it is shown that their result can also be derived via a hypergraph coloring lemma, which slightly generalizes [1]. In [4] we showed that the Slepian-Wolf source coding theorem can easily be derived from the coding theorem for the DMC via the following lemma

Lemma 2.4 (Covering Lemma) *Fix n and $P \in \mathcal{P}_n$ and let $\mathcal{A} \subset \mathcal{T}_P^n$. Then there exist permutations $\pi_1, \dots, \pi_k \in \mathcal{S}_n$ such that*

$$\bigcup_{i=1}^{k} \pi_i \mathcal{A} = \mathcal{T}_P^n,$$

if $k > |\mathcal{A}|^{-1} \cdot |\mathcal{T}_P^n| \log |\mathcal{T}_P^n|$. (Here $\pi\mathcal{A} = \{\pi x^n | x^n \in \mathcal{A}\}$ and $\pi x^n = (x_{\pi 1}, \dots, x_{\pi n})$ for $x^n = (x_1, \dots, x_n) \in \mathcal{X}^n$.)

Here we first show that by this approach every upper bound on the error probability for the DMC yields immediately an upper bound on the error probability for the DMCS, with one source known to the decoder. This way the random coding bound is transformed into the bound derived by Gallager and Koselev, and the expurgated bound is transformed into the bound found by Csiszár and Körner.

Next we show that, conversely, every lower bound on the error probabilities for the DMC yields also a lower bound on the error probabilities for the DMCS. *Theorem* 2.4 *below shows the immediate connection between the source and channel reliability function. We now give the exact statements.*

For the DMCS $(X_t, Y_t)_{t=1}^{\infty}$ we consider the communication situation "source coding with (full) side information", that is, an encoder observes the source X and he has the task to encode this source reliably for a decoder who can observe the source Y. An n-length block code (f, F) for this problem consists of an encoding function $f : \mathcal{X}^n \to \mathcal{Z}$, where $\mathcal{Z}$ is the range of f, and of a decoding function $F : \mathcal{Z} \times \mathcal{Y}^n \to \mathcal{X}^n$. If $x^n \in \mathcal{X}^n$ is observed by the encoder, he gives $f(x^n)$ to the decoder. Having observed the side information y^n the decoder votes for

$F(f(x^n), y^n) \in \mathcal{X}^n$ as being the output of the X-source. The (average) error probability of this code $\lambda(f, F)$ is given by

$$\lambda(f, F) = \sum_{x^n \in \mathcal{X}^n} \sum_{y^n \in \mathcal{Y}^n} Q^n(x^n, y^n)\eta(x^n, F(f(x^n), y^n)),$$

where

$$Q^n(x^n, y^n) = \prod_{t=1}^{n} Q(x_t, y_t),\ Q(x, y) = \Pr\{X = x, Y = y\},$$

and

$$\eta(x^n, x'^n) = \begin{cases} 1, & \text{for } x^n \neq x'^n \\ 0, & \text{for } x^n = x'^n. \end{cases}$$

For $R > 0$ define $\lambda(n, R) = \min \lambda(f, F)$, where the minimum is taken over all n-length block codes (f, F) satisfying $||f|| \leq \exp\{nR\}$. We are interested in the reliability curve

$$e(R) = -\limsup_{n\to\infty} \frac{1}{n} \log \lambda(n, R)$$

for any rate $R > H(X|Y)$.

The joint distribution $Q_{X,Y}$ of (X, Y) induces a channel W, given by

$$W(y|x) := \Pr(Y = y|X = x), \quad \text{for } x \in \mathcal{X}, y \in \mathcal{Y}.$$

For any ED $P \in \mathcal{P}$ define

$$\bar{\lambda}(n, R, P, W) = \min \bar{\lambda}(\mathcal{C}, W),$$

where the minimum is taken over all n-length block codes for W with codewords from $\mathcal{T}_P^n$ and rate at least R. We denote the distribution of X by Q_1. We establish the following connection between $\lambda(n, R)$ and the numbers $\bar{\lambda}(n, R, P, W)$.

Theorem 2.4 (Ahlswede and Dueck [5]) *For any $\delta > 0$ and $m \geq n_0(|\mathcal{X}|, |\mathcal{Y}|, \delta)$,*

(i) $-\frac{1}{n} \log \lambda(n, R + \delta) \geq \min_{P\in\mathcal{P}_n} [D(P||Q_1) - \frac{1}{n} \log \bar{\lambda}(n, H(P) - R, P, W)] - \delta,$

(ii) $-\frac{1}{n} \log \lambda(n, R) \leq \min_{P\in\mathcal{P}_n} [D(P||Q_1) - \frac{1}{n} \log \bar{\lambda}(n, H(P) - R - \delta, P, W)] + \delta.$

In order to get estimates $e(R)$ we can therefore use the familiar estimates on $\bar{\lambda}(n, H(P) - R, P, W)$ and thus obtain the following corollary.

Corollary 2.1

$$e(R) \geq \min_{P \in \mathcal{P}_n} [D(P||Q_1) + E_r(H(P) - R, P, W)], \tag{2.4.1}$$

$$e(R) \geq \min_{P \in \mathcal{P}_n} [D(P||Q_1) + E_{ex}(H(P) - R, P, W)], \tag{2.4.2}$$

$$e(R) \leq \min_{P \in \mathcal{P}_n} \left[D(P||Q_1) + E_{sp}(H(P) - R, P, W)\right], \tag{2.4.3}$$

where

$$E_{sp}(R, P, W) = \min_{\substack{\tilde{W} \in \mathcal{W} \\ I(P, \tilde{W}) \geq R}} D(\tilde{W}||W|P).$$

Remark Equations (2.4.1) and (2.4.3) were obtained in a different form via Chernoff bounds by Gallager [17] and Koselev [20]. Equation (2.4.2) was proved by Csiszár and Körner [8]. In the present form (2.4.1) can be found in [8] and (2.4.3) in [9].

Proof of Theorem 2.4. (i) Fix $R > 0$, $\delta > 0$, and $n \geq n_0(|\mathcal{X}|, |\mathcal{Y}|, \delta)$, $P \in \mathcal{P}_n$. Recall the definition of $\lambda(n, R, P, W)$ and note that any (n, N) code $\mathcal{C} = \{(u_i, \mathcal{D}_i) : i = 1, \dots, N\}$ for W contains at least $N/2$ codewords u_i such that $W^n(\mathcal{D}_i^c|u_i) \leq 2\bar{\lambda}(\mathcal{C}, W)$.

We conclude that for any fixed $P \in \mathcal{P}_n$ there is an (n, N_P) code $\mathcal{C}_P = (u_i^P, \mathcal{D}_i^P)_{i=1}^{N_P}$ for the induced channel W such that

$$\mathcal{U}_P = \left\{u_1^P, \dots, u_{N_P}^P\right\} \subset \mathcal{T}_P^n$$

and

$$N_P \geq \frac{1}{2}\exp\{n(H(P) - R)\}, \tag{2.4.4}$$

and

$$\lambda_{\max}(P, W) \leq 2\bar{\lambda}(n, H(P) - R, P, W). \tag{2.4.5}$$

(It is important here to have a good *maximal* error code.) From these "best" channel codes, constructed for every $P \in \mathcal{P}_n$, we form a source code as follows.

By Lemma 2.4 there exist permutations $\pi_1^P, \dots, \pi_{k_P}^P \in \mathcal{S}_n$ such that

$$\bigcup_{i=1}^{k_P} \pi_i^P \mathcal{U}_P = \mathcal{T}_P^n, \tag{2.4.6}$$

and

$$k_P = \lceil N_P^{-1} \cdot |\mathcal{T}_P^n| \log |\mathcal{T}_P^n| \rceil. \tag{2.4.7}$$

For every $P \in \mathcal{P}_n$ we partition the set $\mathcal{T}_P^n$ into the sets

$$\mathcal{A}_{i,P} = \pi_i^P \mathcal{U}_P - \bigcup_{j=1}^{i-1} \pi_j^P \mathcal{U}_P.$$

We now define an n-length block code (f, F): for every $x^n \in \mathcal{X}^n$ set

$$f(x^n) = (i, P_{x^n}), \quad \text{if } x^n \in \mathcal{A}_{i,P_{x^n}}, \tag{2.4.8}$$

and for every $P \in \mathcal{P}_n$, $i \in \{1, \ldots, k_P\}$, and $y^n \in \mathcal{Y}^n$ set

$$F(i, P, y^n) = \pi_i^P u_j^P, \quad \text{if } y^n \in \pi_i^P \mathcal{D}_j^P. \tag{2.4.9}$$

Next we compute the rate and the error probability of the source code (f, F).

$$\begin{aligned}
||f|| &\leq |\mathcal{P}_n| \cdot \max_{P \in \mathcal{P}_n} k_P \\
&\leq (n+1)^{|\mathcal{X}|} \cdot \max_{P \in \mathcal{P}_n} \left[2\exp\{-n(H(P) - R) + nH(P)\} \cdot n \cdot H(P) + 1\right] \\
&\leq \exp\{n(R + \delta)\}
\end{aligned}$$

for $n \geq n_0(|\mathcal{X}|, |\mathcal{Y}|, \delta)$, where the steps are justified by (2.4.4), (2.4.7), (2.2.1), and (2.2.5). Further,

$$\begin{aligned}
\lambda(f, F) &= \sum_{x^n, y^n} Q^n(x^n, y^n) \cdot \eta(x^n, F(f(x^n), y^n)) \\
&= \sum_{P \in \mathcal{P}_n} \sum_{i=1}^{k_P} \sum_{x^n \in \mathcal{A}_{i,P}} \sum_{y^n} Q_1(x^n) \cdot W^n(y^n|x^n) \cdot \eta(x^n, F(f(x^n), y^n)) \\
&= \sum_{P \in \mathcal{P}} \exp\{-n(D(P||Q_1) + H(P))\} \cdot \sum_{i=1}^{k_P} \sum_{\pi_i^P u_j^P \in \mathcal{A}_{i,P}} W^n\left(\left(\pi_i^P \mathcal{D}_j^P\right)^c | \pi_i^P u_j^P\right) \\
&\leq 2 \sum_{P \in \mathcal{P}_n} \exp\{-n(D(P||Q_1) + H(P))\} \cdot |\mathcal{T}_P^n| \cdot \bar{\lambda}(n, H(P) - R, P, W)
\end{aligned}$$

(by (2.4.5) and (2.2.7)).

Hence, by (2.2.5) and (2.2.1)

$$\begin{aligned}
\lambda(f, F) &\leq 2 \sum_{P \in \mathcal{P}_n} \exp\{-nD(P||Q_1)\} \cdot \bar{\lambda}(n, H(P) - R, P, W) \\
&\leq 2(n+1)^{|\mathcal{X}|} \max_{P \in \mathcal{P}_n} \left[\exp\{-nD(P||Q_1)\} \cdot \bar{\lambda}(n, H(P) - R, P, W)\right],
\end{aligned}$$

and Theorem 2.4 (i) follows.

(ii) Let any code (f, F) of block length n be given. Let $\mathcal{Z} = \{z_1, \ldots, z_{||f||}\}$ be the range of f. For any $z \in \mathcal{Z}$ and every ED $P \in \mathcal{P}_n$ define

$$\mathcal{A}_{P,z} = \left\{x^n : P_{x^n} = P, f(x^n) = z\right\}.$$

For fixed $P \in \mathcal{P}_n$, $z \in \mathcal{Z}$, $x^n \in \mathcal{A}_{P,z}$ define

$$\mathcal{D}_{x^n,P,z} = \{y^n : F(z, y^n) = x^n\}.$$

We now consider for any $P \in \mathcal{P}_n$, $z \in \mathcal{Z}$ the system

$$\mathcal{C}_{P,z} = \left\{\left(x^n, \mathcal{D}_{x^n,P,z}\right) : x^n \in \mathcal{A}_{P,z}\right\} \tag{2.4.10}$$

as a code for the induced channel W. Clearly, $|\mathcal{T}_P^n|/2$ sequences in $\mathcal{T}_P^n$ are contained in sets $\mathcal{A}_{P,z}$ satisfying

$$\left|\mathcal{A}_{P,z}\right| \geq \frac{1}{2}\left|\mathcal{T}_P^n\right| \cdot ||f||^{-1}. \tag{2.4.11}$$

For any $P \in \mathcal{P}_n$ let $Z'(P)$ be the set of those elements in $\mathcal{Z}$ which satisfy (2.4.11). We analyze now the relation between $\lambda(f, F)$ and the error probabilities of the codes in (2.4.10). We get

$$\begin{aligned}
\lambda(f, F) &= \sum_{x^n}\sum_{y^n} Q^n(x^n, y^n)\eta(x^n, F(f(n), y^n)) \\
&= \sum_{P\in\mathcal{P}_n}\sum_{z\in\mathcal{Z}}\sum_{x^n\in\mathcal{A}_{P,z}}\sum_{y^n} Q_1^n(x^n)\cdot W^n(y^n|x^n)\eta(x^n, F(f(x), y^n)) \\
&= \sum_{P\in\mathcal{P}_n} \exp\{-n\,(D(P||Q_1) + H(P))\}\cdot\sum_{z\in\mathcal{Z}}\sum_{x^n\in\mathcal{A}_{P,z}} W^n(\mathcal{D}^c_{x^n,P,z}|x^n) \\
&\geq \max_{P\in\mathcal{P}^n} \exp\{-n\,(D(P||Q_1) + H(P))\}\cdot\sum_{z\in Z'(P)}\sum_{x^n\in\mathcal{A}_{P,z}} W^n(\mathcal{D}^c_{x^n,P,z}|x^n),
\end{aligned}$$

where we have applied (2.2.7). Furthermore,

$$\begin{aligned}
\sum_{x^n\in\mathcal{A}_{P,z}} W^n(\mathcal{D}^c_{x^n,P,z}|x^n) &= |\mathcal{A}_{P,z}|\bar{\lambda}(\mathcal{C}_{P,z}, W) \\
&\geq |\mathcal{A}_{P,z}|\bar{\lambda}\left(n, \frac{1}{n}\log|\mathcal{A}_{P,z}|, P, W\right).
\end{aligned}$$

Now use (2.4.11) to obtain

$$\sum_{z\in Z'(P)} \left|\mathcal{A}_{P,z}\right| \geq \frac{1}{2}\left|\mathcal{T}_P^n\right|$$

and continue again by using (2.4.11) and (2.2.5) to obtain

$$\lambda(f, F) \geq \max_{P \in \mathcal{P}_n} \exp\{-n(D(P||Q_1) + H(P))\}$$
$$\cdot \frac{1}{2}|\mathcal{T}_P^n|\bar{\lambda}\left(n, \frac{1}{n}\log\left(\frac{1}{2} \cdot |\mathcal{T}_P^n| \cdot ||f||^{-1}\right), P, W\right)$$
$$\geq \max_{P \in \mathcal{P}_n} \exp\left\{-nD(P||Q_2) + \log\bar{\lambda}(n, H(P) - R - \delta, P, W) - n\delta\right\}$$

for $n \geq n_0(|\mathcal{X}|, |\mathcal{Y}|, \delta)$. □

2.5 An Iterative Code Construction Achieving the Random Coding and the Expurgated Bound

Theorem 2.5 (Ahlswede and Dueck [5]) *For any* $R > 0, \delta > 0, n \geq n_0(|\mathcal{X}|, |\mathcal{Y}|, \delta)$ *and every ED* $P \in \mathcal{P}_n$ *the following is true.*

(i) Let $\mathcal{C} = \{(u_i, \mathcal{D}_i : i = 1, \ldots, N\}$ *be an* (n, N) *code such that* $(1/2)\log N \leq R$ *and* $u_i \in \mathcal{T}_P^n$ *for* $i = 1, \ldots, N$. *Then there exist an n-sequence* $u_{N+1} \in \mathcal{T}_P^n$ *and proper decoding sets* $\mathcal{E}_1, \ldots, \mathcal{E}_{N+1}$ *such that the enlarged* $(n, N+1)$ *code*

$$\mathcal{C}' = \{(u_i, \mathcal{E}_i) : i = 1, \ldots, N+1\}$$

satisfies for any channel $W \in \mathcal{W}$ *the inequality*

$$\bar{\lambda}(\mathcal{C}', W) \leq \frac{1}{N+1}\left(N \cdot \bar{\lambda}(\mathcal{C}, W) + 2exp\{-n(E_r(R, P, W) - \delta)\}\right). \quad (2.5.1)$$

In particular, if $\bar{\lambda}(\mathcal{C}, W)$ *is less than* $2exp\{-n(E_r(R, P, W) - \delta)\}$, *then also* $\bar{\lambda}(\mathcal{C}', W)$ *is smaller than this quantity.*

(ii) Furthermore, if we prolong the (n, N) *code* $\mathcal{C}$ *to* $\mathcal{C}'$ *by choosing* u_{n+1} *at random according to the equidistribution on* $\mathcal{T}_P^n$, *then the probability of selecting an* u_{N+1} *for which*

$$\bar{\lambda}(\mathcal{C}', W) \leq \frac{1}{N+1}\left(N \cdot \bar{\lambda}(\mathcal{C}, W) + 2exp\{-n(E_r(R+\xi, P, W) - \delta)\}\right). \quad (2.5.2)$$

holds for any $W \in \mathcal{W}$ *is larger than*

$$1 - exp\left\{-n\left(\frac{\delta}{2} + \xi\right)\right\}.$$

Theorem 2.6 (Ahlswede and Dueck [5]) *For any* $R > 0$, $n \geq n_0(|\mathcal{X}|, |\mathcal{Y}|, \delta)$ *and every ED* $\mathcal{P} \in \mathcal{P}_n$ *the following is true*

(i) Let $u_1, \dots, u_N \in \mathcal{T}_P^n$ *be arbitrary sequences,* $N \leq exp\{nR\}$*. For every* $W \in \mathcal{W}$ *let* $\mathcal{C}^W = \{(u_i, \mathcal{D}_i^W) : i = 1, \dots, N\}$ *be the ML code with respect to W for the codewords* $u_1, \dots, u_N$*. Then there exists an n-sequence* $u_{N+1} \in \mathcal{T}_P^n$ *such that for every* $W \in \mathcal{W}$ *the ML code with respect to W satisfies*

$$\bar{\lambda}(\mathcal{C}'^W, W) \leq (1/(N+1)) \left(N \cdot \bar{\lambda}(\mathcal{C}^W, W) + 2exp\{-nE_{ex}(R+\delta, P, W)\}\right). \tag{2.5.3}$$

Again, if $\bar{\lambda}(\mathcal{C}^W, W)$ *is less than* $2exp\{-nE_{ex}(R+\delta, P, W)\}$*, then also* $\bar{\lambda}(\mathcal{C}'^W, W)$ *is smaller than this quantity.*

(ii) If the additional codeword u_{N+1} *is chosen according to the equidistribution on* $\mathcal{T}_P^n$*, then the probability that (2.5.3) can be fulfilled is larger than* $1 - exp\{-(\delta/2)n\}$.

Remark Since for $N = 1$ $\bar{\lambda}(\mathcal{C}, W) \leq 2\exp\{-n(E_r(R, P, W) - \delta)\}$ and $\bar{\lambda}(\mathcal{C}^W, W) \leq 2\exp\{-nE_{ex}(R + \delta, P, W) - \delta)\}$ are obviously achievable, Theorem 2.1 (resp. Theorem 2.2) are immediate consequences of Theorem 2.5 (resp. Theorem 2.6).

Proof of Theorem 2.5. Suppose we are given $\delta > 0$ and an (n, N) code

$$\mathcal{C} = \{(u_i, \mathcal{D}_i) : i = 1, \dots, N\},$$

where $(1/n)\log N \leq R$ and $u_i \in \mathcal{T}_P^n$ for $i = 1, \dots, N$. By Lemma 2.2 (i) there exists a $u_{N+1} \in \mathcal{T}_P^n$ such that

$$g_{\tilde{W},\tilde{\tilde{W}}}(u_{N+1}) \leq \exp\left\{n\left\{H(\tilde{W}|P) - \left[I(P, \tilde{\tilde{W}}) - R\right]^+ + \frac{3}{4}\delta\right\}\right\} \tag{2.5.4}$$

holds for any pair $\tilde{W}, \tilde{\tilde{W}} \in \mathcal{W}$. We show that with such a choice of u_{N+1} (2.5.1) can be fulfilled, so that Theorem 2.5 (i) will follow. It is clear that then Theorem 2.5 (ii) follows directly from this proof and from Lemma 2.2 (ii).

First we define new decoding sets

$$\mathcal{E}_i = \mathcal{D}_i - \left\{y_n : I(u_{N+1} \wedge y^n) > I(u_i \wedge y^n)\right\}, \quad \text{for } i \in \{1, \dots, N\}$$

and

$$\mathcal{E}_{N+1} = \left\{y_n : I(u_{N+1} \wedge y^n) > I(u_i \wedge y^n), \quad \text{for all } i \in \{1, \dots, N\}\right\}.$$

Obviously, the $\mathcal{E}_i$ are disjoint subsets of $\mathcal{Y}^n$. Set $\mathcal{C}' = \{(u_i, \mathcal{E}_i) : i = 1, \dots, N+1\}$. For these codes, $\mathcal{C}$ and $\mathcal{C}'$, we show (2.5.1). We estimate $\bar{\lambda}(\mathcal{C}', W)$ for any $W \in \mathcal{W}$. Now

$$
\begin{aligned}
\bar{\lambda}(\mathcal{C}', W) &= \frac{1}{N+1}\sum_{i=1}^{N+1} W^n(\mathcal{E}_i^c|u_i)\\
&= \frac{1}{N+1}\left(\sum_{i=1}^{N}(W^n(\mathcal{D}_i^c|u_i) + W^n(\mathcal{D}_i - \mathcal{E}_i|u_i)) + W^n(\mathcal{E}_{N+1}^c|u_{N+1})\right)\\
&= \frac{1}{N+1}\Bigg(N\cdot\bar{\lambda}(\mathcal{C}, W) + \sum_{i=1}^{N} W^n(\mathcal{D}_i - \mathcal{E}_i|u_i)\\
&\quad + W^n(\mathcal{E}_{N+1}^c|u_{N+1})\Bigg). \qquad (2.5.5)
\end{aligned}
$$

First we bound the error probability of u_{N+1} from above.

$$
\begin{aligned}
W^n(\mathcal{E}_{N+1}^c : u_{N+1}) &= W^n\left(\{y^n|I(u_{N+1}\wedge y^n) \le I(u_i \wedge y^n),\ \text{for some } 1 \le i \le N\}|u_{N+1}\right)\\
&= \sum_{\substack{\tilde{W},\tilde{\tilde{W}}\in\mathcal{W}_n(P)\\ I(P,\tilde{W})\le I(P,\tilde{\tilde{W}})}} W^n\left(\mathcal{T}_{\tilde{W}}^n(u_{N+1} \cap \bigcup_{i=1}^{N}\mathcal{T}_{\tilde{\tilde{W}}}^n(u_i)|u_{N+1}\right)\\
&= \sum_{\substack{\tilde{W},\tilde{\tilde{W}}\in\mathcal{W}_n(P)\\ I(P,\tilde{W})\le I(P,\tilde{\tilde{W}})}} g_{\tilde{W},\tilde{\tilde{W}}}(u_{N+1})\cdot\exp\left\{-n(D(\tilde{W}||W|P) + H(\tilde{W}|P))\right\}
\end{aligned}
$$

by (2.2.8) and the definition of $g_{\tilde{W},\tilde{\tilde{W}}}$. Observing that $I(P,\tilde{W}) \le I(P,\tilde{\tilde{W}})$ implies $[I(P,\tilde{W}) - R]^+ \le [I(P,\tilde{\tilde{W}}) - R]^+$ we obtain with (2.5.4)

$$
\begin{aligned}
W^n(\mathcal{E}_{N+1}^c|u_{N+1}) &\le \sum_{\substack{\tilde{W},\tilde{\tilde{W}}\in\mathcal{W}_n(P)\\ I(P,\tilde{W})\le I(P,\tilde{\tilde{W}})}} \exp\left\{-n\left(D(\tilde{W}||W|P) + [I(P,\tilde{W}) - R]^+ - \frac{3}{4}\delta\right)\right\}\\
&\le |\mathcal{W}_n(P)|^2\cdot\max_{\tilde{W}\in\mathcal{W}}\exp\left\{-n\left(D(\tilde{W}||W|P) + [I(P,\tilde{W}) - R]^+ - \frac{3}{4}\delta\right)\right\} \qquad (2.5.6)\\
&\le \exp\{-nE_r(R,P,W) - \delta)\} \qquad (2.5.7)
\end{aligned}
$$

for $n \ge n_0(|\mathcal{X}|, |\mathcal{Y}|, \delta)$, because of (2.2.2). Further

$$
\begin{aligned}
\sum_{i=1}^{N} W^n(\mathcal{D}_i - \mathcal{E}_i|u_i) &= \sum_{i=1}^{N} W^n\left(\{y^n \in \mathcal{D}_i : i(u_{N+1}\wedge y^n) > I(u_i \wedge y^n)\}|u_i\right)\\
&= \sum_{\substack{\tilde{W},\tilde{\tilde{W}}\in\mathcal{W}_n(P)\\ I(P,\tilde{W})\le I(P,\tilde{\tilde{W}})}}\sum_{i=1}^{N} W^n\left(\mathcal{D}_i \cap \mathcal{T}_{\tilde{W}}^n(u_i) \cap \mathcal{T}_{\tilde{\tilde{W}}}^n(u_{N+1})|u_i\right) \qquad (2.5.8)
\end{aligned}
$$

By (2.2.8),

$$\begin{aligned} &W^n\left(\mathcal{D}_i \cap \mathcal{T}^n_{\tilde{W}}(u_i) \cap \mathcal{T}^n_{\tilde{\tilde{W}}}(u_{N+1})|u_i\right) \\ &= \exp\left\{-n(D(\tilde{W}||W|P) + H(\tilde{W}|P))\right\} \cdot \left|\mathcal{D}_i \cap \mathcal{T}^n_{\tilde{W}}(u_i) \cap \mathcal{T}^n_{\tilde{\tilde{W}}}(u_{N+1})\right|. \end{aligned} \tag{2.5.9}$$

Since the sets $\mathcal{D}_i$ are disjoint we get

$$\begin{aligned} &\sum_{i=1}^N \left|\mathcal{D}_i \cap \mathcal{T}^n_{\tilde{W}}(u_i) \cap \mathcal{T}^n_{\tilde{\tilde{W}}}(u_{N+1})\right| \\ &\leq \left|\mathcal{T}^n_{\tilde{\tilde{W}}}(u_{N+1}) \cap \bigcup_{i=1}^N \mathcal{T}^n_{\tilde{W}}(u_i)\right| = g_{\tilde{W},\tilde{\tilde{W}}}(u_{N+1}). \end{aligned} \tag{2.5.10}$$

Combining (2.5.8), (2.5.9), and (2.5.10) we obtain as before with (2.5.5)

$$\begin{aligned} \sum_{i=1}^N W^n(\mathcal{D}_i - \mathcal{E}_i|u_i) \leq &\sum_{\substack{\tilde{W},\tilde{\tilde{W}} \in \mathcal{W}_n(P) \\ I(P,\tilde{W}) \leq I(P,\tilde{\tilde{W}}), P\tilde{W} = P\tilde{\tilde{W}}}} \exp\left\{-n(D(\tilde{W}||W|P) + H(\tilde{W}|P))\right\} \\ &\cdot \exp\left\{n\left(H(\tilde{\tilde{W}}|P) - \left[I(P,\tilde{W}) - R\right]^+ + \frac{3}{4}\delta\right)\right\}. \end{aligned}$$

Since $I(P,\tilde{W}) \leq I(P,\tilde{\tilde{W}})$ and $P\tilde{W} = P\tilde{\tilde{W}}$ (by assumption) imply $H(\tilde{\tilde{W}},P) \leq H(\tilde{W},P)$, we conclude (as previously for (2.5.6) and (2.5.7)) that

$$\sum_{i=1}^N W^n(\mathcal{D}_i - \mathcal{E}_i|u_i) \leq \exp(-n(E_r(R,P,W) - \delta)\}, \tag{2.5.11}$$

for $n \geq n_0(|\mathcal{X}|, |\mathcal{Y}|, \delta)$. □

For the proof of Theorem 2.6 we shall need an auxiliary result, which is an analogy to Lemma 2.2 for the expurgated bound.

Fix $R > 0$, $\delta > 0$, $P \in \mathcal{P}_n$, and let $\{u_1, \dots, u_N\} \subset \mathcal{T}^n_P$, $N \leq \exp\{nR\}$ be given.

For any $V \in \mathcal{V}$ we define the function f_V on $\mathcal{X}^n$ by

$$f_V(u) = |\{i : u \in \mathcal{T}^n_V(u_i)\}|, \quad \text{for } u \in \mathcal{X}^n. \tag{2.5.12}$$

$f_V(u)$ measures the V-relationship of u to the given codeword system $\{u_1, \dots, u_N\}$. Note that $f_V(u) = 0$ if $V \notin \mathcal{V}_n(P)$, because in this case $\mathcal{T}^n_V(u_i) = \emptyset$ for all $i = 1, \dots, N$.

Lemma 2.5 *Let U be a RV equidistributed in $\mathcal{T}^n_P$. Then for any $V \in \mathcal{V}$*

(i) $Ef_V(U) \leq (n+1)^{|\mathcal{X}|} \cdot \exp\{n(R - I(P,V))\}$,
(ii) $Pr\{f_V(U) \geq \exp\{n(R - I(P,V) + 3/4\delta)\}$, *for some* $V \in \mathcal{V}\} \leq \exp\{-n(\delta/2)\}$.

Proof

$$
\begin{aligned}
Ef_V(U) &= \sum_{u \in \mathcal{T}_P^n} \Pr(U = u) f_V(u) \\
&\leq \exp\{-nH(P)\} \cdot (n+1)^{|\mathcal{X}|} \sum_{u \in \mathcal{T}_P^n} f_V(u) \\
&= \exp\{-nH(P)\} \cdot (n+1)^{|\mathcal{X}|} \sum_{i=1}^{N} |\{u \in \mathcal{T}_P^n : u \in \mathcal{T}_V^n(u_i)\}| \quad (2.5.13) \\
&= \exp\{-nH(P)\} \cdot (n+1)^{|\mathcal{X}|} \sum_{i=1}^{N} |\mathcal{T}_V^n(u_i)| \\
&\leq N \cdot \exp\{n(H(V|P) - H(P))\} \cdot (n+1)^{|\mathcal{X}|} \quad (2.5.14) \\
&\leq (n+1)^{|\mathcal{X}|} \exp\{n(R + H(V|P) - H(P))\}.
\end{aligned}
$$

The first inequality follows from (2.2.5) and the fact that U is equidistributed. (2.5.13) is obtained by counting and (2.5.14) is a consequence of (2.2.6).

Now let PV be the distribution on $\mathcal{X}$ given by

$$PV(\tilde{x}) = \sum_{x} P(x)V(\tilde{x}|x), \quad \text{for } \tilde{x} \in \mathcal{X}.$$

Then from the definition of f_V it is clear that for $u \in \mathcal{T}_P^n$ $f_V(u) = 0$ if $PV \neq P$. Therefore, we can assume that $PV = P$. Then, however, $H(V|P) - H(P) = H(V|P) - H(PV) = -I(P, V)$. Hence, in any case

$$Ef_V(U) \leq (n+1)^{|\mathcal{X}|} \exp\{n(R - I(P, V))\}.$$

Part (ii) follows by Chebyshev's inequality. □

Proof of Theorem 2.6. Let $\delta, R, P \in \mathcal{P}_n, u_1, \ldots, u_N \in \mathcal{T}_P^n$ be given. Then by Lemma 2.5 there exists a u_{N+1} satisfying

$$f_V(u_{N+1}) \leq (n+1)^{|\mathcal{X}|} \exp\{n(R - I(P, V))\} \quad (2.5.15)$$

for any $V \in \mathcal{V}$.

For any $W \in \mathcal{W}$ we consider the ML codes $\mathcal{C}^W = \{(u_i, \mathcal{D}_i^W) : i = 1, \ldots, N\}$ and $C'^W = \{(u_i, \mathcal{E}_i^W) : i = 1, \ldots, N+1\}$. We estimate for every $W \in \mathcal{W}$

$$\bar{\lambda}(\mathcal{C}'^W, W) = \frac{1}{N+1} \sum_{i=1}^{N+1} W^n \left((\mathcal{E}_i^W)^c \, |u_i \right). \quad (2.5.16)$$

First we bound the error probability for u_{N+1}.

$$W^n\left(\left(\mathcal{E}_{N+1}^W\right)^c|u_{N+1}\right) \leq \sum_{i=1}^{N} \sum_{y^n:W^n(y^n|u_i)>W^n(y^n|u_{N+1})} W^n(y^n|u_{N+1}) \quad (2.5.17)$$

$$\leq \sum_{i=1}^{N} \sum_{y^n \in \mathcal{Y}^n} \sqrt{W^n(y^n|u_i) \cdot W^n(y^n|u_{N+1})}.$$

Now recall the definition of the function d in Theorem 2.2 and observe that

$$\sum_{y^n \in \mathcal{Y}^n} \sqrt{W^n(y^n|u_i) \cdot W^n(y^n|u_{N+1})} = \exp\{-nEd(X, \tilde{X})\}, \quad (2.5.18)$$

where $\tilde{X}$, X are RV's on $\mathcal{X}$ of joint distribution $P_{u_i,u_{N+1}}$.

Now we count how often every sum of the form (2.5.18) occurs in (2.5.17). We use (2.5.15). Note that in (2.5.15) $f_V(u_{N+1})$ is a positive integer so that $f_V(u_{N+1}) = 0$, if $R + (3/4)\delta < I(P, V)$. Hence, we get

$$W^n\left(\left(\mathcal{E}_{N+1}^W\right)^c|u_{N+1}\right) \leq |\mathcal{V}_n| \cdot \exp\Bigg\{-n \cdot \min_{\substack{I(X\wedge\tilde{X})\leq R+(3/4)\delta \\ X,\tilde{X}\, P-\text{distributed}}} \left[Ed(X\tilde{X}) + I(X \wedge \tilde{X}) - R - \frac{3}{4}\delta\right]\Bigg\}$$

$$\leq (n+1)^{|\mathcal{X}|\cdot|\mathcal{Y}|}\exp\left\{-nE_{ex}\left(R + \frac{3}{4}\delta, P, W\right)\right\}$$

$$\leq \exp\{-nE_{ex}(R + \delta, P, W)\}. \quad (2.5.19)$$

for $n \geq n_0(|\mathcal{X}|, |\mathcal{Y}|, \delta)$. Since the code $\mathcal{C}'^W$ is an enlarged version of $\mathcal{C}^W$ and since both $\mathcal{C}'^W$ and $\mathcal{C}^W$ are ML codes, obviously

$$\mathcal{E}_i^W \subset \mathcal{D}_i^W \text{ resp. } \left(\mathcal{E}_i^W\right)^c \supset \left(\mathcal{D}_i^W\right)^c, \quad \text{for } i = 1, \dots, N.$$

Therefore we can write for $i = 1, \dots, N$

$$W^n\left(\left(\mathcal{E}_i^W\right)^c|u_i\right) = W^n\left(\left(\mathcal{D}_i^W\right)^c|u_i\right) + W^n(\mathcal{D}_i^W - \mathcal{E}_I^W|u_i), \quad (2.5.20)$$

where

$$\mathcal{D}_i^W - \mathcal{E}_i^W = \left\{y^n \in \mathcal{D}_i^W : W^n(y^n|u_{N+1}) > W^n(y^n|u_i)\right\}$$

is a subset of $\mathcal{E}_{N+1}^W$.

Using (2.5.15), by the same arguments as above, we get the estimates

$$\sum_{i=1}^{N} W^n\left(\mathcal{D}_i^W - \mathcal{E}_I^W | u_i\right) = \sum_{i=1}^{N} \sum_{y^n \in \mathcal{D}_i^W - \mathcal{E}_i^W} W^n(y^n | u_i) \tag{2.5.21}$$

$$\leq \sum_{i=1}^{N} \sum_{y^n \in \mathcal{Y}^n} \sqrt{W^n(y^n | u_{N+1}) W^n(y^n | u_i)}$$

$$\leq \exp\{-n E_{ex}(R + \delta, P, W)\}$$

for $n \geq n_0(|\mathcal{X}|, |\mathcal{Y}|, \delta)$.

Summarizing we obtain by (2.5.16) and (2.5.19)–(2.5.21)

$$\bar{\lambda}(\mathcal{C}'^W, W) \leq \frac{1}{N+1}\left(\sum_{i=1}^{N} W^N\left(\left(\mathcal{D}_i^W\right)^c |u_i\right) + 2\exp\{-n E_{ex}(R + \delta, P, W)\}\right)$$

$$= \frac{1}{N+1}\left(N \cdot \bar{\lambda}(\mathcal{C}^W, W) + 2\exp\{-n E_{ex}(R + \delta, P, W)\}\right)$$

for $n \geq n_0(|\mathcal{X}|, |\mathcal{Y}|, \delta)$. Theorem 2.6 (i) is proved. Part (ii) follows directly from this proof and Lemma 2.5 (ii). □

2.6 Good Codes Are Highly Probable

In the standard Shannon random coding method [23] one derives bounds on the *expected* average error probability and then concludes that at least one code must be as good as the ensemble average. For high rates this leads to asymptotically optimal results ($E_r(R, W) = E_{sp}(R, W)$ for rates near capacity, see [16]) and therefore in this case "most" codes in the ensemble must be close to the optimum. In the study of complex channel systems such as arbitrarily varying channels ([3]) it is necessary to have estimates on the proportion of codes in the ensemble which are good. Also, if random selection is of any practical use, one would like to have bounds on the probability with which a good code can be found. First steps in this direction were taken by Dobrushin and Stambler [11], and independently in [2] and [3]. The papers [11] and [2] consider the average and the paper [3] the maximal error probability.

Here we show considerably more. Whereas in all those papers the error probability was kept constant we allow here λ to meet the random coding bound and still show that for a random selection the probability of not meeting those bounds is double exponentially small. Moreover, we obtain estimates in the double exponential function.

We first state the result. Theorem 2.7 estimates the probability that randomly selected and expurgated codes are "good". Theorem 2.8 gives a result for nonexpurgated codes. In order to formulate Theorem 2.7 we have to introduce some notation concerning the expurgation of a code.

Let $n, \delta > 0$, and $P \in \mathcal{P}_n$ be given. $U_1, \dots, U_N$ are independent RV's equidistributed on $\mathcal{T}_P^n$, $N = \exp\{nR\}$. For outcomes $u_1, \dots, u_N \in \mathcal{T}_P^n$ of $U_1, \dots, U_N$ we define the functions $F(u_1, \dots, u_N)$ and $G(u_1, \dots, u_N)$ by

1. $F(u_1, \dots, u_N) = 1$ if there exist $u_{j_1}, \dots, u_{j_M} \in \{u_1, \dots, u_N\}$ and suitable decoding sets $\mathcal{D}_{j_1}, \dots, \mathcal{D}_{j_M}$ such that $M \geq N/2$ and for $\mathcal{C} = \{(u_{j_k} | \mathcal{D}_{j_k}) : k = 1, \dots, M\}$
$$\bar{\lambda}(\mathcal{C}, W) \leq 2\exp\{-n(E_r(R, P, W) - \delta)\}, \tag{2.6.1}$$
for every $W \in \mathcal{W}$. $F(u_1, \dots, u_N) = 0$ otherwise. Similarly,
2. $G(u_1, \dots, u_N) = 1$ if there exist $u_{j_1}, \dots, u_{j_M} \in \{u_1, \dots, u_N\}$ such that $M \geq N/2$ and such that for every $W \in \mathcal{W}$ the corresponding ML code
$$\mathcal{C}^W = \left\{\left(u_{j_k}, \mathcal{D}_{j_k}\right) : k = 1, \dots, M\right\}$$
satisfies
$$\bar{\lambda}(\mathcal{C}^W, W) \leq 2\exp\{-nE_{ex}(R + \delta, P, E)\}.$$
$G(u_1, \dots, u_N) = 0$ otherwise.

Theorem 2.7 (Ahlswede and Dueck [5]) *In the notation above for $n \geq n_0(|\mathcal{X}|, |\mathcal{Y}|, \delta)$*

$$Pr(F = 0) \leq exp\{-(n \cdot \delta/4 - \log 2)exp\{nR\}, \tag{2.6.2}$$
$$Pr(G = 0) \leq exp\{-(n \cdot \delta/4 - \log 2)exp\{nR\}, \tag{2.6.3}$$

that is, the procedures fail to achieve the random coding bounds (resp. expurgated bounds) uniformly for every $W \in \mathcal{W}$ with double exponentially small error probabilities. Moreover, the exponent R is optimal.

By somewhat more refined calculations we obtain the next theorem.

Theorem 2.8 (Ahlswede and Dueck [5]) *For any $\delta > 0$, $R > 0$, $n \geq n_0(|\mathcal{X}|, |\mathcal{Y}|, \delta)$, and $P \in \mathcal{P}_n$ the following is true.*

Let $U_1, \dots, U_N$ be independent RV's equidistributed on $\mathcal{T}_P^n$ and for any W let $\mathcal{C}^W(U_1, \dots, U_N)$ be the ML code for the codewords $U_1, \dots, U_N$. Then

$$Pr\left(\bar{\lambda}(\mathcal{C}^W(U_1, \dots, U_N), W) \geq 2exp\{-n(E_r(R, P, W) - 2\delta)\}\right)$$
$$\leq exp\{-exp\{n(R - E_r(R, P, W))\}\}$$

for all $W \in \mathcal{W}$.

Remark This result shows that for $R > E_r(R, P, W)$ codes achieving the random coding bound can hardly be missed by random selection. Notice that for $R < E_r(R, P, W)$ the probability to select a code with P-typical codewords *not* achieving the random coding bound is larger than the probability that in a selected code there are two equal codewords. Since the latter probability is at least exponentially small, for $R < E_r(R, P, W)$ we cannot get any double exponential estimate.

As a *new problem* in the area of error bounds we propose to find the exact exponent for all rates $R > E_r(R, P, W)$.

Proof of Theorem 2.7. Fix $\delta > 0$, $R > 0$. Let $n \geq n_0(|\mathcal{X}|, |\mathcal{Y}|, \delta)$ such that Theorems 2.5 and 2.6 hold.

Let $U_1, \dots, U_N$ be independent RV's equidistributed on $\mathcal{T}_P^n$, $N = \exp\{nR\}$.

Consider the following "expurgated codes": Set $\mathcal{C}_{ex}(U_1) = \{(U_1, \mathcal{D}_{U_1})\}$ with the decoding set $\mathcal{D}_{U_1} = \mathcal{Y}^n$. Clearly, $\bar{\lambda}(\mathcal{C}_{ex}(U_1), W) = 0$ for every $W \in \mathcal{W}$. For $i = 2, \dots, N$ we define the codes $\mathcal{C}_{ex}(U_1, \dots, U_i)$ by extending $\mathcal{C}_{ex}(U_1, \dots, U_{i-1})$. Suppose $i \geq 2$ and assume that $\mathcal{C}_{ex}(U_1, \dots, U_{i-1}) = \{(U_{j1}, \mathcal{D}_{j1}), \dots, (U_{jk}, \mathcal{D}_{jk})\}$ with k codewords $U_{j1}, \dots, U_{jk} \in \{U_1, \dots, U_{i-1}\}$ has been defined. Then we prolong this code by the codeword U_i to the new code

$$\mathcal{C}_{ex}(U_1, \dots, U_{i-1}|U_i) = \left\{\left(U_{j1}, \mathcal{E}_{j1}\right), \dots, \left(U_{jk}, \mathcal{E}_{jk}\right), (U_i, \mathcal{E}_i)\right\},$$

where, for $l = 1, \dots, k$, $\mathcal{E}_{jl} = \mathcal{D}_{jl} - \{y^n : I(U_i \wedge y^n) > I(U_{jl} \wedge y^n)\}$ and where

$$\mathcal{E}_i = \left\{y^n : I(U_1 \wedge y^n) > I(U_{jl} \wedge y^n), \text{ for all } l = 1, \dots, k\right\}.$$

If for all $W \in \mathcal{W}$

$$\bar{\lambda}(\mathcal{C}_{ex}(U_1, \dots, U_{i-1}|U_i), W) \leq 2\exp\{-n(E_r(R, P, W) - \delta)\},$$

then we define

$$\mathcal{C}_{ex}(U_1, \dots, U_i) = \mathcal{C}_{ex}(U_1, \dots, U_{i-1}|U_i).$$

If this is not the case we set

$$\mathcal{C}_{ex}(U_1, \dots, U_i) = \mathcal{C}_{ex}(U_1, \dots, U_{i-1}).$$

In this way we gave a formal definition of the expurgation of a given code with codewords $U_1, \dots, U_N$.

Now let S_i be a RV on $\{0, 1\}$ such that $S_i = 0$ if and only if $\mathcal{C}_{ex}(U_1, \dots, U_i) \neq \mathcal{C}_{ex}(U_1, \dots, U_{i-1})$, that is, $S_i = 0$ if and only if the codeword U_i was not expurgated. We observe

$$\Pr(F = 0) \leq \Pr\left(\sum_{i=1}^{N} S_i \geq \frac{N}{2}\right), \tag{2.6.4}$$

and

$$\Pr(S_i = 1|S_{i-1} = s_{i-1}, \dots, S_1 = s_1) \leq \exp\left\{-n\frac{\delta}{2}\right\} \tag{2.6.5}$$

for any values $s_{i-1}, \dots, s_1 \in \{0, 1\}$. Equation (2.6.4) follows from the definition of the functions F and $S_1, \dots, S_N$. Equation (2.6.5) is a direct application of Theorem 2.5 (ii). Hence, we only have to estimate $\Pr(\sum_{i-1}^{N} S_i \geq N/2)$. This can be done by using Bernstein's trick.

For any $\alpha > 0$

$$\Pr\left(\sum_{i=1}^{N} S_i \geq \frac{N}{2}\right) \leq \exp\left\{-\alpha \cdot \frac{N}{2}\right\} \cdot E \prod_{i=1}^{N} \exp\{\alpha S_i\}.$$

Now we apply (2.5.13) to estimate the expected value on the RHS. Thus we obtain

$$\Pr\left(\sum_{i=1}^{N} S_i \geq \frac{N}{2}\right) \leq \exp\left\{-\alpha \frac{N}{2}\right\} \cdot \left[\exp\left\{-n\frac{\delta}{2}\right\} \cdot \exp\{\alpha\} + \left(1 - \exp\left\{-n\frac{\delta}{2}\right\}\right)\right]^N.$$

Choose in particular

$$\alpha = \log \frac{1 - \exp\left\{-n\frac{\delta}{2}\right\}}{\exp\left\{-n\frac{\delta}{2}\right\}},$$

which is positive for $n \geq n_0(\delta)$. Then,

$$\Pr\left(\sum_{i=1}^{N} S_i \geq N/2\right) \leq \exp\left\{-D((1/2)||\exp\{-n(\delta/2)\}) \cdot N\right\},$$

where $D(p||\lambda)$ denotes the relative entropy between the probability vectors $(p, 1-p)$ and $(\lambda, 1-\lambda)$.

We can estimate this quantity:

$$D\left(\frac{1}{2}||\exp\left\{-n\frac{\delta}{2}\right\}\right) = \log 2 - \frac{1}{2}\log\exp\left\{-n\frac{\delta}{2}\right\} - \frac{1}{2}\log\left(1 - \exp\left\{-n\frac{\delta}{2}\right\}\right)$$
$$\leq -\log 2 + n \cdot \frac{\delta}{4}.$$

Thus, $\Pr(F = 0) \leq \exp\{-(n(\delta/4) - \log 2) \cdot \exp\{nR\}$. This proves the first part of Theorem 2.7. The proof of the second part is completely analogous.

We have to show that the exponent R is best possible. For this, choose any codeword $u \in \mathcal{T}_P^n$, $P \in \mathcal{P}_N$. Define $\mathcal{C}$ as a code with N codewords $u_1, \dots, u_N$; $u_i = u$ for all $i = 1, \dots, N$. We make two observations: $\mathcal{C}$ is a "bad" code, even if one expurgates $\mathcal{C}$. On the other hand, the probability to choose $\mathcal{C}$ at random is of the order $\exp\{-O(n)\exp\{nR\}\}$. □

Proof of Theorem 2.8. Fix $\delta > 0$ and $n \geq n_0(|\mathcal{X}|, |\mathcal{Y}|, \delta)$ such that Theorems 2.5 and 2.6 hold and choose $N = \exp\{nR\}$.

Let $U_1, \dots, U_n$ be independent RV's equidistributed on $\mathcal{T}_P^n$ and let $W \in \mathcal{W}$. We consider the ML codes $\mathcal{C}(U_1, \dots, U_k)$, $k = 1, \dots, N$, that is, codes with codeword set $\{U_1, \dots, U_k\}$ and maximum likelihood decoding with respect to the given channel W. We define the RV's $T_1, \dots, T_N$ on $[0, 1]$ as follows: $T_1 = \bar{\lambda}(\mathcal{C}(U_1), W) = 0$, and for $k = 1, \dots, N-1$ the RV T_{k+1} is defined by

$$\bar{\lambda}(\mathcal{C}(U_1, \dots, U_{k+1}), W) = \frac{1}{k+1}(k \cdot \bar{\lambda}(\mathcal{C}(U_1, \dots, U_k), W) + T_{k+1}).$$

Observe that with this definition

$$\bar{\lambda}(\mathcal{C}(U_1, \dots, U_k), W) = \frac{1}{k}\sum_{i=1}^{k} T_i$$

for any $k = 1, \dots, N$.

Using this notation Theorem 2.5 says that for any $\xi \geq 0$ and for any valuyes $t_1, \dots, t_k$ of the RV's $T_1, \dots, T_k$ we have

$$\begin{aligned}\Pr(T_{k+1} > 2\exp\{-n(E_r(R+\xi, P, W) - \delta)\} | t_1 = t_1, \dots, T_k = t_k) \\ \leq \exp\left\{-n\left(\frac{\delta}{2} + \xi\right)\right\}, \quad k = 1, \dots, N-1.\end{aligned} \tag{2.6.6}$$

For any $\xi \geq 0$ we define RV's S_i, $i = 1, \dots, N$, on $\{0, 1\}$ such that $S_{i,\xi} = 1$ if and only if

$$T_i > 2\exp\{-n(E_r(R+\xi, P, W) - \delta)\}.$$

Thus $\sum +i = 1^N S_{i,\xi}$ counts the number of T_i of a certain size. Note that $|T_i| \leq 1$ and $E_r(R + |\mathcal{X}|, P, W) = 0$ since $E_r(C, P, W) = 0$, where C is the capacity of W. We express the error probability of the code $\mathcal{C}(U_1, \dots, U_N)$ with the help of the "counting variables" $\sum_{i=1}^N S_{i,\xi}$. Let m be a positive integer, $1/m < \delta/2$. Then

$$\begin{aligned}\bar{\lambda}(\mathcal{C}(U_1, \dots, U_N), W) &= \frac{1}{N}\sum_{i=1}^{N} T_i \\ &\leq \frac{1}{N}\sum_{j=1}^{m\cdot|\mathcal{X}|}\left(\sum_{i=1}^{N} S_{i,j/m}\right) \cdot 2\exp\left\{-n\left(E_r\left(R + \frac{j+1}{m}, P, W\right) - \delta\right)\right\}.\end{aligned} \tag{2.6.7}$$

Here we have counted those T_i which lie in intervals of the form

$$\left[2\exp\left\{-n\left(E_r\left(R + \tfrac{j}{m}, P, W\right) - \delta\right)\right\}, 2\exp\left\{-n\left(E_r\left(R + \tfrac{j+1}{m}, P, W\right) - \delta\right)\right\}\right].$$

Therefore $\bar{\lambda}(\mathcal{C}(U_1, \ldots, U_N), W)$ becomes large, if the expressions $\sum_{i=1}^{N} S_{i,j/m}$ become large.

We show that for any $\xi \geq 0$

$$\Pr\left\{\sum_{i=1}^{N} S_{i,\xi} \geq \exp\{n(R - (E_r(R, P, W) - E_r(R + \xi, P, W)))\}\right\}$$
$$\leq \exp\left\{-\left(n\frac{\delta}{2}\right) \cdot \exp\{n(R - (E_r(R, P, W) - E_r(R + \xi, P, W)))\}\right\}. \quad (2.6.8)$$

Again we use Bernstein's trick. Abbreviate $\tau = 1 - \exp\{-n(E_r(R, P, W) - E_r(R + \xi, P, W))\}$. Then for any $\alpha > 0$:

$$\Pr\left\{\sum_{i=1}^{N} S_{i,\xi} \geq N \cdot (1 - \tau)\right\} \leq \exp\{-\alpha \cdot N(1 - \tau)\} \cdot \prod_{i=1}^{N} \exp\left\{\alpha S_{i,\xi}\right\}. \quad (2.6.9)$$

In order to estimate the expectation on the RHS it is necessary to have estimates on conditional probabilities of the $S_{i,\xi}$. Now observe that from the definition of the $S_{i,\xi}$ and because of (2.6.6) we have for any $\xi \geq 0$ and for any values $s_1, \ldots, s_{i=1} \in \{0, 1\}$

$$\Pr(S_{i,\xi} = 1 | S_{1,\xi} = s_1, \ldots, S_{i-1,\xi} = s_{i=1}) \leq \exp\left\{-n\left(\frac{\delta}{2} + \xi\right)\right\}. \quad (2.6.10)$$

We get from (2.6.9) and (2.6.10)

$$\Pr\left\{\sum_{i=1}^{N} S_{i,\xi} \geq N(1 - \tau)\right\}$$
$$\leq \exp\{-\alpha N(1 - \tau)\}\left[\exp\left\{-n\left(\frac{\delta}{2} + \xi\right) + \alpha\right\} + 1 - \exp\left\{-n\left(\frac{\delta}{2} + \xi\right)\right\}\right]. \quad (2.6.11)$$

Set

$$\alpha = \log\left(\frac{1 - \exp\left\{-n\left(\frac{\delta}{2} + \xi\right)\right\}}{\exp\left\{-n\left(\frac{\delta}{2}\right)\right\}} \cdot \frac{1 - \tau}{\tau}\right).$$

Since $E_r(R, P, W) - E_r(R + \xi, P, W) \leq \xi$ for all $\xi \geq 0$, the number α is positive for $n \geq n_0(\delta)$. We obtain from (2.6.11) with this choice of α:

$$\Pr\left\{\sum_{i=1}^{N} S_{i,\xi} \geq N \cdot (1 - \tau)\right\} \leq \exp\left\{-D\left(1 - \tau || \exp\left\{-n\left(\frac{\delta}{2} + \xi\right)\right\}\right) \cdot N\right\},$$

where

$$D\left(1-\tau||\exp\left\{-n\left(\frac{\delta}{2}+\xi\right)\right\}\right) \geq \tau\log\tau+(1-\tau)\log(1-\tau)+n\left(\frac{\delta}{2}+\xi\right)\cdot(1-\tau).$$

From the fact that $\log(1-x) \geq -2x$ for small positive x we conclude that $\log\tau \geq -2(1-\tau)$ for n sufficiently large.

Hence, for large n,

$$\begin{aligned} D\left(\tau||\exp\left\{-n\left(\frac{\delta}{2}+\xi\right)\right\}\right) &\geq \left[n\left(\frac{\delta}{2}+\xi\right) - (E_r(R,P,W)-E_r(R+\xi,P,W))\right] \\ &\quad \cdot(1-\tau)-2(1-\tau) \\ &\geq \left(n\frac{\delta}{2}-2\right)(1-\tau) \qquad (2.6.12) \\ &= \left(n\frac{\delta}{2}-2\right)\cdot\exp\{-n(E_r(R,P,W)-E_r(R+\xi,P,W))\}, \end{aligned}$$

where (2.6.8) is true, because $E_r(R,P,W) - E_r(R+\xi,P,W) \leq \xi$. Equation (2.6.8) is proved. Finally we have to show (2.6.8) and (2.6.7) imply Theorem 2.8.

From (2.6.8) we conclude first that for all $\xi \geq 0$

$$\begin{aligned} &\Pr\left\{\sum_{i=1}^{N} S_{i,\xi} \geq \exp\{n(R-(E_r(R,P,W)-E_r(R+\xi,P,W)))\}\right\} \\ &\leq \exp\left\{-\left(n\frac{\delta}{2}-2\right)\cdot\exp\{n(R_(R,P,W))\}\right\}, \quad \text{if } n \text{ is large.} \qquad (2.6.13) \end{aligned}$$

Suppose now that

$$\sum_{i=1}^{N} S_{i,j/m} \leq \exp\left\{n\left(R-\left(E_r(R,P,W)-E_r\left(R+\frac{j+1}{m},P,W\right)\right)\right)\right\}, \quad j=1,\dots,m\cdot|\mathcal{X}|.$$

Then we can continue with (2.6.7):

$$\begin{aligned} \bar{\lambda}(\mathcal{C}(U_1,\dots,U_N),W) &\leq 2\sum_{j=1}^{m\cdot|\mathcal{X}|}\exp\left\{-n\left(E_r(R,P,W)-E_r\left(R+\frac{j}{m},P,W\right)\right)\right. \\ &\qquad \left. -n\left(E_r\left(R+\frac{j+1}{m},P,W\right)-\delta\right)\right\} \\ &\leq 2\cdot m\cdot|\mathcal{X}|\cdot\exp\left\{-n\left(E_r(R,P,W)+\frac{1}{m}+\delta\right)\right\} \\ &\leq 2\cdot\exp\{-n(E_r(R,P,W)-2\delta)\}, \qquad (2.6.14) \end{aligned}$$

for n sufficiently large. Now (2.6.14), (2.6.13), and (2.6.7) yield

$$\Pr\{\bar{\lambda}(\mathcal{C}(U_1, \dots, U_N), W) \leq 2\exp\{-n(E_r(R, P, W) - 2\delta)\}\}$$
$$\leq \exp\left\{-\left(n\frac{\delta}{2} - 2\right)\exp\{n(R - E_r(R, P, W))\right\}$$

□

References

1. R. Ahlswede, Channel capacities for list codes. J. Appl. Prob. **10**, 824–836 (1973)
2. R. Ahlswede, Elimination of correlation in random codes for arbitrarily varying channels. Z. Wahrscheinlichkeitstheorie verwandte Gebiete **44**, 159–175 (1978)
3. R. Ahlswede, A method of coding and its application to arbitrarily varying channels. J. Comb. Inf. Syst. Sci. **5**(1), 10–35 (1980)
4. R. Ahlswede, Coloring hypergraphs: a new approach to multi-user source coding, Part II. J. Comb. Inf. Syst. Sci. **5**(3), 220–268 (1980)
5. R. Ahlswede, G. Dueck, Good codes can be produced by a few permutations. IEEE Trans. Inf. Theory **IT–28**(3), 430–443 (1982)
6. S. Arimoto, On the converse to the coding theorem for the discrete memoryless channels. IEEE Trans. Inf. Theory **IT–19**, 357–359 (1973)
7. R.E. Blahut, Hypothesis testing and information theory. IEEE Trans. Inf. Theory **IT–20**, 405–417 (1974)
8. I. Csiszár, J. Körner, Graph decomposition: a new key to coding theorems. IEEE Trans. Inf. Theory **IT–27**, 5–12 (1981)
9. I. Csiszár, J. Körner, *Information Theory: Coding Theorems for Discrete Memoryless Systems* (Academic Press, New York, 1981)
10. I. Csiszár, J. Körner, K. Marton, *A new look at the error exponent of a discrete memoryless channel (preprint), in IEEE International Symposium on Information Theory* (Ithaca, NY, 1977)
11. R.L. Dobrushin, S.Z. Stambler, Coding theorems for classes of arbitrarily varying discrete memoryless channels. Probl. Peredach. Inf. **11**, 3–22 (1975)
12. G. Dueck, J. Körner, Reliability function of a discrete memoryless channel at rates above capacity. IEEE Trans. Inf. Theory **IT–25**, 82–85 (1979)
13. R.M. Fano, *Transmission of Information: A Statistical Theory of Communication* (Wiley, New York, 1961)
14. A. Feinstein, A new basic theorem of information theory. IRE Trans. Inf. Theory **4**, 2–22 (1954)
15. R.G. Gallager, A simple derivation of the coding theorem and some applications. IEEE Trans. Inf. Theory **IT–11**, 3–18 (1965)
16. R.G. Gallager, *Information Theory and Reliable Communication* (Wiley, New York, 1968)
17. R.G. Gallager, *Source coding with side information and universal coding (preprint), in IEEE International Symposium on Information Theory* (Ronneby, Sweden, 1976)
18. V.D. Goppa, Nonprobabilistic mutual information without memory. Prob. Contr. Inf. Theory **4**, 97–102 (1975)
19. A. Haroutunian, Estimates of the error exponent for the semi-continuous memoryless channel. Probl. Peredach. Inf. **4**, 37–48 (1968)
20. V.N. Koselev, On a problem of separate coding of two dependent sources. Probl. Peredach. Inf. **13**, 26–32 (1977)
21. J.K. Omura, A lower bounding method for channel and source coding probabilities. Inform. Contr. **27**, 148–177 (1975)
22. C.E. Shannon, A mathematical theory of communication. Bell Syst. Tech. J. **27**(379–423), 632–656 (1948)

23. C.E. Shannon, Certain results in coding theory for noisy channels. Inform. Contr. **1**, 6–25 (1957)
24. C.E. Shannon, R.G. Gallager, E.R. Berlekamp, Lower bounds to error probability for coding on discrete memoryless channels I-II. Inf. Contr. **10**(65–103), 522–552 (1967)
25. D. Slepian, J.K. Wolf, Noiseless coding of correlated information sources. IEEE Trans. Inf. Theory **IT–19**, 471–480 (1973)
26. J. Wolfowitz, The coding of messages subject to chance errors. Illinois J. Math. **1**, 591–606 (1957)

Further Readings

27. R. Ahlswede, Coloring hypergraphs: a new approach to multi-user source coding, Part I. J. Comb. Inf. Syst. Sci. **1**, 76–115 (1979)
28. R.E. Blahut, Composition bounds for channel block codes. IEEE Trans. Inf. Theory **IT–23**, 656–674 (1977)

Chapter 3
Results for Classical Extremal Problems

3.1 Antichains

3.1.1 *Kraft's Inequality and the LYM-property*

In order to prove Kraft's ([7] inequality for prefix codes the codewords were regarded as vertices in a rooted tree. For any rooted tree it is possible to define a relation $\preceq$, say, on the vertices of the tree by $x \preceq y$, if and only if there exists a path from the root through x to y. This relation has the following properties ($\mathcal{X}$ denotes the set of vertices of the tree)

(i) *reflexivity*: $x \preceq x$ for all $x \in \mathcal{X}$
(ii) *antisymmetry*: $x \preceq y, x \preceq y \Rightarrow x = y$ for all $x, y \in \mathcal{X}$
(iii) *transitivity*: $x \preceq y, y \preceq z \Rightarrow x \preceq z$ for all $x, y, z \in \mathcal{X}$

Definition 3.1 A reflexive, antisymmetric and transitive relation $\preceq$ on a set $\mathcal{X}$ is denoted as **order relation**. The pair $(\mathcal{X}, \preceq)$ is called *partially ordered set* or short *poset*.

Further examples for posets are

1. $(\mathbb{N} \cup \{0\}, \leq)$, the non-negative integers with the canonical order,
2. $(\mathbb{N}, |)$, the positive integers with the divisor order ($x \preceq y$, if x divides y),
3. $\big(\mathcal{P}(\{1, \dots, n\}), \subseteq\big)$. Here $\mathcal{P}(\{1, \dots, n\}) \triangleq \{S : S \subseteq \{1, \dots, n\}\}$ is the family of all subsets of an n-elementary set and the inclusion $\subseteq$ determines the order relation.

In $(\mathbb{N} \cup \{0\}, \leq)$ any two elements $n, m \in \mathbb{N}$ are *comparable*, i.e. $n \leq m$ or $m \leq n$ holds for all pairs (n, m). In most posets this is not the case, e.g. in $\mathcal{P}(\{1, 2, 3\})$ the subsets $\{1, 2\}$ and $\{2, 3\}$ are not comparable under inclusion.

A. Ahlswede et al. (eds.), *Combinatorial Methods and Models*,
Foundations in Signal Processing, Communications and Networking 13,
DOI 10.1007/978-3-319-53139-7_3

Definition 3.2 A **chain** (or total order) is a poset in which all elements are comparable.

An **antichain** is a poset in which no two elements are comparable.

Of course, these notions can also be applied to *subposets* $(\mathcal{X}', \preceq')$ of a poset $(\mathcal{X}, \preceq)$. Here $\mathcal{X}' \subseteq \mathcal{X}$ and for $x, y \in \mathcal{X}'$ it is $x \preceq' y$ if and only if $x \preceq y$ in $(\mathcal{X}, \preceq)$. The following examples may serve as an illustration.

1. In the tree a chain consists of elements lying on a path from the root to a leaf. An antichain corresponds to a prefix code.
2. In $(\mathbb{N} \cup \{0\}, \leq)$ any subset $S \subseteq \mathbb{N} \cup \{0\}$ determines the chain $(S, \leq)$, whereas an antichain can only consist of one element.
3. In $(\mathbb{N}, |)$ the powers $\{1, n, n^2, n^3, \dots\}$ of an integer $n \in \mathbb{N}$ form a chain. The prime numbers are an antichain.
4. In $\mathcal{P}(\{1, \dots, n\})$ a possible chain is $\big\{\emptyset, \{1\}, \{1, 2\}, \{1, 2, 3\}, \dots, \{1, 2, \dots, n\}\big\}$, whereas e.g. $\big\{\{1\}, \{2\}, \{3\}\big\}$ is an antichain.

Definition 3.3 A chain (antichain) $(S, \preceq')$, $S \subset \mathcal{X}$ in a poset $(\mathcal{X}, \preceq)$ is said to be **saturated**, if it is not possible to add a further element $x \in \mathcal{X}$ to S without destroying the structure (chain or antichain, respectively).

In our last example 4 the chain is saturated, whereas the antichain $\big\{\{1\}, \{2\}, \{3\}\big\}$ for $n \geq 4$ is not saturated, because $\big\{\{1\}, \{2\}, \{3\}, \{4\}\big\}$ is also an antichain.

Definition 3.4 For the posets presented in the above examples it is also possible to introduce a **rank function** $r : (\mathcal{X}, \preceq) \to \mathbb{N} \cup \{0\}$, which is recursively defined by

(i) $r(x) = 0$ for the *minimal elements* $x \in \mathcal{X}$ (x is minimal, if there is no $y \in \mathcal{X}$ with $y \preceq x$)
(ii) $r(x) = r(y) + 1$, when x is a direct successor of y (i.e. $y \preceq x$ and there is no $z \in \mathcal{X}$ such that $y \preceq z \preceq x$)

In our examples the following rank functions arise.

1. Tree: $r(x) =$ length of the path from the root to x
2. $(\mathbb{N} \cup \{0\}, \leq)$: $r(n) = n$.
3. $(\mathbb{N}, |)$: $r(n) = \sum_{i=1}^{r} k_i$, when $n = p_1^{k_1} \cdot \dots \cdot p_r^{k_r}$ is the prime factorization of n
4. $\big(\mathcal{P}(\{1, \dots, n\}), \subseteq\big)$: $r(S) = |S|$, the cardinality of $S \subseteq \{1, \dots, n\}$.

Not every poset can be equipped with a rank function as the following example demonstrates ($x \preceq y$, if x and y are on a (directed) path from x_0 to x_4):

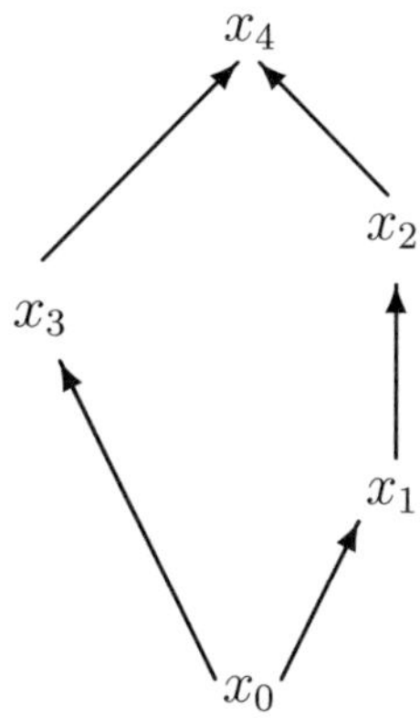

Definition 3.5 If the poset $(\mathcal{X}, \preceq)$ has a rank function r, then the set $\{x \in \mathcal{X} : r(x) = i\}$ is defined as the ith **level** of $(\mathcal{X}, \preceq)$. The size of the ith level is denoted as the ith **Whitney number** $W(i)$.

Many posets have an additional structure. An element $z \in \mathcal{X}$ is said to be lower (upper) bound of x and $y \in \mathcal{X}$, if $z \preceq x$ and $z \preceq y$ ($x \preceq z$ and $y \preceq z$). The greatest lower bound (*infimum*) and the least upper bound (*supremum*) of x and y are denoted as $x \wedge y$ and $x \vee y$, respectively.

Definition 3.6 A **lattice** is a poset $(\mathcal{X}, \preceq)$, where to each pair (x, y) there exist the infimum $x \wedge y$ and the supremum $x \vee y$.

Observe that a tree with the associated order does not yield a lattice. For example in

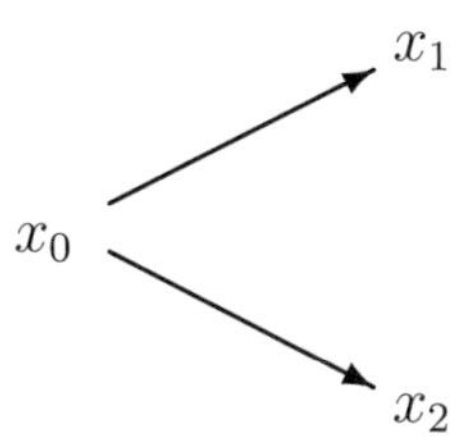

$x_1 \vee x_2$ does not exist.

In the other examples infimum and supremum are given by

1. $(\mathbb{N} \cup \{0\}, \leq)$: $n \wedge m = \min\{n, m\}$, $n \vee m = \max\{n, m\}$.
2. $(\mathbb{N}, |)$: $n \wedge m = gcd(n, m)$ (greatest common divisor), $n \vee m =$ least common multiple of n and m
3. $(\mathcal{P}(\{1, \dots, n\}), \subseteq)$: $S \wedge T = S \cap T$, $S \vee T = S \cup T$ for all S and $T \subseteq \{1, \dots, n\}$.

The existence of a prefix code to given lengths $L(1), L(2), \dots$ is guaranteed by Kraft's inequality. We are now able to interpret this theorem in words of Order Theory. The codewords of a prefix code form an antichain in the order imposed by the tree introduced and the length $L(x)$, $x \in \mathcal{X}$, is the length of the path from the root to $c(x)$. But the length of this path is just the rank of $c(x)$.

The LYM-inequality (LYM: Lubell, Yamamoto, Meshalkin) ([8, 9, 14]) is the analogue to Kraft's inequality for the poset $\big(\mathcal{P}(\{1,\ldots,n\}),\subseteq\big)$. It is often helpful to assign a $\{0,1\}$-sequence of length n to a subset $S \subseteq \{1,\ldots,n\}$, where the ith position in the sequence is 1, exactly if $i \in S$. This obviously defines a bijection $\mathcal{P}(\{1,\ldots,n\}) \leftrightarrow \{0,1\}^n$. For example the sequence $(1, 1, 0, 1, 0)$ is assigned to the subset $\{1,2,4\} \subset \{1,2,3,4,5\}$. The order relation may be demonstrated by a directed graph. Here the subsets are the vertices and there is an edge (S,T), if and only if $T = S \cup \{i\}$ for some $i \in \{1,\ldots,n\}$. Example $\mathcal{P}(\{1,2,3\})$:

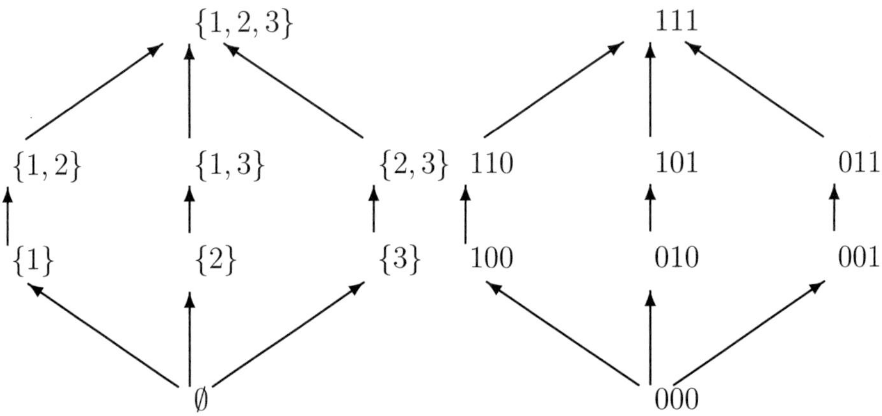

Theorem 3.1 (LYM-inequality) *For an antichain $\mathcal{A} \triangleq \{A_1, A_2, \ldots, A_t\} \in \mathcal{P}(\{1,\ldots,n\})$ the following inequality holds*

$$\sum_{i=1}^{t} \frac{1}{\binom{n}{|A_i|}} \le 1,$$

or equivalently (with $\mathcal{A}_k = \{A_i : |A_i| = k\}$)

$$\sum_{k=0}^{n} \frac{|\mathcal{A}_k|}{\binom{n}{k}} \le 1.$$

Proof The idea of the proof is to count all saturated chains passing through the antichain $\mathcal{A}$. First observe that a saturated chain in $\mathcal{P}(\{1,\ldots,n\})$ is of the form $\big\{\varnothing, \{x_1\}, \{x_1,x_2\}, \ldots, \{x_1,x_2,\ldots,x_{n-1}\}, \{1,\ldots,n\}\big\}$.

Since there are n possible choices for the first element $x_1 \in \{1,\ldots,n\}$, $n-1$ possible choices for x_2, etc., there exist $n!$ saturated chains, all of which have length $n+1$.

Now let A be a set in the antichain with cardinality i, say. Every saturated chain passing through A can be decomposed into a chain $\{\varnothing, \{x_1\}, \{x_1, x_2\}, \ldots, \{x_1, \ldots, x_i\} \triangleq A\}$ and a chain $\{A, A \cup \{x_{i+1}\}, A \cup \{x_{i+1}, x_{i+2}\}, \ldots, \{1, \ldots, n\}\}$.

With the same argumentation as above there are $|A|!(n - |A|)!$ saturated chains passing through A.

Since all the sets in the antichain are disjoint, no saturated chain passing through $\mathcal{A}$ is counted twice, and hence

$$\sum_{i=1}^{t} |A_i|!(n - |A_i|)! \leq n!$$

Division by $n!$ yields the desired result

$$\sum_{i=1}^{t} \frac{1}{\binom{n}{|A_i|}} \leq 1.$$

□

The LYM-inequality was originally used to prove the following theorem.

Theorem 3.2 (Sperner [11]) *The maximum cardinality of an antichain in* $\mathcal{P}(\{1, \ldots, n\})$ *is* $\binom{n}{\lfloor \frac{n}{2} \rfloor} = \binom{n}{\lceil \frac{n}{2} \rceil}$.

Proof In order to find a large antichain, the denominators $\binom{n}{|A_i|}$ have to be chosen as large as possible. This is obviously the case for $|A_i| = \lfloor \frac{n}{2} \rfloor$ or $\lceil \frac{n}{2} \rceil$. It is also possible to construct an antichain to these given cardinalities, since the $\lfloor \frac{n}{2} \rfloor$th level of $\mathcal{P}(\{1, \ldots, n\})$ obviously is an antichain. □

Since for even n $\lfloor \frac{n}{2} \rfloor = \lceil \frac{n}{2} \rceil = \frac{n}{2}$, it is clear that in this case an antichain of maximum cardinality, also denoted as *Sperner set*, consists of all the sets with $\frac{n}{2}$ elements, hence it is just the $\frac{n}{2}$th level in the poset. It can also be shown that for odd n a Sperner set is either the $\lfloor \frac{n}{2} \rfloor$th level or the $\lceil \frac{n}{2} \rceil$th level of the poset $\mathcal{P}(\{1, \ldots, n\})$. Hence it is not possible to find an antichain of cardinality $\binom{n}{\lfloor \frac{n}{2} \rfloor}$ consisting of sets of both levels.

Remark An analogue to the LYM-inequality and Kraft's inequality is not valid in every poset. If a similar inequality is valid, we say that the poset has the *LYM-property*.

3.1.2 Ahlswede–Zhang Identity

In the following sharpening of the LYM-inequality of [2] for every $X \in \mathcal{P} \triangleq \mathcal{P}(\{1, \ldots, n\})$ and for every family $\mathcal{A} \subset \mathcal{P}$ of subsets we define

$$X_{\mathcal{A}} \triangleq \bigcap_{X \supset A \in \mathcal{A}} A \text{ and } W_{\mathcal{A}}(X) = |X_{\mathcal{A}}|.$$

Theorem 3.3 (Ahlswede and Zhang) *For every family $\mathcal{A}$ of non-empty subsets of $\{1, \ldots, n\}$*

$$\sum_{X \in \mathcal{P}} \frac{W_{\mathcal{A}}(X)}{|X| \cdot \binom{n}{|X|}} = 1.$$

Proof Note first that only the minimal elements in $\mathcal{A}$ determine $X_{\mathcal{A}}$ and therefore matter. We can assume therefore that $\mathcal{A}$ is an antichain. Recall that in the proof of the LYM-inequality all saturated chains passing through members of $\mathcal{A}$ are counted. Now we also count the saturated chains not passing through $\mathcal{A}$.

The key idea is to assign to $\mathcal{A}$ the upset

$$\mathcal{U} \triangleq \{X \in \mathcal{P} : X \supset A \text{ for some } A \in \mathcal{A}\}$$

and to count saturated chains according to their exits in $\mathcal{U}$. For this we view $\mathcal{P}(\{1, \ldots, n\})$ as a directed graph with an edge between vertices T, S exactly if $T \supset S$ and $|T \smallsetminus S| = 1$. Observe that in our example for $\mathcal{P}(\{1, 2, 3\})$ we only have to change the direction of the edges.

Since $\varnothing \notin \mathcal{A}$, clearly $\varnothing \notin \mathcal{U}$. Therefore every saturated chain starting in $\{1, \ldots, n\} \in \mathcal{U}$ has a last set, say exit set, in $\mathcal{U}$. For every $U \in \mathcal{U}$ we call $e = (U, V)$ an exit edge, if $V \in \mathcal{P} \smallsetminus \mathcal{U}$ and we denote the set of exit edges by $\varepsilon_{\mathcal{A}}(U)$. So if e.g. in $\mathcal{P}(\{1, 2, 3\})$ we choose $\mathcal{A} = \{011, 100\}$, then $\mathcal{U} = \{111, 110, 101, 011, 100\}$ and $\varepsilon_{\mathcal{A}}(111) = \emptyset$, $\varepsilon_{\mathcal{A}}(110) = \{(110, 010)\}$, $\varepsilon_{\mathcal{A}}(101) = \{(101, 001)\}$, $\varepsilon_{\mathcal{A}}(011) = \{(011, 010), (011, 001)\}$, $\varepsilon_{\mathcal{A}}(100) = \{(100, 000)\}$.

The number of saturated chains leaving $\mathcal{U}$ in U is then

$$(n - |U|)! \cdot |\varepsilon_{\mathcal{A}}(U)| \cdot (|U| - 1)!$$

Therefore

$$\sum_{U \in \mathcal{U}} (n - |U|)! \cdot |\varepsilon_{\mathcal{A}}(U)| \cdot (|U| - 1)! = n!$$

and since $\varepsilon_{\mathcal{A}}(X) = \varnothing$ for $X \in \mathcal{P} \smallsetminus \mathcal{U}$, also

$$\sum_{X \in \mathcal{P}} \frac{|\varepsilon_{\mathcal{A}}(X)|}{|X|\binom{n}{|X|}} = 1.$$

Now just verify that $|\varepsilon_{\mathcal{A}}(X)| = W_{\mathcal{A}}(X)$, as the above example suggests. □

For generalizations and applications see [1].

3.1.3 Sperner's Lemma and Its Original Proof

Let be given a finite set $\mathcal{M}$ of n elements. Let $\mathcal{U} \subset \mathcal{V}$ be subsets both. We recall that the number of elements of a subset is called order. A system S of subsets is called antichain if no subset in S is contained in another subset of S. The number of subsets in S is called *degree of S*.

Theorem 3.4 *If S is an antichain of subsets of $\mathcal{M}$, the degree of S is smaller or equal to $\binom{n}{\lfloor \frac{n}{2} \rfloor}$ with $\mathcal{M}$ having order n.*

Equality holds for

(i) even n, only if S contains all subsets of order $\frac{n}{2}$,
(ii) odd n just in the following two cases:

 (a) S consists of all subset of order $\frac{n+1}{2}$,
 (b) S consists of all subset of order $\frac{n-1}{2}$.

In the last two cases it is obvious that S is an antichain of subsets really having degree $\binom{n}{\lfloor \frac{n}{2} \rfloor}$. It is left to show that for any other antichain the degree is less than $\binom{n}{\lfloor \frac{n}{2} \rfloor}$.

To proof this we need the following lemma:

Lemma 3.1 *m distinct subsets of order k of $\mathcal{M}$ have at least $m + 1$ distinct subsets of order $k - 1$, if $k > \frac{n+1}{2}$.*

Each of the m subsets of order k consists k subsets of order $k-1$, hence all m subsets consists of $k \cdot m$ subsets which are not distinct, in general case. Obviously, a subset of order $k - 1$ appears at most $(n - k + 1)$ times, cause a subset of order $k - 1$ have at most $(n - k + 1)$ uppersets of order k. If r is the number of distinct subsets of these $m \cdot k$ subsets, it holds:

$$r \geq \frac{m \cdot k}{n - k + 1}.$$

Because of $k > \frac{n+1}{2}$ it holds

$$k > n - k + 1,$$

and so

$$\frac{k}{n - k + 1} > 1.$$

By this,

$$r \geq m + 1,$$

and the lemma is proved. □

For even n the orders $\frac{n}{2} + 1, \frac{n}{2} + 2, \ldots, n$ satisfy the condition $k > \frac{n+1}{2}$. For odd n the orders $\frac{n+3}{2}, \frac{n+5}{2}, \ldots, n$ satisfy the same condition. We claim that the lemma is still true for odd n and $k = \frac{n+1}{2}$ if $m < \binom{n}{\lfloor \frac{n}{2} \rfloor}$.

In this case again

$$r \geq \frac{m \cdot k}{n - k + 1},$$

and cause of

$$k = n - k + 1$$

it holds

$$r \geq m.$$

Let be $\mathcal{V}_1, \mathcal{V}_2, \dots \mathcal{V}_m$ the m subsets of order $\frac{n+1}{2}$. Then there exist m subsets $\mathcal{U}_1, \mathcal{U}_2, \dots \mathcal{U}_m$ of order $\frac{n-1}{2}$ which are distinct. The minimal case $r = m$ is only possible if every $\mathcal{U}_i (i = 1, \dots, m)$ appears exactly k times as a subset of some $\mathcal{V}_i (i = 1, \dots, m)$. Now, there is a subset $\mathcal{U}'$ of order $\frac{n-1}{2}$ which differs only in one element from some $\mathcal{U}_i (i = 1, \dots, m)$ but distinct from all other $\mathcal{U}_i (i = 1, \dots, m)$.

First, because there are some subsets of order $\frac{n-1}{2}$ which are distinct from all $\mathcal{U}_i (i = 1, \dots, m)$, because $m < \binom{n}{\lfloor \frac{n}{2} \rfloor}$; moreover it is possible to get any subset of order $\frac{n-1}{2}$ by taking any $\mathcal{U}_i (i = 1, \dots, m)$ and changing only one element step by step.

Let be $\mathcal{U}_k$ $(0 < k \leq m)$ the subset which differs from $\mathcal{U}'$ only by one element. Hence, $\mathcal{U}_k$ and $\mathcal{U}'$ have the same upperset $\mathcal{V}'$ of order $\frac{n+1}{2}$. If now $\mathcal{V}' \subset \{\mathcal{V}_1, \dots, \mathcal{V}_m\}$, then

$$r > m,$$

because $\mathcal{U}'$ is subset of $\mathcal{V}'$.

But, if $\mathcal{V}' \not\subset \{\mathcal{V}_1, \dots, \mathcal{V}_m\}$, then $\mathcal{U}_k$ can appear as subset of $\mathcal{V}'$ at most $(k - 1)$ times as subset of $\mathcal{V}_i (i = 1, \dots, m)$. But in this case too, the minimal case is not reachable.

Analogously to these arguments it is possible to prove this lemma:

Lemma 3.2 *m distinct subsets of order k of $\mathcal{M}$ have at least $m+1$ distinct uppersets of order $k + 1$, if $k < \frac{n-1}{2}$. This fact still holds for odd n in the case $k = \frac{n-1}{2}$ if $m < \binom{n}{\lfloor \frac{n}{2} \rfloor}$.*

Now, we are able to prove Theorem 3.4.

Let be S an antichain of subsets of $\mathcal{M}$. We exclude the cases that

(i) for even n S exists just of subsets of order $\frac{n}{2}$,
(ii) for odd n
  (a) S exists just of subsets of order $\frac{n+1}{2}$,
  (b) S exists just of subsets of order $\frac{n-1}{2}$.

These are the cases for which the theorem is evident.

We will show that under these conditions there always exists a sequence of systems of subsets $S_0, S_1, \ldots, S_r$ with numbers of degree $g_0, g_1, \ldots, g_r$ and the following properties:

(i) $S = S_0$,

(ii) $r \geq 1$,

(iii) $g_0 < g_1 < \cdots < g_{r-1} \leq g_r$, with $g_0 < g_1$ in the case of $r = 1$,

(iv) every $S_i\,(i = 1, \ldots, r)$ is an antichain,

(v) S_r consists of subsets of same order, for even n of order $\frac{n}{2}$, for odd n of order $\frac{n-1}{2}$.

Because of (iv) and (v) it holds:

$$g_r \leq \binom{n}{\lfloor \frac{n}{2} \rfloor}$$

and hence, cause of (ii) and (iii)

$$g_0 < \binom{n}{\lfloor \frac{n}{2} \rfloor}.$$

By use of (i) the degree of S is g_0. The existence of a sequence of this kind is obvious. Let k be the greatest order of those subsets which are in $S = S_0$. Let m subsets of S_0 have this order. If now $k > \frac{n}{2}$ we replace the m subsets of order k by all their subsets of order $k - 1$. Their number is greater or equal to $m + 1$ by Lemma 3.1. This new system S_1 of subsets is an antichain, too. Moreover, $g_1 > g_0$. Doing the same with S_1 yields to S_2 and so on, while the greatest order is still greater than $\frac{n}{2}$. Let S_l, say, be the greatest system of this kind which contains just subsets of order less or equal than $\frac{n}{2}$. (It is still possible that $S_l = S_0$.) Let be h the smallest order which appears for the subsets in S_l and let t subsets have this order.

Substitution of these t subsets by all their uppersets of order $h + 1$, we get an antichain S_{l+1}. We continue this way, till we get a S_r which contains for even n just subsets of order $\frac{n}{2}$, and for odd n just subsets of order $\frac{n-1}{2}$. (For $l > 0$ it is possible that $r = l$. Only in this case it is possible that $g_{r-1} = g_r$, namely if for odd n S_{l-1} contains of all subsets of order $\frac{n+1}{2}$. But then in the light of the restrictions on S above, $r > 1$.) But so, cause of the restrictions on S it holds $r \geq 1$ and the properties (iii) and (iv) are evident from the construction rule of the sequence. Thus, all is proved.□

Remark A simpler proof of uniqueness follows from the AZ-identity.

3.2 On Independence Numbers in Graphs

Theorem 3.5 (Turán (1941) [12]) *The independence number of any simple undirected graph* $\mathcal{G}(V, E)$ *satisfies the inequality*

$$\alpha[\,\mathcal{G}(V, E)\,] \geq \frac{|V|^2}{|V| + 2|E|}, \tag{3.2.1}$$

where $|V|$ *and* $|E|$ *denote the cardinalities of* V *and* E. *Furthermore, there exist graphs for which this bound is tight: equality in (3.2.1) holds if and only if all connected components of* $\mathcal{G}(V, E)$ *are cliques having the same cardinality.*

We will use an auxiliary result formulated below; its proof is given after we complete the proof of the theorem.

Lemma 3.3 *Let* $\mathcal{G}_{n,k}$ *be the simple graph that consists of* k *disjoint cliques, of which* r *have* q *vertices and* $k - r$ *have* $q - 1$ *vertices, where*

$$q = \lfloor (n-1)/k \rfloor + 1, \quad r = n - k(q-1).$$

Then every graph $\mathcal{G}(V, E)$ *such that* $|V| = n$ *and* $\alpha[\,\mathcal{G}(V, E)\,] \leq k$ *that has the minimum possible number of edges is isomorphic to* $\mathcal{G}_{n,k}$.

Proof of Theorem 3.5. Let $|V| = n$, where n is expressed as $k(q-1)+r$ (see Lemma 3.3), and $\alpha[\,\mathcal{G}(V, E)\,] = k$. Then using Lemma 3.3 we conclude that $|E| \geq m_{n,k}$, where

$$\begin{aligned} m_{n,k} &= r \cdot \frac{q(q-1)}{2} + (k-r) \cdot \frac{(q-1)(q-2)}{2} \\ &= \frac{(n-r)(n-k+r)}{2k} \\ &\leq \min_{1 \leq r \leq k} \frac{(n-r)(n-k+r)}{2k} \\ &= \frac{(n-k)n}{2k}. \end{aligned}$$

Therefore,

$$k \geq \frac{n^2}{n + 2m_{n,k}} \geq \frac{|V|^2}{|V| + 2|E|}.$$

If $\mathcal{G}(V, E)$ consists of p cliques with cardinality n_0, then $\alpha[\,\mathcal{G}(V, E)\,] = p$ and

$$\frac{|V|^2}{|V| + 2|E|} = \frac{p^2 n_0^2}{pn_0 + pn_0(n_0 - 1)} = p = \alpha[\,\mathcal{G}(V, E)\,].$$

□

Proof of Lemma 3.3. The statement obviously holds when $n = k + 1, ..., 2k$. Let us fix an integer q, suppose that it holds for $n = qk + 1, ..., (q + 1)k$, and prove that is also the case when $n = (q + 1)k + r$ for all $r = 1, ..., k$.
Let $\mathcal{G}(V, E)$ be a graph with $|V| = n$ and $\alpha[\mathcal{G}(V, E)] \leq k$ that has a minimum number of edges. Hence, $\alpha[\mathcal{G}(V, E)] = k$. Let $S = \{s_1, ..., s_k\}$ be an independent subset. Then each vertex included into $V \backslash S$ is adjacent to S (otherwise, $\alpha[\mathcal{G}(V, E)] > k$). Subgraph $\mathcal{G}(V \backslash S, E')$, where $E' \subseteq E$ is the set of edges belonging to $V \backslash S$, has $n-k$ vertices and its independence number at most k; hence, by the induction hypothesis,

$$|E'| \geq m_{n-k,k}.$$

Since $\mathcal{G}_{n,k}$ can be formed from $\mathcal{G}_{n-k,k}$ by adding a vertex to each of the disjoint cliques in $\mathcal{G}_{n,k}$,

$$m_{n,k} - m_{n-k,k} = n - k.$$

Furthermore, since $|E| \leq m_{n,k}$, it follows that

$$n - k = |V \backslash S| \leq |E| - |E'| \leq m_{n,k} - m_{n-k,k}.$$

Hence,

$$|E| = m_{n,k}, \quad |E'| = m_{n-k,k},$$

i.e., we obtain that $\mathcal{G}(V, E)$ consists of k disjoint cliques.

□

3.3 A Combinatorial Partition Problem: Baranyai's Theorem

Obviously, a set of edges $\mathcal{E} \in \mathcal{E}$ which is simultaneously a covering of $\mathcal{H}$ and a packing of $\mathcal{H}$, is exactly a partition of $\mathcal{H}$.

Theorem 3.6 (Baranyai [3]) *Let* $\mathcal{H} = (\mathcal{V}, \mathcal{E}) = ([n], \binom{[n]}{k})$. *If* k *divides* n, *the set of all* $\binom{n}{k}$ k*-subsets of an* n*-set may be partitioned into disjoint parallel classes* $\mathcal{A}_i$, $i = 1, 2, \ldots, \binom{n-1}{k-1}$.

Proof The usual use of the term "partition" forbids the empty set, in general. But in this sense it is allowed here to occur, perhaps with a multiplicity, so that the total number of subsets is m. So, we use the term "m-partition" of a set $\mathcal{X}$ for a multiset $\mathcal{A}$ of m pairwise disjoint subsets, some of them may be empty, of $\mathcal{X}$ whose union is $\mathcal{X}$.

In order to get an inductive proof, we prove a statement seemingly stronger than the original statement. Let n and k with k divides n be given, and let $m := n/k,\ M := \binom{n-1}{k-1}$. We assert that for any integer l, $0 \le j \le n$, there exists a set

$$\mathcal{A}_1, \mathcal{A}_2, \ldots, \mathcal{A}_M$$

of m-partitions of $\{1, 2, \ldots, l\}$ with the property that each subset $S \subset \{1, 2, \ldots, l\}$ occurs in exactly

$$\binom{n-l}{k-|S|} \tag{3.3.1}$$

of the m-partitions $\mathcal{A}_i$. The binomial coefficient is interpreted as zero if $|S| > k$, of course, and for $S = \emptyset$, the m-partitions containing $\emptyset$ are to be counted with multiplicity equal to the number of times the empty set appears. Now, we prove our assertion by induction on l. We remark that it is true for $l = 0$ where each $\mathcal{A}_i$ will consist of m copies of the empty set. Also notice that the case $l = n$ will prove the theorem, since the binomial coefficient in (3.3.1) is then

$$\binom{0}{k-|S|} = \begin{cases} 1 & \text{if } |S| = k, \\ 0 & \text{otherwise.} \end{cases}$$

Remark This statement is not really more general, but would follow easily from the theorem. If M parallel classes exist as in the statement of the theorem, then for any set L of l points of $\mathcal{X}$, the intersections of the members of the parallel classes with L will provide m-partitions of L with the property above.

For some value of $l < m$ we assume that m-partitions $\mathcal{A}_1, \mathcal{A}_2, \ldots, \mathcal{A}_M$ exist with the required property. We form a transportation network as follows. There is to be a source vertex σ, another named $\mathcal{A}_i$ for each $i = 1, 2 \ldots, M$, another named S for every subset $S \subset \{1, 2, \ldots, l\}$, and a sink vertex τ. Moreover, there is to be a directed edge from σ to each $\mathcal{A}_i$ with capacity 1. There are to be directed edges from $\mathcal{A}_i$ to the vertices corresponding to members of $\mathcal{A}_i$. For this, use j edges to $\emptyset$, if $\emptyset$ occurs j times in $\mathcal{A}_i$. These may have any integral capacity greater or equal to 1. There is to be a directed edge from the vertex corresponding to a subset S to τ of capacity

$$\binom{n-l-1}{k-|S|-1}.$$

Now, we demonstrate a flow in this network constructed above: Assign a flow value of 1 to the edges leaving σ, a flow value of $(k - |S|)/(n - l)$ to the edges from $\mathcal{A}_i$ to each of its members S, and a flow value of $\binom{n-l-1}{k-|S|-1}$ to the edge from S to τ. This must be a flow as easily checked, because the sum of the value on edges leaving a vertex $\mathcal{A}_i$ is

$$\sum_{S \in \mathcal{A}_i} \frac{k - |S|}{n - l} = \frac{1}{n - l}(m\,k - \sum_{S \in \mathcal{A}_i} |S|) - \frac{1}{n - l}(m\,k - l) = 1.$$

The sum of the values on the edges into a vertex S is

$$\sum_{i:S\in\mathcal{A}_i} \frac{k-|S|}{n-l} = \frac{k-|S|}{n-l}\binom{n-l}{k-|S|} = \binom{n-l-1}{k-|S|-1}.$$

This is a maximum flow and has strength M, because all edges leaving σ are saturated. The edges into τ are also saturated in this, and therefore in any flow.

Using the theorem which says, that if all the capacities in a transportation network are integers, then there is a maximum strength flow f for which all values $f(e)$ are integers, this network admits an integral-valued maximum flow f, too. All edges leaving σ will be saturated, so it is clear for each i, f assigns the value 1 to one of the edges leaving $\mathcal{A}_i$ and 0 to all others. Say f assigns 1 to the edge from $\mathcal{A}_i$ to its members S_i. For each subset S, the number of values of i such that $S_i = S$ is $\binom{n-l-1}{k-|S|-1}$.

For completing the induction step, we finally obtain a set of m-partitions $\mathcal{A}'_1, \mathcal{A}'_2, \dots, \mathcal{A}'_M$ of the set$\{1, 2, \dots, l+1\}$ by letting $\mathcal{A}'_i$ be obtained from $\mathcal{A}_i$ by replacing the distinguished member S_i by $S_i \cup \{l+1\}$, $i = 1, \dots, M$. At last, we have to check that each subset T of $\{1, 2, \dots, l+1\}$ occurs exactly

$$\binom{n-(l+1)}{k-|T|}$$

times among $\mathcal{A}'_1, \mathcal{A}'_2, \dots, \mathcal{A}'_M$. But this is done easily.

Now, we are going to prove the part that "k does not divide n".

Let $\mathcal{H} = (\mathcal{X}, \mathcal{E})$ be a hypergraph with vertex set $\mathcal{X}$ and edge set $\mathcal{E}$. A (vertex) *p-coloring* of $\mathcal{H}$ is a partition $\mathcal{C} = \{C_i : i \leq p\}$ of $\mathcal{X}$ into p (possibly empty) subsets (colors). We consider three successively stronger requirements on the coloring.

(i) $\mathcal{C}$ is called *good* if each edge $E \in \mathcal{E}$ has as many colors as it can possibly have, i.e., $|\{i : E \cap C_i \neq \emptyset\}| = \min(|E|, p)$.

(ii) $\mathcal{C}$ is called *fair* if on each edge E the colors are represented as fairly as possible, i.e.,

$$\lfloor\frac{|E|}{p}\rfloor \leq |E \cap C_i| \leq \lceil\frac{|E|}{p}\rceil \text{ for } i = 1, \dots, p.$$

(iii) $\mathcal{C}$ is called *strong* if on each edge E all colors are different, i.e., $|E \cap C_i| \leq 1$ for $i = 1, \dots, p$.

(This is just the special case of a good coloring with p colors when $p \geq \max\{|E| : E \in \mathcal{E})$.)

Theorem 3.7 *Let $\mathcal{H} = K_n^k$ (the complete k-uniform hypergraph) and write $N = \binom{n}{k}$, the number of edges of $\mathcal{H}$. Then*

(i) $\mathcal{H}$ has a good edge p-coloring iff it is not the case that

$$N/\lceil \frac{n}{k} \rceil < p < N/\lfloor \frac{n}{k} \rfloor,$$

i.e. iff

$$\frac{N}{p} \leq \lfloor \frac{n}{k} \rfloor \text{ or } \frac{N}{p} \geq \lceil \frac{n}{k} \rceil.$$

(ii) The strong edge-coloring number of hypergraph $\mathcal{H}$ equals $\lceil N/\lfloor \frac{n}{k} \rfloor \rceil$.

Proof Proof of the necessity:

This part of the proof will be valid for any regular k-uniform hypergraph on n points with N edges. Let $\mathcal{C}$ be any edge p-coloring of $\mathcal{H}$ and define for $x \in \mathcal{X}$

$$c(x) := |\{i : \mathcal{E}_x \cap C_i \neq \emptyset\}|,$$

the number of colors found at point x.

(i) $p < N/\lfloor \frac{n}{k} \rfloor$, i.e., $\lfloor \frac{n}{k} \rfloor < \frac{N}{p}$ means that there exist two non-disjoint edges with the same color i.e., $c(x) < \delta(x) = \Delta$ for some x.
$p > N/\lceil \frac{n}{k} \rceil$, i.e., $\lceil \frac{n}{k} \rceil > \frac{N}{p}$ means that not every color occurs at each point i.e., $c(x) < p$ for some x.

(ii) That the strong edge-coloring number of hypergraph $\mathcal{H}$ is greater or equal than $\lceil N/\lfloor \frac{n}{k} \rfloor \rceil$ immediately follows from (i). □

(i) and (ii) can be formulated more generally as follows. For a regular hypergraph $\mathcal{H} = (\mathcal{X}, \mathcal{E})$ let $\nu(\mathcal{H})$ be the maximum cardinality of a set of pairwise disjoint edges in $\mathcal{H}$, and let $\rho(H)$ be the minimum cardinality of a set of edges covering all vertices.

(i) can be stated as: if

$$\nu(\mathcal{H}) < \frac{|\mathcal{E}|}{p} < \rho(\mathcal{H}),$$

then $\mathcal{H}$ does not have a good edge p-coloring.

(ii) can be stated as:
The strong edge-coloring number of $\mathcal{H}$ is greater or equal to $\lceil \frac{|\mathcal{E}|}{\nu(\mathcal{H})} \rceil$.

Concerning the sufficiency half of Theorem 3.7 we shall in fact prove slightly more, since we need it later. Let s be a positive integer, and $\mathcal{H} = (\mathcal{X}, \mathcal{E})$ be a hypergraph. Then define $s\mathcal{H} = (\mathcal{X}, s\,\mathcal{E})$ to be the hypergraph with the same vertices as $\mathcal{H}$, but with each edge from $\mathcal{H}$ taken with multiplicity s. Obviously $\nu(s\,\mathcal{H}) = \nu(\mathcal{H})$ and $\rho(s\,\mathcal{H}) = \rho(\mathcal{H})$. A coloring of $s\,\mathcal{H}$ with p colors is sometimes called *fractional coloring* of $\mathcal{H}$ with $q = p/s$ colors. We show here that $s\,K_n^k$ has a good edge p-coloring iff p satisfies the condition (i), where now $N = s\binom{n}{k}$.

A hypergraph $(\mathcal{X}, \mathcal{E})$ is called *almost regular* if for all $x, y \in \mathcal{X}$ we have $|\delta(x) - \delta(y)| \leq 1$. Now we have

Theorem 3.8 *Let $a_1, \dots, a_t$ be natural numbers such that $\sum_{i=1}^{t} a_i = N := \binom{n}{k} s$. Then the edges of $s\,K_n^k$ can be partitioned in almost regular hypergraphs $(\mathcal{X}, \mathcal{E}_j)$ such that $|\mathcal{E}_j| = a_j$ with $1 \leq j \leq t$.*

It is easy verified that Theorem 3.7 follows from Theorem 3.8:

(i) If $p \leq N/\lceil \frac{n}{k} \rceil$ then use Theorem 3.8 with $s = 1, t = p$ and $a_1, \cdots, a_{t-1} = \lceil \frac{n}{k} \rceil$, $a_t = N - (t-1)\lceil \frac{n}{k} \rceil$.

If $p \geq N/\lfloor \frac{n}{k} \rfloor$ then use Theorem 3.8 with $t = \lceil N/\lfloor \frac{n}{k} \rfloor \rceil$ and $a_1, \cdots, a_{t-1} = \lfloor \frac{n}{k} \rfloor$, $a_t = N - (t-1)\lfloor \frac{n}{k} \rfloor$.

This also proves (ii).

(ii) Write $f_0 = \lceil \lfloor \frac{\Delta}{k} \rfloor \frac{n}{k} \rceil$ and $f_1 = \lfloor \lceil \frac{\Delta}{k} \rceil \frac{n}{k} \rfloor$. If $p\,f_0 \leq N \leq p\,f_1$ then use Theorem 3.8 with $s = 1, t = p$ and $a_1, \cdots, a_g = \lfloor \frac{N}{p} \rfloor + 1$ and $a_{g+1} = \cdots = a_t = \lfloor \frac{N}{p} \rfloor$ where $g = N - p\lfloor \frac{N}{p} \rfloor$. For all i $f_0 \leq a_i \leq f_1$ guarantees that we get fair coloring.

Theorem 3.8 will be proved in the following as a consequence of much more general theorems.

Baranyai proved a large number of very general theorems all to the effect that if certain matrices exist then hypergraphs exist of which the valency pattern and cardinalities are described by those matrices. An example is the following theorem.

Theorem 3.9 *Let $|\mathcal{X}| = n, \mathcal{H} = (\mathcal{X}, \mathcal{E})$ where $\mathcal{E} = \sum_{i=1}^{s} \binom{\mathcal{X}}{k_i}$ (the k_i not necessarily different). Let $A = (a_{ij})$ be an $s \times t-$ matrix with nonnegative integral entries such that for its row sums $\sum_{j=1}^{t} a_{ij} = \binom{n}{k_i}$ holds. (For $k < 0$ or $k > n$ we read $\binom{n}{k} = 0$.)*

Then there exist hypergraphs $\mathcal{H}_{ij} = (\mathcal{X}, \mathcal{E}_{ij})$ such that

(i) $|\mathcal{E}_{ij}| = a'_{ij}$,

(ii) $\binom{\mathcal{X}}{k_i} = \sum_{j=1}^{t} \mathcal{E}_{ij}$ with $1 \leq j \leq s$,

(iii) $(\mathcal{X}, \sum_{i=1}^{s} \mathcal{E}_{ij})$ is almost regular with $1 \leq j \leq t$.

Note that for $k_1 = \cdots = k_s = k$ this implies Theorem 3.8. If l is an integer, let $l \approx d$ and $d \approx l$ denote that either $l = \lfloor d \rfloor$ or $l = \lceil d \rceil$ holds. We first give some lemmas.

Lemma 3.4 *For integral A we have*

$$\left\lfloor \frac{A}{n} \right\rfloor = \left\lfloor \frac{A - \lceil A/n \rceil}{n-1} \right\rfloor \text{ and } \left\lceil \frac{A}{n} \right\rceil = \left\lceil \frac{A - \lfloor A/n \rfloor}{n-1} \right\rceil.$$

Lemma 3.4 is an easy exercise in calculus.

Lemma 3.5 *Let $\mathcal{H} = (\mathcal{X}, \mathcal{E})$ and $a \in \mathcal{X}$. Then $\mathcal{H}$ is almost regular iff $\mathcal{H}_{\mathcal{X}\setminus\{a\}}$ is almost regular and $\delta_{\mathcal{H}}(a) \approx \frac{1}{n}\sum_{E\in\mathcal{E}} |E|$.*

This can be proved by using Lemma 3.4.

Lemma 3.6 *Let (ϵ_{ij}) be a matrix with real entries. Then there exists a matrix (e_{ij}) with integral entries such that*

(i) $e_{ij} \approx \epsilon_{ij}$ *for all* i, j,

(ii) $\sum_i e_{ij} \approx \sum_i \epsilon_{ij}$ *for all* j,

(iii) $\sum_j e_{ij} \approx \sum_j \epsilon_{ij}$ *for all* i,

(iv) $\sum_{i,j} e_{ij} \approx \sum_{i,j} \epsilon_{ij}$.

Proof This follows straightforwardly from Ford and Fulkerson's Integer Flow Theorem. □

Proof of Theorem 3.9. By induction on $n = |\mathcal{X}|$. If $n = 0$ the theorem is true. The induction step consists of one application of Lemma 3.6. We may suppose that for $i \le s$ we have $0 \le k_i \le n$. Let $\epsilon_{ij} = \frac{k_i}{n} a_{ij}$, the average degree of the hypergraph $(\mathcal{X}, \mathcal{E}_{ij})$ we want to construct.

By Lemma 3.6 there exist nonnegative integers e_{ij} with $\sum_j e_{ij} = \binom{n-1}{k_i-1}$, $\sum_j (a_{ij} - e_{ij}) = \binom{n-1}{k_i}$ and $\sum_i e_{ij} \approx \frac{1}{n}\sum_i k_i a_{ij}$.

Let $a \in \mathcal{X}$ and apply the induction hypothesis to $\mathcal{X}' = \mathcal{X} \setminus \{a\}$ with $s' = 2s$, $t' = t$, $k'_i = k_i$, $k'_{i+s} = k_i - 1 (1 \le i \le s)$, $a'_{ij} = a_{ij} - e_{ij}$, $a'_{(i+s)\,j} = e_{ij}$.

(That this is the proper thing to do is seen by reasoning backward:

When we have $\mathcal{E}_{ij}$ and then remove the point a, $\mathcal{E}_{ij}$ is split up into the class of edges that remain of size k_i and the class of edges that have now size $k_i - 1$. The latter class has cardinality ϵ_{ij} on the average.)

By the induction hypothesis we find hypergraphs $\mathcal{F}_{ij}$ and $\mathcal{G}_{ij}$ such that

$$|\mathcal{F}_{ij}| = a_{ij} - e_{ij}, \quad |\mathcal{G}_{ij}| = e_{ij},$$

$$\sum_j \mathcal{F}_{ij} = \binom{\mathcal{X}}{k_i}, \quad \sum_j \mathcal{G}_{ij} = \binom{\mathcal{X}}{k_i - 1},$$

$$\sum_i \mathcal{F}_{ij} + \mathcal{G}_{ij} \text{ is almost regular.}$$

Defining $\mathcal{E}_{ij} = \mathcal{F}_{ij} \cup \{G \cup \{a\} : G \in \mathcal{G}_{ij}\}$ we are done (using Lemma 3.5). □

(The given theorems and proofs are due to A. Schrijver and A.E. Brouwer (see [4]) and from J.H. van Lindt and R.M. Wilson (see [13]).)

3.4 More on Packing: Bounds on Codes

3.4.1 Plotkin's Bound

In this section we present the bound due to Plotkin [10].

Theorem 3.10 *For an* (n, M, d) *code* $\mathcal{U}$ *and* $n < 2d$ *holds:*

$$M \le 2\lceil \frac{d}{2d-n} \rceil. \tag{3.4.1}$$

Proof We compute the summation $S = \sum_{u \in \mathcal{U}} \sum_{v \in \mathcal{U}} d_H(u, v)$ in two ways. Cause $d_H(u, v) \ge d$ for $u \ne v$ we get

$$S \ge M \cdot (M-1) \cdot d, \tag{3.4.2}$$

on the other hand look at $m \times n$ matrix $U = \begin{pmatrix} u_1 \\ u_2 \\ \vdots \\ u_M \end{pmatrix}$.

Let f_t be the number of zeros in the tth column. It holds

$$S = \sum_{t=1}^{n} \sum_{i,j=1}^{M} d_H(u_{it}, u_{jt}) = \sum_{t=1}^{n} 2 f_t (M - f_t). \tag{3.4.3}$$

For even $M \cdot \sum_{t=1}^{n} 2 f_t (M - f_t)$ is maximal for $f_t = \frac{1}{2} M$ for all t and $S \le \frac{1}{2} n M^2$.

With (3.4.2) follows $M(M-1)d \le \frac{nM^2}{2}$ or $M(d - \frac{n}{2}) \le d$ or $M \le \frac{2d}{2d-n} \le 2\lceil \frac{d}{2d-n} \rceil$.

For odd M yields

$$S \leq n \cdot 2 \cdot \frac{M-1}{2}\frac{M+1}{2} = n \cdot 2 \cdot (\frac{M^2}{4} - \frac{1}{4}) = n \cdot (\frac{M^2-1}{2})$$

or $M(M-1)d \leq n \cdot (\frac{M^2-1}{2})$ or $Md \leq n \cdot \frac{M+1}{2}$ or $M(2d-n) \leq n$ or $m \leq \frac{n}{2d-n} = \frac{2d}{2d-n} - 1 \leq \lceil \frac{2d}{2d-n} \rceil - 1 \leq 2\lceil \frac{d}{2d-n} \rceil$. □

3.4.2 Johnson's Bounds

In this section we present the bound due to Johnson [6]

Theorem 3.11

$$A(n, 2\delta, w) \leq \lceil \frac{\delta n}{w^2 - wn + \delta n} \rceil \text{ } if\, w^2 - w^n + \delta n > 0.$$

Proof Let $\mathcal{U}$ be an $(n, M, 2\delta)$-code with constant weight w and let $|\mathcal{U}| = A(n, 2\delta, w)$. Then $T = \sum_{i \neq j} \sum (u_i, u_j)$. $T \leq (w - \delta)M(M-1)$. Moreover, $T = \sum_{t=1}^{n} \sum_{i \neq j} \sum (u_{it}, u_{jt})$.

Let g_t be the number of ones in the tth column of matrix $U = \begin{pmatrix} u_1 \\ u_2 \\ \vdots \\ u_M \end{pmatrix}$. Then

$$T = \sum_{t=1}^{n} g_t(g_t - 1) \leq (w - \delta)M(M-1). \tag{3.4.4}$$

But $\sum_{t=1}^{n} g_t = wM$. The summation $\sum_{t=1}^{n} g_t^2$ is minimal for $g_t = wM/n$ for all t. This is $\frac{w^2M^2}{n}$. Using (3.4.3) it follows that

$$\frac{w^2M^2}{n} - wM \leq (w - \delta)M(m-1)$$

from which the assertion follows. □

Cause g_t is an integer, we can formulate the next

Theorem 3.12 *For $M = A(n, 2\delta, w)$ and given parameters k, s with $wM = nk + \delta$ for $0 \leq s < n$ it holds*

$$n \cdot k \cdot (k-1) + 2 \cdot k \cdot s \leq (w - \delta)M(M-1).$$

Proof The minimum of $\sum_{t=1}^{n} g_t^2$ with restrict to $\sum_{t=1}^{n} g_t = wM$ is reached for $g_1 = \cdots = g_s = k+1, g_{s+1} = \cdots = g_n = k$. Its value having these parameters is

$$s(k+1)^2 + (n-s)k^2.$$

Using (3.4.3) one has

$$s(k+1)^2 + (n-s)k^2 - (nk+s) \leq (w-\delta)M(M-1).$$

This completes the proof. □

Theorem 3.13

$$A(n, 2\delta, w) \leq \lceil \frac{n}{w} A(n-1, 2\delta, w-1) \rceil.$$

Proof Throwing out all co-ordinates having an one in the tth position we get a code of length $n-1$ with distance $\geq 2\delta$ and weight $w-1$. The number of these code words is less or equal $A(n-1, 2\delta, w-1)$. The number of ones in the original code is $wA(n, 2\delta, w) \leq n \cdot A(n-1, 2\delta, w-1)$. □

Corollary 3.1

$$A(n, 2\delta, w) \leq \lceil \frac{n}{w} \lceil \frac{n-1}{w-1} \cdots \lceil \frac{n-w+\delta}{\delta} \rceil \cdots \rceil.$$

Proof Iteration of Theorem 3.13 and using the equality $A(n, 2\delta, \delta) = \lceil \frac{n}{\delta} \rceil$ yields the result. □

3.4.3 *Basic Methods of Proving Gilbert-Type Bounds on the Cardinality of a Code*

Let us consider the following problem: we are given a code length n and a value of d. What is the upper bound on the cardinality of a binary code having the minimal distance not less than d?

Maximal coding (Gilbert bound [5]).

Since d is the minimal distance of a code, we have an evident inequality

$$M < \frac{2^n}{S_d} \sim 2^{n(1-h(\delta))}, \tag{3.4.5}$$

where $h(x) = -x \log x - (1-x)\log(1-x)$ is the binary entropy function and $\delta = d/n$.

Selection of a random code

Suppose, we want to construct a code with M codewords. select the codewords at random. There are 2^{nM} codes. Let us fix the m-th codeword. Then the number of choices of all other codewords such that at least one of them is located at the Hamming distance d or less form the m-th codeword is not greater than

$$(M-1)2^{n(M-1)}S_{d-1}.$$

Therefore, the number of 'bad' codes(the codes with the minimal distance less than d) is not greater than

$$M(M-1)2^{n(M-1)}S_{d-1}.$$

If this expression is less than the total number of codes, i.e.,

$$M(M-1)2^{n(M-1)}S_{d-1} < 2^{nM},$$

then there exists at least one code with the desired property. Direct calculations show that it is possible if

$$M^2 < \frac{2^n}{S_{d-1}}.$$

Hence, the exponent of our upper bound is twice less than the exponent we get in (3.4.5). The method that can be used to improve the result is known as *expurgation.* Note that the probability to select a bad i-th codeword is upper-bounded by

$$\frac{(M-1)S_{d-1}}{2^n}.$$

Thus, the average number of the bad words is upper-bounded by

$$M\frac{(M-1)S_{d-1}}{2^n}.$$

Let us expurgate a half of these words. Then, constructing a new code that contains only remaining codewords, we get the inequality

$$M < \frac{1}{2}\frac{2^n}{S_{d-1}},$$

which is only twice less than the Gilbert bound (the exponent of the bound is the same as the exponent of Gilbert bound in the ratewise sense).

Selection of clouds of random codes

Suppose that we want to construct (M) clouds such that each cloud consists of k codewords. The minimal distance between every codeword of some cloud and any codewords belonging to different clouds should be equal to not less than d.

A generalization of the previous considerations lead to the following inequality

$$M(k(M-1)S_{d-1})^k 2^{nk(M-1)} < 2^{nkM}.$$

If we set $k = n$ then this inequality can be rewritten as follows:

$$M^{1/k}(k(M-1)S_{d-1}) < 2^n.$$

As a result, we obtain

$$M \sim \frac{1}{2n}\frac{2^n}{S_{d-1}},$$

i.e., the *constructions based on the clouds of codewords instead of one codeword assigned to each message lead to approximately the same result as expurgation.*

References

1. R. Ahlswede, V. Blinovsky, *Lectures on Advances in Combinatorics* (Springer, Berlin, 2008)
2. R. Ahlswede, Z. Zhang, An identity in combinatorial extremal theory. Adv. Math. **80**(2), 137–151 (1990)
3. Z. Baranyai, On the Factorization of the Complete Uniform Hypergraph, in *Infinite and Finite Sets*, ed. by A. Hajnal, R. Rado, V.T. Sos (Amsterdam, 1975), pp. 91–108
4. A. Brouwer, A. Schrijver, Uniform hypergraphs, packing and covering in combinatorics. Math. Centre Tracts **106**, 39–73 (1979)
5. E.N. Gilbert, A comparison of signalling alphabets. Bell Syst. Tech. J. **31**, 504–522 (1952)
6. S.M. Johnson, A new upper bound for error-correcting codes. IRE Trans. Inf. Theory **8**, 203–207 (1962)
7. L.G. Kraft, A device for quantizing, grouping, and coding amplitude modulated pulses, MS Thesis, Cambridge, 1949
8. D. Lubell, A short proof of Sperner's lemma. J. Comb. Theory **1**, 299 (1966)
9. L.D. Meshalkin, Generalization of Sperner's theorem on the number of subsets of a finite set. Theory Probab. Appl. **8**, 203–204 (1963)
10. M. Plotkin, Binary codes with specified minimum distance. IRE Trans. Inf. Theory **6**, 445–450 (1960)
11. E. Sperner, Ein Satz über Untermengen einer endlichen Menge. Mathematische Zeitschrift (in German) **27**(1), 544–548 (1928)
12. P. Turan, Eine Extremalaufgabe aus der Graphentheorie. Mat. Fiz. Lapok. **48**, 436–452 (1941)
13. J.H. van Lint, R.M. Wilson, *A Course in Combinatorics*, 2nd edn. (Cambridge University Press, Cambridge, 2001)
14. K. Yamamoto, Logarithmic order of free distributive lattice. J. Math. Soc. Jpn. **6**, 343–353 (1954)

Part II
Combinatorial Models in Information Theory

Chapter 4
Coding for the Multiple-Access Channel: The Combinatorial Model

4.1 Coding for Multiple-Access Channels

4.1.1 Basic Definitions

The model of multiple-access channels (MACs) is one of the simplest generalizations of the channels with one sender and one receiver: we assume that there are several senders connected with the same receiver.

We will consider discrete memoryless MACs defined by the crossover probabilities

$$W = \{W(z|x, y), (x, y) \in \mathcal{X} \times \mathcal{Y}, z \in \mathcal{Z}\},$$

where $\mathcal{X}$, $\mathcal{Y}$, and $\mathcal{Z}$ are finite sets. The probability to receive $z^n \in \mathcal{Z}^n$ when $(x^n, y^n) \in \mathcal{X}^n \times \mathcal{Y}^n$ was sent is defined as

$$W(z^n|x^n, y^n) = \prod_{t=1}^{n} W(z_t|x_t, y_t).$$

The MAC is said to be *deterministic* if all crossover probabilities are equal to either zero or one; in this case, the output of the channel can be presented as a given function of inputs.

The crossover probabilities can be defined by the $(|\mathcal{X}| \cdot |\mathcal{Y}|) \times |\mathcal{Z}|$ matrix $\mathbf{W}$ whose rows correspond to all possible inputs and columns correspond to all possible outputs. We will consider the special case that $\mathcal{X} = \mathcal{Y} = \{0, 1\}$ and suppose that the first row corresponds to the pair $(0, 0)$, the second row corresponds to the pair $(0, 1)$, the third row corresponds to the pair $(1, 0)$, and the fourth row corresponds to the pair $(1, 1)$. When $\mathcal{Z} = \{0, ..., K-1\}$ for some $K > 1$, we suppose that the first column corresponds to the output 0, etc., the last column corresponds to the output $K - 1$.

A. Ahlswede et al. (eds.), *Combinatorial Methods and Models*,
Foundations in Signal Processing, Communications and Networking 13,
DOI 10.1007/978-3-319-53139-7_4

Examples

1. The adder channel is a deterministic MAC defined by the sets $\mathcal{X} = \mathcal{Y} = \{0, 1\}$, $\mathcal{Z} = \{0, 1, 2\}$, and the function
$$z = x + y,$$
where the addition is performed in the ring of integers. This channel can be also represented by the matrix
$$\mathbf{W} = \begin{pmatrix} 1\,0\,0 \\ 0\,1\,0 \\ 0\,1\,0 \\ 0\,0\,1 \end{pmatrix}.$$
2. The binary symmetric adder channel with crossover probability p is a non-deterministic MAC defined by the sets $\mathcal{X} = \mathcal{Y} = \{0, 1\}$, $\mathcal{Z} = \{0, 1, 2\}$, and the matrix
$$\mathbf{W} = \begin{pmatrix} q^2 & 2pq & p^2 \\ pq & p^2+q^2 & pq \\ pq & p^2+q^2 & pq \\ p^2 & 2pq & q^2 \end{pmatrix},$$
where $q = 1-p$. This channel is obtained if both input symbols are independently corrupted in binary symmetric channels with crossover probabilities p and the results are added.
3. The binary OR channel is a deterministic MAC defined by the sets $\mathcal{X} = \mathcal{Y} = \mathcal{Z} = \{0, 1\}$ and the function
$$z = x \vee y.$$
This channel can be also represented by the matrix
$$\mathbf{W} = \begin{pmatrix} 1\,0 \\ 0\,1 \\ 0\,1 \\ 0\,1 \end{pmatrix}.$$
4. The binary switching channel is a deterministic MAC defined by the sets $\mathcal{X} = \mathcal{Y} = \{0, 1\}$, $\mathcal{Z} = \{0, 1, 2\}$, and the matrix
$$\mathbf{W} = \begin{pmatrix} 0\,0\,1 \\ 1\,0\,0 \\ 0\,0\,1 \\ 0\,1\,0 \end{pmatrix},$$
i.e., if $y = 1$, then $z = x$. Otherwise, $z = 2$ regardless of x.

Remark We consider in the sequel the channels under 1, 3, and 4. There is significant work on the channel under 2. Here unique decodability gives zero rates only and next to average error probability (error concept 3) here only maximal error probability (error concept 2, several authors use the name δ-codes).

Definition 4.1 A code for the MAC is a collection

$$(\mathcal{U}, \mathcal{V}, \{\mathcal{D}_{uv},\ (u, v) \in \mathcal{U} \times \mathcal{V}\}),$$

where

$$\mathcal{U} \subseteq \mathcal{X}^n,$$
$$\mathcal{V} \subseteq \mathcal{Y}^n,$$
$$\mathcal{D}_{uv} \subseteq \mathcal{Z}^n, \quad \text{for all } (u, v) \in \mathcal{U} \times \mathcal{V},$$

and the sets $\mathcal{D}_{uv}$, $(u, v) \in \mathcal{U} \times \mathcal{V}$, are disjoint, i.e.,

$$(u, v) \neq (u', v') \implies \mathcal{D}_{uv} \bigcap \mathcal{D}_{u'v'} = \emptyset.$$

The pair of rates of the code is

$$(R_1, R_2) = \left(\frac{\log |\mathcal{U}|}{n}, \frac{\log |\mathcal{V}|}{n}\right).$$

There are 3 natural criteria that can be used when we can construct codes for MACs:

1. The code should be uniquely decodable (UD): every $z^n \in \mathcal{Z}^n$ can be generated by not more than one pair of codewords $(u, v) \in \mathcal{U} \times \mathcal{V}$, i.e.,

$$\begin{aligned} &W(z^n|u, v) > 0,\ (u, v) \in \mathcal{U} \times \mathcal{V} \\ &\quad \implies W(z^n|u', v') = 0, \text{ for all } (u', v') \in \mathcal{U} \times \mathcal{V} \setminus \{(u, v)\}. \end{aligned} \tag{4.1.1}$$

2. The maximal error probability

$$\lambda_{max} = \max_{(u,v)\in\mathcal{U}\times\mathcal{V}} W(\mathcal{D}^c_{uv}|u, v) \tag{4.1.2}$$

   should not exceed some given λ.
3. The average error probability

$$\overline{\lambda} = \frac{1}{|\mathcal{U}| \cdot |\mathcal{V}|} \sum_{(u,v)\in\mathcal{U}\times\mathcal{V}} W(\mathcal{D}^c_{uv}|u, v) \tag{4.1.3}$$

   should not exceed some given λ.

At present, very few facts are known when we have to construct codes for a MAC under the criterion $\lambda_{max} < \lambda$. However, if the MAC is deterministic, then the requirement that the maximal error probability should be small is equivalent to the requirement that it should be equal to zero, i.e., the code should be uniquely decodable (the conditional probabilities at the right hand side of (4.1.2) for deterministic MACs are equal to either 0 or 1, and if $\lambda_{max} < \lambda$ then they are equal to 0). The criterion that $(\mathcal{U}, \mathcal{V})$ should be a UD code with the 'maximal' possible pair of rates relates to the problem of finding the zero-error capacity of the single-user channel. This problem is very hard and it does not become easier if more than one sender is involved into the transmission process. Nevertheless, there exist interesting approaches to this problem for the specific MACs.

4.1.2 Achievable Rate Region Under the Criterion of Arbitrarily Small Average Decoding Error Probability

Definition 4.2 The set $\overline{\mathcal{R}}$ of pairs (R_1, R_2) is known as achievable rate region for a MAC under the criterion of arbitrarily small average decoding error probability if, for all $\lambda \in (0, 1)$, there exists an $\varepsilon_n(\lambda) \to 0$, as $n \to \infty$, such that one can construct a code $(\mathcal{U}, \mathcal{V})$ of length n with the pair of rates $(R_1 - \varepsilon_n(\lambda), R_2 - \varepsilon_n(\lambda))$ and the average decoding error probability less than λ.

Theorem 4.1 (Ahlswede (1971), [1, 2])

$$\overline{\mathcal{R}} = co\ \mathcal{R},$$

where 'co' denotes the convex hull and $\mathcal{R}$ is the set consisting of pairs (R_1, R_2) such that there exist PDs P_X and P_Y with

$$\begin{aligned} R_1 &\le I(X \wedge Z|Y), \\ R_2 &\le I(Y \wedge Z|X), \\ R_1 + R_2 &\le I(XY \wedge Z), \end{aligned} \tag{4.1.4}$$

where I is the mutual information function in the ensemble

$$\mathcal{A} = \{\mathcal{X}\mathcal{Y}\mathcal{Z},\ P_X(x)P_Y(y)W(z|x, y)\}. \tag{4.1.5}$$

Remark The achievable rate region $\overline{\mathcal{R}}$ is convex because one can apply the time sharing argument: if there are two pairs $(R_1, R_2), (R_1', R_2') \in \overline{\mathcal{R}}$, then we can divide the code length n into two subintervals of lengths αn and $(1-\alpha)n$. The i-th sender transmits one of $M_{1i} = 2^{\alpha n R_i}$ messages within the first interval and one of $M_{2i} = 2^{(1-\alpha)nR_i'}$ messages within the second interval. The total number of messages of the i-th sender is

$$M_{1i} \cdot M_{2i} = 2^{nR_i^*},$$

where

$$R_i^* = \alpha R_i + (1 - \alpha) R_i'.$$

Therefore, the pair $(\alpha R_1 + (1 - \alpha) R_1', \alpha R_2 + (1 - \alpha) R_2')$ is achievable, and this statement is true for any $\alpha \in [0, 1]$. However, the region $\mathcal{R}$ defined by (4.1.4)–(4.1.5) is in general not convex, and the operation $'co'$ is needed if we want to specify $\overline{\mathcal{R}}$. We illustrate this fact in the example below.

Example ([7]) We present now for special channels discussed above average-error capacity regions as special cases of Theorem 4.1. A fortiori they are upper bounds for UD codes. Unfortunately all known constructions are still far away from the capacity bounds.

Let $\mathcal{X} = \mathcal{Y} = \{0, 1, 2\}$, $\mathcal{Z} = \{0, 1\}$, and

$$\mathbf{W} = \begin{pmatrix} 1/2 & 1/2 \\ 1 & 0 \\ 0 & 1 \\ 1 & 0 \\ 1/2 & 1/2 \\ 1/2 & 1/2 \\ 0 & 1 \\ 1/2 & 1/2 \\ 1/2 & 1/2 \end{pmatrix},$$

where the crossover probabilities for the input (x, y) are written in the $(3x+y+1)$-st row.

Let us assign

$$P_X(0) = 1, \quad P_X(1) = P_X(2) = 0;$$
$$P_Y(0) = 0, \quad P_Y(1) = P_Y(2) = 1/2.$$

Then we get

$$I(X \wedge Z|Y) = 0, \quad I(Y \wedge Z|X) = I(XY \wedge Z) = 1.$$

Thus, $(0, 1) \in \overline{\mathcal{R}}$. Interchanging X and Y we also conclude that $(1, 0) \in \overline{\mathcal{R}}$ and (because of the operation $'co'$) that $\overline{\mathcal{R}}$ contains all points belonging to the line $R_1 + R_2 = 1$.

Let us check that the pairs $(R_1, 1 - R_1)$ such that $0 < R_1 < 1$ do not belong to $\mathcal{R}$. Suppose $I(XY \wedge Z) = 1$. Since $H(Z) \leq 1$, this equation is valid only if $H(Z|XY) = 0$, i.e., if Z is a deterministic function of x and y. Thus,

$$P_{XY}(x, y) = 0, \quad (x, y) \in \{(0, 0), (1, 1), (1, 2), (2, 1), (2, 2)\}.$$

However, $P_{XY}(x, y) = P_X(x) \cdot P_Y(y)$, and we obtain that either $P_X(0) = 1$ and $P_Y(0) = 0$, or $P_X(0) = 0$ and $P_Y(0) = 1$. This observation means that either $R_1 = 0$ or $1 - R_1 = 0$.

Note that the mutual information functions at the right hand side of (4.1.4) can be expressed using the entropy functions:

$$\begin{aligned} I(X \wedge Z|Y) &= H(Z|Y) - H(Z|XY), \\ I(Y \wedge Z|X) &= H(Z|X) - H(Z|XY), \\ I(XY \wedge Z) &= H(Z) - H(Z|XY). \end{aligned}$$

For deterministic channels, $H(Z|XY) = 0$, and inequalities (4.1.4) can be simplified as follows:

$$\begin{aligned} R_1 &\le H(Z|Y), \\ R_2 &\le H(Z|X), \\ R_1 + R_2 &\le H(Z). \end{aligned} \tag{4.1.6}$$

To obtain $\overline{\mathcal{R}}$ using Theorem 4.1, one should find the PDs on the input alphabets that give the pairs (R_1, R_2) such that all pairs $(R_1', R_2') \neq (R_1, R_2)$ with $R_1' \ge R_1$ and $R_2' \ge R_2$ do not belong to $\mathcal{R}$. For some channels, using the symmetry, we conclude that these distributions are always uniform.

Example

1. For the adder channel, the optimal input PDs are uniform and

$$\overline{\mathcal{R}} = \{\ (R_1, R_2) :\ R_1, R_2 \le 1,\ R_1 + R_2 \le 3/2\ \},$$

as it follows from (4.1.6). The region $\overline{\mathcal{R}}$ is shown in Fig. 4.1.

2. For the binary symmetric adder channel, the optimal input PDs are also uniform and

$$\begin{aligned} H(Z|XY) &= \frac{1}{2}h(q^2; 2pq; p^2) + \frac{1}{2}h(pq; p^2 + q^2; pq) \\ &= h(p) + h(2pq)/2, \\ H(Z|X) &= h(q/2; 1/2; p/2) \\ &= 1 + h(p)/2, \\ H(Z) &= 3/2, \end{aligned}$$

where

$$h(P_0; ...; P_{K-1}) = -\sum_{k=0}^{K-1} P_k \log P_k$$

denotes the entropy function of the distribution $(P_0, ..., P_{K-1})$ and

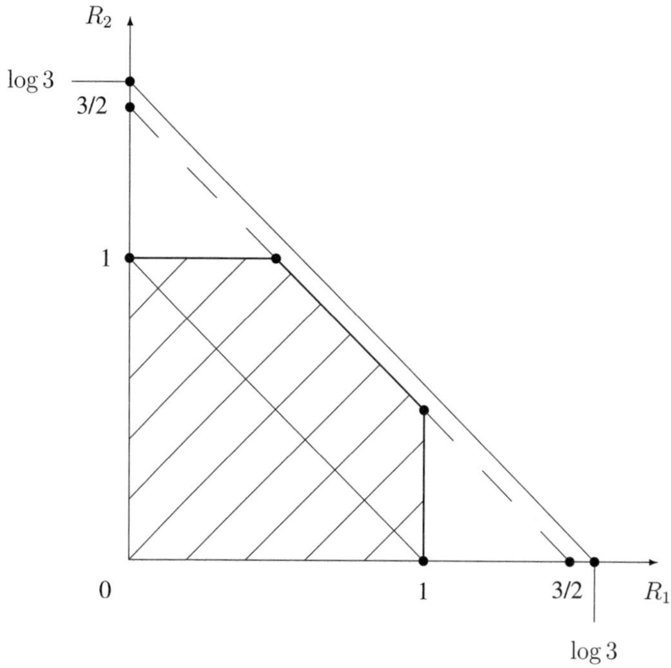

Fig. 4.1 The achievable rate region of the binary adder channel under the criterion of arbitrarily small average decoding error probability. The line $R_1+R_2 = 1$ corresponds to time sharing between the rates (0, 1) and (1, 0). The line $R_1 + R_2 = \log 3$ corresponds to the maximal total rate in the case that one sender uses a channel with the input alphabet $\mathcal{X} \times \mathcal{Y}$ and the crossover probabilities of that channel coincide with the crossover probabilities for the adder channel

$$h(z) = -z \log z - (1 - z) \log(1 - z)$$

is the entropy function of the distribution $(z, 1 - z)$. Therefore

$$\begin{aligned}\overline{\mathcal{R}} = \{\ (R_1, R_2) : \quad & R_1, R_2 \leq 1 - h(p)/2 - h(2pq)/2, \\ & R_1 + R_2 \leq 3/2 - h(p) + h(2pq)/2\ \},\end{aligned}$$

as it follows from (4.1.4).

3. Let us consider the OR channel and suppose that the input PDs are $(p_1, 1 - p_1)$ and $(p_2, 1 - p_2)$. Then

$$\begin{aligned} H(Z|X) &= p_1 \cdot h(p_2), \\ H(Z|Y) &= p_2 \cdot h(p_1), \\ H(Z) &= h(p_1 p_2). \end{aligned}$$

It is easy to see that

$$p_1 \cdot h(p_2) + p_2 \cdot h(p_1) \geq h(p_1 p_2),$$

and if we assign any $p_1 \in [1/2, 1]$ and $p_2 = 1/(2p_1)$, then some point belonging to the line $R_1+R_2 = 1$ will be obtained. This line cannot be lifted since $h(p_1 p_2) \leq 1$ for all p_1 and p_2. On the other hand, this line corresponds to time sharing between the rates (0,1) and (1,0). Hence, a special coding for the OR channel cannot improve the behavior compared to the transmission of uncoded data in a time sharing mode.

4. Let us consider the switching channel. Suppose that $(p_1, 1 - p_1)$ and $(p_2, 1 - p_2)$ are the input PDs. Then

$$\begin{aligned} H(Z|X) &= h(p_2), \\ H(Z|Y) &= p_2 \cdot h(p_1), \\ H(Z) &= p_2 \cdot h(p_1) + h(p_2). \end{aligned}$$

It is easy to see that if $R_1 \in [0, 1/2]$, then we assign $p_1 = p_2 = 1/2$ and obtain that any $R_2 \in [0, 1]$ gives an achievable pair (R_1, R_2). If $R_1 \in (1/2, 1]$, then we assign $p_1 = 1/2$ and $p_2 = R_1$. This choice leads to the inequalities $R_2 \leq h(R_1)$ and $R_1 + R_2 \leq R_1 + h(R_1)$. Hence,

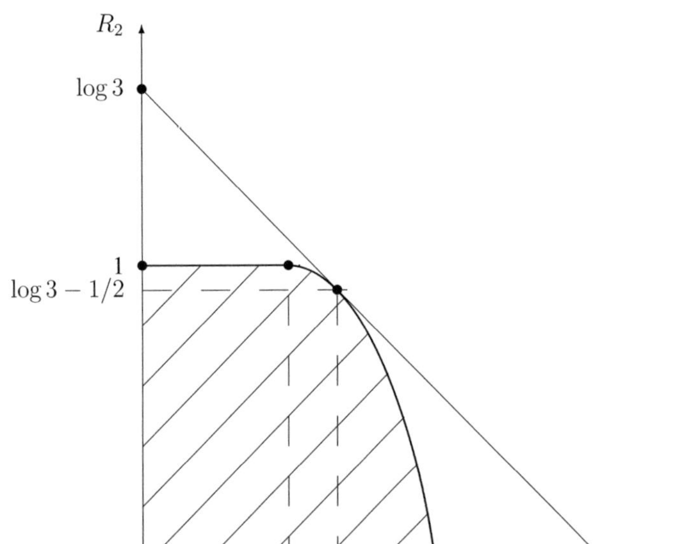

Fig. 4.2 The achievable rate region of the binary switching channel under the criterion of arbitrarily small average decoding error probability. The line $R_1 + R_2 = \log 3$ corresponds to the maximal total rate when one sender uses a channel with the input alphabet $\mathcal{X} \times \mathcal{Y}$ and crossover probabilities of that channel coincide with the crossover probabilities for the switching channel

$$\overline{\mathcal{R}} = \{ (R_1, R_2) : R_1 \in [0, 1/2], R_2 \in [0, 1] \} \cup \{ (R_1, R_2) : R_1 \in [1/2, 1], R_2 \in [0, h(R_1)] \}. \tag{4.1.7}$$

The region $\overline{\mathcal{R}}$ is shown in Fig. 4.2.

4.2 Coding for the Binary Adder Channel

4.2.1 *Statement of the Problem of Constructing UD Codes*

Any deterministic MAC realizes some function of the inputs and instead of {0, 1}-matrix **W** can be defined by the table whose rows correspond to the first input and whose columns correspond to the second input. In particular,

	0	1
0	0	1
1	1	2

(4.2.1)

is the table for the adder channel, and the definition (4.1.1) of UD codes given in the previous section can be reformulated as follows:

$$u + v \neq u' + v', \quad \text{for all } (u, v) \neq (u', v'), \quad (u, v), (u', v') \in \mathcal{U} \times \mathcal{V}. \tag{4.2.2}$$

When both senders may transmit all possible binary n-tuples, we can describe the output vector space taking the n-th Cartesian product of table (4.2.1); if $n = 2$, then we get the following extension:

	00	01	10	11
00	00	01	10	11
01	01	02	11	12
10	10	11	20	21
11	11	12	21	22

(4.2.3)

All ternary vectors, except 00, 02, 20, and 22 are included in the table at least twice, and the construction of a pair of UD codes which attains the maximal achievable pair of rates can be viewed as deleting a minimal number of rows and columns in such a way that all entries of the table are different. For example, table (4.2.3) can be punctured in the following way:

	00	01	10
00	00	01	10
11	11	12	21

(4.2.4)

If the first sender is allowed to transmit one of two codewords, 00 or 11, and the second sender is allowed to transmit one of three codewords, 00, 01 or 10, then one of 6 vectors can be received, and (4.2.4) can be considered as a *decoding table*: the decoder uniquely discovers which pair of codewords was transmitted. Hence, we have constructed a code $(\mathcal{U}, \mathcal{V})$ for the adder channel having length 2 and the pair of rates $(1/2, (\log 3)/2)$. Note that $1/2 + (\log 3)/2 \approx 1.292 > 1$, i.e., these rates give the point above the time sharing line between the rates (0,1) and (1,0). Obviously, this code can be used for any even n if the first user represents his message as a binary vector of length $n/2$ and the second user represents his message as a ternary vector of length $n/2$; after that the first encoder substitutes 00 for 0 and 11 for 1, the second encoder substitutes 00 for 0, 01 for 1, and 10 for 2.

Table (4.2.4) defines the code pair that leads to better characteristics of data transmission systems compared to time sharing and stimulates a systematic study of UD codes for the adder channel.

Definition 4.3 The set $\mathcal{R}_u$ of pairs (R_1, R_2) is an achievable rate region of UD codes for the adder channel if there exists an $\varepsilon_n \to 0$, as $n \to \infty$, such that one can construct a UD code $(\mathcal{U}, \mathcal{V})$ of length n with the pair of rates $(R_1 - \varepsilon_n, R_2 - \varepsilon_n)$.

Finding the region $\mathcal{R}_u$ *is one of the open problems of information theory*, and we present some known results characterizing $\mathcal{R}_u$. Note that this problem can be also considered under *additional restrictions* on available codes $\mathcal{U}$ and $\mathcal{V}$. *One* of these restrictions is *linearity of the codes*. The linear codes will be considered in the next section.

In conclusion of this section we give two statements which are widely used by the procedures that construct codes for the adder channel,

$$u + v = u' + v' \implies u \oplus u' = v \oplus v', \tag{4.2.5}$$
$$\text{consequently } u \oplus u' \neq v \oplus v' \implies u + v \neq u' + v'.$$

These statements can be easily checked for $n = 1$ substituting 16 possible vectors $(u, v, u', v') \in \{0, 1\}^4$, and the case $n \geq 1$ follows.

4.2.2 Rates of UD Codes $(\mathcal{U}, \mathcal{V})$ when $\mathcal{U}$ and $\mathcal{V}$ are Linear Codes

A binary linear (n, k)-code of rate $R = k/n$ is defined by a binary $k \times n$ matrix $\mathbf{G}$, whose rows $\mathbf{g}_1, ..., \mathbf{g}_k \in \{0, 1\}^n$ are linearly independent. This matrix is known as *generator matrix*. The message generated by the source is represented by a binary vector $m \in \{0, 1\}^k$, and the corresponding codeword is defined as $m\mathbf{G}$.

The restriction to linear codes may be motivated by several reasons:

- if the code is linear then, as a rule, the encoding and decoding complexity can be essentially reduced compared to a general case;
- the total number of linear (n, nR)-codes is $n^2 R$, while the total number of all (n, nR)-codes is $n2^{nR}$;
- asymptotic characteristics of the class of linear codes are not worse than the similar characteristics of the whole class of codes for data transmission systems with one sender, one receiver, and a memoryless channel.

In this section we assume that $\mathcal{U}$ and $\mathcal{V}$ are linear (n, k_1)- and (n, k_2)-codes, and denote their generator matrices by $\mathbf{G}_1$ and $\mathbf{G}_2$; the rows of these matrices are distinguished as $\mathbf{g}_{1,1}, ..., \mathbf{g}_{1,k_1}$ and $\mathbf{g}_{2,1}, ..., \mathbf{g}_{2,k_2}$ respectively.

Proposition 4.1 *If $\mathcal{U}$ and $\mathcal{V}$ are linear codes having the rates R_1 and R_2, then the pair $(\mathcal{U}, \mathcal{V})$ can be uniquely decodable if and only if $R_1 + R_2 \leq 1$.*

Proof Suppose that $R_1 + R_2 > 1$ and join the generator matrix $\mathbf{G}_2$ of the code $\mathcal{V}$ to the generator matrix $\mathbf{G}_1$ of the code $\mathcal{U}$. A new matrix

$$\mathbf{G} = \begin{bmatrix} \mathbf{G}_1 \\ \mathbf{G}_2 \end{bmatrix}.$$

has the dimension $n(R_1 + R_2) \times n$, and at least $n(R_1 + R_2) - n$ rows are linearly dependent. For example, suppose that the first row can be expressed as a linear combination of t other rows of $\mathbf{G}_1$ and s rows of $\mathbf{G}_2$, i.e., there exist $i_1, ..., i_t \in \{2, ..., k_1\}$ and $j_1, ..., j_s \in \{2, ..., k_2\}$ such that

$$\mathbf{g}_{1,1} = \mathbf{g}_{1,i_1} \oplus ... \oplus \mathbf{g}_{1,i_t} \oplus \mathbf{g}_{2,j_1} \oplus ... \oplus \mathbf{g}_{2,j_s}.$$

Then

$$\mathbf{g} = \mathbf{g}' \neq 0^n,$$

where

$$\mathbf{g} = \mathbf{g}_{1,1} \oplus \mathbf{g}_{1,i_1} \oplus ... \oplus \mathbf{g}_{1,i_t}$$

is a codeword of $\mathcal{U}$,

$$\mathbf{g}' = \mathbf{g}_{2,j_1} \oplus ... \oplus \mathbf{g}_{2,j_s}$$

is a codeword of $\mathcal{V}$, and 0^n is the all-zero vector of length n. Hence, the decoder gets the same vector in two cases: (1) the first sender sends the all-zero codeword and the second sender sends $\mathbf{g}'$; (2) the first sender sends $\mathbf{g}$ and the second sender sends the all-zero codeword. Therefore any pair of linear codes forms a UD code for the adder channel only if their rates satisfy the inequality $R_1 + R_2 \leq 1$. On the other hand, this bound is achievable by time sharing between the codes of rates 0 and 1 (note that these codes are linear and the resulting code is also linear). The rate region is shown in Fig. 4.3. □

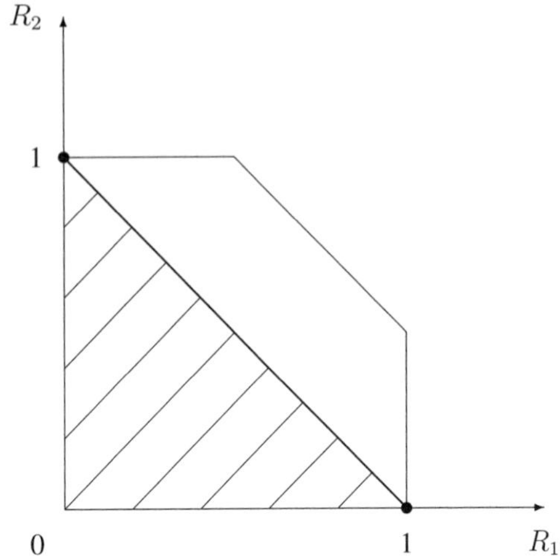

Fig. 4.3 The achievable rate region $\mathcal{R}_u^{(LL)}$ of uniquely decodable codes $(\mathcal{U}, \mathcal{V})$ when $\mathcal{U}$ and $\mathcal{V}$ are linear codes

4.2.3 Rates of UD Codes $(\mathcal{U}, \mathcal{V})$ when $\mathcal{U}$ is a Linear Code

The code $\{00, 11\}$ in table (4.2.4) is a linear (2,1)-code, while the other code $\{00, 01, 10\}$ is non-linear. Note also that the codes of rate 0 and 1 are linear. Therefore, this pair of codes and a possibility to share the time leads to the following statement.

Proposition 4.2 (Weldon (1978), [56]) *There exist UD codes $(\mathcal{U}, \mathcal{V})$ such that $\mathcal{U}$ is a linear code of rate R_1 and $\mathcal{V}$ has the rate*

$$R_2 = \begin{cases} R_1 \log 3, & \text{if } R_1 < 1/2, \\ (1 - R_1) \log 3, & \text{if } R_1 \geq 1/2. \end{cases} \tag{4.2.6}$$

Equation (4.2.6) defines a lower bound on the region of achievable rates $\mathcal{R}_u$ when $\mathcal{U}$ is a linear code. We denote this region by $\mathcal{R}_u^{(L)}$ and write

$$\mathcal{R}_u^{(L)} \supseteq \hat{\mathcal{R}}_u^{(L)},$$

where

$$\hat{\mathcal{R}}_u^{(L)} = \{ (R_1, R_2) : \; R_2 \geq R_1 \log 3, \text{ if } R_1 < 1/2, \\ R_2 \geq (1 - R_1) \log 3, \text{ if } R_1 \geq 1/2 \}.$$

The region $\hat{\mathcal{R}}_u^{(L)}$ is shown in Fig. 4.4. In the following, we abbreviate UD codes linear in $\mathcal{U}$ as LUD codes.

Proposition 4.3 (Weldon (1978), [56]) *Let $\mathcal{U}$ consists of 2^k codewords and have the property that there exists a k-element subset $J \subseteq \{1, ..., n\}$ where the codewords*

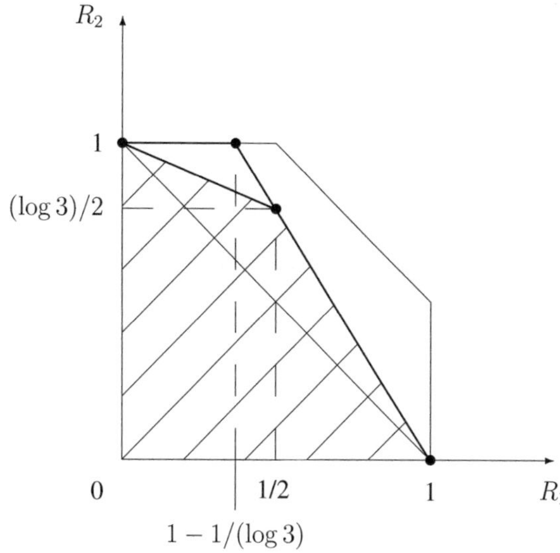

Fig. 4.4 The achievable rate region $\hat{\mathcal{R}}^{(L)}_{u,W}$ of uniquely decodable codes $(\mathcal{U}, \mathcal{V})$ (due to Weldon's lower bound) when $\mathcal{U}$ is a linear code; an upper bound is defined by the inequalities: $R_2 \leq 1$ and $R_2 \leq (1 - R_1) \log 3$

take all possible 2^k values (in particular, all binary linear codes have this property). Then the rate R_2 of any code $\mathcal{V}$ such that $(\mathcal{U}, \mathcal{V})$ is a UD code satisfies the inequality

$$R_2 \leq (1 - R_1) \log 3, \tag{4.2.7}$$

where $R_1 = k/n$.

Proof For all $v \in \mathcal{V}$, there exists an $u \in \mathcal{U}$ such that the vector $u + v$ contains the 1's at positions $j \in J$ (we assign u in such a way that $u_j = v_j \oplus 1$, $j \in J$). Thus, each column of the decoding table contains some vector with 1's at J. There are 3^{n-k} possibilities for the other components, and each vector can be met in the table at most once. Hence, the total number of columns is at most 3^{n-k}. □

Constructions

(a) $\mathbf{R_1 = 0.5}$: $(\mathcal{U}, \mathcal{V}) = (\{00, 11\}, \{00, 01, 10\})$ is a LUD code, which achieves the bound of Proposition 4.3. This construction can be repeated any m times to get codes for $n = 2m$ with $|\mathcal{U}| = 2^m$, $|\mathcal{V}| = 3^m$.

(b) $\mathbf{R_1 > 0.5}$: Now assume that we concatenate r positions to the previous code of length $2m$ to get the length $2m + r$. Obviously, if in the extra r positions the code $\mathcal{U}$ is arbitrary, and if $\mathcal{V}$ has the all zero vector, then $(\mathcal{U}, \mathcal{V})$ for the length $2m + r$ will again be UD.

We thus get $|\mathcal{U}| = 2^{m+r}$, $|\mathcal{V}| = 3^m$, which means that $|\mathcal{V}|$ meets the upper bound (4.2.7). However, if $R_1 > 0.5$ and $R_2 = (1 - R_1) \log_2 3 < 0.5$ it can be shown, that if instead of the code with $R_2 < 0.5$ one takes the linear code with $R_1 < 0.5$, thus we will get a larger rate for the code $\mathcal{V}$. Therefore the construction of LUD codes is of interest with $R_1 < 0.5$. Kasami and Lin [29] obtained an upper bound

$$|\mathcal{V}| \leq \sum_{j=0}^{k} \binom{n-k}{j} 2^j + 2^k \sum_{j=k+1}^{n-k} \binom{n-k}{j}. \tag{4.2.8}$$

This bound comes from the fact, that if the coset of an (n, k) code has maximum and minimum weights $w_{\min}$ and $w_{\max}$, respectively, it can be shown that at most $\min\{2^{n-w_{\max}}, 2^{w_{\min}}\}$ vectors can be chosen from each such coset of the code $\mathcal{V}$.

The upper bound (4.2.8) is an improvement of (4.2.7) for the range $0 \leq R_1 < 0.4$. In an asymptotic form (4.2.8) for that range is:

$R_2 \leq 1$ if $0 \leq R_1 < 1/3$, $R_2 \leq R_1 + (1 - R_1)H(\rho) + O(1)$ if $1/3 \leq R_1 < 2/5$, where $H(\rho)$ is the entropy function, $\rho = R_1/(1-R_1)$, $O(1) \to 0$ when $n \to \infty$. This is the best known upper bound for LUD codes. The best known lower bound is obtained in the work of Kasami, Lin, Wei, and Yamamura in 1983 [30] by using a graph-theoretical approach. The problem of LUD construction had been reduced to the computation of a maximum independent set of an undirected graph. The final result in an asymptotic form is as follows:

$$\begin{aligned} R_2 &\geq 1 - O(1), && \text{if } 0 \leq R_1 < 1/4; \\ R_2 &\geq 1/2(1 + H(2R_1)) - O(1), && \text{if } 1/4 \leq R_1 < 1/3; \\ R_2 &\geq 1/2(\log_2 6) - R_1 - O(1), && \text{if } 1/3 \leq R_1 < 1/2. \end{aligned} \tag{4.2.9}$$

However, the lower bound (4.2.9) is non-constructive, i.e., it does not give a method of an explicit construction of codes.

(c) Construction of LUD codes with $\mathbf{R_1 < 0.5}$:

(1) *Construction of Shannon* [51]. This idea is valid for any UD codes. The idea of construction is simply "time-sharing" between two original UD codes. The users agree to use each of two UD pairs several times to get another UD pair with a longer length. Let $(\mathcal{U}, \mathcal{V})$ and $(\mathcal{U}', \mathcal{V}')$ be UD pairs with rates (R_1, R_2), (R_1', R_2') and lengths n and n' respectively. Them if $(\mathcal{U}, \mathcal{V})$ is used a times, and then $(\mathcal{U}', \mathcal{V}')$ is used b times, the resulting UD pair will have a length $(an + bn')$ and rates $(R_1'', R_2'') = \left(\frac{anR + bn'R_1'}{an+bn'}, \frac{anR_2 + bn'R_2'}{an+bn'}\right)$. This construction will be further referred to as "time-sharing" technique (TS).

Definition 4.4 Two pairs of UD codes P_1 and P_2 will be called equivalent if they can be constructed from each other by TS and this will be denoted by $P_1 \sim P_2$.

It is easy to see, that if one applies TS to different pairs of UD codes with rates (R_1, R_2) and (R_1', R_2'), $R_{\max} = \max\{(R_1, R_2, R_1', R_2')\}$, it is not possible to get an

UD pair (R_1'', R_2''), $R'_{\max} = \max\{R_1'', R_2''\}$, with $R'_{\max} > R_{\max}$. From this observation it is natural to introduce the following partial order between UD pairs:

Definition 4.5 It will be said that an UD pair $P_1 = (R_1, R_2)$ is superior to $P_1' = (R_1', R_2')$ denoted by $P_1 \succeq P_1'$ if $R_1+R_2 \geq R_1'+R_2'$ and $\max\{R_1, R_2\} \geq \max\{R_1', R_2'\}$.

Definition 4.6 It will be said that two different UD pairs P_1, P_2 are incomparable, if they are not equivalent or none of them is superior to the other.

These three definitions give criteria how to compare different UD pairs.

(2) *Construction of Weldon and Yui (1976).* Let $\mathcal{U} = \{0^n, 1^n\}$, $\mathcal{V} = \{(0, 1)^n \setminus 1^n\}$. Then $(\mathcal{U}, \mathcal{V})$ is UD. The proof is obvious, since if the sum vector has at least one "2" then the all one vector 1^n is transmitted by $\mathcal{U}$, otherwise the all zero vector 0^n is transmitted.

Definition 4.7 It is said that a vector $u = (u_1, u_2, \ldots, u_n)$ does not cover a vector $v = (v_1, v_2, \ldots, v_n)$ denoted by $u \nrightarrow v$ if there is at least one i for which $v_i > u_i$.

The following lemma plays an important role for the construction of LUD codes (Figs. 4.5 and 4.6).

Lemma 4.1 (Kasami and Lin 1976, [28]) *The code pair $(\mathcal{U}, \mathcal{V})$ is UD if and only if for any two distinct pairs (u, v) and (u', v') in $\mathcal{U} \times \mathcal{V}$ one of the following conditions holds:*

(i) $u \oplus v \neq u' \oplus v'$
(ii) $u \oplus v = u' \oplus v' \nrightarrow v \oplus v'$

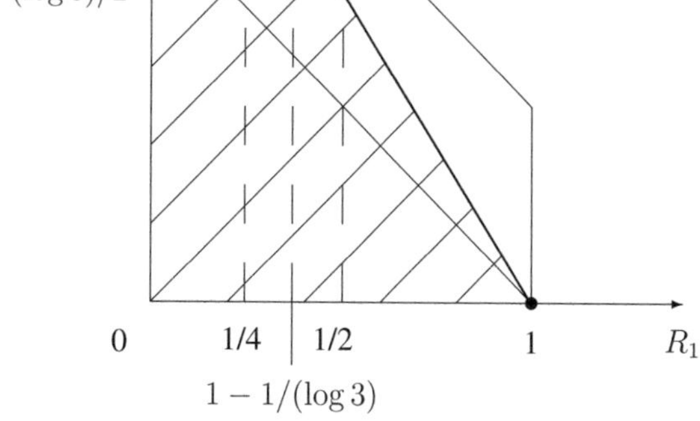

Fig. 4.5 The achievable rate region $\hat{\mathcal{R}}_{u,K}^{(L)}$ of uniquely decodable codes $(\mathcal{U}, \mathcal{V})$ (due to Kasami–Lin–Wei–Yamamura' lower bound) when $\mathcal{U}$ is a linear code; an upper bound is defined by the inequalities: $R_2 \leq 1$ and $R_2 \leq (1 - R_1) \log 3$

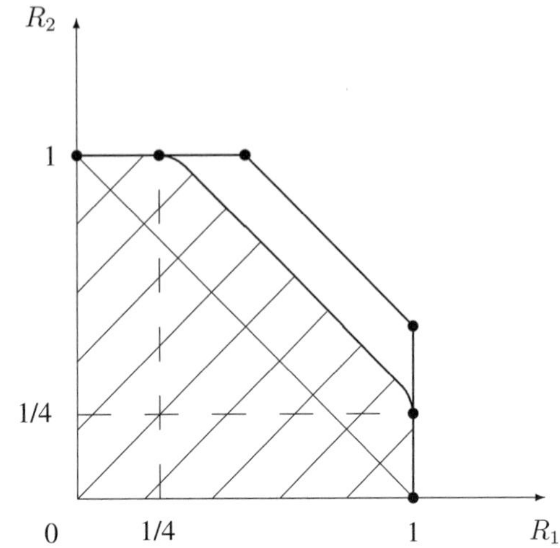

Fig. 4.6 The achievable rate region $\hat{\mathcal{R}}_{u,K}$ of uniquely decodable codes $(\mathcal{U}, \mathcal{V})$ (due to Kasami–Lin–Wei–Yamamura' lower bound); an upper bound is defined by the inequalities: $R_1 \leq 1$, $R_2 \leq 1$ and $R_1 + R_2 \leq 3/2$

Proof Obviously, if two vectors are different modulo 2, they will be different modulo 3, i.e., for the adder channel. Now let us have the second condition, which means, that for some i, $v_i \oplus v_i' = 1$ and $u_i \oplus v_i = 0$ and hence $u_i' \oplus v_i' = 0$. Since $v_i \neq v_i'$, this implies, that $u_i + v_i \neq u_i' + v_i'$ and therefore $u + v \neq u' + v'$. Now let us apply Lemma 4.1 for the construction of LUD codes. If $\mathcal{U}$ is a linear (n, k) code, then evidently code vectors of $\mathcal{V}$ must be chosen from the cosets of $\mathcal{U}$ and the only common vector between $\mathcal{U}$ and $\mathcal{V}$ should be 0^n. □

Lemma 4.2 (Kasami and Lin 1976, [28]) *Let $(\mathcal{U}, \mathcal{V})$ be an LUD pair. Then two vectors v and v' from the same coset can be chosen as code vectors for the code $\mathcal{V}$ if and only if $v \oplus v'$ cannot be covered by any vector of that coset.*

Proof Suppose that $v, v' \in \mathcal{V}$, $u, u' \in \mathcal{U}$, and $u \oplus v = u' \oplus v'$. According to the condition of the lemma, there is some i for which $v_i \oplus v_i' = 1$ and $u_i \oplus v_i = u_i' \oplus v_i' = 0$ and therefore as in Lemma 4.1 $u + v \neq u' + v'$. It is easy to see that the reverse statement of the lemma is also true. □

Lemma 4.2 has been used by G. Khachatrian for the construction of LUD codes.

(3) *Construction of G. Khachatrian, 1982/82,* [31, 32]. In [32] the following general construction of LUD codes is given. It is considered that the generator matrix of $\mathcal{U}$ has the following form.

$$
\begin{array}{cccccccccc}
 & 1 \cdot 10 \cdots\cdot & 01 & 1 & \cdots\cdots & \cdot & 1 \\
 & 0 \cdot 01 \cdot 1000 & \cdot & \cdot & \cdots\cdots & \cdot & \cdot\, r^{(1)} \\
 & \cdots\cdots\cdots & \cdot\, 1 & 1 & \cdots\cdots & \cdot & 1 \\
 & \cdots\cdots\cdots & \cdot\, 0 & 0 & 011 \cdots & \cdot & 1 \\
 & \cdots\cdots\cdots & \cdot\, 0 & 0 & 0 \cdots\cdot & \cdot & \cdot\, r^{(2)} \\
I_k & \cdots\cdots\cdots & \cdot & \cdot & \cdot\, 11 \cdots & \cdot & 1 \\
 & \cdots\cdots\cdots & \cdot & \cdot & \cdots\cdots & \cdot & \cdot \\
 & \cdots\cdots\cdots & \cdot\, 0 & 0 & 00001 & 1 & 1 \\
 & \cdots\cdots\cdots & \cdot & \cdot & \cdots\cdots & \cdot & \cdot\, r^{(m)} \\
 & \cdots\cdot\, 1 \cdot 10 & 0 & 0 & 00001 & 1 & 1 \\
 & l_1 \quad l_2 \quad l_k & r_1^{(1)} & & & r_1^{(m_1)} &
\end{array}
$$

where I_k is an identity matrix, $\sum_{j=1}^{m_1} r^{(j)} = k$, $\sum_{j=1}^{m_1} r_1^{(j)} = n - k - \sum_{i=1}^{k} l_i$. In [33] the following formula for the cardinality of $\mathcal{V}$ is given with the restriction that $l_i = l \ (i = 1 \cdots k)$, $r^{(j)} = r \ (j = 1 \cdots m)$, $r_1^{(j)} = r_1 \ (i = 1 \cdots m)$

$$|\mathcal{V}| = 2^m \sum_{i=1}^{m_1} \sum_{j=0}^{i-1} F_1(i, j) * f(i) \qquad \text{for } i = 0 \to j = 0 \text{ where}$$

$$F_1(i, j) = (-1)^j \times \binom{i}{j} \times \sum_{p=i}^{ir} \binom{(i-j)r}{p} \times (2^{l+1} - 2)^{k-p}$$

$$F(i) = \sum_{j_1=0}^{m-i} \sum_{j_2=j_1+1}^{m-i+1} \cdots \sum_{j_i=j_{i-1}+1}^{m-1} 2^{j_1(r_1-1)} \times 2^{(j_2-j_1)(r_1-1)+1} \times (2^{(m-j_i)(r_1-1)+1} - 1)$$

An analogous formula is obtained in [36] for arbitrary $r^{(1)}, r_1^{(i)}, l_i$, which is more complicated and is not introduced here for the sake of space. The parameters of some codes obtained with the above construction are presented in Table 4.1.

We will relate the condition for a code $(\mathcal{U}, \mathcal{V})$ to be uniquely decodable to an independent set of a graph [30].

Table 4.1 Codes obtained by the construction of G. Khachatrian

n	R_1	R_2	n	R_1	R_2
120	0.125	0.99993	210	0.2	0.99624
120	0.13333	0.99981	156	0.25	0.98458
252	0.14285	0.99974	210	0.2666	0.97957
144	0.1666	0.99896	100	0.3	0.9642
224	0.1875	0.99729	30	0.3333	0.9382
60	0.4	0.8865			

Definition 4.8 Let $\mathcal{G}(V, E)$ be a simple undirected graph (a graph without self loops and multiple edges), where V and E denote the vertex and edge sets respectively.

(i) A set of vertices in $\mathcal{G}$ is said to be an independent set if no two vertices in the set are adjacent (no two vertices in the set are connected by an edge).
(ii) An independent set is said to be maximal if it is not a proper subset of another independent set of the graph.
(iii) An independent set that has the largest number of vertices is called a maximum independent set.
(iv) The number of vertices in a maximum independent set, denoted $\alpha[\mathcal{G}]$, is called the independence number of the graph $\mathcal{G}$.

Note that a maximum independent set is a maximal independent set while the converse is not true.

Definition 4.9 Given $\mathcal{U}, V \subseteq \{0, 1\}^n$, a graph $\mathcal{G}(V, E_{\mathcal{U}})$ whose edge set $E_{\mathcal{U}}$ is defined by the condition:

$$(v, v') \in E_{\mathcal{U}} \Longleftrightarrow u + v = u' + v', \text{ for some } u, u' \in \mathcal{U},$$

is distinguished as a graph associated with $\mathcal{U}$ given a vertex set V.

The following statement reformulates the condition (4.2.2) for UD codes in terms of graph theory.

Proposition 4.4 *Given $\mathcal{U}, V \subseteq \{0, 1\}^n$, there exists a UD code $(\mathcal{U}, \mathcal{V})$ with $\mathcal{V} \subseteq V$ if and only if $\mathcal{V}$ is an independent subset of the graph $\mathcal{G}(V, E_{\mathcal{U}})$; hence, there exists a UD code $(\mathcal{U}, \mathcal{V})$ with $\mathcal{V} \subseteq V$ and*

$$|\mathcal{V}| = \alpha[\mathcal{G}(V, E_{\mathcal{U}})]. \tag{4.2.10}$$

One of the basic results of Graph Theory, known as Turán's theorem, can be used to construct codes for the adder channel. It can be found as Theorem 4 in Sect. 3.2.

Theorem 4.2 (Kasami, Lin, Wei, and Yamamura (1983), [30]) *For all $k \geq 1$ and even $n \leq 2k$, there exists a UD code $(\mathcal{U}, \mathcal{V})$ such that $\mathcal{U}$ is a linear (n, k)-code of rate $R_1 = k/n \leq 1/2$ and $\mathcal{V}$ is a code of rate*

$$\begin{aligned} R_2 &\geq \frac{1}{n} \max_{s=0,\ldots,n/2} \log \left\{ \frac{1}{1 + 2^{k+s-n/2+1}} \cdot \binom{n/2}{s} 2^{n/2} \right\} \\ &= r_2(R_1) - \varepsilon_n, \end{aligned} \tag{4.2.11}$$

where

$$r_2(R_1) = \max_{0 \le \gamma \le 1} \left\{ \frac{1 + h(\gamma)}{2} - \max\{0, R_1 + \gamma/2 - 1/2\} \right\}$$
$$= \begin{cases} 1, & \textit{if } 0 \le R_1 < 1/4, \\ (1 + h(2R_1))/2, & \textit{if } 1/4 \le R_1 < 1/3, \\ (\log 6)/2 - R_1, & \textit{if } 1/3 \le R_1 < 1/2, \end{cases}$$

and $\varepsilon_n \to 0$, *as* $(\log n)/n \to 0$.

Remark The lower bound is non-constructive, i.e., it does not give a method for an explicit construction of codes.

Proof We will represent any binary vector $v \in \{0, 1\}^n$ as a concatenation of two binary vectors of length $n/2$ (n even) and write $v = v_1 v_2$, where $v_1 = (v_{1,1}, \ldots, v_{1,n/2})$ and $v_2 = (v_{2,1}, \ldots, v_{2,n/2})$.

Let us fix a parameter $s \in \{0, \ldots, n/2\}$ and denote the collection of all s-element subsets of the set $\{1, \ldots, n/2\}$ by

$$\mathcal{J}_s = \{\ J \subseteq \{1, \ldots, n/2\} :\ |J| = s\ \}.$$

Denote also

$$V_s = \Big\{\ v = v_1 v_2 \in \{0, 1\}^n :\ v_{1,j} = v_{2,j},\ j \in J,\quad v_{1,j} \ne v_{2,j},\ j \notin J,\quad \text{for some } J \in \mathcal{J}_s\ \Big\}$$

and, for all $v \in V_s$, construct the set $E(v)$ consisting of binary vectors $v' = v_1' v_2' \in V_s$ such that

$$v'_{1,j} v'_{2,j} = \begin{cases} (v_{1,j}, v_{2,j}), & \text{if } v_{1,j} \ne v_{2,j}, \\ (0, 0) \text{ or } (1, 1), & \text{if } v_{1,j} = v_{2,j}, \end{cases}$$

where $j = 1, \ldots, n/2$. It is easy to check that

$$|V_s| = \binom{n/2}{s} 2^{n/2}, \qquad (4.2.12)$$
$$|E(v)| = 2^s, \quad \text{for all } v \in V_s.$$

We will consider V_s as vertex set of a graph $\mathcal{G}(V_s, E_{\mathcal{U}})$ associated with a linear code $\mathcal{U}$ consisting of codewords $(m\mathbf{G}, m\mathbf{G})$, where m runs over all binary vectors of length k and

$$\mathbf{G} = \left[\begin{array}{c|ccc} & g_{1,k+1} & \cdots & g_{1,n/2} \\ & g_{2,k+1} & \cdots & g_{2,n/2} \\ I_k & \cdot & & \cdot \\ & \cdot & & \cdot \\ & g_{k,k+1} & \cdots & g_{k,n/2} \end{array} \right]$$

is a generator matrix of a systematic block $(n/2, k)$-code (I_k denotes the $k \times k$ identity matrix). Since the first half of each codeword of $\mathcal{U}$ coincides with the second half, the vertices v and v' of $\mathcal{G}(V_s, E_{\mathcal{U}})$ can be adjacent only if $v' \in E(v)$. Therefore, using (4.2.5) we write

$$\begin{aligned} |E_{\mathcal{U}}| &= \sum_{v \in V_s} \sum_{v' \in E(v)} 1_{\{\, \exists u, u' \in \mathcal{U}:\, u+v=u'+v' \,\}} \\ &\leq \sum_{v \in V_s} \sum_{v' \in E(v)} 1_{\{\, \exists u, u' \in \mathcal{U}:\, u \oplus u' = v \oplus v' \,\}}. \end{aligned} \tag{4.2.13}$$

Since $\mathcal{U}$ is also a linear code,

$$u, u' \in \mathcal{U} \implies u \oplus u' \in \mathcal{U},$$

and we may rewrite (4.2.13) as follows:

$$|E_{\mathcal{U}}| \leq \sum_{v \in V_s} \sum_{v' \in E(v)} 1_{\{\, v \oplus v' \in \mathcal{U} \,\}}. \tag{4.2.14}$$

Let us introduce an ensemble of generator matrices $\mathbf{G}$ in such a way that the components $g_{i,j}, i = 1, ..., k, j = k + 1, ..., n/2$, of $\mathbf{G}$ are independent binary variables uniformly distributed over $\{0, 1\}$. Then $\mathcal{G}(V_s, E_{\mathcal{U}})$ is a random graph and $|E_{\mathcal{U}}|$ is a random variable. Let the line above denote the averaging over this ensemble. There exist $2^{k(n/2-k)}$ codes $\mathcal{U}$ and a particular non-zero vector whose first half coincides with the second half (the halves have lengths $n/2$) belongs to exactly $2^{(k-1)(n/2-k)}$ codes. Thus, using (4.2.12) and (4.2.14) we obtain

$$\begin{aligned} \overline{|E_{\mathcal{U}}|} &\leq \sum_{v \in V_s} \sum_{v' \in E(v)} \overline{1_{\{\, v \oplus v' \in \mathcal{U} \,\}}} \\ &= \binom{n/2}{s} \cdot 2^{n/2} \cdot 2^s \cdot \frac{2^{(k-1)(n/2-k)}}{2^{k(n/2-k)}} \\ &= \binom{n/2}{s} \cdot 2^{k+s}. \end{aligned} \tag{4.2.15}$$

Turán's theorem (Sect. 3.2) makes it possible to get

$$\alpha[\,\mathcal{G}(V_s, E_{\mathcal{U}})\,] \geq \frac{\binom{n/2}{s}^2 2^n}{\binom{n/2}{s} 2^{n/2} + 2|E_{\mathcal{U}}|} \tag{4.2.16}$$

for all $\mathcal{U}$ (we substitute $|V_s|$ and $|E_{\mathcal{U}}|$ for $|V|$ and $|E|$ respectively). The independence number $\alpha[\,\mathcal{G}(V_s, E_{\mathcal{U}})\,]$ is also a random variable in our code ensemble and its expectation is upper-bounded by the value of the expression at the right hand side of (4.2.16). Let us use the following auxiliary inequality: for any constant a and random variable X with the PD P_X, we may write

$$\sum_x P_X(x) \cdot (a + x)^{-1} \geq \left(a + \sum_x P_X(x) \cdot x\right)^{-1}$$

because, due to the Hölder's inequality,

$$\begin{aligned} 1 &= \sum_x P_X(x) \cdot (a + x)^2 (a + x)^{-2} \\ &\leq \left(\sum_x P_X(x) \cdot (a + x)\right)^2 \cdot \left(\sum_x P_X(x) \cdot (a + x)^{-1}\right)^2. \end{aligned}$$

Therefore,

$$\overline{\alpha}[\,\mathcal{G}(V_s, E_{\mathcal{U}})\,] \geq \frac{2^n \cdot \binom{n/2}{s}^2}{2^{n/2} \cdot \binom{n/2}{s} + 2\overline{|E_{\mathcal{U}}|}} \tag{4.2.17}$$

and using (4.2.15) we obtain

$$\overline{\alpha}[\,\mathcal{G}(V_s, E_{\mathcal{U}})\,] \geq \frac{2^{n/2} \cdot \binom{n/2}{s}}{1 + 2^{k+s-n/2+1}}. \tag{4.2.18}$$

There exists at least one generator matrix $\mathbf{G}^*$ that defines a code $\mathcal{U}^*$ such that

$$\alpha[\,\mathcal{G}(V_s, E_{\mathcal{U}^*})\,] \geq \overline{\alpha}[\,\mathcal{G}(V_s, E_{\mathcal{U}})\,].$$

Thus, using (4.2.10) we conclude that there exists a UD code $(\mathcal{U}^*, \mathcal{V}^*)$ with

$$|\mathcal{V}^*| \geq \overline{\alpha}[\,\mathcal{G}(V_s, E_{\mathcal{U}})\,],$$

and (4.2.11) follows from (4.2.18). □

4.2.4 Constructing UD Codes

Construction 1 (P. Coebergh van den Braak and H. van Tilborg, 1985, [11]). The idea of the construction is as follows: Let a code pair $(C, D \cup E)$ of the length n with partitions $C = C^0 \cup C^1$ and $D = D^0 \cup D^1$ be given, which is called a system of basic codes if

(I) $C, D^i \cup E$ is UD for $i = 0, 1$;
(II) $C^i, D \cup E$ is UD for $i = 0, 1$;
(III) $\forall_{(c,d)\in C^0\times D^0}\forall_{(c',d')\in C^1\times D^1}[c + d \neq c' + d']$;
(IV) there is a bijective mapping $\varphi : D^{(0)} \to D^{(1)}$ such that $\forall_{d\in D^0}\forall_{d'\in D^1}[d' = \varphi(d)$ if $\exists_{c,c'\in C}[c + d = c' + d']$;
(V) $D \cap E = \ominus, C^{(0)} \neq \ominus, C^{(1)} \neq \ominus, D^{(1)} \neq \ominus$.

Let Z be a binary code of length s. Now consider a code $\mathcal{I}$ of length ns which is obtained from the code Z by replacing each coordinate of $Z_i, i = 1, \ldots, s$ by the code vector from the code vector $C^{(i)}, i = 0, 1$. $\mathcal{I}$ will be considered to be the first code for the new UD pair of length ns. Now the question is how many vectors from $(D \cup S)^s$ can be included in the second code. The following theorem gives an explicit answer about the cardinalities of both codes.

Theorem 4.3 *Let $(C, D \cup E)$ be a system of basic codes of length n as defined above. Let Z be a code of length s, where $2 \leq w \leq s/2$, and $\mathcal{I}$ be a code of length ns as defined above. Write $s = qw + r$, $0 \leq r \leq w$ and define $N = sn$, $\delta = \max\{r, w - r\}$, $x = |D^{(0)}| \setminus |D^{(0)} \cup E|$ and $y = |C^{(0)}| \setminus |C|$. Then*

(i) $\mathcal{I}$ is a code of length N and size $|\mathcal{I}| = |C|^s \times \sum_{k=0}^{q} \binom{s}{kw} y^{s-kw}(1 - y)^{kw}$;
(ii) there exists a code $\mathcal{P}$ of length N, such that $(\mathcal{I}, \mathcal{P})$ is UD. The code $\mathcal{P}$ has size $|\mathcal{P}| = |D^{(0)} \cup E|^s \times \{w - \sum_{i=0}^{s} \binom{s}{i}(w - i - 1)x^i(1 - x)^{5-i} + \sum_{i=0}^{s} \binom{s}{i}(w - 2 - 2i)x^{5-i}(1 - x)^i + \sum_{i=w-\beta-1}^{s} \binom{s}{i}(\beta - 1 - i)x^{5-i}(1 - x)^i\}$

For the numerical results a system of basic codes given by $C_i^{(0)} = D_i^{(0)} = \{0^{n_i}\}$, $C_i^{(1)} = D_i^{(1)} = \{1^{n_i}\}$, $E_i = \{0, 1\}^{n_i} \setminus \{0^{n_i}, 1^{n_i}\}$ of length n_i is used which is in fact a system of UD codes given by Construction 1. It is interesting to mention, that if Z is a parity check code correcting single erasures with $w = 2$ this construction coincides with the special case of Construction 3, however it does not cover Construction 3 in more general form. The numerical results for the best UD code pairs obtained with this method will be presented in the final table. It is also interesting to mention, that in the paper [11] where the present construction is given it was also mentioned the construction of a UD pair of length 7 and sizes $|C| = 12$ and $|D| = 47$ found by Coebergh van den Braak in an entirely different way. Although no construction principle of that code has been explained it has the best known sum rate, namely $R_1 = 0.5121$, and $R_2 = 0.7935$, $R_1 + R_2 = 1.3056$.

Construction 2 (R. Ahlswede and V. Balakirsky, 1997, [3]).

(a) *Construction of* $\mathcal{U}$. The code length is $N = tn$, $|\mathcal{U}| = \binom{t}{t/2}$. A code is constructed as follows: At first all $\binom{t}{t/2}$ vectors of the length t and weight $t/2$ are taken and each coordinate then is repeated exactly n times resulting in a code of the length tn and cardinality $\binom{t}{t/2}$.
(b) *Construction of* $\mathcal{V}$. The length tn is divided into t blocks of length n. It is obvious that if a block of length n is a vector $G = \{0, 1\} \setminus \{0^n, 1^n\}$, then in these blocks $\mathcal{U}$ and $\mathcal{V}$ can be decoded uniquely (according to Construction 1). In any r blocks where $\mathcal{U}$ has elements from $B = \{0^n, 1^n\}$, $\mathcal{V}$ may have one of the following $(r+1)$ possible vectors $\{\{0^n\}^i, \{1^n\}^{n-i}\}$, $i = 0, \dots, r$, therefore the cardinality of $\mathcal{V}$ is defined by the formula $|\mathcal{V}| = \sum_{r=0}^{t} \binom{t}{r}(2^n - 2)^{n-r}(1+r) = (2^n - 1)^{n-1}(2^n - 1 + n)$. This construction gives relatively good codes with $n = 2$. The best sum rate is achieved with $t = 26$, $n = 2$, $R_1 = 0.4482$, $R_2 = 0.8554$, and $R_1 + R_2 = 1.3036$. Although this construction does not give a significant improvement over previous non-linear UD constructions it gives by our opinion a very fruitful approach to the construction of better UD codes.

Construction 3 (G. Khachatrian, 1997, [33]). The following construction is considered. Let N be the length of the codes $\mathcal{U}$ and $\mathcal{V}$, t is an arbitrary integer, $N = 2t$.

(a) *Construction of* $\mathcal{U}$. We consider two cases, namely when t is odd and even. Vectors of $\mathcal{U}$ have the form $(a_1 a_1 \cdots a_{ii} \cdots)$ where the number of non-zero elements a_i is equal to

   (i) $(t/2) \pm i$, $1 = 0, \dots, r$ if t is even,
   (ii) $(t+1)/2 + i$ or $(t-1)/2 - i$, $i = 0, \dots, r$, if t is odd.

   Therefore the cardinality of $\mathcal{U}$ is equal to

$$|\mathcal{U}| = 2\sum_{j=0}^{r} \binom{t}{t/2 + j}$$

   if t is even and

$$|\mathcal{U}| = 2\sum_{j=0}^{r} \binom{t}{\frac{t+1}{2} + j}$$

   if t is odd.
(b) *Construction of* $\mathcal{V}$. The positions of $\mathcal{V}$ are divided into t subblocks of length 2. Let t_1, $0 \le t_1 \le t$ be the number of subblocks of length 2, where $\mathcal{V}$ may have either (00) or (11), in the rest of $(t - t_1)$ subblocks $\mathcal{V}$ has either (01) or (10). Now let us see what combinations of (00) and (11) specifically $\mathcal{V}$ is allowed to have in these subblocks of length t_1. $\mathcal{V}$ will consist of vectors of type $\{\{0^n\}^j, \{1^n\}^{n-j}\}$ where $j = (2r+1)k$, if t is even and $j = 2(r+1)k$, if t is odd. Therefore, the number of vectors corresponding to those t_1 subblocks is equal to

$N(t_1) = \lceil (t_1+1)/(2r+1) \rceil$ if i is even $N(t_1) = \lceil (t_1+1)/(2(r+1)) \rceil$ if i is odd. We get the following formula for the cardinality of $\mathcal{V}$: $|\mathcal{V}| = \sum_{r=0}^{t} \binom{t}{r} 2^{n-r} N(t_1)$ and we get that $\mathcal{V} \approx 3^{t-1}/2r * (t + 1.5(2r + 1))$. The best code which is obtained according to this construction has the parameters: $t = 19$, $N = 38$, $r = 2$, $R_1 = 0.48305$, $R_2 = 0.82257$, and $R_1 + R_2 = 1.30562$.

We will construct two binary codes, $\mathcal{U}$ and $\mathcal{V}$, of length tn, where t and n are fixed integers, in such a way that $(\mathcal{U}, \mathcal{V})$ is a UD code for the two-user binary adder channel. Each codeword will be represented as a sequence of binary n-tuples having length t; these n-tuples will be regarded as subblocks. *The main point of our considerations is that we do not only prove the statement of an existence type concerning UD codes, but build specific codes for fixed* t and n *in a regular way. The rates of these codes are located above the Kasami–Lin–Wei–Yamamura (KLWY) lower bound* [30] *and these codes can be used in conjunction with simple encoding and decoding procedures.*

The section is organized as follows. We begin with the description of codes $\mathcal{U}$, $\mathcal{V}$ and illustrate the definitions for specific data. Then we prove a theorem which claims that $(\mathcal{U}, \mathcal{V})$ is a UD code and gives expressions for $|\mathcal{U}|$ and $|\mathcal{V}|$. Some numerical results and a discussion about the relationships between our construction and the Coebergh van den Braak - van Tilborg (CT) construction [11] are also presented. After that we describe a simple decoding procedure. Finally, we point out to the possibility of enumerative coding which follows from the regularity of the construction.

4.2.4.1 Code Construction (u)–(v)

Let us fix integers $t, n \geq 1$ in such a way that t is even and construct the codes $\mathcal{U}$ and $\mathcal{V}$ using the following rules.

(u) Let $\mathcal{C}$ denote the set consisting of all binary vectors of length t and Hamming weight $t/2$, i.e.,

$$\mathcal{C} = \left\{ c = (c_1, ..., c_t) \in \{0, 1\}^t : w_H(c) = t/2 \right\}, \tag{4.2.19}$$

where w_H denotes the Hamming weight. Construct a code

$$\mathcal{U} = \bigcup_{c \in \mathcal{C}} \{ (c_1^n, ..., c_t^n) \} \tag{4.2.20}$$

of length tn repeating n times each component of every vector $c \in \mathcal{C}$.

(v) Given an $s \in \{0, ..., t\}$, let

$$\mathcal{J}_s = \{ J \subseteq [t] : |J| = s \}$$

denote the collection consisting of all s-element subsets of the set $[t] = \{1, ..., t\}$, and let

$$\mathcal{A}^{(s)} = \bigcup_{i=0}^{s} \{1^{in} 0^{(s-i)n}\}, \tag{4.2.21}$$

where $1^0 0^{sn} = 0^{sn}$ and $1^{sn} 0^0 = 1^{sn}$. Furthermore, let us introduce an alphabet

$$\mathcal{B} = \{0, 1\}^n \backslash \{0^n, 1^n\}$$

consisting of $2^n - 2$ binary vectors which differ from 0^n and 1^n.

Let $j_1 < ... < j_s$ be the elements of the set $J \in \mathcal{J}_s$ and let $j'_1 < ... < j'_{t-s}$ be the elements of the set

$$J^c = [t] \backslash J.$$

For all $(a, b) \in \mathcal{A}^{(s)} \times \mathcal{B}^{t-s}$, define a vector

$$v(a, b|J) = (v_1, ..., v_t) \in \{0, 1\}^{tn} \tag{4.2.22}$$

in such a way that

$$v_j = \begin{cases} a_k, \text{ if } j = j_k, \\ b_k, \text{ if } j = j'_k, \end{cases} \tag{4.2.23}$$

where $j = 1, ..., t$, and construct a code

$$\mathcal{V} = \bigcup_{s=0}^{t} \bigcup_{J \in \mathcal{J}_s} \bigcup_{a \in \mathcal{A}^{(s)}} \bigcup_{b \in \mathcal{B}^{t-s}} \{ v(a, b|J) \}.$$

Example Let $t = n = 2$. Then $\mathcal{C} = \mathcal{B} = \{01, 10\}$. The code $\mathcal{U}$ consists of two codewords,

$$u_1 = 00\ 11$$
$$u_2 = 11\ 00$$

and the code $\mathcal{V}$ consists of all binary vectors of length 4, except 0011. We construct $\mathcal{V}$ in the following way.

$s = 0.\ \mathcal{J}_s = \emptyset,\ \mathcal{A}^{(s)} = \emptyset,\ \mathcal{B}^{t-s} = \{0101, 0110, 1001, 1010\}.$

$$\begin{aligned} v_1 &= v(-, 0101|\emptyset) = 01\ 01 \\ v_2 &= v(-, 0110|\emptyset) = 01\ 10 \\ v_3 &= v(-, 1001|\emptyset) = 10\ 01 \\ v_4 &= v(-, 1010|\emptyset) = 10\ 10 \end{aligned}$$

$s = 1.\ \mathcal{J}_s = \{\{1\}, \{2\}\},\ \mathcal{A}^{(s)} = \{00, 11\},\ \mathcal{B}^{t-s} = \{01, 10\}.$

$$\begin{aligned}
v_5 &= v(00, 01|\{1\}) = 00\ 01\\
v_6 &= v(00, 10|\{1\}) = 00\ 10\\
v_7 &= v(11, 01|\{1\}) = 11\ 01\\
v_8 &= v(11, 10|\{1\}) = 11\ 10\\
v_9 &= v(00, 01|\{2\}) = 01\ 00\\
v_{10} &= v(00, 10|\{2\}) = 10\ 00\\
v_{11} &= v(11, 01|\{2\}) = 01\ 11\\
v_{12} &= v(11, 10|\{2\}) = 10\ 11
\end{aligned}$$

$s = 2$. $\mathcal{J}_s = \{\{1, 2\}\}$, $\mathcal{A}^{(s)} = \{0000, 1100, 1111\}$, $\mathcal{B}^{t-s} = \emptyset$.

$$\begin{aligned}
v_{13} &= v(0000, -|\{1, 2\}) = 00\ 00\\
v_{14} &= v(1100, -|\{1, 2\}) = 11\ 00\\
v_{15} &= v(1111, -|\{1, 2\}) = 11\ 11
\end{aligned}$$

The pair $(\mathcal{U}, \mathcal{V})$ is optimal in the following sense: any codes $\mathcal{U}$ and $\mathcal{V}$ such that $(\mathcal{U}, \mathcal{V})$ is a UD code for the binary adder channel may contain at most one common codeword; thus

$$|\mathcal{U}| + |\mathcal{V}| \leq 2^{tn} + 1.$$

In our case,

$$|\mathcal{U}| + |\mathcal{V}| = 17 = 2^{tn} + 1.$$

4.2.4.2 Properties of Codes Constructed by (u)–(v)

Theorem 4.4 *The code $(\mathcal{U}, \mathcal{V})$ of length tn defined in (u)–(v) is a UD code for the binary adder channel and*

$$|\mathcal{U}| = \binom{t}{t/2}, \tag{4.2.24}$$

$$|\mathcal{V}| = (2^n - 1)^t \left[\frac{t}{2^n - 1} + 1\right]. \tag{4.2.25}$$

Hence,

$$\begin{aligned}
R_1 &= \frac{1}{n} - \frac{1}{tn} \log\left[2^t \binom{t}{t/2}^{-1}\right],\\
R_2 &= \frac{1}{n} \log(2^n - 1) + \frac{1}{tn} \log\left[\frac{t}{2^n - 1} + 1\right].
\end{aligned}$$

Proof Equation (4.2.24) directly follows from (4.2.19)–(4.2.20). Given an $s \in \{0, ..., t\}$, the set $\mathcal{J}_s$ consists of $\binom{t}{s}$ elements. For each $J \in \mathcal{J}_s$ there are $s + 1$ possibilities for the vector $a \in \mathcal{A}^{(s)}$ and $(2^n - 2)^{t-s}$ possibilities for the vector

$b \in \mathcal{B}^{t-s}$. Therefore,

$$|\mathcal{V}| = \sum_{s=0}^{t} \binom{t}{s} (s+1)(2^n - 2)^{t-s}.$$

It easy to check that this equation can be expressed as (4.2.25).

The proof is complete if we show that $(\mathcal{U}, \mathcal{V})$ is a UD code. Let us introduce an alphabet $\mathcal{B}^*$ consisting of the $2^n - 2$ elements of $\mathcal{B}$ and an element specified as $'*'$, i.e.,

$$\mathcal{B}^* = \mathcal{B} \bigcup \{*\}. \tag{4.2.26}$$

Let $(\mathcal{B}^*)^t$ denote the t-th extension of $\mathcal{B}^*$. For all $b^* \in (\mathcal{B}^*)^t$, we introduce the set

$$\begin{aligned} \mathcal{V}(b^*) = \{\ v = (v_1, ..., v_t) \in \{0,1\}^{tn} : \ & v_j = b_j^*, \ \text{if } b_j^* \neq *, \\ & v_j \in \{0^n, 1^n\}, \ \text{if } b_j^* = *; \\ & \text{for all } \ j = 1, ..., t\ \}, \end{aligned} \tag{4.2.27}$$

note that $\{\mathcal{V}(b^*), b^* \in (\mathcal{B}^*)^t\}$ is a collection of pairwise disjoint sets and get the following

Proposition 4.5 *Suppose that, for all $b^* \in (\mathcal{B}^*)^t$, there are subsets $\hat{\mathcal{V}}(b^*) \subseteq \mathcal{V}(b^*)$ satisfying the following condition:*

$$(\mathcal{U} + v) \bigcap (\mathcal{U} + v') = \emptyset, \quad \text{for all } \ v, v' \in \hat{\mathcal{V}}(b^*).$$

Then $\left(\mathcal{U}, \cup_{b^* \in (\mathcal{B}^*)^t} \hat{\mathcal{V}}(b^*)\right)$ is a UD code.

Furthermore, using (4.2.19)–(4.2.20) and (4.2.26)–(4.2.27) we obtain

Proposition 4.6 *Given $b^* \in (\mathcal{B}^*)^t$ and $v, v' \in \mathcal{V}(b^*)$, the following two statements are equivalent.*

(i) There exist $u, u' \in \mathcal{U}$ such that

$$u + v = u' + v'.$$

(ii) There exist $c, c' \in \mathcal{C}$ such that

$$\begin{aligned} v_j = v_j' &\Longrightarrow c_j = c_j', \\ (v_j, v_j') = (0^n, 1^n) &\Longrightarrow (c_j, c_j') = (1, 0), \\ (v_j, v_j') = (1^n, 0^n) &\Longrightarrow (c_j, c_j') = (0, 1); \ \textit{for all} \ \ j = 1, ..., t. \end{aligned} \tag{4.2.28}$$

Let us fix $b^* \in (\mathcal{B}^*)^t$ and, for all $v, v' \in \mathcal{V}(b^*)$, define

$$t_{01}(v, v') = \sum_{j=1}^{t} \chi\{\, (v_j, v'_j) = (0^n, 1^n) \,\}, \tag{4.2.29}$$

$$t_{10}(v, v') = \sum_{j=1}^{t} \chi\{\, (v_j, v'_j) = (1^n, 0^n) \,\}.$$

Proposition 4.7 *If* $v, v' \in \mathcal{V}(b^*)$ *and*

$$t_{01}(v, v') \neq t_{10}(v, v'), \tag{4.2.30}$$

then there are no $c, c' \in \mathcal{C}$ *such that statement (4.2.28) is true.*

Proof Since all vectors $c, c' \in \mathcal{C}$ have the same Hamming weight, we obtain

$$\sum_{j=1}^{t} \chi\{\, (c_j, c'_j) = (0, 1) \,\} = \sum_{j=1}^{t} \chi\{\, (c_j, c'_j) = (1, 0) \,\}. \tag{4.2.31}$$

If these vectors satisfy (4.2.28) given $v, v' \in \mathcal{V}(b^*)$, then using (4.2.27), (4.2.29), and (4.2.31), we conclude that $t_{01}(v, v') = t_{10}(v, v')$, but this equation contradicts (4.2.30). □

Let us fix $b^* \in (\mathcal{B}^*)^t$, denote

$$J = \left\{\, j \in [\,t\,] :\; b^*_j = * \,\right\}, \quad s = |J|,$$

and suppose that $j_1 < \ldots < j_s$ and $j'_1 < \ldots < j'_{t-s}$ are the elements of the sets J and J^c. Assign

$$\hat{\mathcal{V}}(b^*) = \left\{\, v \in \mathcal{V}(b^*) :\; (v_{j_1}, \ldots, v_{j_s}) \in \mathcal{A}^{(s)} \,\right\},$$

where the set $\mathcal{A}^{(s)}$ is defined in (4.2.21). Then, for all $v, v' \in \hat{\mathcal{V}}(b^*)$, $v \neq v'$, either $t_{01}(v, v') > 0$ and $t_{10}(v, v') = 0$, or $t_{01}(v, v') = 0$ and $t_{10}(v, v') > 0$. Therefore, based on Proposition 4.1, we conclude that, for all $v, v' \in \hat{\mathcal{V}}(b^*)$, there are no $c, c' \in \mathcal{C}$ such that statement (4.2.28) is true, and using Proposition 4.6 obtain that the sets $\mathcal{U} + v$, $v \in \hat{\mathcal{V}}(b^*)$, are pairwise disjoint. Finally, Proposition 4.5 says that $(\mathcal{U}, \cup_{b^* \in (\mathcal{B}^*)^t} \hat{\mathcal{V}}(b^*))$ is a UD code and, as it is easy to see,

$$\bigcup_{b^* \in (\mathcal{B}^*)^t} \hat{\mathcal{V}}(b^*) = \mathcal{V},$$

where $\mathcal{V}$ is defined in (4.2.22)–(4.2.23).

□

The rates (R_1, R_2) of some UD code are given in Table 4.2. For $R_1 \in (1/3, 1/2)$, the pair

$$\left(R_1, \hat{R}_2 = \frac{\log 6}{2} - R_1 \right)$$

Table 4.2 The rates (R_1, R_2) of some uniquely decodable codes defined by (u)–(v), the sum rates $R_1' + R_2'$ for the codes whose existence is guaranteed by the CT-construction, and the differences between R_2 and the values $\hat{R}_2$ defined by the KLWY lower bound on the maximal rate of uniquely decodable codes

tn	t	R_1	R_2	$R_1 + R_2$	$R_1' + R_2'$	$R_2 - \hat{R}_2$
28	14	0.419458	0.881856	1.301315	1.299426	0.008833
32	16	0.426616	0.875699	1.302315	1.301048	0.009834
36	18	0.432480	0.870463	1.302943	1.302071	0.010462
40	20	0.437382	0.865946	1.303328	1.302714	0.010847
44	22	0.441549	0.862002	1.303550	1.303109	0.011069
48	24	0.445141	0.858521	1.303662	1.303339	0.011181
52	**26**	**0.448272**	**0.855424**	**1.303696**	**1.303457**	**0.011215**
56	28	0.451030	0.852646	1.303676	1.303497	0.011195
60	30	0.453480	0.850138	1.303618	1.303482	0.011137
64	32	0.455672	0.847861	1.303533	1.303428	0.011052
68	34	0.457646	0.845783	1.303428	1.303347	0.010947
72	36	0.459434	0.843876	1.303311	1.303248	0.010829
76	38	0.461063	0.842121	1.303184	1.303134	0.010702
80	40	0.462553	0.840498	1.303051	1.303012	0.010570

belongs to the KLWY lower bound. We show the difference $R_2 - \hat{R}_2$ and the values of the sum rates $R_1' + R_2'$ of the codes $(\mathcal{U}', \mathcal{V}')$ whose existence is guaranteed if we use the CT-construction with given t and n. The sum rates of all codes presented in Table 4.2 are greater than $R_1' + R_2'$ and the points (R_1, R_2) are located above the curve obtained using the KLWY lower bound.

Remark on the CT-construction The authors of [11] described a rather general construction which "almost" contains the Ahlswede/Balakirsky construction (u)–(v) when $t \geq 4$, meaning that we fix the Hamming weight of each element of the set $\mathcal{C}$, while this weight should be divisible by $t/2$ in the CT-construction (if we consider the case $q = 2, r = 0$ - [11], p. 8). Then the expressions for the cardinalities of the codes given in Theorem 4.4 are reduced (in our notations) to

$$|\mathcal{U}'| = 2 + \binom{t}{t/2}$$

$$|\mathcal{V}'| = (2^n - 1)^t \left[\frac{t}{2} - \sum_{i=0}^{t/2-2} \binom{t}{i} (t/2 - i - 1)\pi^i (1-\pi)^{t-i} + \sum_{i=0}^{t/2-2} \binom{t}{i} (t/2 - i - 1)\pi^{t-i} (1-\pi)^i \right],$$

where $\pi = 1/(2^n - 1)$ and t is even. The difference in the code rate between $\mathcal{U}$ and $\mathcal{U}'$ vanishes when t is not very small. For example, consider the case $t = 4$ and set (in the notations of [11])

$$n = s = 2, \quad D^{(0)} = \{00\}, \quad D^{(1)} = \{11\}, \quad E = \{01, 10\},$$

$$\mathbf{y} = (\,00\,00\,01\,01\,), \quad \mathbf{d} = (\,00\,00\,), \quad \mathbf{d}' = (\,11\,11\,).$$

Then,

$$w^*(\mathbf{d}) = w^*(\mathbf{d}') = \gamma(\mathbf{d}, \mathbf{d}') = 0,$$

and the vectors $(\,00\,00\,01\,01\,)$, $(\,11\,11\,01\,01\,)$ cannot simultaneously belong to $\mathcal{V}'$. Nevertheless, it is possible for the code $\mathcal{V}$.

4.2.4.3 Decoding Algorithm

The codes derived in (u)–(v) can be used with a simple decoding procedure. Let $z = (z_1, ..., z_t) \in \{0, 1, 2\}^{tn}$ denote the received vector, where $z_j \in \{0, 1, 2\}^n$ for all $j = 1, ..., t$. We will write $0 \in z_j$ and $2 \in z_j$ if the received subblock z_j has 0 and 2 as one of components, respectively.

Since $u_j \in \{0^n, 1^n\}$ for all $j = 1, ..., t$, each received subblock cannot contain both 0 and 2 symbols. Thus, the decoder knows u_j if z_j contains either 0 or 2. The number of subbocks 1^n in u corresponding to the received subblocks 1^n can be found using the fact that the total Hamming weight of u is fixed to be $tn/2$. These remaining subblocks can be discovered based on the structure of the sets $\mathcal{A}^{(0)}, ..., \mathcal{A}^{(t)}$. A formal description of the decoding algorithm is given below.

(1) Set

$$J_1 = \{\, j \in [\,t\,] : \, z_j = 1^n \,\}, \quad J_1^c = [\,t\,] \backslash J_1.$$

(2) For all $j \in J_1^c$, set

$$u_j = \begin{cases} 0^n, \text{ if } 0 \in z_j, \\ 1^n, \text{ if } 2 \in z_j, \end{cases}$$

and

$$w' = \left|\{\, j \in J_1^c \, : \, 2 \in z_j \,\}\right|.$$

(3) Set

$$w = t/2 - w'$$

and represent the elements of J_1 in the increasing order, i.e.,

$$|J_1| = k, \; j_1, ..., j_k \in J_1 \implies j_1 < ... < j_k.$$

Set

$$u_j = \begin{cases} 0^n, \text{ if } \; j \in \{j_1, ..., j_{k-w}\}, \\ 1^n, \text{ if } \; j \in \{j_{k-w+1}, ..., j_k\}. \end{cases}$$

(4) Set

$$v = (z_1, ..., z_t) - (u_1, ..., u_t).$$

Example Let $t = n = 2$ (see the previous example). If the first received subblock contains 0 then the codeword u_1 was sent by the first sender, and if it contains 2 then this codeword was u_2. Similarly, if the second received subblock contains 0 or 2 then the decoder makes a decision u_2 or u_1. The codeword $v \in \mathcal{V}$ is discovered in these cases after the decoder subtracts u from the received vector. At last, if the received vector consists of all 1's then there are two possibilities: $(u, v) = (u_1, 1100)$ and $(u, v) = (u_2, 0011)$. However $0011 \notin \mathcal{V}$, and the decoder selects the first possibility.

4.2.4.4 Enumerative Coding

Enumerative procedures were developed in source coding to make the storage of a code book unnecessary at the both sides of communication link and essentially reduce computational efforts [5, 12, 48]. In this case, the encoder having received a message calculates corresponding codeword, and the decoder calculates the inverse function. Our decoder does not use the code book to decode transmitted codewords, and an enumerative algorithm for messages completely escapes the storage of code books. We present this algorithm below.

First, we construct one-to-one mappings

$$\begin{aligned} f(m) &\to \mathcal{U}, \\ f_1^{(s)}(m_J) &\to \mathcal{J}_s, \\ f_2^{(s)}(m_a) &\to \mathcal{A}^{(s)}, \\ f_3^{(s)}(m_b) &\to \mathcal{B}^{t-s}, \end{aligned}$$

where m, m_J, m_a, and m_b are integers taking values in the corresponding sets: $m \in \{1, ..., |\mathcal{U}|\}$, $m_J \in \{1, ..., |\mathcal{J}_s|\}$, etc., and $s = 0, ..., t$. The structure of the possible mappings $f_2^{(s)}(m_a)$ and $f_3^{(s)}(m_b)$ is evident; the mappings $f(m)$ and $f_1^{(s)}(m_J)$ are based on the enumeration procedures for binary vectors having a fixed Hamming weight [12].

Let (m, m') be the message to be transmitted over the binary adder channel, where $m \in \{1, ..., |\mathcal{U}|\}$ and $m' \in \{1, ..., |\mathcal{V}|\}$. Encoding and decoding of the message m are obvious: we assign

$$f(m) = u, \quad f^{-1}(u) = m.$$

Let us consider encoding and decoding of the message m'. Denote

$$K_0 = 0,$$
$$K_{s+1} = K_s + \binom{t}{s}(s+1)(2^n - 2)^{t-s}, \quad s = 0, ..., t-1,$$

and

$$M_a^{(s)} = s + 1, \quad M_b^{(s)} = (2^n - 2)^{t-s},$$

for all $s = 0, ..., t$. Furthermore, for all integers $q \geq 0$ and $Q \geq 1$, introduce the function

$$\Delta(q, Q) = q - Q\lfloor q/Q \rfloor.$$

Enumerative encoding procedure is given below.

1. Find the maximal value of $s \in \{0, ..., t-1\}$ such that $m' > K_s$, denote $m_s = m' - K_s - 1$, and set
$$m_J = \lfloor m_s/(M_a^{(s)} M_b^{(s)}) \rfloor + 1,$$
$$m_a = \lfloor \Delta(m_s, M_a^{(s)} M_b^{(s)})/M_b^{(s)} \rfloor + 1,$$
$$m_b = \Delta(\,\Delta(m_s, M_a^{(s)} M_b^{(s)}), M_b^{(s)}\,) + 1.$$
2. Set
$$J = f_1^{(s)}(m_J), \quad a = f_2^{(s)}(m_a), \quad b = f_3^{(s)}(m_b).$$
3. Construct the vector $v(a, b|J)$ in accordance with (4.2.22)–(4.2.23).

Enumerative decoding procedure goes in an opposite direction.

1. Find J, a, and b from v. Denote $s = |J|$.
2. Set
$$m_J = \left(f_1^{(s)}\right)^{-1}(J), \quad m_a = \left(f_2^{(s)}\right)^{-1}(a), \quad m_b = \left(f_3^{(s)}\right)^{-1}(b).$$
3. Set
$$m' = K_s + (m_J - 1)M_a^{(s)} M_b^{(s)} + (m_a - 1)M_b^{(s)} + (m_b - 1) + 1. \quad (4.2.32)$$

Example Let $t = n = 2$ (see first example of this section). Then

$$K_0 = 0,$$
$$K_1 = 0 + \binom{2}{0}(0+1)2^{2-0} = 4,$$
$$K_2 = 4 + \binom{2}{1}(1+1)2^{2-1} = 12.$$

Let $m' = 11$. Then $s = 1$ since $11 > K_1$ and $11 \leq K_2$. Therefore,

$$\begin{aligned} m_1 &= 11 - 4 - 1 = 6, \\ m_J &= \lfloor 6/(2 \cdot 2) \rfloor + 1 = 2, \\ m_a &= \lfloor \Delta(6,4)/2 \rfloor + 1 = 2, \\ m_b &= \Delta(\,\Delta(6,4), 2\,) + 1 = 1, \end{aligned}$$

since $M_a^{(s)} = M_b^{(s)} = 2$ and

$$\begin{aligned} \Delta(6,4) &= 6 - 4\lfloor 6/4 \rfloor = 2, \\ \Delta(2,2) &= 2 - 2\lfloor 2/2 \rfloor = 0. \end{aligned}$$

Suppose that

$$\begin{aligned} f_1^{(1)} &: (1,2) \to (\{1\}, \{2\}), \\ f_2^{(1)} &: (1,2) \to ((00), (11)), \\ f_3^{(1)} &: (1,2) \to ((01), (10)). \end{aligned} \tag{4.2.33}$$

Then we assign

$$J = f_1^{(1)}(2) = \{2\}, \quad a = f_2^{(1)}(2) = (11), \quad b = f_3^{(1)}(1) = (01),$$

and construct the codeword using (4.2.22)–(4.2.23):

$$v(a, b|J) = (\,01,\; 11\,).$$

Let us consider decoding of the message m' when $v = (\,11,\; 10\,)$. We discover that

$$J = \{1\}, \quad a = (11), \quad b = (10).$$

Hence, $s = |J| = 1$ and

$$\begin{aligned} m_J &= \left(f_1^{(1)}\right)^{-1}(\{1\}) = 1, \\ m_a &= \left(f_2^{(1)}\right)^{-1}((11)) = 2, \\ m_b &= \left(f_2^{(1)}\right)^{-1}((10)) = 2, \\ m' &= 4 + (1-1) \cdot 2 \cdot 2 + (2-1) \cdot 2 + (2-1) + 1 = 8, \end{aligned}$$

where (4.2.32), (4.2.33) were used.

In conclusion of this section we show an example of a UD code, which is probably the best known one concerning the value of the sum rate.

Example ([11]) Let $n = 7$, $|\mathcal{U}| = 12$, $|\mathcal{V}| = 47$. The codes $\mathcal{U}$ and $\mathcal{V}$ are given below in decimal notation.

$$\begin{aligned}\mathcal{U} = \{ &1, 4, 10, 19, 28, 31, 96, 99, 108, 117, 123, 126 \},\\ \mathcal{V} = \{ &6, 7, 9, 10, 13, 14, 15, 16, 18, 21, 22, 24, 25, 26,\\ &38, 39, 41, 42, 45,\\ &46, 47, 48, 50, 53, 54, 56, 57, 58, 61, 70, 71, 73,\\ &74, 77, 78, 79, 80,\\ &82, 85, 86, 88, 89, 90, 93, 109, 118, 121 \}.\end{aligned}$$

Then

$$(R_1, R_2) = (\log 12/7, \log 47/7) \approx (0.512138, 0.793513)$$

and $R_1 + R_2 \approx 1.305651$ (the KLWY lower bound claims that for all $R_1 \in (1/2, \log 3/2)$ there exist codes with the sum rate at least 1.292481).

4.2.5 *Coding for the T-User Binary Adder Channel*

A model of the T-user binary adder channel generalizes a model of the two-user binary adder channel.

Definition 4.10 The T-user binary adder channel is a channel with T binary inputs $x_1, ..., x_T$ and one output $z \in \{0, ..., T\}$ defined as the arithmetic sum of the inputs,

$$z = x_1 + ... + x_T.$$

A code $(\mathcal{U}_1, ..., \mathcal{U}_T)$, where $\mathcal{U}_t$ is a binary block code of length n and rate $R_t = \log |\mathcal{U}_t|/n$, $t = 1, ..., T$, is uniquely decodable if and only if

$$u_1 + ... + u_T \neq u'_1 + ... + u'_T, \quad \text{for all} \quad (u_1, ..., u_T) \neq (u'_1, ..., u'_T),$$
$$(u_1, ..., u_T), (u'_1, ..., u'_T) \in \mathcal{U}_1 \times ... \times \mathcal{U}_T.$$

A T-tuple $(R_1, ..., R_T)$ is regarded as achievable rate vector for UD codes if there exists a UD code with rates $R_1, ..., R_T$. The set $\mathcal{R}_{uT}$ consisting of all achievable rates is known as achievable rate region for UD codes. The sum rate of the T-user code is defined as

$$R_{sum}(T) = R_1 + ... + R_T.$$

The achievable rate region for the T-user binary adder channel under the criterion of arbitrarily small average decoding error probability gives an outer bound on $\mathcal{R}_{uT}$ and a direct extension of Theorem 4.1 leads to the following statement.

Proposition 4.8

$$\mathcal{R}_{uT} \subseteq \overline{\mathcal{R}}_T,$$

where $\overline{\mathcal{R}}_T$ consists of all permutations of the T-tuples belonging to the set

$$\overline{\mathcal{R}}'_T = \{ (R_1, ..., R_T) : 1 \geq R_1 \geq R_2 \geq ... \geq R_T \geq 0, \sum_{l=1}^{L} R_l \leq h(B_L), \text{ for all } L = 1, ..., T \},$$

where $h(B_L)$ denotes the entropy of the binomial distribution

$$B_L = (b_L(0), ..., b_L(L)); \ b_L(l) = \binom{L}{l} 2^{-L}, \ l = 0, ..., L,$$

i.e.,

$$h(B_L) = -\sum_{l=0}^{L} b_L(l) \log b_L(l).$$

Achievable rate region of the three-user binary adder channel under the criterion of arbitrarily small average decoding error probability is shown in Fig. 4.7.

The following result is obtained using the Stirling's approximation for the binomial coefficients.

Proposition 4.9 (Chang and Weldon 1979, [9]) *If $(R_1, ..., R_T) \in \mathcal{R}_{uT}$, then*

$$R_{sum}(T) \leq h(B_T),$$

where

$$\frac{1}{2} \log \frac{\pi T}{2} \leq h(B_T) \leq \frac{1}{2} \log \frac{\pi e(T+1)}{2}. \tag{4.2.34}$$

An important special case is obtained if we set $R_1 = ... = R_T = R$; in particular, if each code entering the T-tuple $(\mathcal{U}_1, ..., \mathcal{U}_T)$ consists of two codewords, i.e., $R = 1/n$.

We will present a construction of UD codes $(\mathcal{U}_1, ..., \mathcal{U}_T)$, where each code $\mathcal{U}_t$ consists of two codewords. At first, we reformulate the condition when the code is uniquely decodable.

Lemma 4.3 *A code $(\mathcal{U}_1, ..., \mathcal{U}_T)$, where each code $\mathcal{U}_t$ consists of two codewords, i.e., $\mathcal{U}_t = (u_t^{(0)}, u_t^{(1)})$, is uniquely decodable if and only if*

$$m\mathbf{D} = 0^n \implies m = (0, ..., 0),$$

where $\mathbf{D}$ is a $T \times n$ matrix (known as a difference matrix) whose t-th row, $d_t \in \{-1, 0, 1\}^n$ is defined by the equation $d_t = u_t^{(0)} - u_t^{(1)}$, $m \in \{0, 1\}^T$ is some binary vector, and 0^n is the all-zero vector of length n.

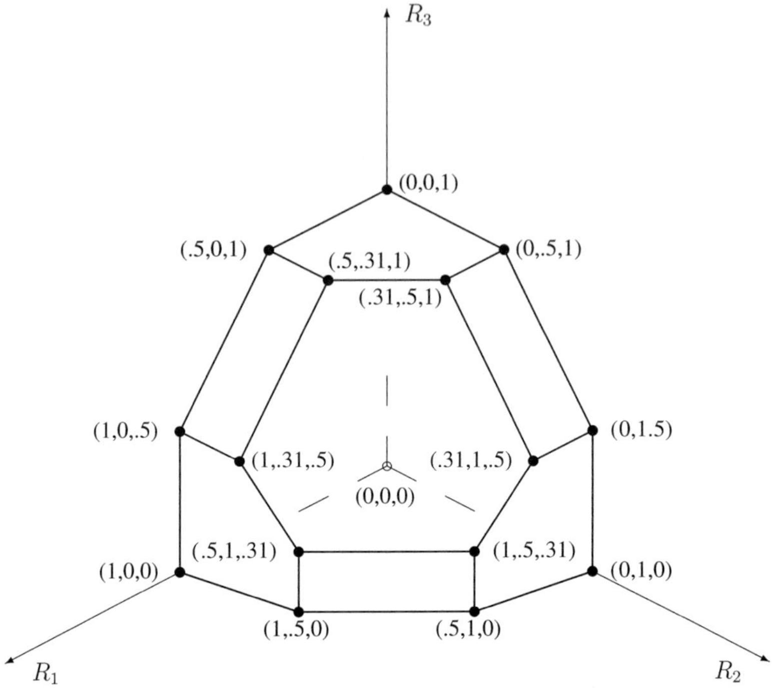

Fig. 4.7 Achievable rate region of the three-user binary adder channel under the criterion of arbitrarily small average decoding error probability

Obviously, if some $T \times n$ matrix matrix $\mathbf{D}$ with the entries -1, 0, and 1 is given, then we can construct codes $\mathcal{U}_1, \ldots, \mathcal{U}_T$ such that $\mathbf{D}$ coincides with their difference matrix.

Example 4.1 Let $T = 3$, $n = 2$, and

$$\mathbf{D}_1 = \begin{bmatrix} 1 & 1 \\ 1 & -1 \\ 1 & 0 \end{bmatrix}. \tag{4.2.35}$$

Then we construct the codes

$$\mathcal{U}_1 = (11, 00), \ \ \mathcal{U}_2 = (10, 01), \ \ \mathcal{U}_3 = (10, 00)$$

because

$$\begin{bmatrix} 11 \\ 10 \\ 10 \end{bmatrix} - \begin{bmatrix} 00 \\ 01 \\ 00 \end{bmatrix} = \mathbf{D}_1.$$

We will construct T-user UD codes such that the sum rate asymptotically achieves $h(B_T)$. Let

$$\mathbf{D}_0 = [1], \quad \mathbf{0}_0 = [0], \quad \mathbf{1}_0 = [1].$$

Then we note that the matrix defined in (4.2.35) satisfies the equation

$$\mathbf{D}_1 = \begin{bmatrix} \mathbf{D}_0 & \mathbf{D}_0 \\ \mathbf{D}_0 & -\mathbf{D}_0 \\ \mathbf{1}_0 & \mathbf{0}_0 \end{bmatrix}.$$

The following theorem claims that this iterative construction can be efficiently used for any $j \geq 1$.

Theorem 4.5 (Chang and Weldon 1979, [9]) *For any integer $j \geq 1$, the matrix*

$$\mathbf{D}_j = \begin{bmatrix} \mathbf{D}_{j-1} & \mathbf{D}_{j-1} \\ \mathbf{D}_{j-1} & -\mathbf{D}_{j-1} \\ \mathbf{1}_{j-1} & \mathbf{0}_{j-1} \end{bmatrix}. \tag{4.2.36}$$

defines a UD code of length

$$n_j = 2^j \tag{4.2.37}$$

for the binary adder channel with

$$T_j = (j+2)2^{j-1} \tag{4.2.38}$$

users, where $\mathbf{1}_j$ is the $2^j \times 2^j$ identity matrix and $\mathbf{0}_j$ is the $2^j \times 2^j$ zero matrix.

Proof We use the induction on j. For $j = 0$, $\mathbf{D}_j = [1]$, - this setting specifies a trivial single-user code of length 1. Assume that $\mathbf{D}_{j-1}$ defines a T_{j-1}-user UD code of length n_{j-1}. Note that $T_j = 2T_{j-1} + n_{j-1}$ and introduce a vector $m = (m_1, m_2, m_3) \in \{-1, 0, 1\}^{T_j}$, where $m_1, m_2 \in \{-1, 0, 1\}^{T_{j-1}}$ and $m_3 \in \{-1, 0, 1\}^{n_{j-1}}$, in such a way that

$$m\mathbf{D}_j = 0^{n_j}.$$

Then using (4.2.36) we have

$$\begin{aligned} m_1\mathbf{D}_{j-1} + m_2\mathbf{D}_{j-1} + m_3 &= 0^{n_{j-1}}, \\ m_1\mathbf{D}_{j-1} - m_2\mathbf{D}_{j-1} &= 0^{n_{j-1}}. \end{aligned}$$

Hence, the vector (m_1, m_2, m_3) consists of all zeroes and, by Lemma 4.3, $\mathbf{D}_j$ is a difference matrix of a T_j-user UD code of length n_j. □

Using (4.2.34) and (4.2.37)–(4.2.38) we conclude that

$$R_{sum}(T_j) = \frac{T_j}{n_j} = 1 + \frac{j}{2}$$

and

$$\frac{1}{2}\log\frac{\pi(j+2)2^{j-1}}{2} \le h(B_{T_j}) \le \frac{1}{2}\log\frac{\pi e((j+2)2^{j-1}+1)}{2}.$$

Hence,

$$\lim_{j\to\infty}\frac{R_{sum}(T_j)}{h(B_{T_j})} = 1$$

and we get the following statement.

Proposition 4.10 (Chang and Weldon (1979), [9]) *The T_j-user UD code specified by Theorem 4.5 has a sum rate asymptotically equal to the maximal achievable sum rate as T_j increases.*

Although this result looks very elegant, the coding problem of the adder channel is rather interesting for the case when the number of users is fixed. The real goal would be the following: to get asymptotically optimal UD codes for fixed T as the length of the codes goes to infinity.

The construction given by Theorem 4.5 was generalized in the work by Ferguson [20] in 1982, where it was shown that instead of $(I_{j-1}\, 0_{j-1})$ in D_i, any $(A\; B)$, if $\overline{A+B}$ is an invertible binary matrix (in which the overbar refers to reduction *modulo* 2), can be used. The construction described in Theorem 4.5 gives codes with length $N = 2^i$.

In 1984, Chang [8] proposed a shortening technique, which allows to construct binary UD codes of arbitrary length. This result was improved in 1986 by Martirossian [42] and best known binary UD codes were found.

Theorem 4.6 *Let $\mathbf{m} = (m_1, m_2, \ldots, m_T)$ be an arbitrary vector with $m_1 \in \{0, 1, -1\}$. Then $\mathcal{U}_1, \mathcal{U}_2, \ldots, \mathcal{U}_T$ is a UD code with T users if the condition $\mathbf{m}\widetilde{D} = 0^n$ holds iff for $\mathbf{m} = 0^T$, where 0^n is the n-dimensional all-zero vector.*

For the code of length n we will denote the difference matrix (DM) of a UD code $\mathcal{U}_1, \mathcal{U}_2, \ldots, \mathcal{U}_T$ by $\widetilde{D}_n = \{d_1^n, d_2^n, \ldots, d_n^n\}$ and the number of users by T_n, respectively.

Theorem 4.7 *If $\widetilde{D}_u$ and $\widetilde{D}_v$ are the DM of binary UD codes of length u and v ($u \leq v$), respectively, then the matrix*

$$\widetilde{D}_{u+v} = \left\| \begin{array}{c|c|c} -\widetilde{D}_v^u & \widetilde{D}_v & \\ \widetilde{D}_u & \widetilde{D}_u & \widetilde{A} \\ \widetilde{I}_u & \widetilde{0}_u & \widetilde{B} \end{array} \right\| \left| \begin{array}{cc|cccc} -d_1^v - d_2^v & \cdots d_u^v & d_1^v d_2^v & \cdots & & d_v^u \\ d_1^u d_2^u & \cdots d_u^u & d_1^u d_2^u & \cdots & d_u^u & \widetilde{A} \\ e_1 e_2 & \cdots e_u & 00 & \cdots & 0 & \widetilde{B} \end{array} \right|, \tag{4.2.39}$$

where $\widetilde{D}_v^u$ consists of the first columns of the matrix $\widetilde{D}_v$, $\widetilde{I}_u$ is the $u \times u$ identity matrix, and $\widetilde{A}$, $\widetilde{B}$ are any two matrices with elements from $\{0, 1, -1\}$, is the DM of a UD of length $u + v$.

Theorem 4.7 allows us to construct D_u from the given $D_{u_1}, D_{u_2}, \ldots, D_{u_s}$, where $u = u_1 + u_2 + \cdots + u_s$ for any s.

Now we will present n as $n = \sum_{k=0}^{s} n_k 2^k$, $s = [\log_2^n]$, $n_k \in \{0, 1\}$ and denote $n^{(j)} = \sum_{k=0}^{l} n_k 2^k$.

Thus, using Theorem 4.7 for the lengths $u = n_j n^{(j)}$, $v = \sum_{r=j+1}^{s} n_r 2^r$ and setting j equal to $s-1, s-2, \ldots, 1, 0$ successively, we will reduce the construction of $\widetilde{D}_n$ to one of constructing $\widetilde{D}_{2^0}, \widetilde{D}_{2^1}, \ldots, \widetilde{D}_{2^s}$. For this case the number of users is obtained successively from the relation $T_{u+v} = T_u + T_v + u$, i.e.,

$$\begin{aligned} T_n &= T_{2^s} + T_{n^{(s-1)}} + n^{(s-1)} \\ &= T_{2^s} + n_{s-1} T_{2^{s-1}} + T_{n^{(s-2)}} + n_{s-1} n^{(s-1)} + n^{(s-2)} \\ &= T_{2^s} + n_{s-1} T_{2^{s-1}} + \cdots + n_0 T_{s^0} + n_{s-1} n^{(s-1)} + n_{s-2} n^{(s-2)} + \cdots + n_0 \\ &= \sum_{k=0}^{s} n_k T_{2^k} + \sum_{l=0}^{s-1} n_l \sum_{k=0}^{1} n_k 2^k \end{aligned}$$

or as $T_{2^k} = (k+2)2^{k-1}$ (see [3], the same result is also obtained from Theorem 4.7) then

$$T_n = \sum_{k=0}^{s} n_k (k+2) 2^{k-1} + \sum_{l=0}^{s-1} n_l \sum_{k=0}^{1} n_k 2^k. \tag{4.2.40}$$

Let us denote the number of users of the code of length n constructed in [8] by T_n'. If we express n as $n = 2^l - j, 0 < j < 2^{l-1}$, then it will be given by the formula

$$T_n' = (i+1)2^{l-1} - j - \sum_{k=0}^{i-2} j_k (k+2) 2^{k-1}, \tag{4.2.41}$$

where $j - 1 = \sum_{k=0}^{i-2} j_k 2^k$, $j_k \in \{0, 1\}$.

Lemma 4.4 *Comparing (4.2.40) and (4.2.41), we have*

$$T_n - T'_n = \sum_{l=0}^{\alpha-1} n_l \sum_{k=0}^{1} n_k 2^k \geq 0. \tag{4.2.42}$$

Now we will introduce the results obtained by Khachatrian and Martirossian. These results were reported during the First Armenian–Japanese Colloquium on Coding Theory, Diligan, Armenia, September 1986 and are finally published in [35]. A construction of non-basic UD codes constructed from binary UD codes given in some special way is represented here. This construction is based on the following.

Lemma 4.5 *Let $\mathcal{U}_1, \ldots, \mathcal{U}_T$ be a UD set, and $\{\{u_1\}, \ldots, \{u_{T_1}\}\}$ be a split of this set into T_1 non-empty subsets. Then the system $\{\mathcal{U}_1^1, \ldots, \mathcal{U}_T^1\}$ will also be a UD, where $\mathcal{U}_i^1$ is the set of all binary vectors that belong to the set of all possible sums*

$$\left(s_{r_1^{(i)}} + x_{r_2^{(i)}} + \cdots + x_{r_{|u_i|}^{(i)}}\right)$$

where

$$x_{r_j^i} \in \mathcal{U}_{r_1^{(i)}} \qquad j = \{1, \ldots, |u_i|\}, \{u_i\} = \left\{\mathcal{U}_{r_1^{(i)}}, \ldots, \mathcal{U}_{r_{|u_i|}^{(i)}}\right\}$$

and $|u_i|$ is the cardinality of the set u_i

$$\sum_{i=1}^{T} |u_i| = T.$$

The proof of the lemma follows directly from the definition of a UD system. The obtained T_1-user UD system will be called to be a T_1-*conjugate system* in respect to T-user $\{\mathcal{U}_1, \ldots, \mathcal{U}_T\}$ (in short $(T_1 - T)$ system).

The following two corollaries are deduced from Lemma 4.5.

Corollary 4.1 *Let $\mathcal{U}_1, \ldots, \mathcal{U}_T$ be a UD set and let*

$$\{\mathcal{U}_{i_1} \cap \mathcal{U}_{i_2} \cap \cdots \cap \mathcal{U}_{i_r}\} \neq \emptyset.$$

Then the $(T - r + 1)$-user system $(\mathcal{U}_0, \mathcal{U}_{j_1}, \ldots, \mathcal{U}_{j_{T-r}})$, $j_1, j_2, \ldots, j_{T-r} \in \{i_1, i_2, \ldots, i_r\}$ is also a UD, where $\mathcal{U}_0 = \mathcal{U}_{i_1} \cup \mathcal{U}_{i_2} \cup \cdots \cup \mathcal{U}_{i_r}$.

Corollary 4.2 *Let*

$$D = [d_{i_1}\, d_{i_2}\, \cdots\, d_{i_k}]^T$$

be the submatrix of the DM for a binary UD system. If each column in D has no more than one non-zero element then the corresponding $\mathcal{U}_{i_1}, \ldots, \mathcal{U}_{i_k}$ codes can be combined into one code with the cardinality equal to 2^k such that the obtained $(T - k + 1)$-user system is also UD.

The last corollary allows us to construct $(T-k+1)$-user UD codes from T-user ones with the same sum rate, which is obviously more favorable since we have the same sum rate for a smaller number of users.

The UD codes will be constructed on the basis of some initial binary UD codes and Lemma 4.5. Now we will try to explain the problem if initial binary UD codes of what kind are constructed. Two cases will be considered here.

First case ($n=2^k$). The construction is implemented iteratively on k. On the kth step 2^{k-1} matrices $D^1_{2^k}, \dots, D^{2^{k-1}}_{2^k}$ are constructed.

At the first step ($k=1$)

$i=1$

$$D^1_{2^1} = \left|\begin{array}{cc} 1 & 1 \\ 1 & -1 \\ \hline 1 & 0 \end{array}\right| = \left|\begin{array}{c} A^1_{2^1} \\ \\ \hline B^1_{2^1} \end{array}\right| = \left|\begin{array}{c} a_{21} \\ a_{22} \\ \hline b_{21} \end{array}\right|$$

At the second step ($k=2$) there are two matrices.

$i=1$

$$D^1_{2^2} = \left|\begin{array}{cccc} 1 & 1 & 1 & 1 \\ -1 & -1 & 1 & 1 \\ \hline 1 & -1 & 0 & 0 \\ 0 & 0 & 1 & -1 \\ 1 & 1 & 0 & 0 \\ 1 & 0 & 0 & 0 \\ 0 & 0 & 1 & 0 \end{array}\right| = \left|\begin{array}{c} A^1_{2^2} \\ \\ \hline \\ \\ B^1_{2^2} \\ \\ \\ \end{array}\right| = \left|\begin{array}{cc} a_{21} & a_{21} \\ -a_{21} & a_{21} \\ \hline a_{22} & 0 \\ 0 & a_{22} \\ a_{21} & 0 \\ B^1_2 & 0 \\ 0 & B^1_2 \end{array}\right|$$

$i=2$

$$D^1_{2^2} = \left|\begin{array}{cccc} 1 & 1 & 1 & 1 \\ 1 & -1 & 1 & -1 \\ -1 & -1 & 1 & 1 \\ -1 & 1 & 1 & -1 \\ \hline 1 & 1 & 0 & 0 \\ 0 & 0 & 1 & -1 \\ 1 & 0 & 0 & 0 \\ 0 & 0 & 1 & 0 \end{array}\right| = \left|\begin{array}{c} \\ A^2_{2^2} \\ \\ \\ \hline \\ B^2_{2^2} \\ \\ \\ \end{array}\right| = \left|\begin{array}{cc} a_{21} & a_{21} \\ a_{22} & a_{22} \\ -a_{21} & a_{21} \\ a_{22} & a_{22} \\ \hline a_{21} & 0 \\ 0 & a_{22} \\ B^1_2 & 0 \\ 0 & B^1_2 \end{array}\right|$$

At the kth step $i=1,2,\dots,2^{k-1}$

$$
D^i_{2^k} = \begin{array}{c|c|c|}
 & a_{2^{k-1}1} & a_{2^{k-1}1} \\
1 & \vdots & \vdots \\
 & a_{2^{k-1}i} & a_{2^{k-1}i} \\
\hline
 & -a_{2^{k-1}1} & a_{2^{k-1}1} \\
2 & \vdots & \vdots \\
 & -a_{2^{k-1}i} & a_{2^{k-1}i} \\
\hline
 & a_{2^{k-1}(i+1)} & 0 \\
3 & \vdots & \vdots \\
 & a_{2^{k-1}2^{k-1}} & 0 \\
\hline
 & 0 & a_{2^{k-1}(i+1)} \\
4 & \vdots & \vdots \\
 & 0 & a_{2^{k-1}2^{k-1}} \\
\hline
 & a_{2^{k-1}1} & 0 \\
5 & \vdots & \vdots \\
 & a_{2^{k-1}\lfloor\frac{i+1}{2}\rfloor^*} & 0 \\
\hline
 & 0 & a_{2^{k-1}\lfloor\frac{i+3}{2}\rfloor} \\
6 & \vdots & \vdots \\
 & 0 & 0 \\
\hline
7 & B^{2^k-2}_{2^{k-1}} & 0 \\
\hline
8 & 0 & B^{2^k-2}_{2^{k-1}}
\end{array}
= \begin{array}{|c|}
A^i_{2^k} \\
\hline
B^i_{2^k}
\end{array}
\tag{4.2.43}
$$

For the sake of convenience the rows of the matrix $D^i_{2^k}$ are split into eight blocks and numbered. Let us denote the number of rows in $D^i_{2^k}$ (the number of users) by $T^i_{2^k}$. It is easy to see that for the matrices constructed by (4.2.43) the following recurrence relation holds:

$$
T^i_{2^k} = T^{2^k-2}_{2^{k-1}} + T^{2^k-2}_{2^{k-1}} + i. \tag{4.2.44}
$$

It follows, particularly, frm (4.2.44) that

$$
T^{2^k-2}_{2^k} = (k+2)2^{k-1} \quad \text{and} \quad T^i_{2^k} = (k=1)2^{k-1} + i.
$$

Theorem 4.8 (Khachatrian and Martirossian 1998, [35]) *For all k and i, $1 \le i \le 2^{k-1}$, the matrix $D^i_{2^k}$ is a DM for a binary UD set of codes.*

Now new binary UD codes can be constructed by regrouping the rows of the matrix (see [35]). The results are summarized by

Theorem 4.9 (Khachatrian and Martirossian 1998, [35]) *For UD codes $R_{SUM}(T)$ satisfies the following relation:*

$$(i)\ R_{SUM}(T) \geq \frac{(k+2)2^k + 2s}{2^{k+1}}, \qquad r = 0$$

$$(ii)\ R_{SUM}(T) \geq \frac{(k+2)2^k + 2s - 1 + \log_2 3}{2^{k+1}}, \qquad r = 1$$

$$(iii)\ R_{SUM}(T) \geq \frac{(k+2)2^k + 2s - 1 - k + \log_2(2^{k+1}+1)}{2^{k+1}}, \quad r = 2$$

The table below gives the best known T-user UD codes based on the results in [35].

T	$\sum R_i$	n	T	$\sum R_i$	n	T	$\sum R_i$	n	T	$\sum R_i$	n
2	1.2924	2	14	2.5680	16	25	3.0183	32	37	3.2683	32
3	1.5283	3	15	2.6250	16	26	3.0326	32	38	3.2826	32
4	1.6666	3	16	2.6666	12	27	3.0625	32	39	3.3125	32
5	1.8305	4	17	2.6930	16	28	3.0808	32	40	3.3308	32
6	2.0000	4	18	2.7500	16	29	3.0951	32	41	3.3451	32
7	2.0731	8	19	2.7586	16	30	3.1250	32	42	3.3750	32
8	2.1666	6	20	2.8180	16	31	3.1433	32	43	3.3933	32
9	2.2500	8	21	2.8750	16	32	3.1666	24	44	3.4076	32
10	2.3231	8	22	2.9116	16	33	3.1875	32	45	3.4375	32
11	2.3962	8	23	2.9430	16	34	3.2058	32	46	3.4358	32
12	2.5000	8	24	3.0000	16	35	3.2201	32	47	3.4701	32
13	2.5366	16				36	3.2500	32	48	3.5000	32

4.3 On the T-User q-Frequency Noiseless Multiple-Access Channel without Intensity Information

A specific noiseless multiple-access channel model, the T-user q-frequency multiple-access channel without intensity information is studied in this section. Information-theoretic bounds on the transmission rate for this model is presented. Constructive coding schemes are given for the channel which achieve zero error probability and whose rate sum is close to the information-theoretic bounds. Although the problem is formulated in terms of frequencies, the results are applicable to any signaling scheme where q orthogonal signals are used in each signaling interval including time partitioning of the interval.

4.3.1 Introduction

In this section a specific class of T-user noiseless multiple-access channels is studied. This class contains the T-user noiseless binary adder channels as a special case [9, 21, 28]. Both, information-theoretic bounds on the achievable rate sum and constructive coding schemes, are presented.

The following description of a general T-user multiple-access communication system assumes block coding. Every n time units, the channel accepts T input n-vectors $X_1, X_2, \ldots, X_T$ provided by the encoders. The channel emits a single n-vector Y in accordance with some prescribed conditional probability distribution $P_{Y|X_1,X_2,\ldots,X_T}(y|x_1, x_2, \ldots, x_T)$. The information to be transmitted over the channel by these vectors is a set of integers $m_1, m_2, \ldots, m_T$.

The ith user provides the ith integer m_i which is chosen from a uniform distribution over the set $\{1, 2, \ldots, 2^{nR_i}\}$. R_i is called the *rate* for the ith user. The T users choose their integers independently of one another. The ith encoder seeing only m_i produces the vector x_i which is one of 2^{nR_i} codewords, one codeword for each possible value of m_i. R_i, the rate for the ith code, is measured in units of bits per channel use. The decoder upon observing Y must decide which set of integers was produced by the T users. The estimate of m_i that the decoder generates will be denoted $\hat{m}_i$ The probability of error for the system, P_e, is defined to be

$$P_e = 1 - P(\hat{m}_1 = m_1, \hat{m}_2 = m_2, \ldots, \hat{m}_t = m_t). \tag{4.3.1}$$

In the specific model we consider for this channel, the aim is to find codewords for each of the T users and a decoding rule such that the probability of error is negligibly small (or better still zero). We measure the goodness of the system by the set of rates $R_1, R_2, \ldots, R_T$ for these codes, and in particular we wish to make the rate sum $R_{\text{sum}} = R_1 + R_2 + \cdots + R_T$ as large as possible.

For ease of future reference, the T-user q-frequency multiple-access channel without intensity knowledge will be referred to as the *A channel*. Each component of each of the T input vectors X_i is chosen from the common alphabet $\{f_1, f_2, \ldots, f_M\}$. For the A channel, the output Y at each time instant is a symbol which identifies which subset of frequencies occurred as inputs to the channel at that time instand but *not* how many of each frequency occurred. One representation for the output symbol at each time instant is an q-dimensional vector. For the A channel this vector has binary components (0, 1), the jth component being a one if and only if one or more channel inputs are equal to f_j, $j = 1, 2, \ldots, q$. The table below shows the outputs of the A channel using this representation for $T = 3$ and $q = 2$.

Inputs			Output A Channel
X_{1i}	X_{2i}	X_{3i}	(f_1, f_2)
f_1	f_1	f_1	(1, 0)
f_1	f_1	f_2	(1, 1)
f_1	f_2	f_1	(1, 1)
f_1	f_2	f_2	(1, 1)
f_2	f_1	f_1	(1, 1)
f_2	f_1	f_2	(1, 1)
f_2	f_2	f_1	(1, 1)
f_2	f_2	f_2	(0, 1)

A three-input two-frequency model

In this model we assume that there are no errors due to noise, phase cancellation of signals, etc. A more sophisticated model taking such errors into account could easily be developed. One method would be to use a noisy channel in cascade with out noiseless channel. It is our contention, however, that although the details are different, the basic ideas are the same in the noisy and noiseless cases. Thus in this section we pursue the noiseless model because of its simplicity. In this model we insist that the probability of decoding error be equal to zero for our code constructions. Thus the resulting output vectors must be uniquely decodable.

Although the problem has been formulated in terms of frequencies, the results are applicable to any signaling scheme where q orthogonal signals are used in each signaling interval. Thus the results apply to pulse position modulation (PPM) where the signaling interval is partitioned into q time slots.

The format of the section is as follows. Sect. 4.3.2 discusses information-theoretic bounds for the channel model. Sect. 4.3.3 is concerned with constructive coding schemes for the A channel.

4.3.2 *Information-Theoretic Bounds*

The capacity region for a multiple-access channel is the set of rate points $(R_1, R_2, \ldots, R_T)$ for which codes exist that lead to negligibly small error probability. Although information-theoretic expressions are known for the outer surface of this region, the region is in general a complicated T-dimensional convex body which is difficult to envision and somewhat complicated to describe. One aspect of this capacity region is that the sum of the rates,

$$R_{\text{sum}} = R_1 + R_2 + \cdots + R_T, \tag{4.3.2}$$

is upper bounded by the joint mutual information

$$R_{\text{sum}} \le \max I(X_1, X_2, \ldots, X_T; Y) \triangleq C_{\text{sum}}(T, q), \tag{4.3.3}$$

where the maximum is taken over all product distributions on the input RV's $X_1, X_2, \ldots, X_T$. Since the mutual information can be written as

$$I(X_1, X_2, \ldots, X_T; Y) = H(Y) - H(Y|X_1, X_2, \ldots, X_T), \tag{4.3.4}$$

and since $H(Y|X_1, X_2, \ldots, X_T) = 0$ for the A channel. $C_{\text{sum}}(T, q)$ can be written as

$$C_{\text{sum}}(T, q) = \max H(Y), \tag{4.3.5}$$

where again the maximum is taken over the same set of input distributions. Our aim is to calculate $C_{\text{sum}}(T, q)$ for the A channel. We use a superscript $C_{\text{sum}}^{(A)}$ to indicate this.

It is tempting to guess that because of the symmetry of the channel each user should use a uniform distribution over the q frequencies in order to maximize $H(Y)$. This line of thought is easily shown to be incorrect by considering the T-user 2-frequency A channel. There are three outputs for for this channel, two of which occur with probability $(1/2)^T$, a quantity which approaches zero as T approaches infinity for a uniform distribution on the outputs. However, it is clear that $C_{\text{sum}}^{(A)}(T, 2) \geq 1$ for all $T \geq 1$, since one can always achieve the output entropy $H(Y) = 1$ by letting one user use a uniform distribution while all other users use a probability distribution which puts all the mass on one of the frequencies (say f_1). Thus an integral part of the calculation of $C_{\text{sum}}(T, q)$ is concerned with the question of finding the input product distribution which maximizes the output entropy. Unfortunately Chang and Wolf [10] were not able to find a general analytic solution for the optimizing distribution and had to resort to a computer search to obtain some of their results.

The following are the results that have been obtained by Chang and Wolf [10] concerning the quantity $C_{\text{sum}}^{(A)}$. The results which were arrived at by a computer search are prefaced by the word (computer). The other results are analytic in nature.

Theorem 4.10 *For the T-user 2-frequency A channel, all users utilize the same probability distribution to maximize the output entropy.*

Proof Let P_{i1} be the probability that symbol f_1 is chosen by the ith user, $i = 1, 2, \ldots, T$. Then by definition

$$C_{\text{sum}}^{(A)}(T, 2) = \max[(-\log_2 e)A_0],$$

where

$$A_0 = A_1 \ln A_1 + A_2 \ln A_2 + (1 - A_1 - A_2) \ln(1 - A_1 - A_2),$$

$$A_1 = \prod_{i=1}^{T} P_{i1}, \qquad \text{and} \qquad A_2 = \prod_{i=1}^{T} (1 - P_{i1}).$$

By differentiating with respect to P_{j1}, for $j = 1, 2, \ldots, T$ we obtain

$$
\begin{aligned}
0 = \frac{\partial A_0}{\partial P_{j1}} &= \frac{A_1 \ln A_1}{P_{j1}} + \frac{A_1}{P_{j1}} - \frac{A_2 \ln A_2}{1 - P_{j1}} - \frac{A_2}{1 - P_{j1}} \\
&+ \left[-\frac{A_1}{P_{j1}} + \frac{A_2}{1 - P_{j1}} \right] \ln(1 - A_1 - A_2) - \frac{A_1}{P_{j1}} + \frac{A_2}{1 - P_{j1}} \\
&= \frac{A_1 \ln A_1 - A_1 \ln(1 - A_1 - A_2)}{P_{j1}} - \frac{A_2 \ln A_2 - A_2 \ln(1 - A_1 - A_2)}{1 - P_{j1}}.
\end{aligned}
$$

Therefore

$$
\frac{P_{j1}}{1 - P_{j1}} = \frac{A_2 \ln A_2 - A_2 \ln(1 - A_1 - A_2)}{A_1 \ln A_1 - A_1 \ln(1 - A_1 - A_2)} = D,
$$

which implies

$$
P_{j1} = \frac{D}{1 + D}, \qquad \text{for } j = 1, 2, \ldots, T.
$$

□

Theorem 4.11 *For the 2-user q-frequency A channel,*

$$
C_{sum}^{(A)}(2, q) = 2 \log_2 q + \frac{1}{q} - 1, \qquad \textit{for } q \geq 2. \tag{4.3.6}
$$

Proof Let P_{ij} be the probability that the ith user ($i = 1, 2$) uses the jth frequency ($j = 1, 2, \ldots, q$). Then the entropy of the output is

$$
H(Y) = -\sum_{j=1}^{q} P_{1j} P_{2j} \log P_{1j} P_{2j} - \sum_{1 \leq k \neq j \leq q} (P_{1k} P_{2j} + P_{1j} P_{2k}) \cdot \log(P_{1k} P_{2j} + P_{1j} P_{2k}).
$$

Using Lagrange multipliers we find the extremum of the quantity

$$
H(Y) - \lambda_1 \left(\sum_{j=1}^{q} P_{1j} - 1 \right) - \lambda_2 \left(\sum_{j=1}^{q} P_{2j} - 1 \right)
$$

to be $P_{1j} = P_{2j} = (1/q)$, for all $j = 1, 2, \ldots, q$. Substituting this result into the expression for $H(Y)$ we obtain the desired result. □

Theorem 4.12 *For the 3-user q-frequency A channel,*

$$
C_{sum}^{(A)}(3, q) = 3 \log_2 q + \log_2 6 \left(\frac{1}{q^2} - 1 \right), \qquad \textit{for } q \geq 3. \tag{4.3.7}
$$

For the T-user q-frequency A channel, the maximum output entropy is achieved when all users utilize a common distribution. For $T \leq q$, this is the uniform distribution. For $T > q$, a non-uniform distribution yields the maximum output entropy.

The non-uniform distribution places heavier weight on one frequency and distributed the remaining weight evenly among the other frequencies. For fixed q, $C_{\text{sum}}^{(A)}(T, q)$ increases with increasing T until it reaches its maximum value at a value of T which is an integer close to $q \cdot \ln 2$. The maximum value of $C_{\text{sum}}^{(A)}(T, q)$ is greater than or equal to $q - 1/2$ and less than q. As T further increases, $C_{\text{sum}}^{(A)}(T, q)$ decreases until, for very large T, $C_{\text{sum}}^{(A)}(T, q)$ asymptotically approaches $q - 1$.

Theorem 4.13 *For $T \leq q$ (where the computer results indicate that the optimizing distribution is the uniform distribution for all users),*

$$C_{sum}^{(A)}(T, q) = \sum_{i=1}^{T} \frac{\binom{q}{i} a_i}{q^T} \log_2 \frac{q^T}{a_i}, \tag{4.3.8}$$

where

$$a_i = i^T - \sum_{j=1}^{i-1} \binom{i}{j} a_j, \qquad i \geq 2, \quad a_1 = 1, \quad a_T = T!,$$

and

$$\sum_{i=1}^{T} \binom{q}{i} a_i = q^T.$$

Proof There are $\binom{q}{i}$ ways in which exactly i of the q frequencies can be received. Let a_i be the number of possible distinct inputs that correspond to a particular output in which i frequencies were received. Then

$$a_i = i^T - \sum_{j=1}^{i-1} \binom{i}{j} a_j, \qquad i \geq 2,$$

since of the i^T possible input patterns that could be generated by T users sending one of i frequencies we must delete those input patterns that result in strictly less than i received frequencies. Also $a_1 = 1$ and $a_T = T!$. The result follows from the fact that each possible input pattern occurs with probability q^{-T}. □

Remark The 2-user binary adder channel is identical to the 2-user 2-frequency A channel.

4.3.3 Construction of Codes for the A Channel

Very simple code constructions achieve rate sums close to $C_{\text{sum}}^{(A)}(T, q)$ for a wide range of values of T and q. As previously mentioned, all constructions in this section yield zero probability of error. The proof of this fact is by displaying the output vectors for all combinations of inputs and showing that they are unique. These proofs are omitted.

4.3.3.1 Construction (A-1)

The first construction, (A-1), is applicable to any values of (T, q) for which $T \geq q-1$. It results in a rate sum of $q - 1$ which is very close to $C_{\text{sum}}^{(A)}(T, q)$. The construction is first explained for the case of $T = q - 1$, then for arbitrary $T = n(q - 1)$, n a positive integer, and then for arbitrary $T \geq q - 1$.

$T = q - 1$. Let the ith code, $i = 1, 2, \ldots, T$, consist of two codewords of block length 1, f_1 and $f_{i=1}$. The output of the channel clearly identifies which codeword was sent by each user.

$T = n(q - 1)$. Each user has two codewords of block length n. One codeword for each user is the symbol $f - 1$ repeated n times. The other codeword consists of $n - 1$ repetitions of f_1 and one component from the set $(f_2, f_3, \ldots, f_q)$. The position and value of this one special component identify the individual user. More specifically, identifying each of the $n(q - 1)$ users by the pair (j, k), $j = 1, 2, \ldots, q - 1$, $k = 1, 2, \ldots, n$ and denoting the code for the (j, k)th user by $\mathcal{U}_{j,k}$ we have

$$
\begin{aligned}
\mathcal{U}_{1,1} &= \{(f_1, f_1, \ldots, f_1), (f_1, f_1, \ldots, f_2)\}, \\
\mathcal{U}_{1,2} &= \{(f_1, f_1, \ldots, f_1), (f_1, f_1, \ldots, f_2, f_1)\}, \\
&\vdots \\
\mathcal{U}_{1,n} &= \{(f_1, f_1, \ldots, f_1), (f_2, f_1, \ldots, f_1)\}, \\
&\vdots \\
\mathcal{U}_{q-1,1} &= \{(f_1, f_1, \ldots, f_1), (f_1, f_1, \ldots, f_q)\}, \\
\mathcal{U}_{q-1,2} &= \{(f_1, f_1, \ldots, f_1), (f_1, f_1, \ldots, f_q, f_1)\}, \\
&\vdots \\
\mathcal{U}_{q-1,n} &= \{(f_1, f_1, \ldots, f_1), (f_q, f_1, \ldots, f_1)\}.
\end{aligned}
$$

For $(n - 1)(q - 1) < T \leq n(q - 1)$, we can combine codes in $\{\mathcal{U}_{1,j_1} : 1 \leq j_1 \leq n\}$, $\{\mathcal{U}_{2,f_2} : 1 \leq j_2 \leq n\}, \ldots, \{\mathcal{U}_{q-1,f_{q-1}} : 1 \leq j_{q-1} \leq n\}$, so that the total number of codes will decrease, but the total rate sum does not change.

An example of code construction (A-1) follows for the case of $q = 3$ and $T = 2, 3, 4$. Here $\mathcal{U}_i$ denotes the codewords for the ith user.

$T = 2:$

$$\mathcal{U}_1 = \{(f_1), (f_2)\}, \qquad \mathcal{U}_2 = \{(f_1), (f_3)\}.$$

$T = 4:$

$$\mathcal{U}_1 = \{(f_1, f_1), (f_1, f_2)\}, \qquad \mathcal{U}_1 = \{(f_1, f_1), (f_2, f_1)\},$$
$$\mathcal{U}_3 = \{(f_1, f_1), (f_1, f_3)\}, \qquad \mathcal{U}_4 = \{(f_1, f_1), (f_3, f_1)\}.$$

$T = 3$: (obtained by combining $\mathcal{U}_3$ and $\mathcal{U}_4$, when $T = 4$)

$$\mathcal{U}_1 = \{(f_1, f_1), (f_1, f_2)\}, \qquad \mathcal{U}_1 = \{(f_1, f_1), (f_2, f_1)\},$$
$$\mathcal{U}_3 = \{(f_1, f_1), (f_1, f_3), (f_3, f_1), (f_3, f_3)\}.$$

4.3.3.2 Construction (A-2)

The next construction, (A-2), holds for $T = 2$ and any $q \geq 1$. The rate sum for this construction is given by

$$R_{\text{sum}} = \frac{1}{2} \log_2[q(q^2 - q + 1)].$$

Both users utilize codes of block length equal to 2. The first user's code consists of the pairs $(f_1, f_1), (f_2, f_2), \ldots, (f_q, f_q)$. The second user's code consists of all pairs (f_i, f_j) where $i \neq j$ except for $i = j = 1$.

The following is an example of this construction with $q = 3$.

$$\mathcal{U}_1 = \{(f_1, f_1), (f_2, f_2), (f_3, f_3)\},$$
$$\mathcal{U}_2 = \{(f_1, f_1), (f_1, f_2), (f_1, f_3), (f_2, f_1), (f_2, f_3), (f_3, f_1), (f_3, f_2)\}.$$

4.3.3.3 Construction (A-3)

Another construction, (A-3), applies to the $T = 2$ user case for arbitrary $M \geq 2$. The rate sum for this construction is

$$R_{\text{sum}} = \begin{cases} 2\log_2 \dfrac{(q+1)}{2}, & q \text{ odd}, \\[2ex] \log_2 \dfrac{(q+2)q}{4}, & q \text{ even}. \end{cases}$$

This construction uses codes of block length 1. The codewords are

$$\mathcal{U}_1 = \{(f_1), (f_2), \ldots, (f_{\lfloor (q+1)/2 \rfloor})\},$$
$$\mathcal{U}_2 = \{(f_1), (f_{\lfloor (q+3)/2 \rfloor}), \ldots, (f_q)\}.$$

Note that Construction (A-2) gives a greater rate sum than Construction (A-3) if and only if $a \leq 10$. Furthermore, the ratio of the rate sum of Construction (A-3) to $C_{\text{sum}}^{(A)}(2, q)$ approaches 1 as q gets large.

4.3.4 Evaluation of the Asymptotics of the Summarized Capacity of a T-User q-Frequency Noiseless Multiple-Access Channel

In this section the best known estimates for the asymptotics of the summarized capacity of an A channel are given. It is shown that the uniform input distribution is asymptotically optimal for a unique value of the parameter λ, $T = \lambda q$, $0 < \lambda < \infty$, namely, $\lambda = \ln 2$, and is not such in all other cases.

An input of an A channel consists of T, $T \geq 2$ independent users. At each time instant (time is discrete) each of the users transmits a symbol from the alphabet $\{1, 2, \ldots, q\}$, $q \geq 2$ using his own probability distribution. An output of an A channel is a binary sequence of length q whose mth position contains the symbol 0 if and only if none of the users transmits the symbol m. We will also refer to alphabet symbols as frequencies, and to users, as stations.

Denote by $X = (X_1, \ldots, X_T)$ a q-ary sequence of length T at the channel input at a fixed time instant and by $Y = (Y_1, \ldots, Y_q)$ a binary sequence at the channel output. Then the summarized capacity of an A channel is

$$C_{\text{sum}}(T, q) = \max H(Y), \tag{4.3.9}$$

where the maximum is taken over all channel input distributions of independent RV's $X_1, \ldots, X_T$:

$$P_X = P_{X_1} \cdots P_{X_T}. \tag{4.3.10}$$

The output distribution corresponding to P_X is denoted by Q_Y. By definition,

$$H(Y) = -\sum_{y \in \{0,1\}^q} Q_y \log Q_y.$$

Put

$$C_{\text{sum}}(\lambda) = \lim_{q \to \infty} \frac{C_{\text{sum}}(\lambda q, q)}{q}, \qquad 0 < \lambda < \infty.$$

The existence of the limit and the convexity of the function $C_{\text{sum}}(\lambda)$, i.e.

$$C_{\text{sum}}(\alpha\lambda_1 + \beta\lambda_2) \geq \alpha C_{\text{sum}}(\lambda_1)\beta C_{\text{sum}}(\lambda_2), \qquad \alpha + \beta = 1, \quad \alpha, \beta \geq 0,$$

can easily be proved by the corresponding frequency division multiplex (for instance, to prove the convexity, it suffices to consider the case where the first $\lambda_1 \alpha q$ stations transmit in the optimal way the first αq frequencies, and the last $\lambda_2 \beta q$ stations, the last βq frequencies). The cases $\lambda = 0$ and $\lambda = \infty$ are described at the end of the section.

In [10], a formula was given for the entropy $H_{\text{unif}}(Y)$ of the output distribution under the uniform distribution of all $X_1, \ldots, X_T$. In [57], the asymptotic behavior of this entropy was found, i.e., for $T = \lambda q$, $0 < \lambda < \infty$, the quantity $H_{\text{unif}}(\lambda) = \lim_{q\to\infty} \frac{H_{\text{unif}}(Y)}{q}$ was computed:

$$H_{\text{unif}}(\lambda) = h(1 - e^{-\lambda}), \qquad h(u) = -u \log u - (1-u)\log(1-u). \tag{4.3.11}$$

It was also shown there that for $\lambda = \ln 2$ the equality

$$C_{\text{sum}}(\ln 2) = H_{\text{unif}}(\ln 2) = 1$$

holds.

An attempt to compute $H_{\text{unif}}(\lambda)$ was taken in [23], but formula (4.3.14) obtained there and, therefore, Theorem 1.2 are wrong (a mistake was due to the incorrect use of the approximation (4.2.12) for binomial coefficients).

Also, in [10], for $T \geq q - 1$ an example is given of an input distribution such that the entropy of the output distribution equals $q - 1$, namely, $P_{(X_t = t)} = P_{(X_t = q)} = 1/2$, $t = 1, \ldots, q - 1$, and $P_{(X_t = q)} = 1$, $t = q, \ldots, T$. Even this example shows that for $T > q$ the uniform distribution is obviously bad; so, it was suggested to use a (common) distribution distorted in favor of one distinguished frequency and uniform on the others.

In [22], for fixed q (and, hence, for $\lambda = infty$), a specific distorted distribution is considered, which was introduced in [25] for the study of another parameter of an A channel, namely,

$$P_{(X_T = q)} = 1 - \frac{(q-1)\ln 2}{q}, \qquad P_{(X_t = m)} = \frac{\ln 2}{T} \tag{4.3.12}$$

for all m from 1 to $q - 1$, $t = 1, \ldots, T$.

Denote the entropy Y for this distribution $H_{\text{distort}}(Y)$ (note that for $T = q \ln 2$ the distorted and the uniform distributions coincide, and the distorted distribution is defined for $T \geq q \ln 2$ only). One can easily compute the asymptotic behavior of this entropy, i.e., find the quantity $H_{\text{distort}}(\lambda) = \lim_{q\to\infty} \frac{H_{\text{distort}}(Y)}{q}$ as $T = \lambda q$.

Proposition 4.11 *We have the equality*

$$H_{distort}(\lambda) = 1, \qquad \ln 2 \leq \lambda < \infty. \tag{4.3.13}$$

Proof This statement, as well as many other similar statements given below, are proved using the same scheme described in [57]. Therefore, we only present a complete proof for an input distribution which was not considered before (see Proposition 4.13). Here, we only explain why one should expect the answer (4.3.13). Indeed by (4.3.12), for a distorted distribution, the mean number of stations that send the frequency q to the channel equals $T - (q - 1)\ln 2$, and the other stations use frequencies different from q equiprobable, i.e., for this $T' = (q - 1)\ln 2$ stations and $q' = q - 1$ frequencies, we dwell on the uniform distribution, which, as we know from [57], gives the sought answer when $\lambda = \ln 2$, $T' = q' \ln 2$. Surely, this is only an explanation why the distorted distributions give the desired asymptotic answer, but a formal proof can also easily be performed taking into account only those input sequences $(X_1, \ldots, X_T)$ for which the deviation from the mean number of users that utilize the frequency q is small as compared to this mean. □

If we confine ourselves to equal distributions at the stations only (i.e., $P_1 = \cdots = P_T$ in (4.3.10)), then the asymptotic behavior of the RHS of (4.3.9) under this restriction on input distributions (denote the corresponding quantity by $H_{\text{com}}(\lambda)$) is completely determined by the uniform and distorted distributions.

Proposition 4.12 *We have the equality*

$$H_{com}(\lambda) = \begin{cases} H_{unif}(\lambda) = h(1 - e^{-\lambda}) & \textit{for } 0 < \lambda < \ln 2, \\ H_{distort}(\lambda) = 1 & \textit{for } \ln 2 \le \lambda < \infty. \end{cases}$$

Proof We have only to prove the inequality

$$H_{\text{com}}(\lambda) \le H_{\text{unif}}(\lambda) = h(1 - e^{-\lambda}) \qquad \text{for } 0 < \lambda \le \ln 2,$$

which is a consequence of the following two facts:

(i) If we consider equal distributions at the stations only, the mean number of units at the output is maximal with the uniform distribution, i.e., $\max \sum_{m=1}^{q}(1 - (1 - p_m)^T)$ on condition that $\sum_{m=1}^{q} p_m = 1$ is attained at $p_m = 1/q$, $m = 1, \ldots, q$ (here, p_m denotes the probability that a station utilizes the frequency m, i.e., $p_m = P_{(X_t=m)}$, $t = 1, \ldots, T$). Moreover, this mean is asymptotically not greater than $q/2$ (since $\lambda \le \ln 2$).

(ii) The probability of a significant deviation from the mean number of units is small, therefore, the entropy of the output distribution is asymptotically not greater than the logarithm of the number of binary sequences of length q with as many units as this mean number

□

Remark Many researchers believed (see, for example, [10, 23]) that the uniform distribution is asymptotically optimal for $\lambda \le 1$. Computations (see, e.g., [22]) have not corroborated this, and Proposition 4.12 shows that they could not, since the uniform distribution is necessarily not asymptotically optimal for $\lambda > \ln 2$. However, for $\lambda = \ln 2 = 0.693\ldots$ it is so, and the expectation of Bassalygo and Pinsker [6] (apparently, as well as that of other researchers) was that this should hold for all smaller values of λ, $0 < \lambda \le \ln 2$. Therefore, we were rather surprised when it was found that, for smaller λ, the best answer is obtained with the following input distribution (certainly, it is different for different users, $t = 1, \ldots, T$, $T < q$):

$$P_{(X_t=m)} = \begin{cases} \frac{1}{2} & \text{for } m = t, \\ \frac{1}{2(q-T)} & \text{for } m > T, \\ 0 & \text{otherwise.} \end{cases} \tag{4.3.14}$$

Denote the entropy of the output distribution for this input distribution by $H^*(Y)$ and denote bu $H^*(\lambda)$ the corresponding asymptotic parameter.

Proposition 4.13 *We have*

$$H^*(\lambda) = \lambda + (1-\lambda)h\left(1 - e^{-\frac{\lambda}{2(1-\lambda)}}\right) \quad \textit{for } 0 < \lambda \leq \frac{2\ln 2}{1+2\ln 2} = 0.581\ldots$$

Proof Although we will perform the proof in detail, let us first explain why one should expect this answer. The distribution (4.3.14) generates at each station its own frequency with probability $1/2$ and $q-T$ common frequencies equiprobable. Therefore, the entropy (in asymptotic representation) for the first $T=\lambda q$ frequencies equals λ and, in fact, is determined by the cases where the number of stations that transmit their own frequencies differs from $T/2$ a little, and, hence, the conditional entropy for the other $q-T$ frequencies coincides with the entropy for the transmission of $q-T=(1-\lambda)q$ stations with the uniform distribution at these stations. By (4.3.11), this entropy (in asymptotic representation) equals $(1-\lambda)h(1-e^{-\frac{\lambda}{2(1-\lambda)}}$.

Now, let us proceed to the formal proof. Denote by $U=(Y_1,\ldots,Y_T)$ the first T components of the sequence Y, and by $V=(Y_{T+1},\ldots,Y_q)$, the remaining $q-T$ components. Then $H^*(Y)=H^*(U)+H^*(V|U)$. Since the components $Y_1,\ldots,Y_T$ are independent and assume the values 0 and 1 with probability $1/2$, we have $H^*(U)=T$.

Now, we have to compute the asymptotic behavior of the conditional entropy. It is clear that the output conditional probabilities $Q^*(v|u)$ depend on the weights $w(u)$ and $w(v)$ of sequences u and v only (where u and v are values of the RV's U and V), i.e., on the number of units in them. Of course, one could write explicit formulas for these probabilities using formula (4.3.5) from [57], which describes the output probability distribution of $q-T$ frequencies for $T-w(u)$ users with uniform input distribution. However, it suffices to know two conditional probabilities only $(t,t'>T, t\neq t')$,

$$q_0(w) \triangleq Q^*(Y_t=0|(y_1,\ldots,y_T), w(y_1,\ldots,y_T)=w) = \left(1-\frac{1}{q-T}\right)^{T-w} \tag{4.3.15}$$

and

$$q_{00}(w) \triangleq Q^*)Y_t{=}Y_{t'}=0|(y_1,\ldots,y_T), w(y_1,\ldots,y_T){=}w) = \left(1-\frac{2}{q-T}\right)^{T-w}. \tag{4.3.16}$$

Given (4.3.15) and (4.3.16), one can easily compute conditional expectation and variance of the RV $w(V)=Y_{T+1}+\cdots+Y_M$:

$$\mathbf{E}(w(V)|u, w(u) = w) \;=\; (q - T)(1 - q_0(w)) = M(1 - \lambda)(1 - q_0(w)), \tag{4.3.17}$$

$$\begin{aligned} \mathbf{D}(w(V)|u, w(u) = w) =\; & (q - T)q_0(w)(1 - q_0(w)) \\ & + (q - T)(q - T - 1)(q_{00}(w) - q_0^2(w)) \quad (4.3.18) \\ \le\; & q(1 - \lambda)q_0(w)(1 - q_0(w)). \end{aligned}$$

For $w = wT$, $0 < w < 1$, and $q \to \infty$, we have

$$\mathcal{E}(\lambda, w) \triangleq \mathbf{E}(w(V)|u, w(u) = wT) \sim q(1 - \lambda)\left(1 - e^{-\frac{(1-w)\lambda}{1-\lambda}}\right) \tag{4.3.19}$$

(here and in what follows, $f(n) \sim g(n)$ means that $\lim \frac{f(n)}{g(n)} = 1$ as $n \to \infty$). Note that one can also easily compute the asymptotic behavior of the conditional variance (4.3.18) but, to apply the Chebyshev inequalities, it suffices to have an upper estimate for the variance (by the way, let us correct the formula for the variance σ_w^2 of the analogous parameter $w^{(n)}$ in [57] – it should be $qe^{-\lambda}(1 - e^{-\lambda} - \lambda e^{-\lambda})$ instead of $qe^{-\lambda}(1 - e^{-\lambda})$ – though this has no effect on the result).

Using the relations

$$Q^*(u) = \frac{1}{2^T},$$

$$1 = \sum_{w=0}^{T} \binom{T}{w} 2^{-T} \sim \sum_{w=\frac{T}{2}-T^{\frac{1}{2}+\varepsilon}}^{\frac{T}{2}+T^{\frac{1}{2}+\varepsilon}} \binom{T}{w} 2^{-T}, \tag{4.3.20}$$

and

$$q_0(w) \sim e^{-\frac{\lambda}{2(1-\lambda)}} \qquad \text{for all } w,\ \frac{T}{2} - T^{\frac{1}{2}+\varepsilon} \le w \le \frac{T}{2} + T^{\frac{1}{2}+\varepsilon}$$

(here and in what follows, ε is a small positive number), one easily obtains the required upper estimate for the conditional entropy, namely,

$$\begin{aligned} H^*(V|U) \le \sum_{t=T+1}^{q} H^*(Y_t|U) &= \sum_{t=T+1}^{q} \sum_{u\in\{0,1\}^T} Q^*(u)\left(-\sum_{y_t\in\{0,1\}} Q^*(y_t|u)\log Q^*(y_t|u)\right) \\ &= (q - T)\sum_{w=0}^{T} \binom{T}{w} w^{-T} h(1 - q_0(w)) \sim q(1 - \lambda)h\left(1 - e^{-\frac{\lambda}{2(1-\lambda)}}\right). \end{aligned}$$

To obtain a lower estimate which asymptotically coincides with the upper one, we need, together with (4.3.20), additional relations:

$$\sum_{j=\mathcal{E}(\lambda,\frac{1}{2})-q^{\frac{1}{2}+\varepsilon}}^{\mathcal{E}(\lambda,\frac{1}{2})+q^{\frac{1}{2}+\varepsilon}} \sum_{\substack{v\in\{0,1\}^{q-T}\\ w(v)=j}} Q^*(v|u) \sim 1$$

and

$$\left(\min_{\mathcal{E}(\lambda,\frac{1}{2})-q^{\frac{1}{2}+\varepsilon}\le j\le \mathcal{E}(\lambda,\frac{1}{2})+q^{\frac{1}{2}+\varepsilon}} \log\binom{q-T}{j}\right) \sim q(1-\lambda)h\left(1-e^{-\frac{\lambda}{2(1-\lambda)}}\right)$$

(the first relation follows from the Chebyshev inequality and estimates (4.3.17) and (4.3.18).

Thus,

$$\begin{aligned}
H^*(V|U) &= -\sum_{u\in\{0,1\}^T} Q^*(u) \sum_{v\in\{0,1\}^{q-T}} Q^*(v|u)\log Q^*(v|u)\\
&\ge -\sum_{w=\frac{T}{2}-T^{\frac{1}{2}+\varepsilon}}^{\frac{T}{2}+T^{\frac{1}{2}+\varepsilon}} \sum_{\substack{u\in\{0,1\}^T\\ w(u)=w}} Q^*(u)\\
&\quad\times \sum_{j=\mathcal{E}(\lambda,\frac{1}{2})-q^{\frac{1}{2}+\varepsilon}}^{\mathcal{E}(\lambda,\frac{1}{2})+q^{\frac{1}{2}+\varepsilon}} \sum_{\substack{v\in\{0,1\}^{q-T}\\ w(v)=j}} Q^*(v|u)\log Q^*(v|u)\binom{q-T}{j}\binom{q-T}{j}^{-1}\\
&\ge \left(\min_{\mathcal{E}(\lambda,\frac{1}{2})-q^{\frac{1}{2}+\varepsilon}\le j\le \mathcal{E}(\lambda,\frac{1}{2})+q^{\frac{1}{2}+\varepsilon}} \log\binom{q-T}{j}\right) \sum_{w=\frac{T}{2}-T^{\frac{1}{2}+\varepsilon}}^{\frac{T}{2}+T^{\frac{1}{2}+\varepsilon}} \sum_{\substack{u\in\{0,1\}^T\\ w(u)=w}} Q^*(u)\\
&\quad\times \sum_{j=\mathcal{E}(\lambda,\frac{1}{2})-q^{\frac{1}{2}+\varepsilon}}^{\mathcal{E}(\lambda,\frac{1}{2})+q^{\frac{1}{2}+\varepsilon}} \sum_{\substack{v\in\{0,1\}^{q-T}\\ w(v)=j}} Q^*(v|u)\\
&\sim q(1-\lambda)h\left(1-e^{-\frac{\lambda}{2(1-\lambda)}}\right).
\end{aligned}$$

(in the latter inequality, we used that $Q^*(v|u)\binom{q-T}{j} \le 1$ if $w(v)=j$). □

According to Proposition 4.13, if $\frac{\lambda}{2(1-\lambda)} = \ln 2$, i.e., $\lambda = \lambda^* \triangleq \frac{2\ln 2}{1+2\ln 2}$, then $H^*(\lambda^*) = 1$. From this and Proposition 4.11, taking into account that the function $C_{\text{sum}}(\lambda)$ is convex, the theorem below immediately follows

Theorem 4.14 *We have*

$$C_{sum} = 1 \quad \textit{for } \lambda = \lambda^* \triangleq \frac{2\ln 2}{1+2\ln 2} = 0.583\ldots$$

Remark Unfortunately, we do not know the exact value of $C_{\text{sum}}(\lambda)$ for $0 < \lambda < \lambda^*$; perhaps, the reason is that we have no non-trivial upper bound. The only estimate known,

$$C_{\text{sum}}(\lambda) \leq \begin{cases} 1 & \text{for } \frac{1}{2} \leq \lambda < \infty, \\ h(\lambda) & \text{for } 0 < \lambda < \frac{1}{2}, \end{cases}$$

coincides with the summarized capacity of an A channel considered as an ordinary channel but not a multiple-access one (in the case of an ordinary channel, the maximum on the RHS of (4.3.9) is taken over all possible input distributions P_X but not independent only).

As to the lower bound, the best answer so far is derived a natural generalization of Proposition 4.13. Consider the following input distribution depending on the parameter γ, $0 < \gamma \leq 1/2$:

$$P_{(X_t=m)} = \begin{cases} \gamma & \text{for } m = t, \\ \frac{1-\gamma}{(q-T)} & \text{for } m > T, \\ 0 & \text{otherwise.} \end{cases} \tag{4.3.21}$$

Denote the entropy of the output distribution for this input distribution by $H_\gamma(Y)$ and denote by $H_\gamma(\lambda)$ the corresponding asymptotic parameter.

Proposition 4.14 *For any* γ, $0 \leq \gamma \leq 1/2$, *we have*

$$H_\gamma(\lambda) = \lambda h(\gamma) + (1-\lambda) h\left(1 - e^{-\frac{\lambda(1-\gamma)}{1-\lambda}}\right) \quad \text{for } 0 < \lambda \leq \frac{\ln 2}{1 + \ln 2 - \gamma}.$$

Proof The proof of this statement repeats that of Proposition 4.13 with the replacement of $1/2$ with γ (for $\gamma = 1/2$, the distribution (4.3.21) gives the distribution (4.3.14)). To obtain the best lower bound for a given λ, one should maximize $H_\gamma(\lambda)$ over γ. □

Theorem 4.15 *For* $0 < \lambda \leq \lambda^*$, *we have*

$$C_{sum}(\lambda) \geq \max_{\max\{0, 1+\ln 2 - \lambda^{-1} \ln 2\} \leq \gamma \leq 1/2} \lambda h(\gamma) + (1-\lambda) h\left(1 - e^{-\frac{\lambda(1-\gamma)}{(1-\lambda)}}\right).$$

Remark In the interval $0 < \lambda \leq \lambda^*$, Theorem 4.15 necessarily gives a better answer than the uniform distribution since $H_\lambda(\lambda) > H_{\text{unif}}(\lambda)$.

Thus, Theorems 4.14 and 4.15 together with the upper bound from the remark following Theorem 4.14 provide the best known estimates of the asymptotic summarized capacity of an A channel for all λ between 0 and ∞, and it only remains to consider its asymptotic behavior at two boundary points.

I. $\lambda = 0$, i.e., $\frac{T}{q} \to 0$ as $q \to \infty$. Then

$$C_{\text{sum}}(T, q) \sim T \log \frac{q}{T}.$$

One can easily check that, asymptotically, this answer is as well obtained, for instance, with the uniform input distribution.

II. $\lambda = \infty$, i.e., $\frac{T}{q} \to \infty$ as $T \to \infty$. Then

$$C_{\text{sum}}(T, q) \sim \begin{cases} q & \text{if } q \to \infty, \\ q - 1 & \text{if } q \text{ is fixed.} \end{cases}$$

For $q \to \infty$, this answer was obtained in [10], and for fixed q, in [22], where it was shown that it is attained with the distorted distribution (4.3.12).

Remark K.S. Zigangirov stated that, for practical purposes, more interesting is the case of a partial multiple access, where the number of simultaneously operating stations is significantly less than their total number.

4.4 Nearly Optimal Multi-user Codes for the Binary Adder Channel

4.4.1 Introduction

A *central* problem in multi-user coding is to assign codes to a collection of senders so that they can communicate simultaneously with a single receiver through a shared multiple-access channel. Multi-user Information Theory provides the prime motivation for studying this problem by revealing that, for many channels of practical interest (e.g., [14], p. 379), multi-user coding can achieve a higher total rate of transmission (sum-rate) than traditional channel multiplexing techniques such as time-division. Unfortunately, despite more than three decades of intensive research on multi-user coding, the performance gains promised by Information Theory remain elusive. There are few non-trivial, multiple-access channels for which explicit code constructions exist and approach the information-theoretic limits (the collision channel without feedback [44], the Galois field adder channel [54], and the codes in [9] are noteworthy exceptions).

Perhaps the most extensively investigated multiple-access channel is the *binary adder channel*, described as follows. T users communicate with a single receiver through a common discrete-time channel. At each time epoch, user i selects an input $X_i \in \{0, 1\}$ for transmission. The channel output is

$$Y \triangleq \sum_{i=1}^{T} X_i \tag{4.4.1}$$

where summation is over the real numbers. A variety of coding problems have been investigated using this model. These variations include user feedback [21, 47], asynchronism [15, 61], jamming [17], superimposed codes [18], and codes for T active users out of M potential users [45, 47]. Here we focus on the oldest and best understood of these problems: the channel is noiseless; there is no feedback; and all users are synchronous, active at all times, and collaborate in code design. Thus a *T-user code* $\mathcal{U} \triangleq (\mathcal{U}_1, \mathcal{U}_2, \ldots, \mathcal{U}_T)$, is a collection of T sets of *codewords* of length n, $\mathcal{U}_i \subset \{0, 1\}^n$. The *rate* of the code is $\mathbf{R} \triangleq (R_1, \ldots, R_T)$ and the *sum-rate* is

$$R_{\text{sum}}(T) \triangleq R_1 + R_2 + \cdots + R_t$$

where $R_i \triangleq (1/n) \log_2 |\mathcal{U}_i|$ is the rate of user i's code.

Chang and Weldon [9] showed that the capacity region of the T-user binary adder channel is the set of all non-negative rates $R_1, \ldots, R_T)$ satisfying

$$\begin{array}{l} 0 \leq R_i \leq H_1, \\ 0 \leq R_i + R_j \leq H_2, \\ \vdots \quad \vdots \quad \vdots \\ 0 \leq R_1 + \cdots + R_T \leq H_T \end{array} \tag{4.4.2}$$

where

$$H_m \triangleq -\sum_{i=1}^{m} \binom{m}{i} 2^{-m} \log_2 \binom{m}{i} 2^{-m}. \tag{4.4.3}$$

(The special case $T = 2$ was derived earlier by Liao [37] (p. 48) in the guise of the *noiseless multiple-access binary erasure channel.*) In particular, observe that the largest achievable sum-rate is $C_{\text{sum}}(T) \triangleq H_T$, which is called the *sum-capacity*.

Most work on code constructions for the binary adder channel has focused on the special case $T = 2$. Farrell [19] has written an excellent survey of the literature up to 1980; more recent constructions can be found in [11, 30, 32, 38, 52, 53]. While many of these two-user codes achieve higher sum-rates than time-division, none approaches the sum-capacity $C_{\text{sum}}(2)$. It was therefore a significant advance when Chang and Weldon [9] presented, for $T > 2$, a family of multi-user codes which are asymptotically optimal in the sense that $R_{\text{sum}}/C_{\text{sum}} \to 1$ as $T \to +\infty$. In their construction, each user's code consists of only two codewords which are defined recursively (so $R_1 = R_2 = \cdots = R_T$). This basic construction has been generalized in several ways [8, 20, 34, 59], and alternate constructions have been proposed based on coin weighing designs [43] and results from additive number theory [27].

Chang and Weldon's construction shows how to approach one point on the boundary of the T-user capacity region. Similarly, all subsequent work for $T > 2$ has focused on the symmetric case $|\mathcal{U}_1| = \cdots = |\mathcal{U}_T| = 2$, except for [34] where $|\mathcal{U}_1| = \cdots = |\mathcal{U}_{T-1}| = 2$ but $|\mathcal{U}_T| > 2$. It is natural to ask, however, whether pther points in the capacity region can be approached by a similar construction.

The goal of section is to present the construction of mixed-rate, multi-user codes for the binary adder channel by Hughes and Cooper [26]. In Sect. 4.4.2, we present two recursive multi-user code constructions. The codewords contained in these codes are equivalent, up to an affine transformation, to those in [9, 43]; however, the recursions are adapted in order to distribute these codewords among *as few users as possible*. As a result, codes with a wide range of information rates are obtained. In Sect. 4.4.3, we then show that these basic codes can be combined by time-sharing to achieve most rates in the capacity region. Specifically, for every T, all rates in the polytope

$$\begin{aligned} 0 \le R_i &\le H_1 - \epsilon_1, \\ 0 \le R_i + R_j &\le H_2 - \epsilon_2, \\ \vdots \quad \vdots \quad &\vdots \\ 0 \le R_1 + \cdots + R_T &\le H_T - \epsilon_T \end{aligned} \tag{4.4.4}$$

can be approached, where $0 \le \epsilon_m < 1.090$ bits per channel use, $1 \le m \le T$. In particular, Hughes and Cooper constructed a family of T-user codes with $R_{\text{sum}}(T) \ge C_{\text{sum}}(T) - 0.547$ bits per channel use, which exceeds the sum-rate of all codes previously reported in [8, 9, 20, 27, 34, 43, 59] for almost every T. In Sect. 4.4.4, we discuss extensions to a T-user, q-frequency adder channel. Finally, the main conclusions of Hughes and Cooper are summarized in Sect. 4.4.5.

4.4.2 Two Multi-user Codes

4.4.2.1 Preliminaries

Earlier work on coding for the T-user binary adder channel has focused almost exclusively on multi-user codes that assign only two codewords to each user. As a consequence, basic definitions have been formulated in terms of the difference between these two codewords (e.g., [9]). However, because our interest is in larger codes, we must extend these basic definitions to a broader class of codes.

Definition 4.11 An (N, K) **affine code** is a pair $(G, \mathbf{m})$, where G is a real $K \times N$ matrix and $\mathbf{m}$ is a real low vector of length N. The **rate** of this code is $R \triangleq K/N$. The **codeword** associated with the message $\mathbf{u} \in \{0, 1\}^K$ is $\mathbf{u}G + \mathbf{m}$. The code is said to be **binary** if $\mathbf{u}G + \mathbf{m} \in \{0, 1\}^N$ for all $\mathbf{u} \in \{0, 1\}^K$.

Remark Observe that $(G, \mathbf{m})$ is a binary affine code if and only if (a) all components of $\mathbf{m} = (m_1, \ldots, m_N)$ are binary, (b) all components of G are in $\{-1, 0, +1\}$ and no column of G contains more than one non-zero component, and (c) all non-zero components of G satisfy $g_{ij} = 1 - 2m_j$.

Definition 4.12 A T**-user** $(N; K_1, K_2, \ldots, K_T)$ **binary affine code** is a collection

$$\mathcal{U} = \{(G_1, \mathbf{m}_1), \ldots, (G_T, \mathbf{m}_T)\}$$

where $(G_i, \mathbf{m}_i)$ is an (N, K_i) binary affine code. The **rate** of this code is $\mathbf{R} \triangleq (R_1, \ldots, R_T)$ and the **sum-rate** is

$$R_{sum}(\mathcal{U}) \triangleq R_1 + R_2 + \cdots + R_T \tag{4.4.5}$$

where R_i is the rate of $(G_i, \mathbf{m}_i)$.

Definition 4.13 Let $\mathcal{U} = \{(G_1, \mathbf{m}_1), \ldots, (G_T, \mathbf{m}_T)\}$ be a T-user $(N; K_1, \ldots, K_T)$ binary affine code. We say that $\mathcal{U}$ is **uniquely decodable** (UD) if the sum

$$\sum_{i=1}^{T} \mathbf{u}_i G_i + \mathbf{m}_i$$

is distinct for each choice of $\mathbf{u}_i \in \{0, 1\}^{K_i}$, $1 \le i \le T$.

Lemma 4.6 *A T-user binary affine code,*

$$\mathcal{U} = \{(G_1, \mathbf{m}_1), \ldots, (G_T, \mathbf{m}_T)\}$$

is uniquely decodable if and only if, for all $\mathbf{w}_i \in \{-1, 0, +1\}^{K_i}$, $1 \le i \le T$

$$\sum_{i=1}^{T} \mathbf{w}_i G_i = \mathbf{0}_N$$

implies $\mathbf{w}_i = \mathbf{0}_{K_i}$, where $\mathbf{0}_N$ denotes the all-zero N-tuple.

Proof Let $\mathbf{u}_i, \mathbf{u}'_i \in \{0, 1\}^{K_i}$, $1 \le i \le T$, denote any two message sequences. Then $\mathcal{U}$ is UD if and only if

$$\sum_{i=1}^{T} \mathbf{u}_i G_i + \mathbf{m}_i = \sum_{i=1}^{T} \mathbf{u}'_i G_i + \mathbf{m}_i$$

implies $\mathbf{u}_i = \mathbf{u}'_i$ for all $1 \le i \le T$. This holds if and only if

$$\sum_{i=1}^{T} \mathbf{w}_i G_i = \mathbf{0}_N$$

implies $\mathbf{w}_i = \mathbf{0}_{K_i}$ for all $1 \le i \le T$, where $\mathbf{w}_i \triangleq \mathbf{u}_i - \mathbf{u}'_i \in \{-1, 0, +1\}^{K_i}$. This completes the proof of Lemma 4.6. □

4.4.2.2 Construction A

We now present the first of two families of mixed-rate, multi-user codes. The codes given in this subsection are similar to Lindström's coin weighting designs [40]. This similarity can be seen by comparing the construction below with Martirossian and Khachatrian's [43] recursive form of the design matrix. Here, we adapt this recursion in order to assign more than two codewords to each user.

For all $j \geq 1$, denote the jth code in the series by the notation

$$\mathcal{U}_A^j \triangleq \{(G_1^j, \mathbf{m}_1^j), \ldots, (G_T^j, \mathbf{m}_T^j)\}.$$

Let T_j and N_j be the number of users and the block length of $\mathcal{U}_A^j$, respectively. The first code in the series is the trivial single-user code $\mathcal{U}_A^1 \triangleq \{(G_1^1, \mathbf{m}_1^1)\}$ with $T_1 = N_1 \triangleq 1$ and

$$G_1^1 \triangleq 1, \qquad \mathbf{m}_1^1 \triangleq 0.$$

Now, for each $j \geq 1$, the code $\mathcal{U}_A^{j+1}$ is constructed from $\mathcal{U}_A^j$ by the recursion

$$\begin{aligned}
G_1^{j+1} &\triangleq \begin{bmatrix} I_{N_j} & O_{N_j} & \mathbf{0}'_{N_j} \\ \mathbf{0}_{N_j} & \mathbf{1}_{N_j} & 1 \end{bmatrix}, & \mathbf{m}_1^{j+1} &\triangleq [\,\mathbf{0}_{N_j} \;\; \mathbf{0}_{N_j} \;\; 0\,] \\
G_{2i}^{j+1} &\triangleq [\; G_i^j \;\; G_i^j \;\; \mathbf{0}'_{N_j} \;], & \mathbf{m}_{2i}^{j+1} &\triangleq [\,\mathbf{m}_i^j \;\; \mathbf{m}_i^j \;\; 0\,] \\
G_{2i+1}^{j+1} &\triangleq [\; G_i^j \;\; -G_i^j \;\; \mathbf{0}'_{N_j} \;], & \mathbf{m}_{2i+1}^{j=1} &\triangleq [\,\mathbf{m}_i^j \;\; \overline{\mathbf{m}}_i^j \;\; 1\,]
\end{aligned} \tag{4.4.6}$$

for $i = 1, \ldots, T_j$. Here, I_N is the identity matrix of order N, O_N is the square all-zero matrix of order N, $\mathbf{1}_N$ is the all-one N-tuple, $\overline{\mathbf{m}}_i^j \triangleq \mathbf{1}_{N_j} - \mathbf{m}_i^j$, and C' denotes the matrix transpose of C.

For example, $\mathcal{U}_A^2$ is the 3-user (3; 2, 1, 1) code

$$\begin{aligned}
G_1^2 &= \begin{bmatrix} 1 & 0 & 0 \\ 0 & 1 & 1 \end{bmatrix}, & \mathbf{m}_1^2 &= [\,0\;0\;0\,] \\
G_2^2 &= [1 \;\; 1 \;\; 0], & \mathbf{m}_2^2 &= [\,0\;0\;0\,] \\
G_3^2 &= [1 \; -1 \;\; 0], & \mathbf{m}_3^2 &= [\,0\;1\;1\,]
\end{aligned}$$

and $\mathcal{U}_A^3$ is the 7-user (7, 4, 2, 2, 1, 1, 1, 1) code

$$G_1^3 = \begin{bmatrix} 1 & 0 & 0 & 0 & 0 & 0 & 0 \\ 0 & 1 & 0 & 0 & 0 & 0 & 0 \\ 0 & 0 & 1 & 0 & 0 & 0 & 0 \\ 0 & 0 & 0 & 1 & 1 & 1 & 1 \end{bmatrix}, \quad \mathbf{m}_1^3 = [\,0\ 0\ 0\ 0\ 0\ 0\ 0\,]$$

$$G_2^3 = \begin{bmatrix} 1 & 0 & 0 & 1 & 0 & 0 & 0 \\ 0 & 1 & 1 & 0 & 1 & 1 & 0 \end{bmatrix}, \quad \mathbf{m}_2^3 = [\,0\ 0\ 0\ 0\ 0\ 0\ 0\,]$$

$$G_3^3 = \begin{bmatrix} 1 & 0 & 0 & -1 & 0 & 0 & 0 \\ 0 & 1 & 1 & 0 & -1 & -1 & 0 \end{bmatrix}, \quad \mathbf{m}_3^3 = [\,0\ 0\ 0\ 1\ 1\ 1\ 1\,]$$

$$G_4^3 = \begin{bmatrix} 1 & 1 & 0 & 1 & 1 & 0 & 0 \end{bmatrix}, \quad \mathbf{m}_4^3 = [\,0\ 0\ 0\ 0\ 0\ 0\ 0\,]$$

$$G_5^3 = \begin{bmatrix} 1 & 1 & 0 & -1 & -1 & 0 & 0 \end{bmatrix}, \quad \mathbf{m}_5^3 = [\,0\ 0\ 0\ 1\ 1\ 1\ 1\,]$$

$$G_6^3 = \begin{bmatrix} 1 & -1 & 0 & 1 & -1 & 0 & 0 \end{bmatrix}, \quad \mathbf{m}_6^3 = [\,0\ 1\ 1\ 0\ 1\ 1\ 0\,]$$

$$G_7^3 = \begin{bmatrix} 1 & -1 & 0 & -1 & 1 & 0 & 0 \end{bmatrix}, \quad \mathbf{m}_7^3 = [\,0\ 1\ 1\ 1\ 0\ 0\ 1\,].$$

Theorem 4.16 *For all $j \geq 1, \mathcal{U}_A^j$ is a T_j-user $(N_j; K_1^j, \ldots, K_{T_j}^j)$ affine code, where*

$$\begin{aligned} T_j &= N_j = 2^j - 1 \\ K_i^j &= 2^{j-\nu(i)}, \qquad 1 \leq i \leq T_j \end{aligned} \tag{4.4.7}$$

and $\nu(i) \triangleq \lceil \log_2(i+1) \rceil$.

Proof From (4.4.6), observe that the code parameters obey the recursions

$$\begin{aligned} T_{j+1} &= 2T_j + 1, T_1 = 1 \\ N_{j+1} &= 2N_j + 1, N_1 = 1 \\ K_1^{j+1} &= N_j + 1, K_1^1 = 1 \\ K_{2i}^{j+1} &= K_i^j \\ K_{2i+1}^{j+1} &= K_i^j \end{aligned}$$

for all $j \geq 1$ and $1 \leq i \leq T_j$. The expressions for T_j and N_j in (4.4.7) are immediate. From the identities $\nu(2i) = \nu(i) + 1$ and $\nu(2i+1) = \nu(i) + 1$ for all $i \geq 1$, it can be verified by direct substitution that $K_i^j = 2^{j-\nu(i)}$ solves the above recursion, thus completing the proof. □

Note that $\nu(i) = k$ if and only if $2^{k-1} \leq i \leq 2^k - 1$. It follows that the collection $\mathcal{U}_A^j$ contains exactly 2^{k-1} codes of dimension $K_i^j = 2^{j-k}$ for all $1 \leq k \leq j$. The sum of the dimensions of all codes in $\mathcal{U}_A^j$ is therefore $j2^{j-1}$, which yields the following corollary.

Corollary 4.3 *The rate of* $\mathcal{U}_A^j$ *is* $\mathbf{R}_A^j \triangleq (R_A^j, \ldots, R_{AT_j}^j)$, *where*

$$R_{Ai}^j \triangleq \frac{2^{j-\nu(i)}}{2^j - 1}$$

and hence the sum rate is

$$R_{sum}(\mathcal{U}_A^j) = \frac{j2^{j-1}}{2^j - 1}.$$

The next theorem is the main result of this section.

Theorem 4.17 $\mathcal{U}_A^j$ *is a* T_j*-user, uniquely decodable, binary affine code.*

Proof The proof is by induction. The theorem is obvious for $j = 1$. Assuming the theorem holds for $\mathcal{U}_A^j$, we now prove that it also holds for $\mathcal{U}_A^{j=1}$.

First we show that $\mathcal{U}_A^{j+1}$ is binary. The only equation in (4.4.6) with the potential to introduce a non-binary code is

$$G_{2i+1}^{j+1} \triangleq [\, g_i^j \; -G_i^j \; \mathbf{0}'_{N_j} \,], \qquad \mathbf{m}_{2i+1}^{j+1} \triangleq [\, \mathbf{m}_i^j \; \overline{\mathbf{m}}_i^j \; 1 \,].$$

For $\mathbf{u} \in \{0, 1\}^{K_{2i+1}^{j+1}}$, we can write

$$\mathbf{u}G_i^j + \mathbf{m}_{2i+1}^{j+1} = [\, \mathbf{u}G_i^j + \mathbf{m}_i^j, \quad \mathbf{1}_{N_j} - \mathbf{u}G_i^j - \mathbf{m}_i^j, \quad 1 \,].$$

Since $\mathbf{u}G_i^j + \mathbf{m}_i^j$ is binary by assumption, and $a \in \{0, 1\}$ implies $1 - a \in \{0, 1\}$, we conclude that $\mathbf{u}_{2i+1}^{j+1} + \mathbf{m}_{2i+1}^{j+1}$ is also binary. It follows that $\mathcal{U}_A^{j+1}$ is a binary code.

Next we prove that $\mathcal{U}_A^{j+1}$ is uniquely decodable. By Lemma 4.6, it suffices to show that

$$\mathbf{s} \triangleq \sum_{i=1}^{T_{j+1}} \mathbf{w}_i G_i^{j+1} = \mathbf{0}_{N_{j+1}} \Rightarrow \mathbf{w}_i = \mathbf{0}_{K_i^{j+1}} \tag{4.4.8}$$

for all $\mathbf{w}_i \in \{-1, 0, 1\}^{K_i^{j+1}}$, $1 \le i \le T_{j+1}$. To this end, partition $\mathbf{w}_1$ and $\mathbf{s}$ as follows:

$$\mathbf{w}_1 = [\, \overbrace{\hat{\mathbf{w}}_1}^{N_j} \;\; w_1 \,]$$
$$\mathbf{s} = [\, \overbrace{\mathbf{s}_1}^{N_j} \;\; \overbrace{\mathbf{s}_2}^{N_j} \;\; s_3 \,].$$

Using this partition and (4.4.6), we can rewrite $\mathbf{s} = \mathbf{0}_{N_{j+1}}$ as

$$\mathbf{s}_1 = \sum_{i=1}^{T_j} (\mathbf{w}_{2i} G_i^j + \mathbf{w}_{2i+1} G_i^j) + \hat{\mathbf{w}}_1 = \mathbf{0}_{N_j}$$

$$\mathbf{s}_2 = \sum_{i=1}^{T_j} (\mathbf{w}_{2i} G_i^j - \mathbf{w}_{2i+1} G_i^j) + w_1 \mathbf{1}_{N_j} = \mathbf{0}_{N_j} \tag{4.4.9}$$

$$s_3 = w_1 = 0.$$

Hence

$$\mathbf{s}_1 + \mathbf{s}_2 - s_3 \mathbf{1}_{N_j} = 2 \sum_{i=1}^{T_j} \mathbf{w}_{2i} G_i^j + \hat{\mathbf{w}}_1 = \mathbf{0}_{N_j}.$$

This equality implies that all of the components of $\hat{\mathbf{w}}_1$ are even. However, since all components of $\hat{\mathbf{w}}_1$ are in $\{-1, 0, 1\}$, it follows that $\hat{\mathbf{w}}_1 = \mathbf{0}_{N_j}$ and

$$\sum_{i=1}^{T_j} \mathbf{w}_{2i} G_i^j = \mathbf{0}_{N_j}.$$

Since $\mathcal{U}_A^j$ is UD by assumption, Lemma 4.6 implies $\mathbf{w}_{2i} = \mathbf{0}_{K_{2i}^{j+1}}$ for all $1 \le i \le T_j$. Substituting this into (4.4.9), we obtain

$$\mathbf{s}_1 = \sum_{i=1}^{T_j} \mathbf{w}_{2i+1} G_i^j = \mathbf{0}_{N_j}$$

from which it similarly follows that $\mathbf{w}_{2i+1} = \mathbf{0}_{K_{2i+1}^{j+1}}$ for all $1 \le i \le T_j$. Since $s_3 = 0$ implies $w_1 = 0$, the proof of (4.4.8), and hence Theorem 4.17, is complete. □

By the remark preceding Corollary 4.3, there are 2^{j-1} single-user codes in $\mathcal{U}_A^j$ containing only two codewords. For these codes, the next theorem shows that there is no need for a separate bias vector $\mathbf{m}$.

Theorem 4.18 *Let*

$$\mathcal{U}_A^{*j} \triangleq \{(G_1^{*j}, \mathbf{m}_1^{*j}), \ldots, (G_{T_j}^{*j}, \mathbf{m}_{T_j}^{*j})\}$$

be the multi-user code obtained by replacing $(G_i^j, \mathbf{m}_i^j)$ in $\mathcal{U}_A^j$ by $(G_i^j + \mathbf{m}_i^j, \mathbf{0})$ for all i and j satisfying $K_i^j = 1$. Then $\mathcal{U}_A^{*j}$ is a uniquely decodable, binary affine code.

Proof By the remark following Definition 4.11, it is obvious that $(G_i^j + \mathbf{m}_i^j, \mathbf{0})$ is a binary affine code if $K_i^j = 1$. To show that $\mathcal{U}_A^{*j}$ is UD requires only a few changes in the proof of Theorem 4.17. Again, we proceed by induction, assuming that $\mathcal{U}_A^{*j}$ is UD.

Let $A_j \triangleq \{i : K_i^j = 1\}$ and observe that $A_{j+1} = 2A_j \cup (2A_j + 1)$. The decoding equations for $\mathcal{U}_A^{*(j+1)}$ corresponding to (4.4.9) are

$$\mathbf{s}_1 \triangleq \sum_{i=1}^{T_j} (\mathbf{w}_{2i} G_i^j + \mathbf{w}_{2i+1} G_i^j) + \sum_{i \in A_j} (\mathbf{w}_{2i} \mathbf{m}_i^j + \mathbf{w}_{2i+1} \mathbf{m}_i^j) + \hat{\mathbf{w}}_1 = \mathbf{0}_{N_j}$$

$$\mathbf{s}_2 \triangleq \sum_{i=1}^{T_j} (\mathbf{w}_{2i} G_i^j - \mathbf{w}_{2i+1} G_i^j) + \sum_{i \in A_j} (\mathbf{w}_{2i} \mathbf{m}_i^j + \mathbf{w}_{2i+1} (\mathbf{1}_{N_j} - \mathbf{m}_i^j)) + w_1 \mathbf{1}_{N_j} = \mathbf{0}_{N_j}$$

$$s_3 = \sum_{i \in A_j} \mathbf{w}_{2i+1} + w_1 = 0.$$

It follows that

$$\begin{aligned} \mathbf{s}_1 + \mathbf{s}_2 - s_3 \mathbf{1}_{N_j} &= 2 \sum_{i=1}^{T_j} \mathbf{w}_{2i} G_i^j + 2 \sum_{i \in A_j} \mathbf{w}_{2i} \mathbf{m}_i^j + \hat{\mathbf{m}}_1 \\ &\triangleq 2 \sum_{i=1}^{T_j} \mathbf{w}_{2i} G_i^{*j} + \hat{\mathbf{w}}_1 = \mathbf{0}_{N_j}. \end{aligned}$$

As in the proof of Theorem 4.17, this equation implies $\hat{\mathbf{w}}_1 = \mathbf{0}_{N_j}$ and hence

$$\sum_{i=1}^{T_j} \mathbf{w}_{2i} G_i^{*j} = \mathbf{0}_{N_j}.$$

Since $\mathcal{U}_A^{*j}$ is UD by assumption, Lemma 4.6 implies $\mathbf{w}_{2i} = \mathbf{0}_{K_i^j}$ for all $1 \le i \le T_j$. Similarly

$$\mathbf{s}_1 - \mathbf{s}_2 + s_3 \mathbf{1}_{N_j} = 2 \sum_{i=1}^{T_j} \mathbf{w}_{2i+1} G_i^{*j} = \mathbf{0}_{N_j}$$

which implies $\mathbf{w}_{2i+1} = \mathbf{0}_{K_i^j}$ for all $1 \le i \le T_j$. It then follows from $s_3 = 0$ that $w_1 = 0$. This completes the proof of Theorem 4.18. □

We can use Theorem 4.18 to modify $\mathcal{U}_A^j$ in a way that distributes the non-zero codewords among fewer users. In Sect. 4.4.3, we will show that this leads to a better tradeoff between the sum-rate and T. Let $\mathcal{U}_A^{*j}$ be the code defined in Theorem 4.18, where $\mathbf{m}_i^{*j} = \mathbf{0}_{N_j}$ for all single-bit codes. Let $\{\mathbf{0}_{N_j}, \mathbf{a}\}$ and $\{\mathbf{0}_{N_j}, \mathbf{b}\}$ be the codewords of any two such codes. Observe that the T-user code remains UD if these *two* codes are replaced by *one* (non-affine) code comprising the three codewords $\{\mathbf{0}_{N_j}, \mathbf{a}, \mathbf{b}\}$. Thus any two single-bit codes can be replaced by one code with three codewords. For $j \ge 2$, let $\overline{\mathcal{U}}_A^j$ be the code obtained by merging pairs of single-bit codes in $\mathcal{U}_A^{*j}$. Since $\mathcal{U}_A^{*j}$ contains exactly 2^{j-1} single-bit codes, the total number of users in $\overline{\mathcal{U}}_A^j$ is $\overline{T}_j \triangleq T_j - 2^{j-2} = 3 \cdot 2^{j-2} - 1$ and the sum-rate is bounded by

$$R_{\text{sum}}(\overline{\mathcal{U}}_A^j) = \frac{1}{N_j}\left(\sum_{\nu=1}^{j-1} 2^{\nu-1}2^{j-\nu} + 2^{j-2}\cdot\log_2 3\right) > \frac{j-1+\log_2\sqrt{3}}{2}. \quad (4.4.10)$$

4.4.2.3 Construction B

The second family of mixed-rate, multi-user codes considered in this section is based on Chang and Weldon's construction [9]. Here, our aim is to partition their encoding matrix and define biases which permit more than two codewords to be assigned to each user.

Abusing notation slightly, we denote the jth code in the series by

$$\mathcal{U}_B^j \triangleq \{(G_1^j, \mathbf{m}_1^j), \ldots, (G_{T_j}^j, \mathbf{m}_{T_j}^j)\}$$

and we denote the number of users and block length by T_j and N_j, respectively.

Again, the first code in the series is the single-user code $\mathcal{U}_B^1 \triangleq \{G_1^1, \mathbf{m}_1^1)\}$ with $T_1 = N_1 \triangleq 1$ and

$$G_1^1 \triangleq 1, \qquad \mathbf{m}_1^1 \triangleq 0.$$

Now $\mathcal{U}_B^{j=1}$ is recursively contructed from $\mathcal{U}_B^j$ by

$$\begin{aligned} G_1^{j+1} &\triangleq [\ I_{N_j} \quad O_{N_j}\ ], & \mathbf{m}_1^{j+1} &\triangleq [\ \mathbf{0}_{N_j} \quad \mathbf{0}_{N_j}\ ] \\ G_{2i}^{j+1} &\triangleq [\ G_i^j \quad G_i^j\ ], & \mathbf{m}_{2i}^{j+1} &\triangleq [\ \mathbf{m}_i^j \quad \mathbf{m}_i^j\ ] \\ G_{2i+1}^{j+1} &\triangleq [\ G_i^j \quad -G_i^j\ ], & \mathbf{m}_{2i+1}^{j+1} &\triangleq [\ \mathbf{m}_i^j \quad \overline{\mathbf{m}}_i^j\ ] \end{aligned} \quad (4.4.11)$$

for all $j \geq 1$ and $1 \leq i \leq T_j$.

For example, $\mathcal{U}_B^2$ is the 3-user (3; 1, 1, 1) code

$$\begin{aligned} G_1^2 &= [\ 1 \quad 0\ ], & \mathbf{m}_1^2 &= [\ 0 \ \ 0\ ] \\ G_2^2 &= [\ 1 \quad 1\ ], & \mathbf{m}_2^2 &= [\ 0 \ \ 0\ ] \\ G_3^2 &= [\ 1 \ \ -1\ ], & \mathbf{m}_3^2 &= [\ 0 \ \ 1\ ] \end{aligned}$$

and $\mathcal{U}_B^3$ is the 7-user (4; 2, 1, 1, 1, 1, 1, 1) code

$$G_1^3 = \begin{bmatrix} 1 & 0 & 0 & 0 \\ 0 & 1 & 0 & 0 \end{bmatrix}, \quad \mathbf{m}_1^3 = [\,0\ 0\ 0\ 0\,]$$

$$G_2^3 = [1 \quad 0 \quad 1 \quad 0], \quad \mathbf{m}_2^3 = [\,0\ 0\ 0\ 0\,]$$

$$G_3^3 = [1 \quad 0 \ -1 \quad 0], \quad \mathbf{m}_3^3 = [\,0\ 0\ 1\ 1\,]$$

$$G_4^3 = [1 \quad 1 \quad 1 \quad 1], \quad \mathbf{m}_4^3 = [\,0\ 0\ 0\ 0\,]$$

$$G_5^3 = [1 \quad 1 \ -1 \ -1], \quad \mathbf{m}_5^3 = [\,0\ 0\ 1\ 1\,]$$

$$G_6^3 = [1 \ -1 \quad 1 \ -1], \quad \mathbf{m}_6^3 = [\,0\ 1\ 0\ 1\,]$$

$$G_7^3 = [1 \ -1 \ -1 \quad 1], \quad \mathbf{m}_7^3 = [\,0\ 1\ 1\ 0\,]$$

The next theorem, which is the main result of this subsection, gives results for $\mathcal{U}_B^j$ which are analogous to Theorems 4.16 and 4.17. We omit the proof since it is similar to those given in the previous subsection.

Theorem 4.19 *For all $j \geq 1, \mathcal{U}_B^j$ is a uniquely decodable T_j-user $(N_j; K_1^j, \ldots, K_{T_j}^j)$ binary affine code, where*

$$\begin{aligned} T_j &= 2^j - 1 \\ N_j &= 2^{j-1} \\ K_i^j &= \begin{cases} 2^{j-1-\nu(i)}, & 1 \leq i \leq 2^{j-2} - 1 \\ 1, & 2^{j-2} \leq i \leq T_j \end{cases} \end{aligned} \tag{4.4.12}$$

and

$$\nu(i) \triangleq \lceil \log_2(i+1) \rceil.$$

The rate of $\mathcal{U}_B^j$ is

$$\mathbf{R}_B^j \triangleq (R_{B1}^j, \ldots, R_{BT_j}^j)$$

where

$$R_{Bi}^j \triangleq 2^{-\nu(i)} \quad \textit{for } 1 \leq i \leq 2^{j-2} - 1$$

and

$$R_{Bi}^j \triangleq 2^{1-j} \quad \textit{for } 2^{j-2} \leq i \leq 2^j - 1.$$

Hence the sum rate is

$$R_{sum}(\mathcal{U}_B^j) = \frac{j+1}{2}.$$

Remark The recursion in (4.4.11) is obtained by partitioning the encoding matrix of [9]. This partition does not increase the sum-rate of the code; however, it does increase the sum-rate for a given T. This distinction can be most easily seen through an example. For $T = 7$, the best code from [9] has a sum-rate of 1.75-b/channel use, and is obtained by dropping one codebook from the $T = 8$ code with sum-rate 2.)-b/channel use. Our partitioning scheme allows us to reassign the codewords in the $T = 8$ code to only $T = 7$ users (see $\mathcal{U}_B^j$). The sum-rate of this code is not changed by this reassignment, but the sum-rate for $T = 7$ users is now increased from 1.75 to 2.0-b/channel use.

Unfortunately, there is no analog of Theorem 4.18 for $\mathcal{U}_B^j$, which can be seen by observing that replacing $(G_3^2, \mathbf{m}_3^2)$ with $(G_3^2, \mathbf{m}_3^2, 0)$ clearly destroys the unique decodability of $\mathcal{U}_B^2$. However, $\mathcal{U}_B^j$ will always contain two single-bit codes with $\mathbf{m}_i^j = 0$, which arise from the codes $(G_1^2, \mathbf{0})$ and $G_2^2, \mathbf{0})$ through the recursion for G_{2i}^{j+1} in (4.4.11). We can merge these two single-bit codes to get a UD code $\overline{\mathcal{U}}_B^j$ with $2^j - 2$ users and sum-rate

$$R_{\text{sum}}(\overline{\mathcal{U}}_B^j) = \frac{j+1}{2} + \frac{\log_2 3 - 2}{2^{j-1}}. \tag{4.4.13}$$

4.4.3 Performance

From any T-user code $\mathcal{U}$, a variety of other multi-user codes can be constructed by elementary operations. First, we can reorder (i.e., reassign) the single-user codes in $\mathcal{U}$. Second, we can delete codes from $\mathcal{U}$. Third, we can use time-sharing to obtain still other codes. For the sake of brevity, we say that $\mathcal{U}'$ can be *constructed by elementary time-sharing* from $\mathcal{U}$ if it can be obtained by these three basic operations. The aim of this section is to characterize the set of all rates of codes that can be constructed by elementary time-sharing from $\mathcal{U}_A^j$ and $\mathcal{U}_B^j$, and to compare this set with the capacity region of the T-user binary adder channel.

4.4.3.1 Capacity and Majorization

Before examining the performance of the codes constructed in the previous section, it is convenient to introduce a result from the theory of majorization [41]. For any real vector $\mathbf{y} \triangleq (y_1, \ldots, y_n)$, let $y_{[1]} \geq y_{[2]} \geq \cdots \geq y_{[n]}$ denote the components of $\mathbf{y}$ arranged in decreasing order. The real vector $\mathbf{x} \triangleq (x_1, \ldots, x_n)$ is said to be *weakly submajorized* by $\mathbf{y}$, denoted $\mathbf{x} \prec \mathbf{y}$, if

$$\sum_{i=1}^{m} x_{[i]} \leq \sum_{i=1}^{m} y_{[i]}, \qquad m = 1, \ldots, n. \tag{4.4.14}$$

Let $\mathbf{y} \in \mathbb{R}_+^n$, where $\mathbb{R}_+^n$ is the set of non-negative real numbers. The next lemma gives a simple characterization of the set of all non-negative vectors that are weakly submajorized by $\mathbf{y}$.

Lemma 4.7 (Mirsky [41] p. 28) *For any* $\mathbf{y} \in \mathbb{R}_+^n$, *the set* $\{\mathbf{x} \in \mathbb{R}_+^n : \mathbf{x} \prec \mathbf{y}\}$ *is the convex hull of the set of all vectors of the form* $(\eta_1 y_{\pi_1}, \dots, \eta_n y_{\pi_n})$ *where* $(\pi_1, \dots, \pi_n)$ *is a permutation of* $(1, \dots, n)$ *and each* η_i *is* 0 *or* 1.

Lemma 4.7 permits us to give a simple answer to the following question: Given a T-user code of rate $\mathbf{R} \triangleq (R_1, \dots, R_T)$, is it possible to construct by elementary time-sharing from $\mathcal{U}$ another T-user code of rate $\mathbf{R}'$? Observe that, by reassigning or deleting codes in $\mathcal{U}$, it is possible to achieve any rate of the form $(\eta_1 R_{\pi_1}, \dots, \eta_T R_{\pi_T})$ where $(\pi_1, \dots, \pi_T)$ is a permutation of $(1, \dots, T)$ and each η_i us 0 or 1. By time-sharing, any point in the convex hull of these rates can be approached arbitrarily closely. Therefore, by Mirsky's lemma, a code of rate $\mathbf{R}'$ can be constructed by elementary time-sharing from $\mathcal{U}$ if $\mathbf{R}'$ is weakly submajorized by $\mathbf{R}$.

This observation has an important consequence for the capacity region of the T-user binary adder channel. Upon setting

$$\begin{aligned} \mathbf{C}_T &= (C_1, \dots, C_T) \\ &\triangleq (H_1, H_2 - H_1, H_3 - H_2, \dots, H_T - H_{T-1}) \qquad (4.4.15) \\ &\doteq (1, 0.5, 0.311278, 0.219361, 0.167553, \dots) \end{aligned}$$

we can rewrite (4.4.2) in the form

$$0 \le \sum_{i=1}^{m} R_{[i]} \le \sum_{i=1}^{m} C_{[i]} = H_m, \qquad m = 1, \dots, T.$$

Thus a rate $\mathbf{R}$ is in the capacity region if and only if it is weakly submajorized by the vector $\mathbf{C}_T$. It follows from Lemma 4.7 that any rate in the capacity region can be constructed by elementary time-sharing from a multi-user code of rate $\mathbf{C}_T$.

4.4.3.2 Codes Constructed from $\mathcal{U}_A^j$

In this subsection, we show that codes achieving a large portion of the T-user capacity region for any $T \ge 1$ can be constructed from the family of codes $|U_A^j$. We begin by fixing $j \ge 1$ and considering the particular case $T = T_j = 2^j - 1$. By the preceding subsection, a T_j-user code of rate $\mathbf{R} = (R_1, \dots, R_{T_j})$ can be constructed by elementary time-sharing from $\mathcal{U}_A^j$ if $\mathbf{R}$ is weakly submajorized by the rate of $\mathcal{U}_A^j$, i.e., if

$$0 \le \sum_{i=1}^{m} R_{[i]} \le \sum_{i=1}^{m} R_{Ai}^j, \qquad m = 1, \dots, T_j, \qquad (4.4.16)$$

where $\mathbf{R}_A^j \triangleq (R_{A1}^j, \ldots, R_{AT_j}^j)$ is given in Corollary 4.3. To compare this with the capacity region (4.4.2), it suffices to compare the partial sum-rate $\sum_{i=1}^m R_{Ai}^j$ with the corresponding entropy H_m. To this end, the following bounds are useful.

Lemma 4.8 (Chang and Weldon [9]) *For all* $m \geq 1$

$$\frac{1}{2}\log_2 \pi \frac{m}{2} \leq H_m \leq \frac{1}{2}\log_2 \pi e \left\lceil \frac{m}{2} \right\rceil. \tag{4.4.17}$$

Remark Note that $H_m - (1/2)\log_2 \pi e m/2 \to 0$ as $m \to +\infty$. To see this, let $\{X_i\}$ be a Bernoulli sequence with

$$\Pr\{X_i = 0\} = \Pr\{X_i = 1\} = 1/2.$$

Define

$$Z_m \triangleq \frac{1}{\sqrt{m}} \sum_{i=1}^m X_i$$

so that $H_m = H(Z_m)$. By the central limit theorem, Z_m converges in distribution to a Gaussian RV with variance 1/4. We get that $H_m + \log_2(1/\sqrt{m})$ converges to $(1/2)\log_2(\pi e/2)$.

We will also need a bound on the partial sum-rate $\sum_{i=1}^m R_{Ai}^j$, which is given by the following theorem

Theorem 4.20 (Partial sum-rate bounds for $\mathcal{U}_{\mathbf{A}}^{\mathbf{j}}$) *Let*

$$\mathbf{R}_A^j \triangleq (R_{A1}^j, \ldots, R_{AT_j}^j)$$

be given as in Corollary 4.3. Then for all $1 \leq m \leq T_j$

$$\sum_{i=1}^m R_{Ai}^j > \frac{1}{2}\log_2\left(\frac{e\ln 2}{2}(m+1)\right). \tag{4.4.18}$$

When $m = 2^l - 1$ *for some* $l \geq 1$, *this bound can be tightened to*

$$\sum_{i=1}^m R_{Ai}^j > \frac{1}{2}\log_2(m+1). \tag{4.4.19}$$

Remark An examination of the proof reveals that both of the lower bounds in Theorem 4.20 are asymptotically tight, e.g.

$$\sum_{i=1}^{m_k} R_{Ai}^j - \frac{1}{2}\log_2\left(\frac{e\ln 2}{2}(m_k+1)\right) \to 0$$

as $j, k \to +\infty$, where $m_k \triangleq \lceil 2^k \log_2 e \rceil - 1$ and $m_k < T_j = 2^j - 1$.

Proof of Theorem 4.20. For all $m \geq 1$, we can write $m = 2^{k+\theta} - 1$ for some integer $k \geq 1$ and $0 < \theta \leq 1$. Note that $\log_2(m+1) = k + \theta$ and $\nu(m) = k + 1$. The partial sum-rates can then be bounded by

$$\begin{aligned}
\sum_{i=1}^{m} R_{Ai}^{j} &> \sum_{i=1}^{m} 2^{-\nu(i)} \\
&= \sum_{i=1}^{2^k-1} 2^{-\nu(i)} + (m - 2^k + 1)2^{-\nu(m)} \\
&= \sum_{\nu=1}^{k} 2^{\nu-1}2^{-\nu} + (2^{k+\theta} - 1 - 2^k + 1)2^{-k-1} \\
&= \frac{k}{2} + \frac{2^\theta - 1}{2} \\
&= \frac{1}{2}\log_2(m+1) + \frac{1}{2}(2^\theta - 1 - \theta).
\end{aligned} \tag{4.4.20}$$

The second inequality follows by observing that $\nu(i) = \nu$ if and only if $2^{\nu-1} \leq i \leq 2^\nu - 1$, so there are exactly $2^{\nu-1}$ positive integers satisfying $\nu(i) = \nu$. When $m = 2^l - 1$ for some $l \geq 1, \theta = 1$ and the second term on the right vanishes, thereby proving (4.4.19). For other values of m, observe that $2^\theta - 1 - \theta$ is minimized by $\theta = -\log_2 \ln 2$. We can therefore continue the above bound by

$$\begin{aligned}
\sum_{i=1}^{m} R_{Ai}^{j} &> \frac{1}{2}\log_2(m+1) + \frac{1}{2}\left(\frac{1}{\ln 2} - 1 + \log_2 \ln 2\right) \\
&= \frac{1}{2}\log_2\left(\frac{e \ln 2}{2}(m+1)\right)
\end{aligned}$$

since $1/ln2 = \log_2 e$. This completes the proof of Theorem 4.20. □

Now let us define for all $j \geq 1$ and $1 \leq m \leq T_j$ the non-negative quantity

$$\epsilon_{m,j} \triangleq H_m - \sum_{i=1}^{m} R_{Ai}^{j}. \tag{4.4.21}$$

Combining Lemma 4.8 and Theorem 4.20, and observing that $\lceil m/2 \rceil \leq (m+1)/2$, we see that

$$\epsilon_{m,j} < \frac{1}{2}\log_2\left(\frac{\pi e}{2}(m+1)\right) - \frac{1}{2}\log_2\left(\frac{e\ln 2}{2}(m+1)\right)$$
$$= \frac{1}{2}\log_2\left(\frac{\pi}{\ln 2}\right) \doteq 1.090 \text{ b/channel use (b/cu)} \tag{4.4.22}$$

for all $j \geq 1$ and $1 \leq m \leq T_j$. A slightly tighter bound can be obtained for the sum-rate (where $m = T_j = 2^j - 1$) by using (4.4.19)

$$0 \leq C_{\text{sum}}(T_j) - R_{\text{sum}}(\mathcal{U}_A^j) < \frac{1}{2}\log_2\left(\frac{\pi e}{2}\right) \doteq 1.047 \text{ b/cu.} \tag{4.4.23}$$

Thus each supporting hyperplane of the polytope (4.4.16) is within 1./090 b/cu of a corresponding supporting hyperplane of the capacity region! By the remarks following Lemma 4.8 and Theorem 4.20, (4.4.22) and (4.4.23) are asymptotically tight.

Thus far, we have considered only multi-user codes in which the number of users is $T = 2^j - 1$ for some $j \geq 1$. However, it is not difficult to extend these results to an arbitrary number of users. Fix $T \geq 1$ and set $j = \nu(T)$. Observe that the number of users in $\mathcal{U}_A^j$ then satisfies $T_j \geq T$; hence, a T-user code $\mathcal{U}_{A,T}$ can be formed by taking the first T codes in $\mathcal{U}_A^j$. Since the partial sum-rates $\sum_{i=1}^m R_{Ai}^j$ are the same for $\mathcal{U}_{A,T}$ and $\mathcal{U}_A^j$ for all $1 \leq m \leq T$, we can construct, by elementary time-sharing from $\mathcal{U}_{A,T}$, codes with any rate satisfying (4.4.16) with T_j replaced by T. Thus combining (4.4.16) and (4.4.21), we can achieve all rates $(R_1, \ldots, R_T)$ in the region

$$0 \leq \sum_{i=1}^{m} R_{[i]} \leq H_m - \epsilon_{m,\nu(T)}, \qquad m = 1, \ldots, T \tag{4.4.24}$$

for every $T \geq 1$. Analogous codes $\overline{\mathcal{U}}_{A,T}$, $\mathcal{U}_{B,T}$, and $\overline{\mathcal{U}}_{B,T}$ can be obtained from $\overline{\mathcal{U}}_A^j$, $\mathcal{U}_B^j$, and $\overline{\mathcal{U}}_B^j$, respectively.

By modifying $\mathcal{U}_A^j$ slightly, we can obtain a (non-affine) UD code with a sum-rate even closer to the sum-capacity. Let $\overline{\mathcal{U}}_A^j$ be the code defined in the last subsection of Sect. 4.4.2 which was formed by merging pairs of single-bit codes in $\mathcal{U}_A^j$. The total number of users in $\overline{\mathcal{U}}_A^j$ is $\overline{T}_j = 3 \cdot 2^{j-2} - 1$ and hence the sum-rate is bounded by

$$R_{\text{sum}}(\overline{\mathcal{U}}_A^j) > \frac{(j - 1 + \log_2\sqrt{3})}{2} = \frac{1}{2}\log_2\frac{2}{\sqrt{3}}(\overline{T}_j + 1).$$

Thus Lemma 4.8 implies

$$0 \leq C_{\text{sum}}(\overline{T}_j) - R_{\text{sum}}(\overline{\mathcal{U}}_A^j) < \frac{1}{2}\log_2\left(\frac{\sqrt{3}\pi e}{4}\right) \doteq 0.943 \text{ b/cu.}$$

4.4.3.3 Codes Constructed from $\mathcal{U}_B^j$

We now consider the family of multi-user codes that can be constructed by elementary time-sharing from $\mathcal{U}_B^j$ (cf. the last subsection of Sect. 4.4.2). Most of the results of the preceding subsection carry over with little or no change; however, we need to adapt Theorem 4.20.

Theorem 4.21 (Partial sum-rate bounds for $\mathcal{U}_\mathbf{B}^\mathbf{j}$) *Let*

$$R_{Bi}^j \triangleq 2^{-\nu(i)} \qquad \text{for } 1 \le i \le 2^{j-2} - 1$$

and

$$R_{Bi}^j \triangleq 2^{1-j} \qquad \text{for } 2^{j-2} \le i \le 2^j - 1.$$

Then, for all $1 \le m \le 2^j - 1$, *inequalities (4.4.18) and (4.4.19) apply with* R_{Bi}^j *replacing* R_{Ai}^j *and with "*$\ge$*" replacing "*$>$*". Moreover, in the particular case* $m = T_j = 2^j - 1$, *we have the exact expression*

$$R_{sum}(\mathcal{U}_B^j) = \frac{1}{2}\log_2(2(T_j + 1)). \tag{4.4.25}$$

Proof Observe that $R_{Bi}^j \ge 2^{-\nu(i)}$ for all $1 \le m \le 2^j - 1$; hence $\sum_{i=1}^m R_{Bi}^j$ can be bounded below as in (4.4.20) with "$\ge$" replacing "$>$". Now (4.4.25) follows from Theorem 4.19 by observing that

$$R_{\text{sum}}(\mathcal{U}_B^j) = \frac{(j+1)}{2} = \frac{1}{2}(\log_2(T_j + 1) + 1) = \frac{1}{2}\log_2(2(T_j + 1)).$$

This completes the proof of Theorem 4.21. □

Proceeding as in the last subsection, we can show that multi-user codes constructed by elementary time-sharing from $\mathcal{U}_B^j$ can achieve all rates in (4.4.24) with $\epsilon_{m,j}$ replaced by

$$\epsilon'_{m,j} \triangleq H_m - \sum_{i=1}^m R_{Bi}^j.$$

Using Theorem 4.21 and Lemma 4.8, we can also show that $\epsilon'_{m,j} \le 1.090$ b/cu, for all $j \ge 1$ and $1 \le m \le T_j$. However, $\mathcal{U}_B^j$ can actually achieve *higher* sum-rate that $\mathcal{U}_A^j$. From Lemma 4.8 and (4.4.25), we obtain

$$0 \le C_{\text{sum}}(T_j) - R_{\text{sum}}(\mathcal{U}_B^j) \le \frac{1}{2}\log_2\left(\frac{\pi e}{4}\right) \doteq 0.547 \text{ b/cu}.$$

As in the last subsection, we can obtain multi-user codes for any T, say $\mathcal{U}_{B,T}$ and $\overline{\mathcal{U}}_{B,T}$, by taking the first T codes in $\mathcal{U}_B^j$ and $\overline{\mathcal{U}}_B^j$, respectively, for $T_j, \overline{T}_j \ge T$. In terms of sum-rate, $\mathcal{U}_{B.T}$ and $\overline{\mathcal{U}}_{B,T}$ are the most nearly optimal of all codes presented in this section.

4.4.4 The T-User, q-Frequency Adder Channel

The results presented in Sects. 4.4.2 and 4.4.3 have applications to the T-user, q-frequency multiple-access channel introduced by Wolf in [60]. This channel models a communication situation in which T synchronized users employ the same q-ary orthogonal signaling scheme, such as frequency shift keying or pulse position modulation. The channel is defined as follows: T users communicate with a single receiver through a shared discrete-time channel. At each time epoch, user i selects a *frequency* from the set $\{f_1, \ldots, f_q\}$ for transmission over the channel. The channel output consists of the q numbers $(N_1, \ldots, N_q)$, where N_i is the number of users transmitting at frequency f_i.

To make our notation compact, it is convenient to identify each frequency with an element of the set $F \triangleq \{0, 1, x, \ldots, x^{q-2}\}$, where x is an indeterminate variable. With this correspondence, the channel is equivalent to the polynomial adder channel, where i chooses an input $X_i \in F$ and the channel output is

$$Y \triangleq N_2 + N_3 x + \cdots + N_q x^{q-2}.$$

Note that N_1 is redundant because $N_1 = T - (N_2 + \cdots + N_q)$.

Most of the definitions given in Sect. 4.4.2 carry over with minor modifications to the present channel. We say that $(G, \mathbf{m})$ is an (N, K) *affine code* if G is a $K \times N$ matrix and $\mathbf{m}$ is an N-tuple, both with components that are real polynomials in x. The code is *q-ary* if

$$\mathbf{u} \in \{0, 1\}^K \Rightarrow \mathbf{u}G + \mathbf{m} \in \{0, 1, x, \ldots, x^{q-2}\}^N.$$

In particular, note that simultaneous transmission of more than one frequency (e.g., $1+x$) by a single user is not permitted. It is not difficult to show that $(G, \mathbf{m})$ is q-ary if and only if the following conditions are met

(i) $\mathbf{m} = (m_1, \ldots, m_N)$ is q-ary,
(ii) no column of G contains more than one non-zero component, and
(iii) all non-zero components of G take the form $g_{ij} = a - m_j$, for some $a \in F$.

In [10], Chang and Wolf generalized the Chang–Weldon codes to the T-user, q-frequency adder channel (see Sects. 4.3, 4.3.1–4.3.3). The main idea underlying their approach is to construct q-ary codes by "multiplexing" binary codes onto the $q - 1$ non-zero frequencies in F. To illustrate, let $(G, \mathbf{m})$ be any (N, K) binary affine code and consider the $(q - 1)$-user code

$$\mathcal{U}_q \triangleq \{(G, \mathbf{m}), x(G, \mathbf{m}), \ldots, x^{q-2}(G, \mathbf{m})\}.$$

Since $(G, \mathbf{m})$ is binary, the codewords generated by $x^i(G, \mathbf{m})$ take values in $\{0, x^i\}^N$. Thus the non-zero frequencies produced by each code in $\mathcal{U}_q$ are distinct, so user i "sees" a *single-user binary channel* in which the one is mapped into the q-ary symbol x^i. Clearly, $\mathcal{U}_q$ is a uniquely decodable, q-ary affine code.

More generally, if

$$\mathcal{U} = \{(G_1, \mathbf{m}_1), \ldots, (G_t, \mathbf{m}_t)\}$$

is a t-user $(N; K_1, K_2, \ldots, K_t)$ uniquely decodable, binary affine code then

$$\begin{aligned}\mathcal{U}_q \triangleq \{&(G_1, \mathbf{m}_1), x(G_1, \mathbf{m}_1), \ldots, x^{q-2}(G_1, \mathbf{m}_1),\\ &\cdot(G_2, \mathbf{m}_2), x(G_2, \mathbf{m}_2), \ldots, x^{q-2}(G_2, \mathbf{m}_2),\\ &\quad\vdots \qquad\qquad \vdots \qquad\qquad\qquad \vdots\\ &\cdot(G_t, \mathbf{m}_t), x(G_t, \mathbf{m}_t), \ldots, x^{q-2}(G_t, \mathbf{m}_t)\}\end{aligned} \tag{4.4.26}$$

is a T-user $(N; K_1', \ldots, K_T')$ uniquely decodable, q-ary affine code, where $T \triangleq (q-1)t$ and $K'_{(q-1)i+j} \triangleq K_{i+1}$ for all $0 \le i \le t-1$ and $1 \le j \le q-1$.

The same approach can be used to construct mixed-rate, q-ary codes from the binary codes presented in Sect. 4.4.2. Let $\mathcal{U}^j_{A,q}$ and $\mathcal{U}^j_{B,q}$ denote, respectively, the codes obtained by multiplexing $\mathcal{U}^j_A$ and $\mathcal{U}^j_B$ as indicated in (4.4.26). In the remainder of this section, we examine the rates of codes that can be constructed from $\mathcal{U}^j_{A,q}$ and compare them to the information-theoretic limits. For the sake of brevity, we omit a similar treatment of $\mathcal{U}^j_{B,q}$.

The following theorem is an immediate consequence of Theorems 4.16 and 4.17.

Theorem 4.22 *For all $j \ge 1$ and $q \ge 2$, $\mathcal{U}^j_{A,q}$ is a uniquely decodable, T_j-user $(N_j; K_1^j, \ldots, K_{T_j}^j)$ q-ary affine code, where*

$$\begin{aligned}T_j &= (q-1)(2^j - 1)\\ N_j &= 2^j - 1\\ K_i^j &= 2^{j-\nu'(i)}, \qquad 1 \le i \le T_j\end{aligned} \tag{4.4.27}$$

and $\nu'(i) \triangleq \lceil \log_2(i/(q-1)+1) \rceil$. The rate of $\mathcal{U}^j_{A,q}$ is given by $\mathbf{R}^j_q \triangleq (R^j_{q1}, \ldots, R^j_{qT_j})$, where

$$R^j_{qi} \triangleq \frac{2^{j-\nu'(i)}}{2^j - 1}.$$

Hence, the sum rate is

$$R_{sum}(\mathcal{U}^j_{A,q}) = \frac{(q-1)j2^{j-1}}{2^j - 1}. \tag{4.4.28}$$

From (Eqs. 4.4.11 and 4.4.15), it can be inferred that the capacity region of ther T-user, q-frequency adder channel is the set of all non-negative rates $(R_1, \ldots, R_T)$ satisfying

$$0 \leq \sum_{i=1}^{m} R_{[i]} \leq H(q,m), \qquad m = 1, \ldots, T \tag{4.4.29}$$

where

$$H(q,m) \triangleq - \sum_{m_1+\cdots+m_q=m} \binom{m}{m_1, \ldots, m_q} q^{-m} \cdot \log_2 \binom{m}{m_1, \ldots, m-q} q^{-m}. \tag{4.4.30}$$

We now characterize the rates of codes that can be constructed from $\mathcal{U}^j_{A,q}$. Since the arguments used here are similar to the derivation of (4.4.24), we will be brief. Let $T \geq 1$ and $q \geq 2$ be arbitrary, and set $j \triangleq \nu'(T)$. Proceeding as in Sect. 4.4.3, we can construct, by elementary time-sharing from $\mathcal{U}^j_{A,q}$, a uniquely decodable T-user code with any rate $(R_1, \ldots, R_T)$ satisfying

$$0 \leq \sum_{i=1}^{m} R_{[i]} \leq H(q,m) - \epsilon_{q,m,\nu'(T)}, \qquad m = 1, \ldots, T \tag{4.4.31}$$

where

$$\epsilon_{q,m,j} \triangleq H(q,m) - \sum_{i=1}^{m} R^j_{qi} \tag{4.4.32}$$

is defined for all $j \geq 1$, $q \geq 2$, and $1 \leq m \leq (q-1)(2^j - 1)$. Once again, our aim is to bound $\epsilon_{q,m,j}$ by first obtaining individual bounds on $H(q,m)$ and $\sum_{i=1}^{m} R^j_{qi}$.

Chang and Wolf [10] have given an estimate of $H(q,m)$ for large q and m

$$H(q,m) \approx \frac{q-1}{2} \log_2 \frac{2\pi e m}{q^{q/(q-1)}}. \tag{4.4.33}$$

For our purposes, however, it will be more useful to have an upper bound on $H(q,m)$.

Lemma 4.9 *For all $m \geq 1$ and $q \geq 2$*

$$\begin{aligned} H(q,m) &\leq \frac{q-2}{2} \log_2 2\pi e \left(\frac{m}{q} + \frac{1}{12}\right) + \frac{1}{2} \log_2 2\pi e \left(\frac{m}{q^2} + \frac{1}{12}\right) \\ &< \frac{q-1}{2} \log_2 2\pi e \left(\frac{m}{q} + \frac{1}{12}\right). \end{aligned} \tag{4.4.34}$$

Before proceeding with the proof, we need a slight extension of the differential entropy bound. The proof is a straightforward generalization of the one in [14] (p. 235) and so is omitted.

Lemma 4.10 (Differential entropy bound) *Let* $\mathbf{X}$ *be a random vector values in the integer lattice in* $\mathbb{R}^n$. *Then*

$$H(\mathbf{X}) \leq \tfrac{1}{2}\log_2(2\pi e)^n |Cov(\mathbf{X}) + \tfrac{1}{12}I_n| \tag{4.4.35}$$

where $Cov(\mathbf{X})$ *is the covariance matrix of* $\mathbf{X}$, $|A|$ *denotes the absolute value of the determinant of the matrix* A, *and* I_n *is the identity matrix of order* n.

Proof of Lemma 4.9. To prove Lemma 4.9, let $(X_1, \dots, X_q)$ denote a random vector with the multinomial distribution

$$\Pr\{X_1 = m_1, \dots, X_q = m_q\} \triangleq \binom{m}{m_1, \dots, m_q} q^{-m}$$

for all $m_i \geq 0$, $m_1 + \cdots + m_q = m$, and observe that

$$H(q, m) = H(X_1, \dots, X_1).$$

Let $\mathbf{X} \triangleq (X_1, \dots, X_{q-1})$ denote the first $q-1$ components of $(X_1, \dots, X - q)$ and note that $H(q, m) = H(\mathbf{X})$ since $X_q = m - \sum_{i=1}^{q-1} X_i$ is redundant. Since

$$\mathbf{E}(X_i) = m/q$$
$$\mathbf{E}(X_i - m/q)(X_j - m/q) = \begin{cases} m(q-1)/q^2, & i = j \\ -m/q^2, & i \neq j \end{cases}$$

it follows that

$$\mathrm{Cov}(\mathbf{X}) = (m/q)I_{q-1} - (m/q^2)J_{q-1}$$

where J_{q-1} is the square, all-one matrix of order $q-1$. Using the well-known determinant formula

$$|aI_n + bJ_n| = a^{n-1}(a + nb)$$

with $n = q-1$, $a = m/q + 1/12$, and $b = -m/q^2$, we obtain

$$\left|\mathrm{Cov}(\mathbf{X}) + \frac{1}{12}I_{q-1}\right| = \left(\frac{m}{q} + \frac{1}{12}\right)^{q-2}\left(\frac{m}{q^2} + \frac{1}{12}\right). \tag{4.4.36}$$

Applying Lemma 4.10 with $n = q-1$, we obtain the upper bound in Lemma 4.9. □

Remark For $q = 2$, the first bound reduces to $(1/2)\log_2 \pi e(m/2 + 1/6)$, which improves on (4.4.17) for odd m.

Theorem 4.23 (Partial sum-rate bounds for $\mathcal{U}^{\mathbf{j}}_{\mathbf{A},\mathbf{q}}$) *For any* $j \geq 1$ *and* $q \geq 2$, *consider the rate*

$$\mathbf{R}_q^j \triangleq (R_{q1}^j, \dots, R_{qT_j}^j)$$

defined in Theorem 4.22. For all $1 \le m \le (q-2)(2^j-1)$

$$\sum_{i=1}^{m} R_{qi}^{j} > \frac{q-1}{2}\log_2\left[\frac{e\ln 2}{2}\left(\frac{m}{q-1}+1\right)\right]. \tag{4.4.37}$$

When $m = (q-1)(2^l-1)$ *for some* $l \ge 1$, *this bound can be tightened to*

$$\sum_{i=1}^{m} R_{qi}^{j} > \frac{q-1}{2}\log_2\left[\frac{m}{q-1}+1\right]. \tag{4.4.38}$$

Proof The proof is similar to that of Theorem 4.20. For any $m \ge 1$, we can write $m = (q-1)(2^{k+\theta}-1)$ for some $k \ge 0$ and $0 < \theta \le 1$. Note that $\log_2[m/(q-1)+1] = k+\theta$ and $\nu'(m) = k+1$. The partial sum of the rates can then be bounded by

$$\begin{aligned}\sum_{i=1}^{m} R_{qi}^{j} &= \frac{2^j}{2^j-1}\sum_{i=1}^{m} 2^{-\nu'(i)} \\ &> \sum_{i=1}^{(q-1)(2^k-1)} 2^{-\nu'(i)} + [m-(q-1)(2^k-1)]2^{-\nu'(m)} \\ &= (q-1)\left[\frac{k}{2}+\frac{(2^\theta-1)}{2}\right] \\ &= \frac{q-1}{2}\log_2\left(\frac{m}{q-1}+1\right)+\frac{q-1}{2}(2^\theta-1-\theta).\end{aligned}$$

If $m = (q-1)(2^l-1)$ for some $l \ge 1$, then $\theta = 1$ and the second term above vanishes, proving (4.4.38). For other values of m, we can bound $2^\theta - 1 - \theta$ below by its minimum value, which is achieved at $\theta = -\log_2 \ln 2$. This completes the proof of Theorem 4.23. □

Combining the bounds in Lemma 4.9 and Theorem 4.23, we obtain for all q, m, and j

$$\begin{aligned}\epsilon_{q,m,j} &< \frac{q-1}{2}\log_2\left[\frac{4\pi}{\ln 2}\left(\frac{m/q+1/12}{m/(q-1)+1}\right)\right] \\ &< \frac{q-1}{2}\log_2\left(\frac{4\pi}{\ln 2}\right) \doteq 2.047(q-1) \text{ b/cu.}\end{aligned} \tag{4.4.39}$$

Thus each supporting hyperplane of the polytope (4.4.31) is within $2.047(q-1)$ b/cu of the corresponding supporting hyperplane of the capacity region (4.4.29).

It is useful to compare the exact values of $H(q,m)$ and $\sum_{i=1}^{m} R_{qi}^{j}$ for small m and q, where the bound in Lemma 4.9 is loose. To this end, define

$$R(q, m) \triangleq \sum_{i=1}^{m} 2^{-\nu'(i)}. \tag{4.4.40}$$

so that

$$\sum_{i=1}^{m} R_{qi}^{j} = \frac{2^j}{2^j - 1} R(q, m)$$

and hence $\sum_{i=1}^{m} R_{qi}^{j} \doteq R(q, m)$ for large j. The table below gives the values of $H(q, m)$ and $R(q, m)$ for $q = 3, 4, 5$, and even values of m between 2 and 40. For these values of m and large j, we see that $\epsilon_{3,m,j}$, $\epsilon_{4,m,j}$, and $\epsilon_{5,m,j}$ take values in the ranges 2–2.7, 3.1–4.4, 3.8–6.2, respectively. Thus $\epsilon_{q,m,j}$ is significantly smaller than the bound in (4.4.39). However, it can be shown using (4.4.33) that the bound in (4.4.39) can be approached for large j and q by certain values of m.

m	$H(3,m)$	$R(3,m)$	$H(4,m)$	$R(4,m)$	$H(5,m)$	$R(5,m)$
4	3.57950	1.5000	4.81511	1.7500	5.83830	2.0000
6	4.21984	2.0000	5.79298	2.2500	7.13552	2.5000
8	4.66393	2.2500	6.48269	2.7500	8.06979	3.0000
10	5.00146	2.5000	7.00785	3.1250	8.78770	3.5000
12	5.27347	2.7500	7.42940	3.3750	9.36535	4.0000
14	5.50145	3.0000	7.78082	3.6250	9.84639	4.2500
16	5.69783	3.1250	8.08204	3.8750	10.2577	4.5000
18	5.87040	3.2500	8.34568	4.1250	10.6166	4.7500
20	6.02439	3.3750	8.58018	4.3750	10.9349	5.0000
22	6.16344	3.5000	8.79143	4.5625	11.2209	5.2500
24	6.29020	3.6250	8.98367	4.6875	11.4807	5.5000
26	6.40669	3.7500	9.16008	4.8125	11.7186	5.7500
28	6.51446	3.8750	9.32310	4.9375	11.9382	6.0000
30	6.61471	4.0000	9.47463	5.0625	12.1420	6.1250
32	6.70844	4.0625	9.61620	5.1875	12.3323	6.2500
34	6.79644	4.1250	9.74905	5.3125	12.5107	6.3750
36	6.87938	4.1875	9.87419	5.4375	12.6787	6.5000
38	6.95780	4.2500	9.99246	5.5625	12.8374	6.6250
40	7.03217	4.3125	10.1046	5.6875	12.9877	6.7500

Entropies and partial sum-rates for$q$ − ary codes

4.4.5 Concluding Remarks

We have presented two multi-user code constructions of Hughes and Cooper [26] for the binary adder channel. The codewords in these codes are equivalant, up to an affine transformation, to the coin weighing design in [34] and the symmetric-rate multi-user code of Chang and Weldon. The main idea behind their construction is to distribute these codewords among *as few users as possible*. This yields several important benefits. First, we obtain multi-user codes with a variety of information rates. Second, because decreasing the number of users also shrinks the capacity region, we obtain multi-user codes which are more nearly optimal. Third, by time-sharing, we can construct multi-user codes approaching all rates in the polytope (4.4.24), where each supporting hyperplane of the polytope is within 1.090 b/cu of a corresponding hyperplane of the capacity region. Similar results were also presented for the T-user q-frequency adder channel.

In this section, we conclude with several remarks concerning the performance of the codes presented here. First, it is important to recognize that many uniquely decodable, multi-user codes are known with rates that fall outside of the polytope (4.4.24). Specifically, this is true of almost all codes developed for the two-user binary adder channel. It is also true of many codes that can be constructed by elementary time-sharing from the trivial code with rate $\mathbf{R} = (1, 0, \ldots, 0)$. However, for large T, most of the rates in (4.4.24) are new. In particular, the sum-rate of $\overline{\mathcal{U}}_{B,T}$ is higher than that of almost all codes previously reported in [8, 9, 20, 27, 34, 43, 59]. For $T \geq 3$, the only codes with higher sum-rates are the $T = 5$, 10–12, 20–25 codes in [34].

Second, it is interesting to compare the sum-rate of Hughes and Cooper's codes with that of Chang and Weldon's codes [9]. For each $j \geq 1$, Chang and Weldon constructed a uniquely decodable, T'_j-user $(N'_j; 1, \ldots, 1)$ code, where $N'_j = 2^j$ and $T'_j = (j+2)2^{j-1}$. They further showed that this code, which we denote by $\mathcal{U}'^j$, is asymptotically optimal in the sense that the relative difference $(C_{\text{sum}}(T'_j) - R_{\text{sum}}(C'^j))/C_{\text{sum}}(T'_j)$ vanishes as $j \to +\infty$. However, observe that

$$R_{\text{sum}}(C'^j) = (j+2)/2 = \frac{1}{2}\log_2 T'_j - \frac{1}{2}\log_2\left(\frac{j+2}{8}\right).$$

Hence, from the lower bound in Lemma 4.8

$$C_{\text{sum}}(T'_j) - R_{\text{sum}}(C'^j) \geq \frac{1}{2}\log_2\left(\frac{\pi(j+2)}{16}\right) \to +\infty$$

as $j \to +\infty$. Thus while the *relative* distance between $C_{\text{sum}}(T'_j)$ and $R_{\text{sum}}(C'^j)$ vanishes, the *absolute* difference grows without bound. By contrast, the sum-rate of the code $\mathcal{U}_B^j$ defined in the last subsection of Sect. 4.4.2 is not more than 0.547 b/cu from the sum-capacity for any $j \geq 3$.

Finally, recall that the capacity region (4.4.2) consists of all rates for which there exist codes which can achieve an *arbitrarily small* probability of decoding error. However, the unique decodability condition imposed in Definition 4.12 implies that the codes constructed in this section have an error probability of zero. Thus two remarks are in order. First, (4.4.24) constitutes an inner bound on the *zero-error* capacity region of the T-user binary adder channel. Second, since the zero-error capacity region is in general smaller than the arbitrarily small error capacity region, it might not be possible to find T-user uniquely decodable codes achieving all rates in (4.4.2).

4.5 Coding for the Binary Switching Channel

4.5.1 UD Codes for the Binary Switching Channel

The binary switching channel was defined in the first example of Sect. 4.1 in such a way that user 2 'switches' the connection between the user 1 and the receiver on and off by sending zeroes and ones, respectively. Thus, a codeword $v \in \mathcal{V}$ can be considered as an erasure pattern on a codeword $u \in \mathcal{U}$: the received word z equals u, except in those coordinates where v is 0; there the receiver reads the symbol 2. Thus, *the decoder always knows the codeword* v. The problem of sending information from user 1 to the receiver over the channel resembles the coding problem for memories with defects, because a codeword of user 1 can become corrupted when a codeword of user 2 erases some of its symbols. In our case, however, the decoder knows the erased (defect) positions, while the encoder (user 1) does not. Furthermore, user 2 can choose the defect positions by choosing $\mathcal{V}$.

The achievable rate region for the arbitrarily small average decoding error probability was defined by the formula (4.1.7) in Sect. 4.1. We will consider the problem of specifying the achievable rate region of the UD codes and give the following

Definition 4.14 A code $(\mathcal{U}, \mathcal{V})$, where $\mathcal{U}$ and $\mathcal{V}$ are block codes of length n is referred to as a UD code for the binary switching channels if, for all $(u, v), (u', v) \in \mathcal{U} \times \mathcal{V}$ such that $u \neq u'$,

$$u \wedge v \neq u' \wedge v, \tag{4.5.1}$$

where $\wedge$ denotes the binary AND operation.

Let $T(\mathcal{U})$ denote the set consisting of all binary vectors v such that (4.5.1) is satisfied for all $u, u' \in \mathcal{U}$ with $u \neq u'$. The elements of $T(\mathcal{U})$ can be viewed as erasure patterns *tolerated by* $\mathcal{U}$. Obviously, $(\mathcal{U}, \mathcal{V})$ with $\mathcal{V} = T(\mathcal{U})$ is a UD code having the maximal possible cardinality for a given $\mathcal{U}$.

Theorem 4.24 (Vanroose (1988), [55])

(i) If $(\mathcal{U}, \mathcal{V})$ *is a UD code and* $|\mathcal{U}| > 2^{k-1}$, *then*

$$|\mathcal{V}| \leq \sum_{i=0}^{n-k} \binom{n}{i}. \tag{4.5.2}$$

(ii) There exist linear (n,k)-codes $\mathcal{U}$ such that

$$|T(\mathcal{U})| \geq \frac{1}{2} \sum_{i=0}^{n-k-1} \binom{n}{i}. \tag{4.5.3}$$

It is easy to see that (4.5.2) and (4.5.3) asymptotically coincide and we have the following

Corollary 4.4 *For the binary switching channel, all rate pairs (R_1, R_2) such that*

$$R_2 \leq \begin{cases} h(R_1), & \text{if } R_1 \geq 1/2, \\ 1, & \text{if } R_1 \leq 1/2, \end{cases}$$

can be asymptotically achieved with the decoding error probability zero when $\mathcal{U}$ is a linear code.

The average error capacity region coincides with the one for UD codes.

If n is finite, then values obtained from (4.5.2) are less than corresponding values obtained from (4.5.3). We denote

$$\underline{R} = \max_k \left\{ \frac{k-1}{n} + \frac{1}{n} \log \sum_{i=0}^{n-k} \binom{n}{i} \right\},$$

$$\overline{R} = \min_k \left\{ \frac{k}{n} + \frac{1}{n} \log \frac{1}{2} \sum_{i=0}^{n-k-1} \binom{n}{i} \right\},$$

and show $\underline{R}$ and $\overline{R}$ in Table 4.3.

The proof of Theorem 4.24 uses the following auxiliary result.

Table 4.3 The values of the lower and upper bounds, $\underline{R}$ and $\overline{R}$, on the sum rate of uniquely decodable codes for the switching channel; n is the code length

n	$\underline{R}$	$\overline{R}$	n	$\underline{R}$	$\overline{R}$
1	0	1	25	1.435	1.515
2	0.292	1.292	50	1.501	1.541
3	0.667	1.333	100	1.539	1.559
4	0.865	1.365	250	1.564	1.572
5	1.000	1.400	500	1.573	1.577
6	1.077	1.410	1000	1.579	1.581
7	1.143	1.429	2000	1.582	1.583
8	1.192	1.442			
9	1.225	1.447			
10	1.259	1.459			

Proposition 4.15 *Given a linear* (n, k)*-code* $\mathcal{U}$ *with the weight distribution*

$$A_l = \sum_{u \in \mathcal{U} \setminus \{0^n\}} \chi\{\, w_H(u) = l \,\}, \quad l = 0, \ldots, n,$$

we may write

$$|T(\mathcal{U})| \geq \sum_{i=0}^{n-k} \binom{n}{i} - \sum_{i=0}^{n-k-1} A_{n-k-i} \sum_{j=0}^{i} \binom{i+k}{j}. \tag{4.5.4}$$

Proof Let

$$\mathcal{F}_w = \bigcup_{u \in \mathcal{U} \setminus \{0^n\}} \mathcal{F}_w(u),$$

where

$$\mathcal{F}_w(u) = \left\{\, x \in \{0, 1\}^n : \; w_H(x) = w, \; x \wedge u = 0^n \,\right\}.$$

Then

$$T(\mathcal{U}) = \{0, 1\}^n \setminus \bigcup_{w=0}^{n} \mathcal{F}_w$$

and

$$|T(\mathcal{U})| = 2^n - \sum_{w=0}^{n} |\mathcal{F}_w|$$

since we should exclude from $T(\mathcal{U}$ the elements $x \in \{0, 1\}^n$ such that $u \wedge x = u' \wedge x$. If $w < k$ then

$$|\mathcal{F}_w| = \binom{n}{w}.$$

Otherwise, we note that

$$|\mathcal{F}_w(u)| = \binom{n - w_H(u)}{w},$$

because the w ones in a codeword of $\mathcal{F}_w(u)$ have to be placed in those coordinates where u has zeroes, and

$$|\mathcal{F}_w| \leq \sum_{u \in \mathcal{U} \setminus \{0^n\}} \binom{n - w_H(u)}{w} = \sum_{l=1}^{n} A_l \binom{n-l}{w}.$$

Combining these cases we get

$$
\begin{aligned}
|T(\mathcal{U})| &= \sum_{i=0}^{n-k} \binom{n}{i} - \sum_{w=k}^{n} |\mathcal{F}_w| \\
&\geq \sum_{i=0}^{n-k} \binom{n}{i} - \sum_{l=1}^{n} A_l \sum_{w=k}^{n} \binom{n-l}{w} \\
&= \sum_{i=0}^{n-k} \binom{n}{i} - \sum_{l=1}^{n-k} A_l \sum_{w=k}^{n} \binom{n-l}{w},
\end{aligned}
$$

rename the indices: $(i, j) = (n-k-l, n-w-l)$, and obtain (4.5.4). □

4.5.1.1 Proof of Theorem 4.24

If $|\mathcal{U}| > 2^{k-1}$ then $\mathcal{U}$ cannot tolerate more than $n-k$ elements. Consequently, $\mathcal{V} \subseteq \{v : w_H(v) \geq k\}$ and (4.5.2) follows.

To prove (4.5.3) we note that

$$
\begin{aligned}
\sum_{l=1}^{m} \sum_{s=l}^{m} \binom{n}{l}\binom{n-l}{s-l} &= \sum_{s=1}^{m} \sum_{l=l}^{s} \binom{n}{l}\binom{n-l}{s-l} \qquad (4.5.5) \\
&= \sum_{s=1}^{m} \sum_{l=l}^{s} \binom{n}{s}\binom{s}{l} \\
&= \sum_{s=1}^{m} \binom{n}{s}(2^s - 1).
\end{aligned}
$$

It is well-known that the average weight distribution of a linear (n, k)-code satisfies the equations:

$$
\begin{aligned}
\overline{A}_0 &= 1, \\
\overline{A}_l &= 2^{k-n}\left[\binom{n}{l} - \binom{n-k}{l}\right] \leq 2^{k-n}\binom{n}{l}, \quad l = 1, ..., n.
\end{aligned}
$$

If $\mathcal{U}$ is a code with the weight distribution

$$
(A_0, ..., A_n) = (\overline{A}_0, ..., \overline{A}_n),
$$

then

$$
\sum_{i=0}^{n-k-1} A_{n-k-i} \sum_{j=0}^{i} \binom{i+k}{j} \leq \sum_{i=0}^{n-k-1} \binom{n}{i+k} 2^{k-n} \sum_{j=0}^{i} \binom{i+k}{j}.
$$

Therefore using (4.5.4) and (4.5.5) we write:

$$
\begin{aligned}
|T(\mathcal{U})| &\geq \sum_{i=0}^{n-k} \binom{n}{i} - 2^{k-n} \sum_{i=0}^{n-k-1} \binom{n}{i+k} \sum_{j=0}^{i} \binom{i+k}{j} \\
&= \sum_{i=0}^{n-k} \binom{n}{i} - 2^{k-n} \sum_{s=1}^{n-k} \binom{n}{s} (2^s - 1) \\
&= \sum_{i=0}^{n-k} \binom{n}{i} (1 - 2^{k-n+i} - 2^{k-n}) \\
&\geq \sum_{i=0}^{n-k} \binom{n}{i} (1 - 2^{k-n+i}) \\
&\geq \frac{1}{2} \sum_{i=0}^{n-k-1} \binom{n}{i}
\end{aligned}
$$

and complete the proof.

□

4.6 Coding for Interference Channels

4.6.1 Statement of the Coding Problem for Interference Channels

Discrete memoryless interference channels differ from MACs as follows:

- the output alphabet $\mathcal{Z}$ is represented as a Cartesian product of two finite sets, $\mathcal{Z}_1$ and $\mathcal{Z}_2$, i.e., the channel is defined by the crossover probabilities

$$W(z_1, z_2|x, y), \quad (z_1, z_2) \in \mathcal{Z}_1 \times \mathcal{Z}_2, \ (x, y) \in \mathcal{X} \times \mathcal{Y};$$

- there are two receivers; the first receiver gets z_1 and estimates the message of the first user, and the second receiver gets z_2 and estimates the message of the second user.

The definition of the achievable rate region under the criterion of arbitrarily small average decoding error probability can be introduced for the interference channels in the same way as for MACs. However, in general case, only the following result is known [46]: *the achievable rate region under the criterion of arbitrarily small average decoding error probability for the interference channels consists of all pairs* (R_1, R_2) *such that, for some* $n > 1$, *there exist auxiliary random variables* X^n *and* Y^n *with the property:*

$$R_1 \leq \frac{1}{n} I(Z_1^n \wedge X^n), \quad R_2 \leq \frac{1}{n} I(Z_2^n \wedge Y^n).$$

Note that the region defined by these inequalities does not have a single-letter characterization, i.e., we are supposed to increase n up to infinity.

Open problem *There exists the following conjecture: given an $\varepsilon > 0$, one can specify a value $f(\varepsilon) < \infty$ such that the achievable rate region can be found with the distortion less than ε if we restrict considerations to all $n < f(\varepsilon)$. Prove or disprove this conjecture.*

We will deal with the problem of constructing UD codes for a special class of deterministic interference channels.

Definition 4.15 An interference channel will be referred to as $(\vee, \wedge)$-channel if $\mathcal{X} = \mathcal{Y} = \mathcal{Z}_1 = \mathcal{Z}_2 = \{0, 1\}$ and

$$z_1 = x \vee y, \quad z_2 = x \wedge y,$$

where the signs $\vee$ and $\wedge$ stand for the binary OR and AND operations, respectively. In other words,

$$\begin{aligned}
(x, y) = (0, 0) &\Longrightarrow \ (z_1, z_2) = (0, 0),\\
(x, y) = (0, 1) &\Longrightarrow \ (z_1, z_2) = (1, 0),\\
(x, y) = (1, 0) &\Longrightarrow \ (z_1, z_2) = (1, 0),\\
(x, y) = (1, 1) &\Longrightarrow \ (z_1, z_2) = (1, 1).
\end{aligned}$$

Definition 4.16 A pair of rates (R_1, R_2) is the point belonging to the achievable rate region $\mathcal{R}_{\vee,\wedge}$ of UD codes for the $(\vee, \wedge)$-channel if and only if there exist codes $\mathcal{U}$ and $\mathcal{V}$ of rates R_1 and R_2 such that, for all $(u, v), (u', v') \in \mathcal{U} \times \mathcal{V}$,

$$u \vee v = u' \vee v' \Longrightarrow \ u = u', \tag{4.6.1}$$

$$u \wedge v = u' \wedge v' \Longrightarrow \ v = v'. \tag{4.6.2}$$

A partial solution of the problem of finding the region $\mathcal{R}_{\vee,\wedge}$ is specifying the maximal value of the sum rate of UD codes or the maximal product of the cardinalities of UD codes.

Proposition 4.16 *1. There exist codes $(\mathcal{U}, \mathcal{V})$ satisfying (4.6.1)–(4.6.2) such that*

$$|\mathcal{U}| \cdot |\mathcal{V}| \geq 2^n. \tag{4.6.3}$$

2. If a pair $(\mathcal{U}, \mathcal{V})$ satisfies (4.6.1)–(4.6.2), then

$$|\mathcal{U}| \cdot |\mathcal{V}| \leq 3^n. \tag{4.6.4}$$

Proof Let us fix $\alpha \in (0, 1)$ in such a way that αn is an integer and assign

$$\mathcal{U} = \bigcup_{b \in \{0,1\}^{\alpha n}} \{(b, 0^{(1-\alpha)n})\}, \quad \mathcal{V} = \bigcup_{b \in \{0,1\}^{(1-\alpha)n}} \{(1^{\alpha n}, b)\},$$

where $0^{(1-\alpha)n}$ and $1^{\alpha n}$ denote the all-zero vector of length $(1-\alpha)n$ and the all-one vector of length αn, respectively. It is easy to see that $(\mathcal{U}, \mathcal{V})$ satisfy (4.6.1)–(4.6.2) and

$$|\mathcal{U}| \cdot |\mathcal{V}| = 2^n$$

for any α. Hence, (4.6.3) follows.

To prove (4.6.4) let us denote by

$$w = \min_{u \in \mathcal{U}} w_H(u)$$

the minimal Hamming weight of codewords included into $\mathcal{U}$ and let $u_0 \in \mathcal{U}$ be a codeword of weight w. Then using (4.6.2) we obtain that all elements of the set $u_0 \wedge \mathcal{V}$ are different and $|\, u_0 \wedge \mathcal{V} \,| = |\, \mathcal{V} \,|$. However, if $y = u_0 \wedge v$ for some $v \in \{0, 1\}^n$, then $y_j \leq u_j$ for all $j = 1, ..., n$ and $w_H(y) \leq w$. Thus,

$$|\, \mathcal{V} \,| \leq 2^w, \quad |\, \mathcal{U} \,| \leq \sum_{i=w}^{n} \binom{n}{i}$$

and we get

$$|\mathcal{U}| \cdot |\mathcal{V}| \leq 2^w \cdot \sum_{i=w}^{n} \binom{n}{i} \leq \sum_{i=0}^{n} \binom{n}{i} \cdot 2^i = 3^n.$$

□

4.6.2 The Sandglass Conjecture

The problem of constructing UD codes for $(\vee, \wedge)$-channels can be presented for a more general setup when we assume that u and v are some elements of the lattice.

Definition 4.17 Let $\mathcal{L}$ be a set consisting of elements $a, b, ...$ Suppose that there is a binary relation $' \leq '$ defined between pairs of elements of $\mathcal{L}$ in such a way that

$$a \leq a,$$
$$a \leq b, \; b \leq a \Longrightarrow \; a = b,$$
$$a \leq b, \; b \leq c \Longrightarrow \; a \leq c.$$

Then $\mathcal{L}$ is referred to as a partially ordered set by the relation $' \leq'$. A partially ordered set $\mathcal{L}$ is referred to as a lattice if, for any pair $(a, b) \in \mathcal{L}^2$, there exist the elements $a \vee b$ and $a \wedge b$ such that

$$a \leq c,\ b \leq c \Longrightarrow\ a \vee b \leq c,$$
$$c \leq a,\ c \leq b \Longrightarrow\ c \leq a \wedge b.$$

The elements $a \vee b$ and $a \wedge b$ are known as the least upper bound of a and b and the greatest lower bound of a and b, respectively.

Definition 4.18 A pair $(\mathcal{U}, \mathcal{V})$ of subsets of a lattice $\mathcal{L}$ is said to form a sandglass if there exists an element $c \in \mathcal{L}$ that satisfies $c \leq u$ for every $u \in \mathcal{U}$ and $c \geq v$ for every $v \in \mathcal{V}$. A sandglass is full or saturated if adding any new element to $\mathcal{U}$ or $\mathcal{V}$ the new pair is not a sandglass any more.

Note that in a lattice we could equivalently define a sandglass by the property that $v \leq u$ holds for every $(u, v) \in \mathcal{U} \times \mathcal{V}$ (for general partially ordered sets these two possible definitions do not coincide).

Let

$$M(\mathcal{L}) = \max_{\mathcal{U},\mathcal{V}} |\mathcal{U}| \cdot |\mathcal{V}|,$$

where the maximum is taken over all pairs $(\mathcal{U}, \mathcal{V}) \subseteq \mathcal{L}^2$ such that the statements (4.6.1)–(4.6.2) are valid for all $(u, v), (u', v') \in \mathcal{U} \times \mathcal{V}$.

Sandglass Conjecture (Ahlswede and Simonyi (1994), [4]) *Let $\mathcal{L}$ be the product of k finite length chains. Then there exists a saturated sandglass $(\mathcal{U}, \mathcal{V}) \subseteq \mathcal{L}^2$ for which*

$$|\mathcal{U}| \cdot |\mathcal{V}| = M(\mathcal{L}).$$

The Sandglass Conjecture is trivial for $k = 1$. We show that it holds for $k = 2$.

Theorem 4.25 (Ahlswede and Simonyi (1994), [4]) *Let $\mathcal{L}$ be a lattice obtained as the product of two finite length chains. Then $M(\mathcal{L})$ can be achieved by a sandglass.*

First we prove four Lemmas (note that the first three of them are valid for any lattice).

Lemma 4.11 *If $(\mathcal{U}, \mathcal{V})$ is a recovering pair and there exists a pair of elements $(u, v) \in \mathcal{U} \times \mathcal{V}$ with $v \geq u$ then there exists a sandglass $(\mathcal{U}', \mathcal{V}')$ with $|\mathcal{U}'| \geq |\mathcal{U}|$ and $|\mathcal{V}'| \geq |\mathcal{V}|$.*

Proof Using the definition of recovering pairs we get

$$|\mathcal{U}| \leq \min_{v\in\mathcal{V}} |\{\, a \in \mathcal{L} :\ a \geq v \,\}|\,,$$
$$|\mathcal{V}| \leq \min_{u\in\mathcal{U}} |\{\, a \in \mathcal{L} :\ a \leq u \,\}|\,.$$

If there exists a pair of elements $(u, v) \in \mathcal{U} \times \mathcal{V}$ with $v \geq u$ then consider the sandglass

$$\mathcal{U}' = \{\, a \in \mathcal{L} : \; a \geq v \,\},$$
$$\mathcal{V}' = \{\, a \in \mathcal{L} : \; a \leq v \,\}.$$

Since

$$\{\, a \in \mathcal{L} : \; a \leq u \,\} \subseteq \mathcal{V}',$$

we have $|\mathcal{U}'| \geq |\mathcal{U}|$ and $|\mathcal{V}'| \geq |\mathcal{V}|$.

□

We call a recovering pair $(\mathcal{U}, \mathcal{V})$ *canonical* if there are no pairs of elements $(u, v) \in \mathcal{U} \times \mathcal{V}$ with $v \geq u$. It remains to analyze canonical pairs.

Note that the statements (4.6.1)–(4.6.2) are equivalent to the following ones:

$$(u \vee \mathcal{V}) \bigcap (u' \vee \mathcal{V}) = \emptyset, \quad \text{for all} \;\; (u, u') \in \mathcal{U}^2, \; u \neq u', \tag{4.6.5}$$

$$(v \wedge \mathcal{U}) \bigcap (v' \wedge \mathcal{U}) = \emptyset, \quad \text{for all} \;\; (v, v') \in \mathcal{V}^2, \; v \neq v' \tag{4.6.6}$$

and define

$$Max(u, \mathcal{V}) = \bigvee_{a \in u \vee \mathcal{V}} a,$$
$$Min(v, \mathcal{U}) = \bigwedge_{b \in v \wedge \mathcal{U}} b.$$

Lemma 4.12 *If $(\mathcal{U}, \mathcal{V})$ is a recovering pair and there exists a $u_0 \in \mathcal{U}$ such that $Max(u_0, \mathcal{V}) \in u_0 \vee \mathcal{V}$ then the set*

$$\mathcal{U}^+ = \{\mathcal{U} \backslash u_0\} \cup \{Max(u_0, \mathcal{V})\}$$

also forms a recovering pair with $\mathcal{V}$.

Proof Note that, for all $u \in \mathcal{U}$ such that $u \neq u_0$, the values $u \vee v$ and $u \wedge v$ do not change if we substitute $Max(u_0, \mathcal{V})$ for u_0.

Using the definition of $Max(u_0, \mathcal{V})$ we also write

$$Max(u_0, \mathcal{V}) \vee v = Max(u_0, \mathcal{V})$$

for all $v \in \mathcal{V}$. Since $Max(u_0, \mathcal{V})$ is an element of $u_0 \vee \mathcal{V}$, it could not be contained in any other $u \vee \mathcal{V}$ with $u \neq u_0$ and (4.6.5) is satisfied for $\mathcal{U}^+$ and $\mathcal{V}$.

We may also write

$$Max(u_0, \mathcal{V}) \wedge v = v$$

for all $v \in \mathcal{V}$. It is obvious that $v \in v' \wedge \mathcal{U}$ with $v' \neq v$ is impossible unless there exists a $u \in \mathcal{U}$ with $u \geq v$. However, then $u \wedge v = v$, too, contradicting (4.6.6). So, if (4.6.6) is satisfied for $(\mathcal{U}, \mathcal{V})$ then it is also the case for $(\mathcal{U}^+, \mathcal{V})$.

□

Dually, we have another lemma.

Lemma 4.13 *If $(\mathcal{U}, \mathcal{V})$ is a recovering pair and there exists a $v_0 \in \mathcal{V}$ such that $Min(v_0, \mathcal{U}) \in v_0 \vee \mathcal{U}$ then the set*

$$\mathcal{V}^- = \{\mathcal{V} \backslash v_0\} \cup \{Min(v_0, \mathcal{U})\}$$

also forms a recovering pair with $\mathcal{U}$.

The following lemma makes use of the special structure of $\mathcal{L}$ in the theorem.

Lemma 4.14 *If $\mathcal{L}$ is the product of two finite length chains then for any canonical recovering pair $(\mathcal{U}, \mathcal{V})$ containing an incomparable pair (u, v), then*

$$\left[\begin{array}{ll} \textit{either} \; \exists u_0 \in \mathcal{U} : u_0 \neq Max(u_0, \mathcal{V}), & Max(u_0, \mathcal{V}) \in u_0 \vee \mathcal{V}, \\ \textit{or} \quad\; \exists v_0 \in \mathcal{V} : v_0 \neq Min(v_0, \mathcal{U}), & Min(v_0, \mathcal{U}) \in v_0 \wedge \mathcal{U} \end{array}\right.$$

Proof Let the elements of $\mathcal{L}$ be denoted by (a, b) in the natural way, i.e., a is the corresponding element of the first and b is that of the second chain defining $\mathcal{L}$. Note that if two elements, (a, b) and (a', b'), are incompatible, then either $a < a', b > b'$ or $a > a', b < b'$, holds.

Consider all those elements of $\mathcal{U}$ and $\mathcal{V}$ for which there are no incompatible elements in the other set, i.e., define the set $\mathcal{D}$ consisting of all $u \in \mathcal{U}$ and $v \in \mathcal{V}$ such that there exist $u' \in \mathcal{U}$ or $v' \in \mathcal{V}$ with (u, v') or (u', v) incompatible. Choose an element $(u, v) \in \mathcal{D}$ for which (possibly negative) value of $u - v$ is minimal within $\mathcal{D}$. Denote it by (u^*, v^*). We claim that this element can take the role of u_0 and v_0 depending on whether it is in $\mathcal{U}$ or $\mathcal{V}$. Since $(i^*, v^*) \in \mathcal{D}$, it is clearly not equal to both $Max(u^*, \mathcal{V})$ and $Min(v^*, \mathcal{U})$.

Assume $(u^*, v^*) \in \mathcal{U}$. Consider the elements of $\mathcal{V}$ that are incomparable with (u^*, v^*). Let (u', v') be an arbitrary one of them. by the choice of (u^*, v^*) we know that $u^* - v^* \leq u' - v'$. Since (u^*, v^*) and (u', v') are incomparable this implies $u' > u$ and $v' < v$, thus $(u^*, v^*) \vee (u', v') = (u', v^*)$. Since $(\mathcal{U}, \mathcal{V})$ is canonical this implies that every element of $(u^*, v^*) \vee \mathcal{V}$ has the form $(., v^*)$. This means that $(u^*, v^*) \vee \mathcal{V}$ is an ordered subset of $\mathcal{L}$. Thus, it contains its maximum $Max(u^*, \mathcal{V})$.

Similarly, if $(u^*, v^*) \in \mathcal{V}$, then $(u^*, v^*) \wedge \mathcal{U}$ consists of the elements of the form $(u^*, .)$ and so is an ordered subset of $\mathcal{L}$. Thus, it contains its minimum $Max(v^*, \mathcal{U})$.

□

Proof of Theorem 4.25. By Lemma 4.11 it suffices to consider a canonical recovering pair $(\mathcal{U}, \mathcal{V})$. If it contains incomparable pairs then by Lemmas 4.12–4.14, we can modify these sets step by step in such a way that the cardinalities do not change and the modified sets form canonical recovering pairs while the number of incomparable pairs is strictly decreasing at each step. So this procedure ends with a canonical recovering pair $(\mathcal{U}', \mathcal{V}')$, where $\mathcal{U}' = \mathcal{U}$ and $\mathcal{V}' = \mathcal{V}$ and every element of $\mathcal{U}'$ is comparable to every element of $\mathcal{V}'$. Then $(\mathcal{U}', \mathcal{V}')$ is a sandglass.

□

4.7 UD Codes for Multiple-Access Adder Channels Generated by Integer Sets

4.7.1 *Statement of the Problem*

We will consider a multi-user communication system in which there are T users and one receiver. Each user is given a code which is a subset of the set of integers

$$\mathcal{N}_n = \{0, 1, ..., 2^n - 1\},$$

where n is a fixed parameter. We also assume that 0 belongs to all codes and denote the i-th code by $\mathcal{U}_i$, $i = 1, ..., T$. The i-th user transmits some $u_i \in \mathcal{U}_i$ over a multiple-access adder channel, and the receiver gets an integer

$$z = u_1 + u_2 + ... + u_T \in \{0, ..., (2^n - 1)T\}. \tag{4.7.1}$$

The case when the user transmits 0 is interpreted as the situation when he is non-active, while if the user transmits a positive integer we say that he is active. We want to construct codes having the maximal possible cardinalities in such a way that the decoder can uniquely specify *all active users and their codewords*.

Note that a UD code $(\mathcal{U}_1', ..., \mathcal{U}_T')$ for the T-user binary adder channel can generate not a UD code $(\mathcal{U}_1, ..., \mathcal{U}_T)$, where $\mathcal{U}_i = \mathcal{U}_i' \cup \{0\}$, $i = 1, ..., T$, for our multiple-access adder channels generated by integer sets because the decoder does not know which users were active. This conclusion is illustrated in the following example where we show the elements of $\mathcal{U}_1, ..., \mathcal{U}_T$, both as the binary codewords and integers.

Example Let $T = 3, n = 2$, and

$$\mathcal{U}_1' = \{(00), (11)\} \equiv \{0, 3\},$$
$$\mathcal{U}_2' = \{(01), (10)\} \equiv \{1, 2\},$$
$$\mathcal{U}_3' = \{(00), (10)\} \equiv \{0, 2\}.$$

It is easy to check that this code is uniquely decodable for the 3-user binary adder channel. However, the code

$$(\mathcal{U}_1 = \{0, 3\},\ \mathcal{U}_2 = \{0, 1, 2\},\ \mathcal{U}_3 = \{0, 2\}\ )$$

is not uniquely decodable (we include 0 into $\mathcal{U}_2$ since the second user can be non-active): for example, $3 = 3 + 0 + 0 = 0 + 1 + 2$.

Let us denote by

$$R_I(T, n) = R_1 + ... + R_T$$

the sum rate of the code $(\mathcal{U}_1, ..., \mathcal{U}_T)$, where $R_t = \log |\mathcal{U}_i|/n, t = 1, ..., T$.

Proposition 4.17 (Jevtić (1992), [27]) *If $(\mathcal{U}_1, ..., \mathcal{U}_T)$ is a UD code, then the following inequalities are valid,*

$$R_I(T, n) \leq \frac{T}{n} \log\left(\frac{2^n - 1}{T} + 1\right) < T, \tag{4.7.2}$$

$$R_I(T, n) < 1 + \frac{\log T}{n}, \tag{4.7.3}$$

$$R_I(T, n) < G(n), \tag{4.7.4}$$

where

$$G(n) = 1 + \frac{1}{n} \log(n + \log(1 + n + \log n)), \quad n \geq 2.$$

Proof If $(\mathcal{U}_1, ..., \mathcal{U}_T)$ is a UD code, then $\mathcal{U}_i \cap \mathcal{U}_j = \{0\}$ for all $i \neq j$. Thus, the sets $\mathcal{U}_1, ..., \mathcal{U}_T$ partition the set $\{1, ..., 2^n - 1\}$ and the maximal sum rate is attained when these sets have equal cardinalities $(2^n - 1)/T + 1$, and (4.7.2) follows.

Inequality (4.7.3) is a corollary from the evident inequalities

$$|\mathcal{U}_1|, ..., |\mathcal{U}_T| < 2^n.$$

There are 2^T sums $\alpha_1 u_1 + ... + \alpha_T u_T$, where $\alpha_1, ..., \alpha_T \in \{0, 1\}$. If $(\mathcal{U}_1, ..., \mathcal{U}_T)$ is a UD code, then all these sums are distinct. Each sum does not exceed $T2^n$ and we get

$$2^T < T2^n. \tag{4.7.5}$$

UD codes exist only if $T < 2^n$. Hence, using (4.7.5) we conclude that

$$T < 2n \Longrightarrow 2^T < n2^{n+1} \Longrightarrow T < 1 + n + \log n. \tag{4.7.6}$$

Taking logarithms in (4.7.5) and using (4.7.6) we obtain

$$T < n + \log(1 + n + \log n)$$

and combining this inequality with (4.7.3) get (4.7.4).

□

The function $G(n)$ is given in Fig. 4.8. Note that

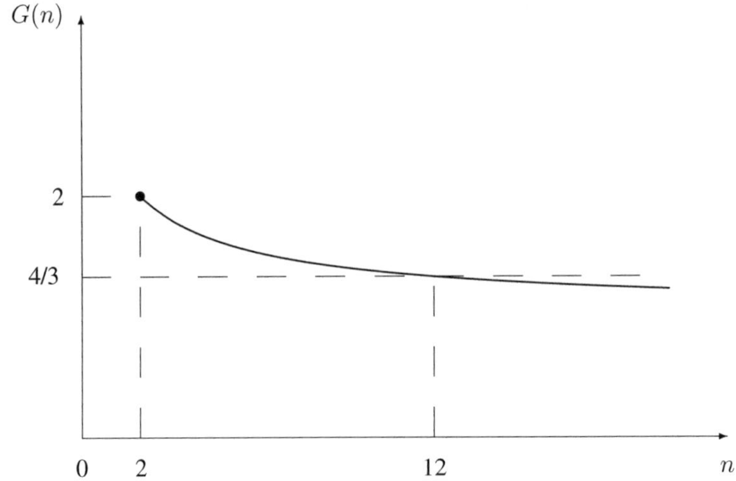

Fig. 4.8 The function $G(n)$

$$G(n) < 4/3, \quad \text{for all} \quad n > 12. \tag{4.7.7}$$

This fact will be used in the further considerations.

Definition 4.19 A set of positive integers

$$\mathcal{U}_{T,n} = \{u_1, ..., u_T\}$$

such that

$$u_1 < ... < u_T < 2^n$$

is referred to as the sum-distinct if all 2^T sums $\alpha_1 i_1 + ... + \alpha_T u_T$, where $\alpha_1, ..., \alpha_T \in \{0, 1\}$, are distinct. The parameter

$$d(\mathcal{U}_{Tn}) = T/n$$

is referred to as the density of $\mathcal{U}_{T,n}$. Given n and an integer c, a class of sum-distinct sets with the density $1 + c/n$ is denoted by $\mathcal{B}_c$.

Example Let $n = 3$. Then $\mathcal{U}_{4,3} = \{3, 5, 6, 7\}$ is a sum-distinct set with the density $4/3 = 1 + 1/3$. Hence, $\mathcal{U}_{4,3} \in \mathcal{B}_1$.

Proposition 4.18 *Let*

$$d_n^* = \max_{c:\mathcal{B}_c \neq \emptyset} (1 + c/n)$$

be the maximal density of sum-distinct subsets in $\mathcal{N}_n$ *and let*

$$d^* = \max_{n \geq 2} d_n^*.$$

Then $d^* = 4/3$.

Proof The example above gives the set with the density 4/3. Thus,

$$d^* \geq 4/3. \tag{4.7.8}$$

Let $(\mathcal{U}_1, ..., \mathcal{U}_T)$ be a UD code such that each $\mathcal{U}_i$ consists two elements, 0 and $u_i < 2^n$. Then $\mathcal{U}_{T,n} = \{u_1, ..., u_T\}$ is a sum-distinct set with the density $T/n = R_I(T, n)$. Hence, using Proposition 4.17 we conclude that $d_n^* < G(n)$. Because of (4.7.7) we can examine only the cases $n \leq 12$. Direct inspection shows that $\mathcal{B}_c = \emptyset$ for $n < 3c$ and $c = 2, 3, 4$. Therefore, inequality (4.7.8) is tight. □

4.7.2 Code Design

An obvious way to design UD codes is to partition sum-distinct integer sets. Since a code $(\mathcal{U}_1, ..., \mathcal{U}_T)$, where $(\mathcal{U}_i = \{0, u_i\}$ and $T = n + c$, has the rate sum $(n + c)/n$ it is desirable to have c as large as possible (note that according to [39] there exists a 23-element sum-distinct set for $n = 21$).

If $n < 3$ then $\mathcal{B}_1 = \emptyset$. If $n = 3$ then only $\mathcal{H}_4 = \{3, 5, 6, 7\}$ belongs to $\mathcal{B}_1$. If $n = 3$ then according to [39] $|\mathcal{B}_1| \geq (32\sqrt{2})^{-1} 2^n$. Elements from $\mathcal{B}_1$ are obtained recursively. The possible designs are

$$\mathcal{H}_{n+1} = \{u\} \bigcup \{\, 2i : i \in \mathcal{H}_n \,\}, \quad n = 4, 5, ...,$$

$$\mathcal{D}_n = \{2^{n-1} - 1\} \bigcup \left(\bigcup_{j=0}^{n-1} \{2^{n-1} - 1 - 2^j\} \right),$$

where u runs over all odd integers less than 2^n.

A sum-distinct subset $\mathcal{X}_n$ is referred to as *an* Ω*-set if any sum of* $t + 1$ *terms from* $\mathcal{X}_n$ *is greater than any sum of* t *terms.* Note that $\mathcal{X}_n$ has to be a 'dense' set since any of its t-element sums are upper-bounded by its minimal $(t + 1)$-element sum. An example of an Ω-set is $\mathcal{D}_n$. Another example is the sequence $\mathcal{X}_n^*$, $n = 1, 2, ...$ of ordered sets $\mathcal{X}_n^* = \{x_1, ..., x_n\}$ defined recursively as $\mathcal{X}_1^* = \{1\}$ and

$$\mathcal{X}_{n+1}^* = \{x\} \bigcup_{z \in \mathcal{X}_n^*} \{x + z\},$$

where $x \in \mathcal{X}_n^*$ is in position $n' = \lceil (n + 1)/2 \rceil$, i.e., $x = z_{n'}$.

Codes with rate sum above one do not necessarily have to be generated by partitioning sum-distinct sets; much less by partitioning a set from $\mathcal{B}_c$. For example, the code $(\{0, 1, ..., 2^n - 2\}, \{0, 2^n - 1\})$ is obtained by partitioning $\{1, ..., 2^n - 1\}$. Call a set of positive integers $\{x_1, ..., x_n\}$ *an* $\mathcal{A}$-set *if its elements satisfy the inequalities* $x_{t+1} > x_1 + ... + x_t$ *for* $t = 1, ..., n-1$. Clearly, by definition, $\mathcal{A}$-set is sum-distinct. Further, it is obvious by induction that $x_n \geq 2^{n-1}$ for any $\mathcal{A}$-set and thus these sets do not belong to $\mathcal{B}_1$. A code $\mathcal{U}_1 = \{0, ..., r-1\}$ and $\mathcal{U}_j = \{0, r2^{j-2}\}$, $j = 2, ..., T$, is uniquely decodable for any $r \geq 2$ since $\{x_{j_1}, ..., x_{j_T}\}$ is an $\mathcal{A}$-set for any choice of $x_{j_i} \in \mathcal{U}_i$ and all such $\mathcal{A}$-sets have different sums. Clearly, the best choice for r is $r = 2^{n-T+1} - 1$. for any choice of T. Then the rate sum is $R_0 = 1 + \log(2 - 2^{T_n-1})/n > 1$ for any $T \geq 2$ and $n \geq T$. Note that this class is enforced by the choice of the first component. This is easily seen by taking $\mathcal{U}_1 = \{0, ..., r-1\}$ and noting that $(\mathcal{U}_1, \{0, r\}, \{0, 2r\})$ has higher rate than $(\mathcal{U}_1, \{0, r, 2r\})$.

Given $T > 3$, any code from $\mathcal{B}_1$ has a higher rate than R_0. However, the code with this rate presented before can be decoded using a rather simple procedure. Note that the requirement of simplicity of the decoding procedure for the codes from $\mathcal{B}_1$ would require a special design of a corresponding sum-distinct set.

4.7.3 UD Codes in $\{0, 1\}^n$

Let us consider the codes $(\mathcal{U}_1, ..., \mathcal{U}_T)$ consisting of binary codewords of length n under the restriction that each component $\mathcal{U}_t$ contains the all-zero vector which will be denoted by 0^n. The maximal sum rate of UD codes given n will be denoted by $R^{(0)}(T, n)$.

Proposition 4.19 (Jevtić (1992), [27]) *Let* $A(n)$ *denote the number of ones in the binary representation of the first n positive integers. Then*

$$\frac{A(n)}{n} \leq R^{(0)}(T, n) \leq \log(1 + n + n^2) \tag{4.7.9}$$

Proof Let $(\mathcal{U}_1, ..., \mathcal{U}_T)$ be a UD code such that $\mathcal{U}_t \neq \{0^n\}$ for all $t = 1, ..., T$. By the arguments of Proposition 4.17 we obtain $R^{(0)}(T, n) \geq T/n$. In particular, if $\mathcal{U}_t = \{0^n, u_t\}$ then $\{u_1, ..., u_T\}$ has to be sum-distinct. A construction of a class of T-element sets with $T \geq A(n)$ is given in [39] and the lower bound in (4.7.9) follows. To establish the upper bound, note that there are at most $(T+1)^n$ values corresponding to the elements of $\mathcal{U} = \mathcal{U}_1 \cup ... \cup \mathcal{U}_T)$ in $\{0, ..., T\}^n$. Hence, $2^T < (T+1)^n$. Using also the inequality $T < 2^n$ we complete the proof.

□

Note in conclusion that the codes considered in this section can be viewed as *signature codes*: we want distribute some document among T participants and having received the sum of the codewords realize who of them 'signed' this document.

4.8 Coding for the Multiple-Access Channels with Noiseless Feedback

4.8.1 Example of an Information Transmission Scheme over the Binary Adder Channel

Note that our work on feedback went in parallel to the work on the MAC [1, 2]. When we wrote [1, 2], it was clear to us that feedback for the MAC makes it possible for the senders to build up cooperation and that therefore certain dependent RV's X and Y enter the characterization of the capacity region. However, we could not establish a general capacity theorem or even find a candidate for the achievable rate region. Therefore we did not write about it. On the other hand we could expand on list code ideas in [2]. The topic of feedback was thern addressed by others. It is well-known that the capacity of a single-input single-output discrete memoryless channel is not increased even if the encoder could observe the output of the channel via a noiseless delayless feedback link [50]. We will present an example discovered by Gaarder an Wolf [21] which shows that it is not the case for the two-user binary adder channel.

As we discussed before, one of the restrictions on the pairs of rates (R_1, R_2) belonging to the achievable rate region for the two-user binary adder channel under the criterion of arbitrarily small decoding error probability is as follows: $R_1 + R_2 \leq 1.5$. Therefore, the pair

$$(R_1, R_2) = (0.76, 0.76)$$

does not belong to this region. We will construct a coding scheme in such a way that this pair belongs to the achievable rate region for the two-user binary adder channel with the noiseless feedback.

Suppose that each encoder observes the sequence of the output symbols from the adder channel. The t-th outputs of the first and second encoders can then depend upon the first $(t-1)$ outputs of the channel as well as the message that should be transmitted. Let n be an integer such that $k = 0.76n$ is also integer. Let $M = 2^k$ be the total number of messages which can be transmitted by each encoder. Let both encoders first transmit their messages uncoded using the channel k times and consider the sequence of output symbols corresponding to this input. If some received symbol is equal to either 0 or 2, then the decoder knows the input symbols. However, for those positions where the output symbol is 1, the decoder knows only that the input symbols were complements to each other. Let n_1 be the number of positions for which the output symbol was 1. Since both encoders observe the output symbols via a noiseless feedback link, the encoders know the positions where 1 occurred and also know the other input sequence exactly. Both encoders can then cooperate to retransmit corresponding symbols of the first encoder at the remaining $n-k$ positions. Since the encoders can cooperate completely in this endeavor, they can send 3^{n-k} different messages using the input pairs (0,0), (0,1) and (1,1). If $2^{n_1} \leq 3^{n-k}$, the decoder will be able to reconstruct the two messages without error. Otherwise, we will declare an error and show that the probability of this event can be made as small as desired by choosing N large enough. Really, the probability of decoding error can be expressed as

$$P_e = \Pr\{n_1 > \log 3^{n-k} = (0.24 \log 3)n\}.$$

However, n_1 is a random variable with mean

$$\overline{n_1} = k/2 = 0.38n$$

and variance

$$\sigma^2 = k/4 = 0.19n.$$

Then

$$\begin{aligned}\Pr\{n_1 > (0.24 \log 3)n\} &< \Pr\{|n_1 - \overline{n_1}| > 0.00039n\}\\ &\leq \frac{\sigma^2}{(0.00039n)^2}\\ &= \frac{0.19}{(0.00039)^2 n},\end{aligned}$$

which can be made as small as desired by choosing n large enough.

4.8.2 Cover–Leung Coding Scheme

We will consider a general model of two-user memoryless MACs with feedback.

Definition 4.20 An $((M_1, M_2), n)$-code for the MAC defined by the input alphabet $\mathcal{X} \times \mathcal{Y}$, output alphabet $\mathcal{Z}$, and crossover probabilities

$$\{\ P(z|x, y),\ (x, y) \in \mathcal{X} \times \mathcal{Y},\ z \in \mathcal{Z}\},$$

where $\mathcal{X}$, $\mathcal{Y}$, and $\mathcal{Z}$ are finite sets, is given by the following:

(i) a collection of encoding functions

$$\begin{aligned}f_t &: \{1, ..., M_1\} \times \mathcal{Z}^{t-1} \to \mathcal{X},\\ g_t &: \{1, ..., M_2\} \times \mathcal{Z}^{t-1} \to \mathcal{Y},\end{aligned}$$

where $t = 1, ..., n$ and $\mathcal{Z}^0 = \emptyset$;

(ii) a decoding function

$$\varphi : \mathcal{Z}^n \to \{1, ..., M_1\} \times \{1, ..., M_2\}.$$

Encoding of the messages $m_1 \in \{1, ..., M_1\}$ and $m_2 \in \{1, ..., M_1\}$ is defined as

$$(f_1(m_1), f_2(m_1, z_1), ..., f_t(m_1, z_1, ..., z_{t-1}), ..., f_n(m_1, z_1, ..., z_{n-1}))$$

and

$$(g_1(m_2), g_2(m_1, z_1), ..., g_t(m_2, z_1, ..., z_{t-1}), ..., g_n(m_2, z_1, ..., z_{n-1})),$$

respectively, and decoding is defined as

$$(\hat{m}_1, \hat{m}_2) = \varphi(\mathbf{z}).$$

An achievable rate region for the MACs with feedback constructed under the criterion of arbitrarily small average decoding probability can be introduced similarly to the corresponding region for the MACs without feedback: we are interested in all pairs of rates (R_1, R_2) such that there exist encoding and decoding providing the arbitrarily small average decoding probability of the event $(\hat{m}_1, \hat{m}_2) \neq (m_1, m_2)$ when $M_1 = 2^{nR_1}$, $M_2 = 2^{nR_2}$, and n tends to infinity. We will describe a coding scheme which allows us to attain the asymptotic characteristics given below.

Theorem 4.26 (Cover and Leung (1981), [13]) *Let U be a discrete random variable which takes values in the set $\{1, ..., K\}$, where*

$$K = \min\{|\mathcal{X}| \times |\mathcal{Y}|, |\mathcal{Z}|\}.$$

Consider the set $\mathcal{P}$ of all joint distributions of the form

$$P_{UXYZ}(u, x, y, z) = P_U(u) P_{X|U}(x|u) P_{Y|U}(y|u) P(z|x, y), \tag{4.8.1}$$

where P is fixed by the MAC. For each $P_{UXYZ} \in \mathcal{P}$, denote by $\mathcal{R}(P_{UXYZ})$ the set of all rate pairs (R_1, R_2) satisfying the inequalities

$$\begin{aligned} R_1 &\leq I(X \wedge Z|Y, U), \\ R_2 &\leq I(Y \wedge Z|X, U), \\ R_1 + R_2 &\leq I(XY \wedge Z), \end{aligned} \tag{4.8.2}$$

where the mutual information functions are computed in accordance with (4.8.1). Then the set

$$conv\left(\bigcup_{P_{UXYZ} \in \mathcal{P}} \mathcal{R}(P_{UXYZ})\right),$$

where 'conv' denotes the convex hull of a set, contains the achievable rate region for the MAC with feedback constructed under the criterion of arbitrarily small average decoding probability.

A complete proof of this result can be found in [13], and we restrict our attention to the description of the coding scheme. The scheme uses a large number B of blocks, each of length n, and it is assumed that, the first encoder has to transmit a sequence of messages

$$(m_{11}, ..., m_{1B}), \quad m_{1b} \in \{1, ..., M_1\}$$

and the second encoder has to transmit a sequence of messages

$$(m_{21}, ..., m_{2B}), \quad m_{2b} \in \{1, ..., M_2\}.$$

In block b, where $b \in \{1, ..., B\}$, the encoders send enough information to the decoder to enable him to resolve any uncertainty left from block $b - 1$. Superimposed on this information is some new independent information which each encoder wishes to convey to the decoder. The rate of this new information is small enough so that each encoder can reliably recover the other's message through the feedback links.

Let us fix a distribution $P_{UXYZ} \in \mathcal{P}$ and introduce a random code in the following way.

1. Given an $\varepsilon > 0$, fix
$$R_0 = I(Y \wedge U)$$
and generate a sequence of 2^{nR_0} i.i.d. random vectors
$$\mathbf{u}(m_0) = (u_1(m_0), ..., u_n(m_0)), \quad m_0 = 1, ..., 2^{nR_0}.$$
The probability of each vector is defined as
$$\prod_{t=1}^{n} P_U(u_t(m_0)).$$
2. For each $\mathbf{u}(m_0)$, $m_0 = 1, ..., 2^{nR_0}$, generate 2^{nR_1} conditionally independent vectors
$$\mathbf{x}(m_0, m_1) = (x_1(m_0, m_1), ..., x_n(m_0, m_1)), \quad m_1 = 1, ..., 2^{nR_1}.$$
in such a way that the conditional probability of each vector given $\mathbf{u}(m_0)$ is defined as
$$\prod_{t=1}^{n} P_{X|U}(x_t(m_0, m_1)|u_t(m_0)).$$
Analogously, generate 2^{nR_2} conditionally independent vectors
$$\mathbf{y}(m_0, m_2) = (y_1(m_0, m_2), ..., y_n(m_0, m_2)), \quad m_2 = 1, ..., 2^{nR_2}.$$

in such a way that the conditional probability of each vector given $\mathbf{u}(m_0)$ is defined as

$$\prod_{t=1}^{n} P_{Y|U}(y_t(m_0, m_2)|u_t(m_0)).$$

The idea of introducing the vectors $\mathbf{u}(m_0)$, $\mathbf{x}(m_0, m_1)$, and $\mathbf{y}(m_0, m_2)$ in the definitions above is as follows. *It is intended that the 'cloud center' $\mathbf{u}(m_0)$ will be correctly decoded during the block in which it was sent. The 'satellite' indices m_1 and m_2 will be decoded correctly by the encoders, but only partially understood by the decoder.*

In the first block no cooperative information is sent: the transmitters and receiver use a predetermined index j_1 and encode $m_{11} \in \{1, ..., 2^{nR_1}\}$ and $m_{21} \in \{1, ..., 2^{nR_2}\}$ into $\mathbf{x}(j_1, m_{11})$ and $\mathbf{y}(j_1, m_{21})$. In the last B-th block the transmitters send no new information and the decoder receives enough information to resolve the residual uncertainty. If B is large, the effective rates over B blocks will be only negligibly affected by the rates in the first and last blocks.

Suppose that j_b is the index which is to be sent to the decoder in block b in order to resolve his residual uncertainty about the new messages that were sent in block $b - 1$. Also, let us denote the two new messages to be sent in block b by $(k^*, m_{2b}) \in \{1, ..., 2^{nR_1}\} \times \{1, ..., 2^{nR_2}\}$. Then the first encoder sends $\mathbf{x}(j_b, m_{1b})$ and $\mathbf{y}(j_b, m_{2b})$. Let $\mathbf{z}_b$ denote the sequence received by the decoder.

(-) The decoder declares that $\hat{j}_b$ was sent iff $(\mathbf{u}(\hat{j}_b), \mathbf{z}_b)$ is a jointly typical pair of vectors (the number of entries (u, z) in $(\mathbf{u}(\hat{j}_b), \mathbf{z}_b)$ is close to $nP_{UZ}(u, z)$ for all $(u, z) \in \mathcal{U} \times \mathcal{Z}$, where P_{UZ} is a probability distribution obtained from P_{UXYZ}).

(-) The first encoder declares that $\hat{m}_{2b}$ was sent by the second encoder iff $(\mathbf{x}(j_b, m_{1b}), \mathbf{y}(j_b, \hat{m}_{2b}), \mathbf{z}_b)$ is a jointly typical triple, and the second encoder declares that $\hat{m}_{1b}$ was sent by the second encoder iff $(\mathbf{x}(j_b, \hat{m}_{1b}), \mathbf{y}(j_b, m_{2b}), \mathbf{z}_b)$ is a jointly typical triple (the definitions of jointly typical triples are similar to the definition of jointly typical pairs given above).

(-) Both encoders construct the set

$$\mathcal{S}_b = \{(m_1, m_2) : (\mathbf{x}(j_b, m_1), \mathbf{y}(j_b, m_2), \mathbf{z}_b) \text{ is a jointly typical triple}\}$$

and number its elements as $1, ..., |\mathcal{S}|$. Then $(m_{1b}, m_{2b}) \in \mathcal{S}_b$ with high probability. The first encoder declares that $\hat{j}_1$ is the index of a vector $\mathbf{u}$ in the next block iff $(m_{1b}, \hat{m}_{2b})$ is numbered by $\hat{j}_{1b}$. The second encoder declares that $\hat{j}_{2b}$ is the index of a vector $\mathbf{u}$ in the next block iff $(\hat{m}_{1b}, m_{2b})$ is numbered by $\hat{j}_{2b}$.

Decoding error takes place after the transmission of the b-th block if one of the following events occur,

(i) $\hat{j}_b \neq j_b$;
(ii) $\hat{m}_{2b} \neq m_{2b}$;
(iii) $\hat{m}_{1b} \neq m_{1b}$;
(iv) $(m_{1b}, m_{2b}) \notin \mathcal{S}_b$;
(v) $|\mathcal{S}_b| > 2^{nR_0}$.

If the parameters R_1 and R_2 satisfy (4.8.2) then probabilities of all these events can be upper-bounded by the functions exponentially decreasing with n [13].

It is known [58] that Theorem 4.26 gives the achievable rate region for the MACs with feedback if the MAC has the following property: at least one of the inputs is completely determined by the output and the other input (alternatively, either $H(X|YZ) = 0$ or $H(Y|XZ) = 0$). Note that the binary adder channel has this property:

$$z = x + y \implies y = z - x,$$

while the OR-channel does not:

$$z = x \vee y,\ x = z = 1 \implies y \text{ is unknown.}$$

A similar statement is also valid for a more general model when three messages, m_0, m_1, and m_2 should be delivered to the decoder in such a way that the first encoder has access to m_0 and m_1, and the second encoder has access to m_0 and m_2 [16].

4.9 Some Families of Zero-Error Block Codes for the Two-User Binary Adder Channel with Feedback

4.9.1 Introduction

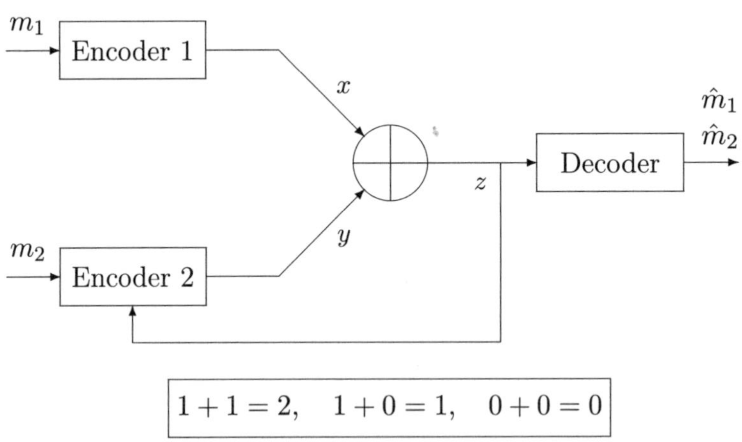

Consider the two-access communication system shown above. Two independent sources wish to send information to the receiver. During a message interval, the messages emanating from the sources are encoded independently with two binary block codes of the same length n. We assume that we have bit and block synchronization.

The two binary input vectors x and y are transformed by the channel into an output vector $z = x + y$, where the plus sign denoted bit-by-bit addition over the reals. This so-called "binary adder channel" is a special case of the multiple access channel. Block coding for this channel has been studied by several authors [1, 24, 28, 30, 52]. Its Shannon capacity region has been determined, and many results have been obtained about uniquely decodable (i.e., zero-error) codes for it. We shall design some uniquely decodable block codes for cases in which z is fed back either to one or to both senders.

If both encoding functions depend on the previous channel outputs, we say we have *full feedback*. If only one of them does, we say we have *partial feedback*. We concentrate mainly on the partial feedback case. One reason for this is that in the full feedback case, variable length codes perform significantly better than block codes [24].

We now describe the encoding procedure for the two-access channel with partial feedback shown above. The informed encoder's encoding function f_k for the kth time slot depends both on the message m_2 it is trung to send and on the channel outputs during the first $k-1$ time slots. That is, $y_k = f_k(m_2, z_1, \dots, z_{k-1})$. The uninformed encoder's output x_k during the kth time slot depends only on the message m_1. In the full feedback case we would have $x_k = g_k(m_1, z_1, \dots, z_{k-1})$.

4.9.2 Two Families of Codes for the Binary Adder Channel with Partial Feedback

Definition 4.21 For any $x \in \{0,1\}^n$, let

$$f(x) = |\{i : x_i \neq x_{i-1}, i > 1\}|$$

denote the number of transitionsin x. Define

$$W(n,k) = \{x \in \{0, 1|^n : t(x) = k\},$$

and note that $|\{W(n,k)| = \binom{n-1}{k}$.

Definition 4.22 The segment $(x_j, \dots, x_{j+s})$ of x is denoted by $l_i(x)$ and called the ith run if

(i) $x_j = \cdots = x_{j+s}$,
(ii) $x_{j-1} = x_{j+s+1} \neq x_j$,
(iii) $|\{s : 1 \leq s < j, x_s \neq x_{s+1}\}| = i - 1$.

4.9.2.1 The First Family of Codes

The strategy of the first encoder, which receives no feedback, will be simply to transform its message m_1 into a word $x \in W(n, k)$. The second encoder which is privy to the feedback, first maps its message m_2 into a word $v \in W(s, t)$ which it then sends in n transmissions as follows. Let

$$v = (v_1, \dots, v_s).$$

Define $f(0) = 0$, $f(1) = 2$. The second encoder keeps sending v_1 until it receives a feedback $f(v_1)$. Then it keeps sending v_2 until it receives feedback $f(v_2)$, and so on. If and when it finishes with v, the second encoder keeps sending v_s.

The decoder receives $z = (z_1, \dots, z_n) \in \{0, 1, 2\}^n$. Denote the indices of the non-1 components of z by $a_1, a_2, a_3, \dots$. If the number of the entries of this sequence is s or bigger, then define the second encoding function by $\hat{v}(z) = \left(f^{-1}(z_{a_1}), \dots, f^{-1}(z_{a_s})\right)$. From $\hat{v}(z)$, the decoder can reconstruct the sequence y transmitted by the second encoder in the manner

$$\hat{y}_j = \begin{cases} \hat{v}_l, & a_{l-1} < j \le a_l, \\ \hat{v}_s, & j > a_s. \end{cases}$$

The second decoding function is then defined by

$$\hat{x}(z) = z - \hat{y}(z).$$

It is easy to see that a necessary and sufficient condition for this code to be uniquely decodable is that, for any $x \in W(n, k)$ and $v \in W(s, t)$, the length of the a-sequence is at least s. This is because in this case, and only in this case, can the second encoder finish sending v within n slots. Note that the second encoder successfully sends its digit y_i only when y_i agrees with the current digit sent by the first encoder. Thus, v_1 is sent successfully at the latest at the first transition in x. After the ith successful transmission, if $v_{i+1} \neq v_i$, then the second encoder will succeed again at the next transition in the first encoder's sequence; but if $y_{i+1} = y_i$, then the second encoder next succeeds either immediately if $x_{i+1} = x_i$ or at the smallest j such that $x_j \neq x_{i+1}$ if $x_{i+1} \neq x_i$. Thus, a sufficient condition that the a-sequence has length at least s is that the number of transitions in x equals or exceeds one plus the number of transitions t in v plus twice the number $(s - t - 1)$ of non-transitions in v. That is, the condition that guarantees unique decodability is

$$1 + t + 2(s - t - 1) = 2s - t - 1 \le k. \tag{4.9.1}$$

4.9.2.2 Rate Pairs and Rate Sum

The rate for the first encoder is

$$R_1 = \frac{1}{n} \log_2 \binom{n-1}{k},$$

and the rate for the second encoder is

$$R_2 = \frac{1}{n} \log_2 \binom{s-1}{t}.$$

For large n, s, and t, let $k/n = p$, $t/s = q$, and $s/n = r$. Then we have

$$R_1 \sim h(p) \qquad R_2 \sim rh(q),$$

where $h(p)$ is the binary entropy function, and becomes

$$2r - rq \leq p. \tag{4.9.2}$$

Using equality in (4.9.2), we have

$$r = p/(2-q).$$

Therefore, the second rate is

$$R_2 = ph(q)/(2-q).$$

If $p = 1/2$, then $R_1 = 1$, in which case the highest rate for the second encoder under the constraint (4.9.2) is $R_2 = 0.347$. The highest rate sum reached by this family of codes is found by equating to zero the derivatives of $R_1 + R_2$ with respect to p and to q. This yields $h'(p) = h'(q)$ which implies that $p = q$. The optimizing p is then seen to satisfy $h(p) + (2-p)h'(p) = 0$, which reduces to $p^2 + p - 1 = 0$, so it is $p^* = (\sqrt{5} - 1)/2$. The resulting maximized rate sum is $-\log(1 - p^*)$. the numerical value of which is

$$\max(R_1 + R_2) = \log_2[2/(3 - \sqrt{5})] = 2\log_2[(1 + \sqrt{5})/2] = 1.3885.$$

4.9.2.3 The Second Family of Codes

For given n and k, we construct a code from the set $W(n, k)$. For any $x \in W(n, k)$, let $|l_i(x)|$ be b_i. Define

$$v(x) = \left(1, \underbrace{\ldots}_{b_1} 1, 0, 0, 1, \underbrace{\ldots}_{b_2} 1, 0, 0, 1, \ldots\right).$$

This is a binary sequence in $W(n + 2k, 2k)$ which consists of k runs of 1's, whose lengths are the b_i, separated from one another by pairs of consecutive 0's. The first encoder sends the sequence $v(x)$ for some $x \in W(n, k)$. The second encoder continually uses the feedback to recover the sequence that the first encoder sends an arbitrary sequence in $\{0, 1\}^{n+k}$ into which it inserts a 0 whenever the feedback indicates that the first encoder has just sent the first of a pair of consecutive 0's in the previous slot. The decoder is able to recover the sequences sent by both encoders because it receives 0's only either in isolation or in runs of length 2. It knows that each of the 0-pairs sent by the first encoder ends either at a received isolated 0 or at the end of a received pair of 0's. Thus, the decoder is able to recover the sequence sent by the first encoder. Using that sequence, the decoder can then recover the sequence transmitted by the second encoder and expunge from it the k extra 0's that the second encoder injected.

The rates of this code are

$$R_1 \sim \frac{1}{n + 2k} \log_2 \binom{n}{k}$$

and

$$R_2 = \frac{n + k}{n + 2k}.$$

Letting $k/n = p$, we have for large n that

$$R_1 \sim h(p)/(1 + 2p)$$

and

$$R_2 = (1 + p)/(1 + 2p).$$

Numerical results show that the best rate sum of any code in this family is 1.375, slightly smaller than the best rate sum of the first code family.

4.9.3 *Codes Generated by Difference Equations*

4.9.3.1 Square Dividing Strategy

Encoding for the binary adder channel with full or partial feedback can also be described by means of square dividing strategies analogous to those used by [49] for binary, two-way channels. Assume that the message sets at the two encoders are C_t, $t = 1, 2$, where $C_1 = \{1, \ldots, a\}$ and $C_2 = \{1, \ldots, b\}$. We consider the set $C_1 \times C_2$. In the first slot for a message in a certain subset of C_1, say $C_1^1(0)$, send 0; for a message in the set $C_1^1(1) = C_1 - C_1^1(0)$, send 1. Similarly, for the second encoder, define $C_2^1(0)$ and $C_2^1(1)$, and send 0 and 1, respectively. Thus, the square $C_1 \times C_2$ is divided into four subsquares with outputs 0, 1, and 2 as shown below

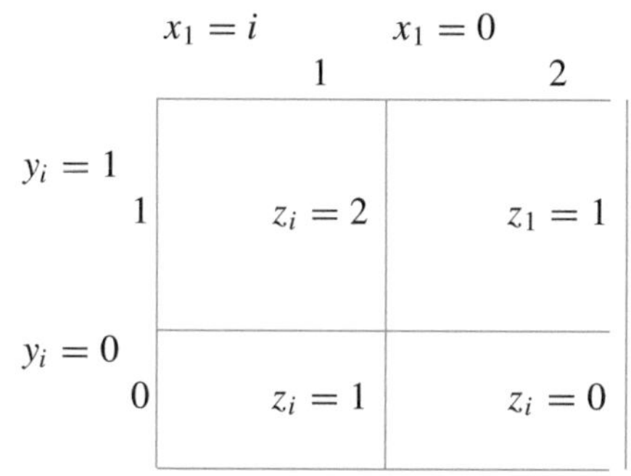

Temporarily confine attention to the full feedback case. In this case, after receiving the first feedback, each encoder can identify which of these four subsquares was employed during the first slot. They then divide it into smaller subsquares during the second slot, and into finer and finer subsquares during subsequent slots. The decoder, on the other hand, knows the input is a message pair that belongs to one of the subsquares that is consistent with the sequence of channel outputs observed thus far. The decoder can make the correct decision provided that eventually there is only one message pair that is consistent with the channel output sequence. We shall continue to consider only cases in which no decoding error is allowed. Kasami and Lin [28] call such zero-error codes "uniquely decodable". In the square-dividing terminology, unique decodability means that eventually the square is divided into $a \cdot b$ subsquares, each of which has a unique channel output sequence.

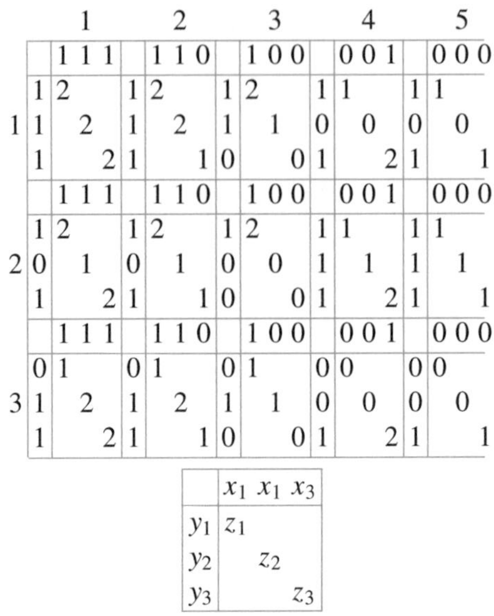

(3, 5) is 3-attainable: $\left(\frac{\log_2 3}{3}, \frac{\log_2 5}{3}\right) = (0.528, 0.774) \in R_0$.

The figure above is an example of a uniquely decodable code with sizes $a = 5$ and $b = 3$. Note that in this example the first encoder always sends either 111 for message one, 110 for message two, 100 for message three, 001 for message four or 000 for message five. Hence, the first encoder need not be privy to the feedback. In slot 1 the second encoder sends 1 for either message one or message two and sends 0 for message three. In the second slot for either message one or message three, a 1 is sent if the larger of the two possible feedback symbols was received and a 0 is sent if the smaller one was received; the opposite is done for message two. In the third and final slot, a 1 is sent unless the feedback pair from the first two slots indicates that the first encoder is trying to send the third of its five possible messages, in which case a 0 is sent. A unique ternary output sequence, shown on the subsquare diagonals, results for each of the 15 possible message pairs. This is but one of many different zero-error coding strategies in this example. The rationale underlying this particular strategy is explained in the next subsection in conjunction with the proof of Theorem 4.28.

A pair of numbers (a, b) is called *k-attainable* (for the full feedback case or the partial feedback case) if there exists a UD code (for the corresponding case) of length k with codeword sets of sizes a and b. Clearly, any pair that is k-attainable for the partial feedback case must be k-attainable for the full feedback case. Of course, if (a, b) is k-attainable, $c \geq a$, and $d \leq b$, then (c, d) is also k-attainable. As shown by the example, (5, 3) is 3-attainable with partial feedback. For small k, it is not difficult to determine whether or not a pair of numbers is k-attainable, but as k grows this task becomes imposing.

We present a method to generate families of attainable pairs. We call the codes we use to reach these pairs *codes generated by difference equations*. As their name implies, they have a recursive construction that makes them easy to encode and decode. Because of their high rates and ease of implementability, they are an interesting family of codes.

4.9.3.2 Fibonacci Codes

The simplest of our codes generated by difference equations will be called *Fibonacci codes*. Let $\{a_i\}$ be the Fibonacci numbers defined by

$$a_i = a_{i-1} + a_{i-2} \tag{4.9.3}$$

with $a_0 = a_1 = 1$. We prove below that (a_i, a_{i+1}) is i-attainable.

A pair of difference equations

$$a_i = \phi(a_j, j < i) \tag{4.9.4}$$

$$b_i = \psi(b_j, j < i) \tag{4.9.5}$$

will be called a pair of *code generating equations* for some initial conditions and a positive integer S if the sequences $\{a_i\}$ and $\{b_i\}$ generated by these equations and initial conditions have the property that (a_i, b_i) is $(S + i)$-attainable for all $i \geq 1$ Thus, we have claimed that the pair of Fibonacci equations $a_i = a_{i-1} + a_{i-2}$ and $b_i = b_{i-1} + b_{i-2}$ are code generating equations for $a_0 = a_1 = b_0 = 1$, $b_1 = 2$, and $S = 0$.

It is easy to find code generating equations, but at present we have no general way of finding ones that possess high rates. We shall show, however, that the aforementioned Fibonacci codes and another set of code generating equations we present in Sect. 4.9.4 do indeed achieve high rates.

We now prove the claim that a pair of Fibonacci equations are code generating fr the full feedback case. Subsequently, we extend this result to the partial feedback case. To facilitate the proof, we introduce the concept of an *attainable cluster*. The union of all subsquares that share the same output sequence is called a cluster. For example, in the $(5, 3)$ code above, after the first step, the 2×2 rectangle in the upper right corner and the 1×3 rectangle of the lower left corner together constitute a cluster. A cluster is k-attainable if after k or fewer further divisions, it can be reduced to single "points" each of which has a distinct output sequence. The cluster comprised of the aforementioned 2×2 and 1×3 rectangles is 2-attainable. These two rectangles are *input-disjoint* in the sense that the user inputs can be chosen independently for these two rectangles. It should be obvious that a cluster composed of two input-disjoint rectangles of sizes 1×2 and 1×1 is 1-attainable.

Theorem 4.27 *A pair of Fibonacci equations with $a_0 = a_1 = b_0 = 1$, $b_1 = 2$, and $S = 0$ are code generating for full feedback.*

Proof First, we define two types of parametrized clusters and prove that, by one step of square dividing, each of them can be reduced to clusters of the same two types with smaller parameter values. The first cluster type is a union of two input-disjoint rectangles with sizes $a_k \times b_{k-1}$ and $a_{k-1} \times b_k$; the second is a rectangle with size $a_k \times b_k$. We denote them, respectively, by

$$\lambda_k = a_k \times b_{k-1} \cup a_{k-1} \times b_k \tag{4.9.6}$$

and

$$\mu_k = a_k \times b_k, \tag{4.9.7}$$

where $a \times b$ denotes an a by b rectangle and $\cup$ denotes the union of input-disjoint rectangles. Note that we can choose the next input digit for the two users as to divide λ_k into three parts,

$$\lambda_k = [\mu_{k-1}]_2 \cup [\lambda_{k-1}]_1 \cup [\mu_{k-1}]_0, \tag{4.9.8}$$

where $[\mu_{k-1}]_2$ means that the set with output 2 is μ_{k-1}, and so on.

For μ_k, we can similarly choose the next input digit so that

$$\mu_k = [\mu_{k-1}]_2 \cup [\lambda_{k-1}]_1 \cup [\mu_{k-2}]_0. \tag{4.9.9}$$

Since the 1-attainability of both $\lambda_1 = 1 \times 1 \times \cup 1 \times 2$ and $\mu_1 = 1 \times 2$ are obvious, the theorem is proved. □

The limiting rates of the Fibonacci code family are

$$R_1 = R_2 = R_f = \lim_{k \to \infty} \frac{1}{k} \log_2 a_k = \log_2 \left[\left(\sqrt{5} + 1 \right) /2 \right] = 0.694.$$

Now we show that the Fibonacci codes actually are implementable in the partial feedback case.

Theorem 4.28 *A pair of Fibonacci equations with $a_0 = a_1 = b_1 = 1$, $b_1 = 2$, and $S = 0$ are code generating for partial feedback.*

Proof We need to prove that the Fibonacci encoding strategy can be implemented with one of the two encoders not having access to the feedback. That is, we must exhibit a technique by means of which the uninformed encoder can correctly divide each of the clusters that appears in the square dividing procedure into 1-subsets and 0-subsets. Note, as shown in the figure below, that the sizes of the horizontal edges of the subsquares after the successive square divisions are:

(i) originally: a_k
(ii) after one division: a_{k-1}, a_{k-2}
(iii) after two divisions: $a_{k-2}, a_{k-3}, a_{k-2}$
(iv) after three divisions: $a_{k-3}, a_{k-4}, a_{k-3}, a_{k-3}, a_{k-4}$
after four divisions: $a_{k-4}, a_{k-5}, a_{k-4}, a_{k-4}, a_{k-5}, a_{k-4}, a_{k-5}, a_{k-4}$

and so on. Observe that, at the ith step, each of the sizes in question is either a_{k-1} or a_{k-i-1}.

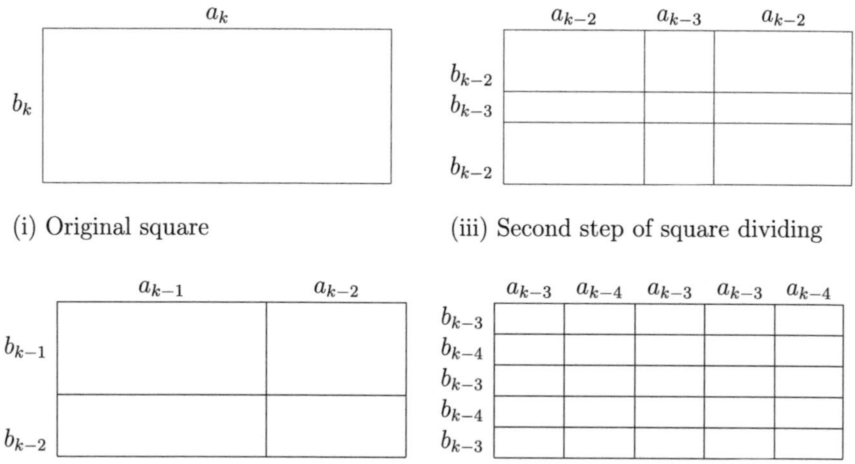

(i) Original square

(ii) First step of square dividing

(iii) Second step of square dividing

(iv) Third step of square dividing

Define $u_i = 1$ if the message to be sent by the encoder without feedback is a member of a subset of size a_{k-i}; otherwise, define $u_i = 0$. The strategy of the encoder without feedback is to send the product $x_i = u_i u_{i-1}$ at the ith step. This strategy is depicted in the figures (i) and (ii) below, the former of which shows the subset sizes arranged on successive levels of a tree and the latter of which shows the corresponding binary transmissions.

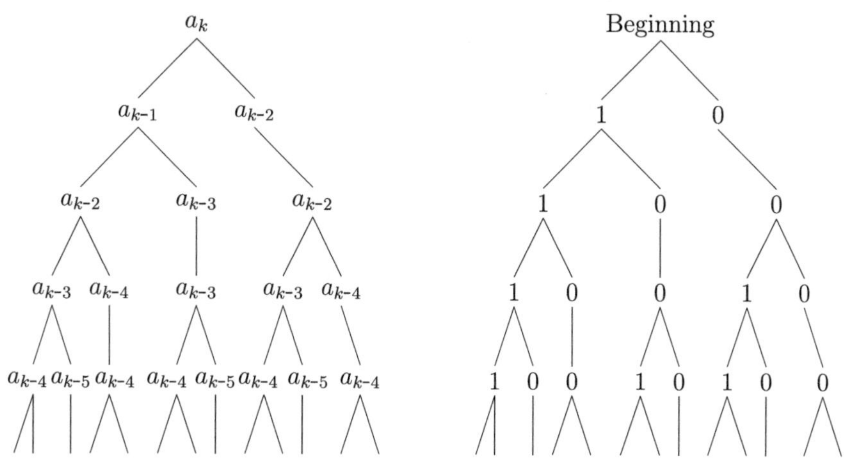

(i) (ii)

(i) Square dividing tree, (ii) Encoding tree of encoder without feedback.

Analogously define the binary function v_i of the sizes of the vertical squares at the ith step by $v_i = 1$ for subsets of size b_{k-i} and $v_i = 0$ for subsets of size $b_{k-i} - 1$. The strategy of the encoder with feedback is to send $y_i = v_i \oplus 1 \oplus u_{i-1}$ at the ith step, which can be done using the past feedback to deduce the value of x_{i-1}, and hence, recursively, the value of u_{i-1}.

Now we prove by induction that these two encoding algorithms achieve the same square dividing strategy we described in the full feedback case. At the first step, this is obvious. Generally, we need to prove that, for the two clusters studied in the proof of Theorem 4.27, the new strategies give precisely the desired dividing. In the case of the first cluster of size $a_k \times b_k$, the two encoders are both sending 1's for the bigger subsets of sizes a_{k-1} and b_{k-1}, respectively, and 0's for the smaller ones of sizes a_{k-2} and b_{k-2}, respectively. It is easy to check that the resulting outputs, shown in (i) below, are precisely the ones we need in the proof of Theorem 4.27. For the second cluster, $a_{k-1} \times b_{k-2} \cup a_{k-2} \times b_{k-1}$, the channel inputs calculated by the two encoders in accordance with the above prescriptions are shown in (ii); note that the resulting outputs again exactly satisfy the requirements of the proof of Theorem 4.27. The next step, shown in (iii), has the (5, 3)-code from above embedded within it. We omit the general step in the induction argument because its validity should be apparent by now.

1	0
1(2)	1(1)
0(1)	0(0)

1	0	0
1(2)	1(1)	0(0)
0(1)	0(0)	1(1)
1(2)	1(1)	0(0)

1	0	0	1	0
1(2)	1(1)	0(0)	1(2)	1(1)
0(1)	0(0)	1(1)	0(1)	0(0)
1(2)	1(1)	0(0)	1(2)	1(1)
1(2)	1(1)	0(0)	1(2)	1(1)
0(1)	0(0)	1(1)	0(1)	0(0)

4.9.3.3 The Inner Bound to the Zero-Error Capacity Region

The three families of zero-error copdes we have presented can be combined by tangent lines representing time-sharing to produce an inner bound to the zero-error capacity region of the binary adder channel with partial feedback. This bound can be mildly improved in the low rate region for uninformed encoder by appealing to an inner bound to the zero-error capacity region derived by Kasami et al. [30] for the case in which there is no feedback to either encoder; clearly, any inner bound for that case is an inner bound for the partial feedback case. That bound and a time-sharing line joining it to the second of our code families completes our overall inner bound. The straight-line portion of this bound has slope -1 and a rate sum of 1.3885.

4.9.4 *Codes Generated by Difference Equations for the Binary Adder Channel with Full Feedback*

4.9.4.1 Refinement of the Fibonacci Code

We call a k-attainable pair (a, b) optimal if $(a+1, b)$ and $(a, b+1)$ are no longer k-attainable. Consider the first few Fibonacci code sizes:

$$(1, 2), (2, 3), (3, 5), (5, 8), (8, 13), (13, 21), \ldots.$$

It is not hard to prove that the first three terms are optimal for $k = 1, 2$, and 3, respectively. It turns out, however, that $(5, 9)$ is 4-attainable and $(8, 14)$ is 5-attainable. This suggests that there may exist code generating equations that generate codes with asymptotically equal rates greater than R_f. We proceed to show that this is indeed the case.

Theorem 4.29 *With* $a_0 = a_1 = b_0 = 1$, $b_1 = 2$, *and* $S = 0$, *the following are code generating equations:*

$$a_k = a_{k-1} + a_{k-2} + 5a_{k-11} \tag{4.9.10}$$

$$b_k = b_{k-1} + b_{k-2} + 5b_{k-11}. \tag{4.9.11}$$

We prove this theorem in Sect. 4.9.5. The $\{a_k\}$ and $\{b_k\}$ of Theorem 4.29 give a limiting rate pair of $(0.717, 0.717)$ which dominates that of the Fibonacci codes. We refer to the associated codes as *refined Fibonacci codes*. It is not yet ascertained whether or not (4.9.10) or (4.9.11) are code generating for the partial feedback case as well.

4.9.4.2 Inner Bound for the Zero-Error Capacity Region of a Binary Adder Channel with Full Feedback

The convex hull of our first family of codes for the partial feedback case is an inner bound for the zero-error capacity region for the full feedback case. (The mirror image of the performance of the first family of codes dominates the performance of the second family of codes. Since either encoder one or encoder two could choose to ignore its feedback, we get a better bound for the full feedback case by using only the first code family.) An additional improvement is obtained by incorporating the point $(0.717, 0.717)$, corresponding to the refined Fibonacci code, and then re-taking the convex hull.

Dueck [16] has derived the exact form of the zero-error full feedback capacity region for a certain class of multiple access channels to which this feedback case belongs. However, numerical evaluation of his capacity region description is fraught with challenging obstacles even in this special case so that this inner bound is still of some interest.

4.9.5 Proof of Theorem 4.29 via Three Lemmas

Lemma 4.15 $(5, 45)$ *is a 6-attainable pair.*

To prove this lemma requires checking the square dividing procedure step by step. We omit this tedious but straightforward task.

Lemma 4.16 *There exists K such that for $k > K$,*

$$a_k/90 \le a_{k-9} \tag{4.9.12}$$

$$b_k/90 \le b_{k-9} \tag{4.9.13}$$

Proof We need to prove only that, if C is the largest eigenvalue of the characteristic equation of the difference equation (4.9.10), then

$$C^9/90 - 1 \le 0. \tag{4.9.14}$$

This is readily verified by calculation. □

Lemma 4.17 *The following clusters are k-attainable:*

$$\begin{aligned}
\lambda_k &= (a_{k-1} - a_k) \times b_k \cup a_k \times 5b_{k-9} \cup a_k \times (b_{k+1} - b_k) \cup 5a_{k-1} \times b_k \\
\beta_k &= a_x \times b_k \cup a_{k+1} \times 5b_{k-8} \cup (a_{k-1} + 5a_{k-10}) \times 5b_{k-9} \\
\delta_k &= (a_{k+1} - a_k) \times (b_{k+1} - b_k) \cup a_{k+2} \times 5b_{k-9} \\
\rho_k &= 5b_{k-6} \times \frac{a_{k+3}}{2} \cup 5a_{k-9} \times b_k \cup a_k \times 5b_{k-9} \\
\pi_k &= (a_{k+1} - a_k) \times (b_{k+1} - b_k) \cup 2a_{k-2} \times b_{k-2} \cup a_{k-2} \times 2b_{k-2} \\
\tau_k &= 2a_{k-1} \times b_{k-1} \cup a_k \times 5b_{k-1} \cup 5a_{k-9} \times b_k.
\end{aligned}$$

It is obvious that Theorem 4.29 is a consequence of Lemma 4.9.6.

Proof We prove the following recursive inequalities in which $'$ denotes a cluster with the roles of a and b exchanged and $\le$ means that the parts after a square dividing are subsets of the sets listed on the right side.

$$\lambda_k \le [\beta_{k-1}]_0 \cup [\lambda_{k-1}]_1 \cup [\beta'_{k-1}]_2 \tag{4.9.15}$$

$$\beta_k \le [\beta_{k-1}]_0 \cup [\lambda_{k-1}]_1 \cup [\delta_{k-1}]_2 \tag{4.9.16}$$

$$\delta_k \le [\beta_{k-1}]_0 \cup [\rho_{k-1}]_1 \cup [\rho_{k-1}]_2 \tag{4.9.17}$$

$$\tau_k \le [\beta_{k-1}]_0 \cup [\beta'_{k-1}]_1 \tag{4.9.18}$$

$$\pi_k \le [\beta_{k-1}]_0 \cup [\tau_{k-1}]_1 \cup [\tau'_{k-1}]_2 \tag{4.9.19}$$

$$\rho_k \le (5, 45) \otimes \{[\beta_{k-7}]_0 \cup [\lambda_{k-7}]_1 \cup [\pi_{k-7}]_2\}, \tag{4.9.20}$$

where the operator "$(\alpha, \beta)\otimes$" multiplies the row and column cardinalities of each code in the succeeding curly bracket by α and by β, respectively. The lemma follows from these inequalities.

Proof of (4.9.15).

$$\begin{aligned}\lambda_k &= [a_{k-1} \times b_{k-1} \cup a_k \times 5b_{k-9} \cup (a_k - a_{k-1}) \times 5b_{k-10}]_0 \\ &\quad \cup [a_{k-1} \times (b_k - b_{k-1}) \cup 5a_{k-10} \times b_{k-1} \cup a_{k-1} \times 5b_{k-10} \cup (a_k - a_{k-1}) \times b_{k-1}]_1 \\ &\quad \cup [a_{k-1} \times b_{k-1} \cup 5a_{k-9} \times b_k \cup (b_k - b_{k-1}) \times 5a_{k-10}]_2 \\ &= [\beta_{k-1}]_0 \cup [\lambda_{k-1}]_1 \cup [\beta'_{k-1}]_2.\end{aligned}$$

Proof of (4.9.16). Since

$$a_k - a_{k-1} > a_{k-2} \geq 2a_{k-4} \geq 4a_{k-6} \geq 8a_{k-8} > 5a_{k-10},$$

we have $a_{k-1} + 5a_{k-10} \leq a_k$, so

$$\begin{aligned}\beta &= [a_{k-1} \times b_{k-1} \cup (a_k + 5a_{k-10}) \times 5b_{k-9}]_0 \\ &\quad \cup [a_{k-1} \times (b_k - b_{k-1}) \cup (a_k - a_{k-1}) \times b_{k-1}]_1 \\ &\quad \cup [(a_k - a_{k-1}) \times (b_k - b_{k-1}) \cup a_{k+1} \times 5b_{k-8}]_2 \\ &\leq [\beta_{k-1}]_0 \cup [\lambda_{k-1}]_1 \cup [\delta_{k-1}]_2.\end{aligned}$$

Proof of (4.9.17). Since

$$b_{k-1} \geq 2b_{k-3} \geq 4b_{k-5} \geq 8b_{k-7} > 5b_{k-10},$$

we have

$$\begin{aligned}\delta_k &= [a_{k-1} \times b_{k-1}]_0 \cup \left[a_{k-1} \times 5b_{k-1} \cup b_{k-1} \times 5a_{k-10} \cup \frac{a_{k+2}}{2} \times 5b_{k-7}\right]_1 \\ &\quad \cup \left[5b_{k-10} \times 5a_{k-10} \cup \frac{a_{k+2}}{2} \times 5b_{k-7}\right]_2 \\ &\leq [\beta_{k-1}]_0 \cup [\rho_{k-1}]_1 \cup [\rho_{k-1}]_2.\end{aligned}$$

Proof of (4.9.18).

$$\begin{aligned}\tau_k &= [a_{k-1} \times b_{k-1} \cup \times 5b_{k-9}]_0 \cup [a_{k-1} \times b_{k-1} \cup b_k \times 5a_{k-9}]_1 \\ &\leq [\beta_{k-1}]_0 \cup [\beta'_{k-1}]_1.\end{aligned}$$

Proof of (4.9.19).

$$\begin{aligned}\pi_k &= [a_{k-1} \times b_{k-1}]_0 \cup [a_{k-1} \times 5b_{k-10} \cup 2a_{k-2} \times b_{k-2} \cup b_{k-1} \times 5a_{k-10}]_1 \\ &\quad \cup [5a_{k-10} \times 5b_{k-10} \cup a_{k-2} \times 2b_{k-2}]_2 \\ &\leq [\beta_{k-1}]_0 \cup [\tau_{k-1}]_1 \cup [\tau'_{k-1}]_2.\end{aligned}$$

Proof of (4.9.20).

$$\begin{aligned}\rho_k &\le (5,45) \otimes \{b_{k-6} \times a_{k-6} \cup a_{k-9} \times 2b_{k-9} \cup 2a_{k-9} \times b_{k-9}\}\\ &\le (5,45) \otimes \{[b_{k-7} \times a_{k-7}]_0 \cup [a_{k-7} \times (b_{k-6} - b_{k-7}) \cup (a_{k-6} - a_{k-7}) \times b_{k-7}]_1\\ &\quad \cup [(a_{k-6} - a_{k-7}) \times (b_{k-6} - b_{k-7}) \cup a_{k-9} \times 2b_{k-1} \cup 2a_{k-9} \times b_{k-9}]_2\}\\ &\le (5,45) \otimes \{[\beta_{k-7}]_0 \cup [\lambda_{k-7}]_1 \cup [\pi_{k-7}]_2\}.\end{aligned}$$

□

Theorem 4.29 gives a limiting rate pair (0.717, 0.717), which dominates that of the Fibonacci codes.

References

1. R. Ahlswede, *Multi-way communication channels, in 2nd International Symposium Information Theory, Armenian SSR, 1971* (Publishing House of the Hungarian Academy of Sciences, Tsahkadzor, 1973), pp. 23–52
2. R. Ahlswede, The capacity region of a channel with two senders and two receivers. Ann. Prob. **2**(5), 805–814 (1974)
3. R. Ahlswede, V.B. Balakirksy, Construction of uniquely decodable codes for the two-user binary adder channel. IEEE Trans. Inf. Theory **45**(1), 326–330 (1999)
4. R. Ahlswede, G. Simonyi, On the optimal structure of recovering set pairs in lattices: the sandglass conjecture. Discrete Math. **128**, 389–394 (1994)
5. V.F. Babkin, A universal encoding method with nonexponential work expediture for a source of independent messages. Problemy Peredachi Informatsii **7**(4), 13–21 (1971)
6. L.A. Bassalygo, M.S. Pinsker, Evaluation of the asymptotics of the summarized capacity of an M-frequency T-user noiseless multiple-access channel, Problemy Peredachi Inf., No. 2, 3–9 (2000); Problems Inf. Transm., **36**(2), 91–97 (2000)
7. M. Bierbaum, H.-M. Wallmeier, A note on the capacity region of the multiple-access channel. IEEE Trans. Inf. Theory **25**, 484 (1979)
8. S.C. Chang, Further results on coding for T-user multiple-access channels. IEEE Trans. Inform. Theory **30**, 411–415 (1984)
9. S.-C. Chang, E.J. Weldon, Coding for T-user multiple-access channels. IEEE Trans. Inf. Theory **25**, 684–691 (1979)
10. S.C. Chang, J.K. Wolf, On the T-user M-frequency noiseless multiple-access channels with and without intensity information. IEEE Trans. Inf. Theory **27**(1), 41–48 (1981)
11. P. Coebergh van den Braak, H. van Tilborg, A family of good uniquely decodable code pairs for the two-access binary adder channel. IEEE Trans. Inf. Theory **31**, 3–9 (1985)
12. T.M. Cover, Enumerative source coding. IEEE Trans. Inf. Theory **19**(1), 73–77 (1973)
13. T.M. Cover, C. Leung, An achievable rate region for the multiple-access channel with feedback. IEEE Trans. Inf. Theory **27**(3), 292–298 (1981)
14. T.M. Cover, J.A. Thomas, *Elements of Information Theory* (Wiley, New York, 1991)
15. M.A. Deaett, J.K. Wolf, Some very simple codes for the nonsynchronized two-user multiple-access adder channel with binary inputs. IEEE Trans. Inf. Theory **24**(5), 635–636 (1978)
16. G. Dueck, The zero error feedback capacity region of a certain class of multiple-access channels. Probl. Control Inf. Theory **14**(2), 89–103 (1985)
17. T. Ericson, The noncooperative binary adder channel. IEEE Trans. Inf. Theory **32**, 365–374 (1986)
18. T. Ericson, L. Györfi, Superimposed codes in R^n. IEEE Trans. Inf. Theory **34**, 877–880 (1988)

19. P.G. Farrell, Survey of channel coding for multi-user systems, in *New Concepts in Multi-User Communications*, ed. by J.K. Skwirrzynski (Alphen aan den Rijn, Sijthoff and Noordhoff, 1981), pp. 133–159
20. T. Ferguson, Generalized T-user codes for multiple-access channels. IEEE Trans. Inf. Theory **28**, 775–778 (1982)
21. N.T. Gaarder, J.K. Wolf, The capacity region of a discrete memoryless multiple-access channel can increase with feedback. IEEE Trans. Inf. Theory **21**(1), 100–102 (1975)
22. P. Gober, A.J. Han Vinck, Note on "On the asymptotical capacity of a multiple-acces channel" by L. Wilhelmsson and K.S. Zigangirov, Probl. Peredachi Inf., **36**(1), 21–25 (2000); Probl. Inf. Trans. 36(1), 19–22 (2000)
23. A.J. Grant, C. Schlegel, Collision-type multiple-user communications. IEEE Trans. Inf. Theory **43**(5), 1725–1736 (1997)
24. T.S. Han, H. Sato, On the zero-error capacity region by variable length codes for multiple channel with feedback, preprint
25. A.J. Han Vinck, J. Keuning, On the capacity of the asynchronous T-user M-frequency noiseless multiple-access channel without intensity information. IEEE Trans. Inf. Theory **42**(6), 2235–2238 (1996)
26. B.L. Hughes, A.B. Cooper, Nearly optimal multiuser codes for the binary adder channel. IEEE Trans. Inf. Theory **42**(2), 387–398 (1996)
27. D.B. Jevtić, Disjoint uniquely decodable codebooks for noiseless synchronized multiple-access adder channels generated by integer sets. IEEE Trans. Inf. Theory **38**(3), 1142–1146 (1992)
28. T. Kasami, S. Lin, Coding for a multiple-access channel. IEEE Trans. Inf. Theory **22**, 129–137 (1976)
29. T. Kasami, S. Lin, Bounds on the achievable rates of block coding for a memoryless multiple-access channel. IEEE Trans. Inf. Theory **24**(2), 187–197 (1978)
30. T. Kasami, S. Lin, V.K. Wei, S. Yamamura, Graph theoretic approaches to the code construction for the two-user multiple-access binary adder channel. IEEE Trans. Inf. Theory **29**, 114–130 (1983)
31. G.H. Khachatrian, Construction of uniquely decodable code pairs for two-user noiseless adder channel, Problemy Peredachi Informatsii (1981)
32. G.H. Khachatrian, On the construction of codes for noiseless synchronized 2-user channel. Probl. Control Inf. Theory **11**(4), 319–324 (1982)
33. G.H. Khachatrian, *New construction of uniquely decodable codes for two-user adder channel, Colloquim dedicated to the 70-anniversary of Prof* (R. Varshamov, Thakhkadzor, Armenia, 1997)
34. G.H. Khachatrian, S.S. Martirossian, Codes for T-user noiseless adder channel. Prob. Contr. Inf. Theory **16**, 187–192 (1987)
35. G.H. Khachatrian, S.S. Martirossian, Code construction for the T-user noiseless adder channel. IEEE Trans. Inf. Theory **44**, 1953–1957 (1998)
36. G.H. Khachatrian, H. Shamoyan, The cardinality of uniquely decodable codes for two-user adder channel. J. Inf. Process. Cybernet. EIK **27**(7), 351–355 (1991)
37. H.J. Liao, Multiple-Access Channels, Ph.D. Dissertation, Dept. of Elect. Eng. University of Hawaii (1972)
38. S. Lin, V.K. Wei, Nonhomogeneous trellis codes for the quasi-synchronous multiple-acces binary adder channel with two users. IEEE Trans. Inf. Theory **32**, 787–796 (1986)
39. B. Lindström, On a combinatorial problem in number theory. Canad. Math. Bull. **8**(4), 477–490 (1965)
40. B. Lindström, Determining subsets by unramified experiments, in *A Survey of Statistical Designs and Linear Models*, ed. by J. Srivastava (North Holland Publishing Company, Amsterdam, 1975), pp. 407–418
41. A.W. Marshall, I. Olken, *Inequalities: Theory of Majorization and its Applications* (Academic Press, New York, 1979)

42. S.S. Martirossian, Codes for noiseless adder channel, in *X Prague Conference on Information Theory*, pp. 110–111 (1986)
43. S.S. Martirossian, G.H. Khachatrian, Construction of signature codes and the coin weighing problem. Probl. Inf. Transm. **25**, 334–335 (1989)
44. J.L. Massey, P. Mathys, The collision channel without feedback. IEEE Trans. Inf. Theory **31**, 192–204 (1985)
45. P. Mathys, A class of codes for a T active users out of M multiple access communication system. IEEE Trans. Inf. Theory **36**, 1206–1219 (1990)
46. Q.A. Nguyen, Some coding problems of multiple-access communication systems, DSc Dissertation, Hungarian Academy of Sciences (1986)
47. E. Plotnick, Code constructions for asynchronous random multiple-access to the adder channel. IEEE Trans. Inf. Theory **39**, 195–197 (1993)
48. J. Schalkwijk, An algorithm for source coding. IEEE Trans. Inf. Theory **18**, 395–399 (1972)
49. J. Schalkwijk, On an extension of an achievable rate region for the binary multiplying channel. IEEE Trans. Inf. Theory **29**, 445–448 (1983)
50. C.E. Shannon, The zero error capacity of a noisy channel. IEEE Trans. Inf. Theory **2**, 8–19 (1956)
51. C.E. Shannon, Two-way communication channels. Proc. 4th Berkeley Symp. Math. Stat. Prob. **1**, 611–644 (1961)
52. H.C.A. van Tilborg, Upper bounds on $|C_2|$ for a uniquely decodable code pair (C_1, C_2) for a two-access binary adder channel. IEEE Trans. Inf. Theory **29**, 386–389 (1983)
53. H.C.A. van Tilborg, An upper bound for codes for the noisy two-access binary adder channel. IEEE Trans. Inf. Theory **32**, 436–440 (1986)
54. R. Urbanke, B. Rimoldi, *Coding for the F -adder channel: Two applications of Reed-Solomon codes* (IEEE International Symposium on Information Theory, San Antonio, United States, 17–22 January 1993)
55. P. Vanroose, Code construction for the noiseless binary switching multiple-access channel. IEEE Trans. Inf. Theory **34**, 1100–1106 (1988)
56. E.J. Weldon, Coding for a multiple-access channel. Inf. Control **36**(3), 256–274 (1978)
57. L. Wilhelmsson, K.S. Zigangirov, On the asymptotical capacity of a multiple-access channel. Probl. Inf. Trans. **33**(1), 12–20 (1997)
58. F.M.J. Willems, The feedback capacity region of a class of discrete memoryless multiple-access channels. IEEE Trans. Inf. Theory **28**, 93–95 (1982)
59. J.H. Wilson, Error-correcting codes for a T-user binary adder channel. IEEE Trans. Inf. Theory **34**, 888–890 (1988)
60. J.K. Wolf, Multi-user communication networks, Communication Systems and Random Process Theory, J.K. Skwirrzynski, Ed., Leyden, The Netherlands, Noordhoff Int., 1978
61. Z. Zhang, T. Berger, J.L. Massey, Some families of zero-error block codes for the two-user binary adder channel with feedback. IEEE Trans. Inf. Theory **33**, 613–619 (1987)

Further Readings

62. E.R. Berlekamp, J. Justesen, Some long cyclic linear binary codes are not so bad. IEEE Trans. Inf. Theory **20**(3), 351–356 (1974)
63. R.E. Blahut, *Theory and Practice of Error Control Codes* (Addison-Wesley, Reading, 1984)
64. E.L. Blokh, V.V. Zyablov, *Generalized Concatenated Codes* (Sviaz Publishers, Moscow, 1976)
65. R.C. Bose, S. Chowla, Theorems in the additive theory of numbers. Comment. Math. Helv. **37**, 141–147 (1962)
66. D.G. Cantor, W.H. Mills, Determination of a subset from certain combinatorial properties. Can. J. Math. **18**, 42–48 (1966)

67. R.T. Chien, W.D. Frazer, An application of coding theory to document retrieval. IEEE Trans. Inf. Theory **12**(2), 92–96 (1966)
68. R. Dorfman, The detection of defective members of large populations. Ann. Math. Stat. **14**, 436–440 (1943)
69. D.-Z. Du, F.K. Hwang, *Combinatorial Group Testing and Its Applications* (World Scientific, Singapore, 1993)
70. A.G. Dyachkov, A.J. Macula, V.V. Rykov, New constructions of superimposed codes. IEEE Trans. Inf. Theory **46**(1), 284–290 (2000)
71. A.G. Dyachkov, V.V. Rykov, A coding model for a multiple-access adder channel. Probl. Inf. Transm. **17**(2), 26–38 (1981)
72. A.G. Dyachkov, V.V. Rykov, Bounds on the length of disjunctive codes. Problemy Peredachi Informatsii **18**(3), 7–13 (1982)
73. A.G. Dyachkov, V.V. Rykov, A survey of superimposed code theory. Probl. Control Inf. Theory **12**(4), 1–13 (1983)
74. P. Erdös, P. Frankl, Z. Füredi, Families of finite sets in which no set is covered by the union of r others. Israel J. Math. **51**(1–2), 70–89 (1985)
75. P. Erdös, A. Rényi, On two problems of information theory. Publ. Math. Inst. Hungarian Academy Sci. **8**, 229–243 (1963)
76. T. Ericson, V.A. Zinoviev, An improvement of the Gilbert bound for constant weight codes. IEEE Trans. Inf. Theory **33**(5), 721–723 (1987)
77. P. Frankl, On Sperner families satisfying an additional condition. J. Comb. Theory Ser. A **20**, 1–11 (1976)
78. Z. Füredi, On r-cover free families. J. Comb. Theory **73**, 172–173 (1996)
79. Z. Füredi, M. Ruszinkó, Superimposed codes are almost big distant ones, in *Proceedings of the IEEE International Symposium on Information Theory*, 118, Ulm (1997)
80. R.G. Gallager, *Information Theory and Reliable Communication* (Wiley, New York, 1968)
81. L. Györfi, I. Vadja, Constructions of protocol sequences for multiple access collision channel without feedback. IEEE Trans. Inf. Theory **39**(5), 1762–1765 (1993)
82. F.K. Hwang, A method for detecting all defective members in a population by group testing. J. Am. Stat. Assoc. **67**, 605–608 (1972)
83. F.K. Hwang, V.T. Sós, Non-adaptive hypergeometric group testing. Studia Scientarium Mathematicarum Hungarica **22**, 257–263 (1987)
84. T. Kasami, S. Lin, Decoding of linear δ-decodable codes for multiple-access channel. IEEE Trans. **24**(5), 633–635 (1978)
85. T. Kasami, S. Lin, S. Yamamura, Further results on coding for a multiple-access channel, in *Conference of the Proceedings Hungarian Colloquium on Information Theory*, Keszthely, pp. 369–391 (1975)
86. W.H. Kautz, R.C. Singleton, Nonrandom binary superimposed codes. IEEE Trans. Inf. Theory **10**, 363–377 (1964)
87. G.H. Khachatrian, Decoding for a noiseless adder channel with two users. Problemy Peredachi Informatsii **19**(2), 8–13 (1983)
88. G.H. Khachatrian, A class of δ-decodable codes for binary adder channel with two users, in *Proceedings of the International Seminar on "Convolutional Codes, Multiuser Communication"*, Sochi, pp. 228–231 (1983)
89. G.H. Khachatrian, New construction of linear δ-decodable codes for 2-user adder channels. Probl. Control Inf. Theory **13**(4), 275–279 (1984)
90. G.H. Khachatrian, Coding for adder channel with two users. Probl. Inf. Transm. **1**, 105–109 (1985)
91. G.H. Khachatrian, Decoding algorithm of linear δ-decodable codes for adder channel with two users, in *Proceedings of the 1st Joint Colloqium of the Academy of Sciences of Armenia and Osaka University (Japan) on Coding Theory*, Dilijan, pp. 9–19 (1986)
92. G.H. Khachatrian, A survey of coding methods for the adder channel, Numbers, Information, and Complexity (Festschrift for Rudolf Ahlswede), Kluwer, pp. 181–196 (2000)

93. E. Knill, W.J. Bruno, D.C. Torney, Non-adaptive group testing in the presence of error. Discr. Appl. Math. **88**, 261–290 (1998)
94. H. Liao, A coding theorem for multiple access communication, in *Proceedings of the International Symposium on Information Theory* (1972)
95. B. Lindström, On a combinatory detection problem I. Publ. Math. Inst. Hungarian Acad. Sci. **9**, 195–207 (1964)
96. N. Linial, Locality in distributed graph algorithms. SIAM J. Comput. **21**(1), 193–201 (1992)
97. J.H. van Lint, T.A. Springer, Generalized Reed-Solomon codes from algebraic theory. IEEE Trans. Inf. Theory **33**, 305–309 (1987)
98. F.J. MacWilliams, N. Sloane, *The Theory of Error-correcting Codes* (North Holland, Amsterdam, 1977)
99. E.C. van der Meulen, The discrete memoryless channel with two senders and one receiver, in *Proceedings of the 2nd International Symposium on Information Theory, Hungarian Academy of Sciences*, pp. 103–135 (1971)
100. Q.A. Nguyen, T. Zeisel, Bounds on constant weight binary superimposed codes. Probl. Control Inf. Theory **17**(4), 223–230 (1988)
101. Q.A. Nguyen, L. Györfi, J.L. Massey, Constructions of binary constrant-weight cyclic codes and cyclically permutable codes. Probl. Control Inf. Theory **38**(3), 940–949 (1992)
102. W.W. Peterson, E.J. Weldon, *Error-correcting Codes* (Mir, Moscow, 1976)
103. V.C. da Rocha, Jr., Maximum distance separable multilevel codes. IEEE Trans. Inf. Theory **30**(3), 547–548 (1984)
104. V. Rödl, On a packing and covering problem. Europ. J. Comb. **5**, 69–78 (1985)
105. M. Ruszinkó, Note on the upper bound of the size of the r-cover-free families. J. Comb. Theory **66**(2), 302–310 (1994)
106. P. Smith, Problem E 2536, Amer. Math. Monthly, Vol. 82, No. 3, 300, 1975; Solutions and comments in Vol. 83, No. 6, 484, 1976
107. M. Sobel, P.A. Groll, Group testing to eliminate efficiently all defectives in a binomial sample. Bell Syst. Tech. J. **38**, 1178–1252 (1959)
108. A. Sterrett, On the detection of defective members in large populations. Ann. Math. Stat. **28**, 1033–1036 (1957)
109. M. Szegedy, S. Vishwanathan, Locality based graph coloring, in *Proceedings of the 25th Annual ACM Symposium on Theory of Computing*, San Diego, pp. 201–207 (1993)
110. H.C.A. van Tilborg, An upper bound for codes in a two-access binary erasure channel. IEEE Trans. Inf. Theory **24**(1), 112–116 (1978)
111. H.C.A. van Tilborg, *A few constructions and a short table of δ-decodable codepair for the two-access binary adder channel* (Univ. of Technology, Technical report, Eindhoven, 1985)
112. M.L. Ulrey, The capacity region of a channel with s senders and r receivers. Inf. Control **29**, 185–203 (1975)
113. J.K. Wolf, Born again group testing: multiaccess communications. IEEE Trans. Inf. Theory **31**(2), 185–191 (1985)
114. K. Yosida, *Functional Analysis*, 4th edn. (Springer, Berlin, 1974)
115. V.A. Zinoviev, Cascade equal-weight codes and maximal packings. Probl. Control Inf. Theory **12**(1), 3–10 (1983)
116. V.A. Zinovev, S.N. Litzin, *Table of best known binary codes* Institute of Information Transmission Problems, Preprint, Moscow, (1984)

Chapter 5
Packing: Combinatorial Models for Various Types of Errors

The following two lectures are based on the papers [23, 24]. They were presented in a series of lectures of Levenshtein when he was guest of Rudolf Ahlswede at the university of Bielefeld.

5.1 A Class of Systematic Codes

In this section (see Siforov [36] and Levenshtein [23]) we consider a class of systematic codes with error detection and correction obtained using one of the code construction algorithms of V.I. Siforov [36]. The size (number of elements) of codes of this class is within the bounds known at present for the maximum size of codes. We investigate certain properties of the codes and also outline a method for decreasing the computational work in their practical construction.

5.1.1 Basic Definitions

We call $\mathcal{X} \triangleq \{0, 1\}$ the *alphabet*. An element $x \in \mathcal{X}$ is called a *letter*. A *word* x^n in the alphabet $\mathcal{X}$ is a finite sequence of elements of $\mathcal{X}$

$$x^n = (x_1, \dots, x_n), \qquad x_i \in \mathcal{X}.$$

The set of all words on the alphabet $\mathcal{X}$ is denoted by $\mathcal{X}^*$ and is equipped with the associative operation defined by the of concatenation of two sequences

$$(x_1, \dots, x_n)(y_1, \dots, y_m) = (x_1, \dots x_n, y_1, \dots, y_m).$$

A. Ahlswede et al. (eds.), *Combinatorial Methods and Models*,
Foundations in Signal Processing, Communications and Networking 13,
DOI 10.1007/978-3-319-53139-7_5

This operation is associative. This allows us to write

$$x^n = x_1 \dots x_n$$

instead of $x^n = (x_1, \dots, x_n)$, by identifying each element $x \in \mathcal{X}$ with the sequence (x). The empty sequence is called the *empty word* and is denoted by 1. It is the neutral element for concatenation. The set of nonempty words on $\mathcal{X}$ is denoted by $\mathcal{X}^+$. Thus we have $\mathcal{X}^+ = \mathcal{X}^* - \{1\}$.

The *length* $|x^n|$ of the word $x^n = x_1 \dots x_n$ with $x_i \in \mathcal{X}$ is the number n of letters in x^n.

We shall use $\oplus$ $(\sum)$, to denote term-by-term addition mod 2 of arbitrary sequences $x^n = x_1 \dots x_n$ and $y^n = y_1 \dots y_n$ of $\mathcal{X}_2^n$, and also for the addition of digits mod 2, and we shall omit the multiplication sign;

$$x^n \oplus y^n = (x_1 \oplus y_1, \dots, x_n \oplus y_n), \quad \lambda x^n = (\lambda x_1, \dots, \lambda x_n).$$

By the *value* $val(x^n)$ of a word $x^n = x_1 \dots x_n$ we mean the integer whose binary representation is the sequence x^n, i.e., $val(x^n) = \sum_{i=1}^{n} x_i 2^{i-1}$.

We define the *weight* $w(x^n)$ of a word x^n as the number of ones in x^n, $w(x^n) = \sum_{i=1}^{n} x_i$. With this definition of weight, the *Hamming distance* [18] between two words x^n and y^n of $\mathcal{X}_2^n$, i.e., the number of digits in which these symbols differ, can be expressed by: $d_H(x^n, y^n) = w(x^n \oplus y^n)$.

For arbitrary words x^n, y^n, and z^n in $\mathcal{X}_2^n$ we have:

Lemma 5.1 $d_H(x^n, y^n) = d_H(x^n \oplus z^n, y^n \oplus z^n)$.

Lemma 5.2 *The inequalities*

$$val(x^n) < val(y^n \oplus z^n), \; val(y^n) < val(x^n \oplus z^n), \; \textit{and} \; val(z^n) < val(x^n \oplus y^n)$$

are incompatible.

A set of words in $\mathcal{X}_2^n$, such that the distance between any two of them is not less than some number d, will be called a *d-code*. A d-code will be called *systematic* if it forms a group under term-by-term addition mod 2. Let us call a d-code, all words of which belong to $\mathcal{X}_2^n$, *saturated* in $\mathcal{X}_2^n$ if it is impossible to adjoin another word of $\mathcal{X}_2^n$ in such a way that it remains a d-code. We call it *maximal* in $\mathcal{X}_2^n$ if there is no d-code with a greater number of words in $\mathcal{X}_2^n$.

5.1.2 Construction of a Maximal d-Code

Let us put all words of the set $\mathcal{X}_2^n$ in a definite order, and let us consider the following algorithm to construct a set S of words possessing some property Γ. As the first element a_0 of S we take the first word with the property Γ in the order in $\mathcal{X}_2^n$. If $a_0, \ldots, a_{i-1}$ have already been chosen we take as a_i the first word (if such exists) in the order in $\mathcal{X}_2^n$, different from those already chosen, such that $a_0, \ldots, a_i$ *have the property* Γ. This algorithm will be called *the trivial algorithm* for constructing S. In particular, in the trivial algorithm for constructing a d-code, we take as a_0 the first element of $\mathcal{X}_2^n$ in the order and, if $a_0, \ldots a_{i-1}$ have already been chosen, we take as a_i the first word of $\mathcal{X}_2^n$ at a distance not less than d from each of $a_0, \ldots, a_{i-1}$, if it exists. It is easy to see that the d-code obtained by the trivial algorithm is maximal in $\mathcal{X}_2^n$. As V.I. Siforov [36] showed, generally speaking, the number of elements in such a d-code depends on *the order in which the words in* $\mathcal{X}_2^n$ *have been put.*

The order in $\mathcal{X}_2^n$ (or $\mathcal{X}_2^*$), in which the words are arranged with their values increasing, will be called the *natural order*. The trivial construction algorithm, in the case when the words of $\mathcal{X}_2^n$ are put in the natural order, will likewise be called the *natural algorithm*. We denote by S_d^n the code obtained from $\mathcal{X}_2^n$ by the natural d-code construction algorithm and we set $S_d^* = \cup_{n=0}^{\infty} S_d^n$.

The basic result is the following

Theorem 5.1 *For any n and d, the codes S_d^* and S_d^n are systematic.*

Proof It is sufficient to show that the words a_i, $i = 0, 1, \ldots,$ successively obtained by the natural d-code construction algorithm, satisfy the relation

$$a_i = \sum_{j=1} \sigma_{i,j} a_{2^{j-1}}, \qquad \text{when } i = \sum_{j=1} \sigma_{i,j} 2^{j-1}. \tag{5.1.1}$$

We shall prove this formula by induction on i. It is trivial for $i = 0$ and also for $i = 2^m$, $m = 0, 1, \ldots$. Hence for the proof of (5.1.1) it is sufficient to show that

$$a_{2^m+r} = a_{2^m} \oplus a_r, \quad 1 \le r \le 2^m - 1, \tag{5.1.2}$$

under the hypothesis that formula (5.1.1) holds for all $i < 2^m + r$. For the proof of (5.1.2) let us suppose the contrary, i.e.,

$$a_{2^m+r} \neq a_{2^m} \oplus a_r. \tag{5.1.3}$$

By Lemma 5.1 and the induction hypothesis it is easy to show that

$$d_H(a_{2^m+r} \oplus a_{2^m}, a_j) = d_H(a_{2^m+r}, a_{2^m+j}) \ge d, 0 \le j < r; \tag{5.1.4}$$

$$d_H(a_{2^m+r} \oplus a_r, a_l) = d_H(a_{2^m+r}, a_{l'}) \ge d, 0 \le l, l', \le 2^m - 1; \tag{5.1.5}$$

$$d_H(a_{2^m} \oplus a_r, a_i) \ge d, i < 2^m + r. \tag{5.1.6}$$

From the definition of the natural algorithm and from inequalities (5.1.3)–(5.1.6) it follows that

$$val(a_r) < val(a_{2^m+r} \oplus a_{2^m}), \ val(a_{2^m}) < val(a_{2^m+r} \oplus a_r),$$

$$\text{and } val(a_{2^m+r}) < val(a_{2^m} \oplus a_r),$$

which are incompatible by Lemma 5.2. This competes the proof of the theorem. □

Two important properties of the codes S_d^n follow from formula (5.1.1).

(i) The sequence a_{2^i}, $i = 1, 2, \dots$ has zeros in the places numbered by $n(j)$, $1 \leq j \leq i$, where $n(j)$ is the position of the last one in the sequence $a_{2^{j-1}}$.
(ii) If $1 \leq r \leq 2^i - 1$, then $a_{2^i+r} = a_{2^i} \oplus a_r$.

Starting from these properties, it is easy to show that one can also construct the code S_d^n by the following algorithm.

For the first word a_0 we take $(0, 0, \dots, 0)$. If $a_0, \dots, a_{2^i-1}$ have already been chosen, we take as a_{2^i} the least value of the elements of $\mathcal{X}_2^n$ having zero in the places numbered by $n(j)$, $1 \leq j \leq i$, and having distance not less than d from each a_l, $0 \leq l \leq 2^i - 1$, if such still exist. The words a_{2^i+r}, $1 \leq r \leq 2^i - 1$, are defined by $a_{2^i+r} = a_{2^i} \oplus a_r$.

This algorithm differs advantageously from the natural algorithm used to define S_d^n by involving considerably less computation, both because of the decrease of length of the selected words and because of the decrease of their number.

In order to formulate a theorem expressing the fact that a natural order is in a certain sense the unique order from which one always gets a systematic code, let us introduce the notation of equivalence of orders. Two orders $b_0, b_1, \dots$ and $b'_0, b'_1, \dots$ of arrangement of the words in $\mathcal{X}_2^*$ are called *equivalent* if $b_j \oplus b_k = b_l$ implies $b'_j \oplus b'_k = b'_l$.

Theorem 5.2 *In order that, for any order of arrangement of the words in $\mathcal{X}_2^*$ equivalent to a given one, the trivial d-code construction algorithm applied to the first 2^n words should give a systematic code for arbitrary n, it is necessary and sufficient that the given order should be equivalent to the natural order.*

5.1.3 Estimation of the Size

In order to estimate the size (number of elements) of the code S_d^n we denote by $m(n, d)$ the number of generators of this code. Then the size of S_d^n will be $2^{m(n,d)}$. One can show that the quantity $m(n, d)$ satisfies the same relations,

$$m(n, 2s + 1) = m(n + 1, 2s + 2);$$

$$\left\lceil log_2 \frac{2^n}{2 + \binom{n-1}{1} + \cdots + \binom{n-1}{2s-1}} \right\rceil \leq m(n, 2s + 1) \leq \left\lfloor log_2 \frac{2^n}{1 + \binom{n}{1} + \cdots + \binom{n}{s}} \right\rfloor,$$

as are known [18, 38] for the number of generators of a maximum systematic code.

It follows in particular that S_3^n and S_4^n are maximum in the class of systematic codes, moreover

$$m(n, 3) = m(n + 1, 4) = \left\lfloor log_2 \frac{2^n}{n + 1} \right\rfloor.$$

Lemma 5.3 *The code S_3^n coincides with the Hamming code [18].*

Proof One can define the Hamming single error correcting code H_n in another way as the set of all words $a = (a_1, \dots, a_n)$ for which

$$e = \sum_{i=1}^{n} a_i e_i = (0, 0, \dots, 0),$$

where $e_i = (\sigma_{i,1}, \dots, \sigma_{i,n})$, when $i = \sum_{j=1} \sigma_{i,j} 2^{j-1}$. First we prove by induction that if $a_l \in S_3^n$ then $a_l \in H_n$. For the word $a_0 = (0, \dots, 0)$ this is obvious. We assume it is true for all words $a_r, 0 \leq r \leq l - 1$, and we show that it is also true for $a_l = (a_{l,1}, \dots, a_{l,n})$.

Let us suppose the contrary, i.e., $e = \sum_{i=1}^{n} a_{l,i} e_i \neq (0, \dots, 0)$. Let the last one in the sequence e have the position number t. Then there exists p such that $a_{l,p} = 1, \sigma_{p,t} = 1$. Hence the word $e \oplus e_p$ is equal to some e_q where $0 \leq q \leq p \leq n$. Let us consider the word $b = (b_1, \dots, b_n)$ where $b_p = 0, b_q = a_{l,q} \oplus 1$ (if $q \neq 0$) and $b_j = a_{i,j}$ otherwise. It is clear that $val(b) < val(a_l)$ and at the same time, as a consequence of the fact that $a_0, \dots, a_{l-1}$ and, as is easy to verify, b belongs to H_n, we have the inequalities $d_H(b, a_i) \geq d, 0 \leq i \leq l - 1$. This contradicts the fact that the word a_l of the code S_3^n is selected by the natural algorithm after the words $a_0, \dots a_{l-1}$. Thus $S_3^n \subset H_n$. On the other hand, since a maximal d-code can not be a proper part of another d-code, $S_3^n = H_3$, and the lemma is proved. □

Table1	Table2
1111	1111
11000111	11000111
1010010011	1010010011
01010010101	01010010101
0110010000011	0110010000011
11010010000101	11010010000101
110101001001001	110101001001001
1011001010010001	1011001010010001
11100110100100001	11100110100100001
1110010000000000011	1110010000000000011
10010010000000000101	1011001000000000101
111000001001000001001	0100010010000000001001
	1100000000010000010001

The statement, asserting that all the codes S_d^n are maximum in the class of systematic codes, turns out to be false. More detailed investigation showed that the codes S_5^n, for example, with $n \leq 21$ are actually maximum in the class of systematic codes with the possible exception of S_5^{18}, and then that S_5^{22} is not. In Table 1 are the generators of the maximal systematic code S_5^{22} and in Table 2 are the generators of a maximum systematic code for $n = 22$ and $d = 5$.

5.1.4 The Practical Construction

For the practical construction of the codes S_d^n it is expedient to take into account the properties (i) and (ii) and to make the computations only for the generators of these codes and only for those digits of the generators whose position numbers are not of the form $n(j)$. One can achieve this by the following device.

Let us denote by C_d^n the set of words which has the property that for any pairwise distinct words $c_1, \dots, c_{d-1}$ of the set we have (cf. [38])

$$w(c_1) \geq d-1,\ w(c_1 \oplus c_2) \geq d-2, \dots, w(c_1 \oplus \cdots \oplus c_{d-1}) \geq 1,$$

which is obtained form $\mathcal{X}_2^n$ by the natural algorithm. Let $C_d^* = \cup_{n=0} C_d^n$. From the first m words $C_i^* = (\gamma_{i,1}, \dots, \gamma_{i,k(i)}), 1 \leq i \leq m$, (where $k(i)$ is the position number of the last one in the sequence c_i) of the set C_d^* we form the word $c_i' = (\gamma_{i,1}', \dots, \gamma_{i,k(i)+i}')$, $1 \leq i \leq m$, setting

$$\gamma_{i,j}' = \begin{cases} 0, & j = k(l) + l, \quad 1 \leq l < i; \\ \gamma_{i,j-l}, & k(l) + l < j < k(l+1) + l + 1,\ 0 \leq l < i\ (k(0) = 0); \\ 1, & j = k(i) + i. \end{cases} \quad (5.1.7)$$

We denote by G_d^m the group under operation of term-by-term addition mod 2 generated by the words $c_1', \dots, c_m'$.

Lemma 5.4 $S_d^n = G_d^m$, *where* $k(m) + m \leq n < k(m+1) + m + 1$, *moreover,* $a_{2^{i-1}} = c_i'$, $1 \leq i \leq m$.

Table3

1111	110011	1010101	0101011	01101001
11010101	11011011	10110111	11101111	111010001
100101001	111000111	1001100001	1011010001	0100101001
1100000101	1100010011	1111101011	0110011111	11011000001
01110100001	00011010001	10001001001		

On the basis of Lemma 5.4 one can reduce the problem of constructing the codes C_d^n to the problem of finding some of the first words in C_d^*. The first 23 words in

C_5^* (Table 3) were found on a computing machine. By means of these according to formula (5.1.7) one can determine the generators of all the codes S_5^n for $n \leq 34$, and by this one can construct these codes.

5.2 Asymptotically Optimum Binary Code with Correction for Losses of One or Two Adjacent Bits

In this section (see Sellers [35] and Levenshtein [26]) a method is presented for the construction of a code permitting correction for the loss of one or two adjacent bits containing at least $\frac{2^{n-1}}{n}$ binary words of length n. On the other hand, for arbitrary $\epsilon > 0$ it is shown that any code having the indicated corrective property contains for sufficiently large n fewer than $(1+\epsilon) \cdot \frac{2^{n-1}}{n}$ binary words of length n.

5.2.1 *Codes with Correction for Losses of l or Fewer Adjacent Bits*

For any binary word $x^n = x_1, \dots, x_n$ we say about every word $x_1, \dots, x_{i-1}, x_{i+l-1}, \dots x_n$, $1 \leq i \leq n+1-l$ that it is obtained from x^n by the loss of l adjacent bits (beginning with the ith one). We say that this loss is *reduced* if for any j, $i < j \leq n+1-l$ the word $x_1, \dots, x_{i-1}, x_{i+l-1}, \dots, x_n \neq x_1, \dots, x_{j-1}, x_{j+l-1}, \dots, x_n$. We call a set of binary words a code with correction for losses of l or fewer adjacent bits if any binary word can be obtained from at most one word of the code by the loss of l or fewer adjacent bits.[1]

We observe that a code which permits correction of all losses of l adjacent bits, in general, does not permit correction of all losses of a smaller number of adjacent bits. On the other hand, a code that permits correction for all reduced losses of l adjacent bits also permits correction for all losses of l adjacent bits.

We denote by $S_l(n)$ the maximum number of words of a code in $\mathcal{X}_2^n$ with correction for losses of l or fewer adjacent bits. The first examples of codes of this type were proposed by Sellers [35]. The construction of Sellers codes are exceedingly simple, but their size, which is equal to 2^k, where

$$n - 2 \cdot \sqrt{(n-l-1) \cdot (l-1)} - l < k < n - 2 \cdot \sqrt{(n-l-1) \cdot (l-1)} + 3 \cdot (l+1),$$

as apparent from below, differs significantly from $S_l(n)$. In [24] the Varshamov–Tenengol'ts code [46] is used to prove the asymptotic equation

[1] A code with correction for gains of l or fewer adjacent bits can be defined analogously. It is easily verified, however, that the two definitions are equivalent.

$$S_1(n) \approx \frac{2^n}{n}.$$

The fundamental result of [26] is the asymptotic equation

$$S_2(n) \approx \frac{2^{n-1}}{n}.$$

5.2.2 *Upper Estimate of the Size of Binary Codes with Correction for Losses of l Adjacent Bits*

We first introduce some notation. Let $x^n = x_1, \dots, x_n$ be any binary word. Word x^n is logically represented in the form of a product of words $u_0, \dots, u_s$, where u_i is nonempty and consists of bits of one type, while the words u_{i-1} and u_{i+1} (if they exist) are formed of bits of the other type. The words u_i are called *strings* of word x^n. For example, the word 01110100 has five strings. The number of strings of word x^n is denoted by $||x^n||$. We denote by $||x^n||_l$ the number of distinct words obtained from word x^n by losses of l adjacent bits. We note that $||x^n||_1 = ||x^n||$. We denote by $y_{i,l}$ $(1 \le i \le l)$ the word $x_i, x_{i+1}, \dots, x_{\lfloor \frac{n-i}{l} \rfloor \cdot l+i}$. It is readily verified by induction on the length n of word x^n that for $n \ge l$

$$||x^n||_l = 1 + \sum_{i=1}^{l} (||y_{i,l}|| - 1). \tag{5.2.1}$$

Lemma 5.5 *For any fixed*[2] *l, $l = 1, 2, \dots$ and $n \to \infty$*

$$S_l(n) \lesssim \frac{2^{n-l+1}}{n}. \tag{5.2.2}$$

Proof Consider an arbitrary code $\mathcal{C}$, maximal in $\mathcal{X}_2^n$, with correction for losses of l adjacent bits. Let $m = \lfloor \frac{n-l}{l} \rfloor$ and let r be arbitrary natural numbers such that $2 \cdot r \le m$. The code $\mathcal{C}$ contains a certain number S'_r of words x^n such that $||y_{i,l}|| > r + 1$ for $1 \le i \le l$ and a certain number S''_r of words x^n such that the relation $||y_{i,l}|| \le r + 1$ holds for at least one i. It follows from (5.2.1) and the corrective properties of code $\mathcal{C}$ that

$$S'_r(l(r+1)+1) \le 2^{n-i}$$

[2]It can be shown, changing only the choice of r in the proof of the lemma, that it is valid when $l = o\left(\frac{n}{(ln(n))^3}\right)$.

and

$$S_r^n \leq \sum_{i_1}^{l} \sum_{j=0}^{r} \binom{\lfloor \frac{n-i}{l} \rfloor}{j} \cdot 2^{n-\lfloor \frac{n-i}{l} \rfloor} \leq l \cdot 2^{n-m} \cdot \sum_{j=0}^{r} \binom{m}{j}.$$

Consequently, for $m \geq 2 \cdot r$,

$$S_l(n) \leq \frac{2^{n-l}}{l \cdot (r+1)+1} + l \cdot 2^{n-m} \cdot \sum_{j=0}^{r} \binom{m}{j}. \tag{5.2.3}$$

We set $r = \left\lfloor \frac{m-\sqrt{2 \cdot m \cdot ln(m)}}{2} \right\rfloor$ and let n (and hence, $m = \lfloor \frac{n-l}{l} \rfloor$) tend to infinity. With this choice of r the relations $2 \cdot r \leq m$ and $\lfloor l \cdot (r+1)+1 \rfloor \approx \frac{n}{2}$ are valid, and by the theorem of large ratios

$$\sum_{j=0}^{r} \binom{m}{j} = O\left(\frac{2^m}{m}\right).$$

But then the asymptotic inequality (5.2.2) follows from (5.2.3).

This proves the lemma. □

5.2.3 A Class of Binary Codes with Correction for Losses of One or Two Adjacent Bits

Let $z = z_0 z_1 \dots z_h$ be an arbitrary binary word. We enumerate the strings of words z from 0 to $||z|| - 1$. We introduce the function $k_z(i)$, which is equal to the index number of the string containing the $(i+1)$th bit z_i of word z, and we put

$$M(z) = \sum_{i=0}^{h} k_z(i).$$

For example, if $z = 01110100$, then $M(z) = 3 \cdot 1 + 1 \cdot 2 + 1 \cdot 3 + 2 \cdot 4 = 16$.

Another function that has an important role in the ensuing discussion is

$$r_z^{\gamma,\delta}(i) = (1+\gamma)k_z(i) + 2(n-i) - (1+\delta),$$

where γ and δ are 0 or 1. We call the number i *interior to the word* $z = z_0 z_1 \dots z_h$ if $0 \leq i \leq h$ and $z_i = z_{i+1}$. We now show that

$$r_z^{\gamma,\delta}(i) \geq r_z^{\gamma,\delta}(j) \text{ for } 0 \leq i < j \leq h,$$

where

$$r_z^{\gamma,\delta}(i) > r_z^{\gamma,\delta}(j) \text{ for } 0 \le i < j \le h, \tag{5.2.4}$$

if the number i is interior to word z. We note for the proof that

$$\begin{aligned} r_z^{\gamma,\delta}(i) - r_z^{\gamma,\delta}(j) &= 2(j-i) - (1+\gamma)(k_z(j) - k_z(i)) \\ &\ge 2((j-i) - (k_z(j) - k_z(i))). \end{aligned} \tag{5.2.5}$$

Analyzing the number of strings of word z that contain the word $z_{i+1} \dots z_j$, we readily note that $j - i \ge k_z(j) - k_z(i)$, where the inequality is definite if the number i is interior to z. This completes the proof of inequalities (5.2.4) and (5.2.5).

We are now in a position to define a class of codes. For arbitrary integers n ($n \ge 1$) and a ($0 \le a \le 2n-1$) let

$$B_n^a = \{x : x \in \mathcal{X}_2^n, M(0x) \equiv a \text{ (mod } 2n)\}.$$

We show that every code B_n^a ($0 \le a \le 2n-1$) permits correction of reduced (and, hence, all) losses of one or two adjacent bits.[3]

Suppose that as a result of the reduced loss of one or two adjacent bits from a word $x = x_1 \dots x_n \in B_r^a$, word $x' = x_1' \dots x_h'$ ($h = n-1$ of $h = n-2$) is obtained. We set $\Delta M = M(0x) - M(0x')$ and $x_0 = x_0' = 0$.

The following six mutually exclusive cases are possible; we proceed with an analysis of each:

I. The lost bit is x_{i+1}, $x_i = x_{i+1}$, $i = 0, \dots, n-1$. Here $x_{i+2} \ne x_i$ when $i + 2 \le n'$, due to the reduced character of the loss. In this case

$$\Delta M = k_{0x}(i+1) = k_{0x'}(i) \le ||0x'|| - 1, \tag{5.2.6}$$

where the number i is not interior to the word $0x'$.

II. The lost bit is x_{i+1}, $x_i \ne x_{i+1}$, $i = 0, \dots, n-1$. Now $x_{i+2} = x_i$ when $i + 2 \le n$, due to the reduced character of the loss.
In this case,

$$\Delta M = k_{0x}(i+1) + 2 \cdot (n-i-1) = k_{0x'}(i) + 2 \cdot (n-i) - 1 = r_{0x'}^{0,0}(i).$$

By virtue of the monotonic decrease (see (5.2.4)) of the function $r_{0x'}^{0,0}(i)$, we have

$$||0x'|| = k_{0x'}(n-1) + 1 = r_{0x'}^{0,0}(n-1) \le r_{0x'}^{0,0}(i) \le r_{0x'}^{0,0}(0) = 2 \cdot n - 1.$$

[3] If we define the code B_n^a as the set of words $x = x_1, x_2, \dots, x_n$ such that $M(x) \equiv a \ (mod\ 2n)$, then such a code, in general, will not guarantee the capability of correcting for several losses of one and two adjacent bits. This was remedied by a method in which for the determination of membership of word x in code B_n^a a fixed bit (say, 0) is assigned to the left of word x.

In this case, therefore,

$$||0x'|| \le \Delta M = r_{0x'}^{0,0}(i) \le 2 \cdot n - 1, \tag{5.2.7}$$

where the number i is interior to the word $0x'$.

III. The lost bits are x_{i+1} and $x_{i+2}, x_i = x_{i+1} = x_{i+2}, 0 \le i \le n-2$. Now $x_{i+3} \ne x_i$ when $i+3 \le n$, due to the reduced character of the loss.
In this case $\Delta M = k_{0x}(i+1) + k_{0x}(i+2) = 2 \cdot k_{0x'}(i)$, so that

$$0 \le \Delta M = 2 \cdot k_{0x'}(i) \le 2 \cdot ||0x'|| - 2, \tag{5.2.8}$$

where the number i is not interior to the word $0x'$.

IV. The lost bits are x_{i+1} and x_{i+2}, $x_i = x_{i+1}$, $x_{i+1} \ne x_{i+2}$, $0 \le i \le n-2$. Now $x_{i+3} \ne x_i$ when $i+3 \le n$, due to the reduced character of the loss.
In this case $\Delta M = k_{0x}(i+1) + k_{0x}(i+2) = 2 \cdot k_{0x'}(i) + 1$; hence,

$$1 \le \Delta M = 2 \cdot k_{0x'}(i) + 1 \le 2 \cdot ||0x'|| - 1, \tag{5.2.9}$$

where the number i is not interior to the word $0x'$.

V. The lost bits are x_{i+1} and x_{i+2}, $x_i \ne x_{i+1}$, $x_{i+1} = x_{i+2}$, $0 \le i \le n-2$. Here $x_{i+3} = x_i$ when $i+3 \le n$, due to the reduced character of the loss.
In this case

$$\Delta M = k_{0x}(i+1) + k_{0x}(i+2) + 2 \cdot (n-i-2)$$

$$= 2 \cdot k_{0x'}(i) + 2 \cdot (n-i) - 2 = r_{0x'}^{1,1}(i).$$

By virtue to the monotonic decrease of the function $r_{0x'}^{1,1}(i)$ we have

$$2 \cdot ||0x'|| = 2 \cdot k_{0x'}(n-2) + 2 = r_{0x'}^{1,1}(n-2)$$

$$\le r_{0x'}^{1,1}(i) \le r_{0x'}^{1,1}(0) = 2 \cdot n - 2.$$

In this case, therefore,

$$2 \cdot ||0x'|| \le \Delta M = r_{0x'}^{1,1}(i) \le 2 \cdot n - 2, \tag{5.2.10}$$

where the number i is interior to the word $0x'$.

VI. The lost bits are x_{i+1} and x_{i+2}; $x_i \ne x_{i+1}, x_{i+1} \ne x_{i+2}$; $0 \le i \le n-2$. Here $x_{i+3} = x_i$ when $i+3 \le n$, due to the reduced character of the loss.
In this case

$$\Delta M = k_{0x}(i+1) + k_{0x}(i+2) + 2 \cdot (n-i-2)$$

$$= 2 \cdot k_{0x'}(i) + 2 \cdot (n-i) - 1 = r^{1,0}_{0x'}(i).$$

By virtue to the monotonic decrease of the function $r^{1,0}_{0x'}(i)$ we have the following:

$$2 \cdot ||0x'|| + 1 = 2 \cdot k_{0x'}(n-2) + 3 = r^{1,0}_{0x'}(n-2)$$

$$\leq r^{1,0}_{0x'}(i) \leq r^{1,0}_{0x'}(0) = 2 \cdot n - 1.$$

In this case, therefore,

$$2 \cdot ||0x'|| \leq \Delta M = r^{1,0}_{0x'}(i) \leq 2 \cdot n - 1, \tag{5.2.11}$$

where the number i is interior to the word $0x'$.

We verify first that it is possible on the basis of the word $x' = x'_1 \dots x'_h$ obtained by the loss of one or two adjacent bits from word x, to determine which of the six possible situations actually occurs. We note, first of all, that the word $0x'$ and the numbers $a, h, ||0x'||$, and $M(0x')$ may be regarded as known. In as much as the number $\Delta M = M(0x) - M(0x')$, according to (5.2.6)–(5.2.11), is always between 0 and $2 \cdot n - 1$ and $M(0x) \equiv a \ (\bmod 2 \cdot n)$, it follows that ΔM is equal to the smallest non-negative residue of the number $a - M(0x') \bmod 2 \cdot n$ and may therefore also be regarded as known. The values of the numbers $h, ||0x'||$, and ΔM enable us right away to determine which of the six possible cases occurs. Thus, the case

I. occurs when $h = n - 1$ and $\Delta M < ||0x'||$;
II. occurs when $h = n - 1$ and $\Delta M \geq ||0x'||$;
III. occurs when $h = n - 2, \Delta M < 2 \cdot ||0x'||$, and ΔM is even;
IV. occurs when $h = n - 2, \Delta M < 2 \cdot ||0x'||$, and ΔM is odd;
V. occurs when $h = n - 2, \Delta M \geq 2 \cdot ||0x'||$, and ΔM is even;
VI. occurs when $h = n - 2, \Delta M \geq 2 \cdot ||0x'||$, and ΔM is odd.

We now show how, once it has been determined which of the six cases occurs, the word $0x'$ can be used to find the word $0x$ and, therefore, word x. Clearly, it is sufficient for this to find the beginning $x'_0 \dots x'_i = x_0 \dots x_i$ of the word $0x'$ and then to insert the beginning of the letter x_i in case *I.*, the letter $\overline{x_i}$ in case *II.*, the word $x_i x_i$ in case *III.*, the word $x_i \overline{x_i}$ in case *IV.*, the word $\overline{x_i x_i}$ in case *V.*, or the word $\overline{x_i} x_i$ in case *VI.*. We verify that in each of these cases the word $x'_0 \dots x'_i$ can be determined from the word $0x'$ and the number ΔM. In fact, it is possible in cases *I.*, *III.*, and *IV* to find (see (5.2.6), (5.2.8), and (5.2.9)) the number $k_{0x'}(i)$, where i is not interior to the word $0x'$ from the number ΔM. Consequently in these cases the word $x'_0 \dots x'_i$ coincides with the word formed by strings with index numbers $0, 1, \dots, k_{0x'}(i)$ of the word $0x'$. In cases *II*, *V*, and *VI* the number ΔM is equal to the value of one of the functions $r^{\gamma,\delta}_{0x'}$

of the argument i, where i is interior to the word $0x' = x'_0 \dots x_h$ or $i = h$. But then, by inequality (5.2.5), the number i is uniquely determined from the number ΔM, and this makes it possible to determine the word $x'_0 \dots x'_h$. This completes the proof of B_n^a as a code with correction for losses of one or two adjacent bits. □

5.2.4 Size of Codes B_n^a

We denote the number of elements of an arbitrary set $K \subseteq \mathcal{X}_2^n$ by $\#K$. For arbitrary $x = x_1 \dots x_n \in \mathcal{X}_2^n$ we set $W(x) = \sum_{i=1}^{n} x_i \cdot i$.

Let $K_{n,m}^a = \{x : x \in \mathcal{X}_2^n, W(x) \equiv a \,(\bmod m)\}$. The codes $K_{n,n+1}^a$ were introduced by Varshamov and Tenengol'ts [46] as codes with correction for unsymmetric substitution (e.g., 0-1) in a single bit. In [24] it was demonstrated that the codes $K_{n,n+1}^a$ are codes with correction for the loss or gain of one bit, and codes $K_{n,2\cdot n}^a$ are codes with correction for the loss, gain, or substitution of one bit. Ginzburg [17] has shown that

$$\#K_{n,n+1}^a = \frac{1}{2\cdot(n+1)} \sum_{\substack{d \text{ odd} \\ d|a; d|(n+1)}} d \sum_{\substack{u \text{ odd} \\ d|u; u|(n+1)}} \mu\left(\frac{u}{d}\right) 2^{\frac{n+1}{u}}, \tag{5.2.12}$$

where code $K_{n,n+1}^0$ is maximum of the codes $K_{n,n+1}^a$ and

$$\#K_{n,n+1}^0 = \frac{1}{2\cdot(n+1)} \sum_{\substack{d \text{ odd} \\ d|(n+1)}} \phi(d) 2^{\frac{n+1}{d}}, \tag{5.2.13}$$

where μ is the Möbius function and ϕ is the Euler function. It can be shown on the basis of the definition of the codes $K_{n,m}^a$ that

$$\#K_{n,n+1}^a = \frac{1}{2}\#K_{n+1,n+1}^a, \; \#K_{n,2\cdot n}^a = \frac{1}{2}\#K_{n,n}^a$$

and, hence,

$$\#K_{n,2\cdot n}^a = \#K_{n-1,n}^a. \tag{5.2.14}$$

It is proved below that the size of code B_n^a is equal to the size of $K_{n,2\cdot n}^a$ and can therefore be found by means of (5.2.14) and (5.2.12).

For an arbitrary word $x = x_1 \dots x_n$ we denote by $\dot{x}$ the word $b_1 \dots b_n$, where $b_i = x_i \oplus x_{i+1}$, $1 \le i \le n-1$, and $b_n = x_n$ (the symbol $\oplus$ denotes addition mod 2). The mapping $x \to \dot{x}$ is a one-to-one mapping of $\mathcal{X}_2^n$, because $x_i = b_i \oplus b_{i+1} \oplus \dots \oplus b_n$, $1 \le i \le n$.

Lemma 5.6 $B_n^a = \{x : x \in \mathcal{X}_2^n, W(\dot{x}) \equiv -a(mod\, 2\cdot n)\}$.

Proof We verify that for any $x \in \mathcal{X}_2^n$

$$M(0x) \equiv -W(\dot{x}) \qquad (\bmod\ 2 \cdot n). \tag{5.2.15}$$

Let word x begin with k_0 ($k_0 \geq 0$) zeros and have s strings, not counting (in the case $k_0 > 0$) the first string of zeros. Let k_i, $i = 1, \ldots, s$ be the number of letters in these strings of word x and $\dot{x} = b_1 b_2 = b_n$. Then

$$M(0x) = \sum_{i=0}^{s} k_i \cdot i = \sum_{i=1}^{s} \sum_{j=1}^{i-1} k_j = ns - \sum_{i=1}^{s} \sum_{j=0}^{i-1} k_j$$

$$= ns - \sum_{i=1}^{n-i} b_i \cdot i = n \cdot (s + x_n) - W(\dot{x}).$$

In as much as $s + x_n$ is always even, relation (5.2.15) holds, and the lemma is proved. □

It follows from Lemma 5.6 and Eqs. (5.2.12)–(5.2.14) that

$$\# B_n^a = \# K_{n,2\cdot n}^{-a} = \# K_{n,2\cdot n}^{a}, \tag{5.2.16}$$

$$\# B_n^a = \frac{1}{2 \cdot n} \sum_{\substack{d \text{ odd} \\ d|a; d|n}} d \sum_{\substack{u \text{ odd} \\ ds|u; u|n}} \mu\left(\frac{u}{d}\right) 2^{\frac{n}{u}}, \tag{5.2.17}$$

where code $\mathcal{X}_2^n$ is the maximum of the codes B_n^a, and

$$|B_n^a| = \frac{1}{2 \cdot n} \sum_{\substack{d \text{ odd} \\ d|n}} \phi(d) 2^{\frac{n}{d}}. \tag{5.2.18}$$

Theorem 5.3 $S_2(n) \approx \frac{2^{n-1}}{n}$.

The theorem is a consequence of Lemma 5.5 and the fact that $\# B_n^0 \geq \frac{2^{n-1}}{n}$.

5.3 Single Error-Correcting Close-Packed and Perfect Codes

This section is based on [29], a paper by Martirossian in Russian language. The translation was organized by Rudolf Ahlswede in the frame of a joint Armenian/German project INTAS.

5.3.1 Introduction

Some methods for construction of optimum or close to optimum classes of q-ary codes that correct the most likely errors on amplitude and phase modulated channels are given.

The purpose of this section is to present in a brief and compact form most of the definitions, statements and conclusions that are general throughout the whole work.

Let the coded information be transmitted (be stored) as a vector

$$\overline{x} = (x_1, x_2, \ldots, x_n), \quad x_i \in \{0, 1, \ldots, q-1\}, \quad q \geq 2.$$

Definition 5.1 We'll say that a single error of the type $\{\varepsilon_1, \varepsilon_2, \ldots, \varepsilon_t\}$, where ε_s $(1 \leq s \leq t)$ are integers, $|\varepsilon_s| \leq q-1$ occurs on the channel, if:

A. ***Amplitude-modulated channel***
Symbol x_l may turn into any of symbols $x_l + \varepsilon_s$ as a result of an error in the lth position, if $0 \leq x_l + \varepsilon_s \leq q-1$.

Ph. ***Phase-modulated channel***
Symbol x_l may turn into any of symbols $(x_l + \varepsilon_s) \mod q$ (here and afterwards $a \mod q$ means the least positive residue of number a modulo q) as a result of an error in the lth position.

It follows from the definition that, in particular, errors of the type $\{\varepsilon_1, \varepsilon_2, \ldots, \varepsilon_{q-1}\}$, where ε_s $(1 \leq s \leq q-1)$ runs through the full set of the least absolute residues modulo q, correspond to single symmetrical errors in Hamming metric on a phase-modulated (**Ph**) channel. And the errors of the type $\{\varepsilon_1, \varepsilon_2, \ldots, \varepsilon_{q-1}\}$, where $(1 \leq s \leq q-1)$ runs through the full set of the least positive residues, correspond to single asymmetrical errors, in the general sense on an amplitude-modulated (**A**) channel.

Denote the code powers and the code sets for channels **A** and **Ph** by $\mathbf{M}_A$, $\mathbf{M}_{Ph}$ and $\mathbf{V}_A$, $\mathbf{V}_{Ph}$, respectively. It is evident that the code capable of correcting errors on **Ph** channel will be also capable of correcting errors of the same type on **A** channel, therefore it is natural to expect $\mathbf{M}_A \geq \mathbf{M}_{Ph}$.

Each vector $\overline{x} \in \mathbf{V}_{Ph}$ may turn into exactly $tn + 1$ different vectors as a result of single errors of the type $\{\varepsilon, \varepsilon_2, \ldots, \varepsilon_t\}$, where $|\varepsilon_i| \leq (q-1)/2$ on **Ph** channel. Therefore, the Hamming upper bound or the close-packed bound holds also for these code powers.

$$\mathbf{M}_{Ph} \leq \frac{q^n}{tn+1}. \tag{5.3.1}$$

Codes which power achieve this bound we call *perfect* or *close-packed.*

Some classes of perfect $(n, n-r)$ codes on the basis of q which unlike the basis of the known perfect codes, is not necessarily a power of a prime number are constructed in this work. The code set $\mathbf{V}_{Ph}$ will be defined as the null space of check matrix of size $r \times n$ $\ H = (\overline{h}_1, \overline{h}_2, \ldots, \overline{h}_n)$ over the ring $\mathbf{Z}_q$ ($\mathbf{Z}_q$ is the ring of residue classes modulo q):

$$\mathbf{V}_{Ph} = \left\{ \overline{x} \mid \overline{x}\, H^T = \overline{0} \right\}.$$

(although we deal with an alphabet here which not a field, we preserve the denomination of linear codes for it).

For a distorted vector $\overline{x}' \in U_x$, $\overline{x}' = (x_1, x_2, \ldots, x_l + \varepsilon_s, \ldots x_n)$, $\overline{x}'\, H' = \varepsilon_s \overline{h}_l$ over the ring $\mathbf{Z}_q$. The quantity $\varepsilon_s \overline{h}_l$ is called *error syndrome* of the quantity ε in the lth position.

Each vector $\overline{x}$ may turn into no more than $tn + 1$ different vectors as a result of single errors of the type $\{\varepsilon_1, \varepsilon_2, \ldots, \varepsilon_t\}$ on **A** channel. For the vector $\overline{x}$ denote

$$U_{\overline{x}} = \left\{ \begin{matrix} \overline{x}' : x' = (x_1, x_2, \ldots, x_l + \varepsilon_s, \ldots, x_n), \; 0 \le x_l + \varepsilon_s \le q - 1, \\ 1 \le l \le n, \qquad\qquad\qquad\qquad 1 \le s \le t \end{matrix} \right\}.$$

Then we can write a bound analogous to that in (5.3.1).

$$\sum_{\overline{x} \in V_A} |U_x| \le q^n. \tag{5.3.2}$$

It follows from this bound that the perfect codes for the **A** channel (that is the codes for which the equals sign stands in (5.3.2)) may be not necessarily of the highest power. Therefore the powers of the codes constructed for **A** channel we compare with the bound (5.3.1).

The presented method for construction of codes for **A** channel is based on the known idea by which the code set $\mathbf{V}_A$ may be defined as the set of all possible solutions for the congruence of the form

$$\sum_{i=1}^{n} f(i)\, x_i \equiv j \mod m, \tag{5.3.3}$$

where $f(i)$ is a numerical function, and h, m, j are natural numbers, or

$$V_A = \left\{ \overline{x} = (x_1, \ldots, x_n) : \sum_{i=1}^{n} f(i)\, x_i - j \equiv 0 \mod m \right\}.$$

The set V_A is capable of correcting single errors of the type $\{\varepsilon_1, \varepsilon_2, \ldots, \varepsilon_t\}$ on **A** channel if and only if all the $tn + 1$ error syndromes $\varepsilon_s f(l)$ $(1 \le s \le t, 1 \le l \le n)$ (including the null syndrome that corresponds to the distorted vector) are different by modulo m. Hence, we have $m \ge tn + 1$. For the codes constructed in this work $m = tn + 1$. Therefore, the value of the code power $\mathbf{M}_A$ just on average will be

$$M_A = \max_j M_A(j) \ge \frac{q^n}{tn + 1}, \tag{5.3.4}$$

where $M_A(j)$ is the number of solutions for the congruence (5.3.3) for $m = tn + 1$.

Comparing (5.3.1) and (5.3.4) for the same parameters of codes we came to the conclusion that the power of codes for **A** channel is over the close-packed bound for **Ph** channel. We call such codes *close-packed.*

Now formulate the following statement that will be helpful in our further study.

5.3.2 *The Criterion of Unique Decodability (UD)*

In order that the codes constructed for **A** and **Ph** channel be capable of correcting single errors of the type $\{\varepsilon_1, \varepsilon_2, \ldots, \varepsilon_t, \}$, it is necessary and sufficient for all the $tn + 1$ error syndromes to be different. The transition probability of a symbol into other symbols is different on both channels. The probability of transition of a symbol into a one with adjacent amplitude (phase) on **A (Ph)** channel is considerable higher than into a symbol with a greatly differing amplitude (phase) (adjacent phases correspond to the symbols 0 and $q - 1$).

Denote the transition probability of the symbol i into the symbol j by p_{ij}. Then it follows from $|i - j| < |i_1 - j_1|$ that $p_{ij} > p_{i_1 j_1}$ for **A** channel and from $\min\{|i - j|, q - |i - j|\} < \min\{|i_1 - j_1|, q - |i_1 - j_1|\}$ that $p_{ij} > p_{i_1 j_1}$ for **Ph** channel. Thus, we have that the most likely errors on both channels are low weight errors of the type $\{1, -1\}$, $\{1, 2\}$ or $\{1, -2\}$, $\{1, -1, 2, -2\}$.

5.3.3 $\{1, -1\}$-*Type Error-Correcting Codes*

Close-packed codes on an arbitrary basis of q and of arbitrary length and perfect codes on bases of odd q's and of the length $n = (q^n - 1)/2$ for **A** and **Ph** channels, respectively are constructed in this section.

A-Channel

Theorem 5.4 *For arbitrary n's and q's the set of all possible solutions* $\mathbf{V}_A$ *for the congruence*

$$\sum_{i=1}^{n} i\, x_i \equiv j \mod 2n + 1, \tag{5.3.5}$$

where $x_i \in \{0, 1, \ldots, q - 1\}$ *is the code capable of correcting errors of the type* $\{1, -1\}$ *on* **A** *channel.*

Proof All the numbers $0, 1, \ldots, n, -1, -2, \ldots, -n$, which differ by modulo $2n + 1$ will be the error syndromes. Wherefrom by u.d. criterion the statement of the theorem is true. □

Example For $q = 3, n = 5, j = 4$ the set of code vectors $\mathbf{V}_A$ includes the following 23 vectors

21000	00221
02000	20021
10100	01021
00010	12002
21120	20102
02120	01102
12210	10012
10220	22212
21201	20222
02201	01222
22011	12122
11111	

The power of the best code satisfies the relation

$$\mathbf{M}_A \geq \frac{q^n}{2n+1}. \tag{5.3.6}$$

The exact formula for the power of these codes has the form

$$\mathbf{M}_A\,(j) = \frac{1}{2n+1} \sum_{u\,|\,2n+1} \left(\frac{q}{u}\right) q^{\frac{2n+1-u}{2u}} \sum_{\substack{t \\ t|(u,k-j)}} \mu\left(\frac{u}{t}\right) t, \tag{5.3.7}$$

where $\left(\frac{q}{u}\right)$ is Jacobs symbol (for $(q, u) \neq 1$, $\left(\frac{q}{u}\right)$ is assumed to be zero); k is any solution of the congruence$16y \equiv -(q-1) \mod 2n+1$ and $\mu\,(\cdot)$ is Möbius function. A more close study of formula (5.3.6) shows that the exact number of solutions of the relation (5.3.5) little differs from the average value of (5.3.6), e.g. for simple modules $2n+1 = p$

$$\mathbf{M}_A\,(j) = \frac{q^n \pm 1}{p} \quad \text{or} \quad \frac{q^n \pm (p-1)}{p}$$

the maximum deviation from average value can be obtained in case when $2n+1$ has a great number of common divisors. For $16_j \equiv -(q-1) \mod 2n+1$ the formula (5.3.7) will be of the form

$$\mathbf{M}_A\,(j) = \frac{1}{2n+1} \sum_{u\,|\,2n+1} \left(\frac{q}{u}\right) q^{\frac{2n+1-u}{2u}} \varphi\,(u),$$

where $\varphi\,(\cdot)$ is Euler's function.

For $2n+1 = q^r$ we have from (5.3.7) that $\mathbf{M}_A\,(j)$ does not depend on j and for every j holds

$$\mathbf{M}_A\,(j) = \frac{q^n}{q^r} = q^{n-r}. \tag{5.3.8}$$

Ph-Channel

Theorem 5.5 *For an odd basis of q the null space of the matrix* $\mathbf{H} = (\overline{h}_1, \overline{h}_2, \ldots \overline{h}\,(q^r - 1)/2,)$*, where* $\overline{h}_i$ $(1 \le i \le (q^r - 1)/2)$ *is all the q-nary vectors of length r having first non-zero components* $1, 2, \ldots, (q-1)/2$*, is a linear* $((q^r - 1)/2,\ (q^r - 1)/2 - r)$ *perfect code capable of correcting errors of the type* $\{+1, -1\}$ *on* **Ph** *channel.*

For $q = 3$ the code given by the Theorem 5.5 corresponds to the Hamming ternary perfect codes with symmetrical error-correcting capabilities, since every transitions between symbols are possible.

Example For $q = 5$, $r = 2$, $n = \left(5^2 - 1\right)/2 = 12$ the check matrix of the linear (12, 10) perfect code will be

$$\mathbf{H} = \begin{bmatrix} 1\,1\,1\,1\,1\,0\,2\,2\,2\,2\,2\,0 \\ 0\,1\,2\,3\,4\,1\,0\,1\,2\,3\,4\,2 \end{bmatrix}.$$

Comparing powers of codes of the lengths $(q^r - 1)/2 < n \le (q^r - 1)/2$ for odd $q's$ constructed in 1, 2 (here n implies the code lengths of shortened codes described in 2), we get

$$\mathbf{M}_A\,(j) \ge q^{\lceil \log_q (2n+1) \rceil - \log_q (2n+1)} \mathbf{M}_{Ph}. \tag{5.3.9}$$

For $n = (q^n - 1)/2$ from (5.3.8) and (5.3.9) we have $\mathbf{M}_A = \mathbf{M}_{Ph}$, and for the remaining n $\left(q^{r-1} - 1\right)/2 < n < (q^r - 1)/2$ $\mathbf{M}_A > \mathbf{M}_{Ph}$ and may exceed it up to q times.

5.3.4 {1, 2}- or {−1, −2}-Type Error-Correcting Codes

The conditions placed on the basis of q and code length n under which the close-packed and perfect codes capable of correcting low weight asymmetrical errors of the type $\{1, 2\}$ or $\{-1, -2\}$ on both **A** and **Ph** channels can exist are given in this section.

In case of existence of such codes, methods for their construction are presented.

A-Channel

Two steps are necessary to construct close-packed codes for **A**channel.

Let $p = 2n + 1$ be a prime number.

Theorem 5.6 *In order that there should exist a function $f\,(i)$ such that the set* $\mathbf{V}_A$ *of all possible solutions for the congruence*

$$\sum_{i=1}^{n} f(x)x_i \equiv j \ mod\ \ 2n + 1 \tag{5.3.10}$$

be capable of correcting errors of the type $\{1, 2\}$ *or* $\{-1, -2\}$ *on* **A** *channel, it is necessary and sufficient that* $2 \mid p\,(2)$.

Proof Sufficiency. Let $2 \mid p\,(2)$. The function $f\,(i)$ is defined as follows

$$f\,(i) = a_{s(i)}\, 2^{2[i - s(i)p(2)/2] - 1}, \tag{5.3.11}$$

where $s(i) = \left\lceil \frac{2i}{p(2)} \right\rceil - 1,\ a_0 = 1$, and $a_{s(i)}$ for $s(i) > 0$ is any integer which satisfies the condition

$$a_{s(i)} \not\equiv a_t\, 2^r \mod p \tag{5.3.12}$$

for all $t < s,\ n - r\ \left(r = \overline{1, p\,(2)}\right)$.

For $s = 0$ error syndromes $p\,(2)$ in the first $p\,(2)\,/2$ positions of the functions $f\,(i)$ and $2f\,(i)$ take the following values, respectively

$$\begin{array}{ll} f\,(i) & 2, 2^3, \ldots, 2^{p(2)-1} \\ & \qquad\qquad\qquad (\text{mod } p) \\ 2f\,(i) & 2^2, 2^4, \ldots, 2^{p(2)}. \end{array} \tag{5.3.13}$$

All the numbers in (5.3.13) form a subgroup of multiplicative group $\mathbf{Z}_p^*$ over the field $\mathbf{Z}_p$ of residue classes of integers modulo p. If take a representative from the cosets of decomposition of the group $\mathbf{Z}_p^*$ with respect to this subgroup as $a_1, a_2, \ldots, a_{(p-1)/p(2)-1}$ then all $2n$ error syndromes correspond to all the elements of the group $\mathbf{Z}_p^*$ and together with null syndrome to all the elements of the field $\mathbf{Z}_p$. This fact together with the u.d. criterion proves the theorem.

Thus, the set $\mathbf{V}_A$ of all possible solutions for the congruence

$$\sum_{s=0}^{\frac{p-1}{p(2)}-1} \sum_{t=1}^{\frac{p(2)}{2}} a_s 2^{2t-1} x_{t+p(2)s/2} \equiv j \mod p$$

is the code capable of correcting errors of the type $\{1, 2\}$ or $\{-1, -2\}$ on **A** channel.

Necessity. Let $2 \nmid p\,(2)$. Without loss of generality, we'll seek the values of the functions among the numbers that are less than $p - 1$. Form the following matrix of size $2 \times p - 1$

$$\begin{array}{r} i \\ 2i \mod p \end{array} \begin{bmatrix} 1\ 2 \cdots (p-1)\,/2\ (p+1)\,/2 \cdots p-1 \\ 2\ 4 \cdots \quad p-1 \qquad\quad 1 \qquad\ \cdots p-2 \end{bmatrix}.$$

In this matrix each number $<p - 1$ appears exactly twice. The problem of finding the values of the function $f\,(i)$ satisfying the u.d. criterion is reduced to the one of choosing that $(p-1)\,/2$ columns from the matrix, all elements in which are distinct. All the numbers in the subgroup $2 \mod p,\ 2^2 \mod p, \ldots, 2^{p(2)} \mod p$ are included in the $p\,(2)$ columns of the matrix. In order to include all these numbers

in the chosen columns at least $(p\,(2)+1)/2$ columns of the matrix should be taken. Then some of these columns should be taken twice. □

Now give an example of Theorem 5.6 application.

Example $n=8$, $2n+1=17$, $17\,(2)=8$.

$$f\,(1)\equiv 2,\ f\,(2)\equiv 2^3\equiv 8,\ f\,(3)\equiv 2^5\equiv 15,\ f\,(4)\equiv 2^7\equiv 9\ \ (\bmod\ 17)\,.$$

Since $3\not\equiv 2^r \bmod 17$ for any r, then assuming $a_1=3$, we get

$$f\,(5)\equiv 3\cdot 2\equiv 6,\ f\,(6)\equiv 3\cdot 2^3\equiv 7,\ f\,(7)\equiv 3\cdot 2^5\equiv 11,\ f\,(8)\equiv 10\ \ (\bmod\ 17)\,.$$

Hence, by Theorem 5.6, the set of all possible solutions for the congruence

$$2x_1+8x_2+15x_3+9x_4+6x_5+7x_6+11x_7+10x_8\equiv j \bmod 17$$

for an arbitrary j is the closed-packed code which corrects errors of the type $\{1,2\}$ or $\{-1,-2\}$ on **A** channel.

The following lemma shows that the set of prime numbers satisfying the condition in Theorem 5.6 is infinite.

Lemma 5.7 *For the prime numbers of the form* $p=8k+3$ *and* $p=8k+5$ $2|$ $p\,(2)$.

Proof Let $p=8k+3$ and $p=8k+5$. Assume $2\dagger p\,(2)$. Then $p-1$ may be represented as $(p-1)=2p\,(2)\,l$ or $(p-1)/2=p\,(2)\,l$. Rasing the congruence $2^{p(2)}\equiv 1 \bmod p$ to the lth power, we have

$$2^{p(2)l}\equiv 2^{\frac{p-1}{2}}\equiv 1 \bmod p \tag{5.3.14}$$

On the other hand, by the theory of quadratic residues for the prime numbers of the form $p=8k+3$ and $p=8k+5$

$$2^{\frac{p-1}{2}}\equiv\left(\frac{2}{p}\right)\equiv -1 \bmod p$$

which is a contradiction to (5.3.14) and thus proves the theorem. □

Remark In order that $2\mid p\,(2)$ the condition of Lemma 5.7 $p=8k+3$ and $p=8k+5$ is not necessary. $2\mid p\,(2)$ holds also for the numbers of the form $p=8k+1$. Thus, e.g. for all prime numbers of the form $p=8k+1=a2^k+1$, where a is an odd number, $k\geq 3$ and $2^a<p$ $2\mid p\,(2)$.

From $17=1\cdot 2^4+1$ and $2^{'}<17$ follows that $2\mid 17\,(2)=10$;
from $41=5\cdot 2^3+1$ and $2^5<41$ follows that $2\mid 41\,(2)=10$ or
from $97=3\cdot 2^5+1$ and $2^3<97$ follows that $2\mid 97\,(2)\ (97\,(2)=48)$.

Now proceed to the case of composite module. Let $2n + 1 = p_1, \dots p_s$.

The following theorem allows to define recurrently from Theorem 5.6 the coefficients $f(i)$ in the congruence (5.3.10) satisfying the u.d. criterion.

Theorem 5.7 *Let $m = 2n + 1 = m_1 m_2$. and for modules $m_1 = 2r + 1$ and $m_2 = 2s + 1$ the coefficients $f(1), f(2), \dots, f(r)$ and $\varphi(1), \varphi(2), \dots, \varphi(s)$, respectively satisfying the u.d. criterion were found before. Then the set of all possible solutions for the congruence*

$$\sum_{i=1}^{r} m_2 f(i)\, x_i + \sum_{k=0}^{m_1-1} \sum_{l=1}^{s} (\varphi(l) + k\, m_2)\, x_{r+ks+l} \equiv j \ mod\ 2n + 1 \tag{5.3.15}$$

is a closed-packed code correcting errors of the type $\{1, 2\}$ or $\{-1, -2\}$ on **A** *channel.*

Proof By the condition of the theorem for $\varepsilon_1, \varepsilon_2 \in \{1, 2\}$ we have

$$\left.\begin{array}{l} \varepsilon_1 f(k) \not\equiv \varepsilon_2 f(l) \mod m_1 \\ \varepsilon_1 \varphi(k) \not\equiv \varepsilon_2 f(l) \mod m_2. \end{array}\right. \quad \text{for } k \neq l \text{ or } \varepsilon_1 \neq \varepsilon_2 \tag{5.3.16}$$

Consider three cases.

Case 1. Let for $\varepsilon_1 \neq \varepsilon_2$ or $i \neq j$ $(i, j = 1, r)$ holds

$$\varepsilon_1 m_2 f(i) \equiv \varepsilon_2 m_2 f(i) \mod m_1 m_2$$

and from the above congruence we have

$$\varepsilon_1 f(i) \equiv \varepsilon_2 f(i) \mod m_1$$

which is a contradiction to (5.3.16).

Case 2. Let for $o \leq k \leq m_1 - 1;\ 1 \leq i \leq r;\ 1 \leq j \leq s$ holds

$$\varepsilon_1 m_2 f(i) \equiv \varepsilon_2 (\varphi(i) + k\, m_2) \mod m_1 m_2.$$

Then

$$m_2 (\varepsilon_1 f(i) - \varepsilon_2 k) \equiv \varepsilon_2 \varphi(j) \mod m_1 m_2$$

which reduces to a contradictionary conclusion.

$$\varepsilon_2 \varphi(j) \equiv 0 \mod m_2.$$

Case 3. Let for $\varepsilon_1 \neq \varepsilon_2$ or $i \neq j$ or $k \neq l_1;\ 0 \leq k, l \leq m_1 - 1;\ 1 \leq i, j \leq s$ holds

$$\varepsilon_1 (\varphi(i) + k m_2) \equiv \varepsilon_2 (\varphi(j) + l\, m_2) \mod m_1 m_2.$$

Then

$$\varepsilon_1\varphi\,(i) - \varepsilon_2\varphi\,(j) \equiv m_2\,(\varepsilon_1 k - \varepsilon_2 l) \mod m_1 m_2. \tag{5.3.17}$$

From which we have

$$\varepsilon_1\varphi\,(i) \equiv \varepsilon_2\varphi\,(j) \mod m_2.$$

This holds only for $\varepsilon_1 = \varepsilon_2, i = j, k \neq l$. Then from (5.3.17) we get

$$\varepsilon_1 k - \varepsilon_1 l \equiv 0 \mod m_1 \qquad \varepsilon_1\,(k - l) \equiv 0 \mod m_1.$$

The latter is impossible in view of $(\varepsilon_1, m_1) = 1, 0 < k - l < m_1$.

By Theorem 5.7 the closed-packed codes correcting errors of the type $\{1, 2\}$ or $\{-1, -2\}$ exist on an arbitrary basis of q and of such lengths n that for all primes $p \mid 2n + 1,\ \ 2 \mid p\,(2)$. The latter condition is also necessary in case of composite module, which follows from Theorem 5.6. Now give an example of Theorem 5.7 application.

Example Let $2n + 1 = m_1 m_2 = 5 \cdot 17 = 85,\ n = 42,\ r = (5 - 1)/2 = 2,\ s = (17 - 1)/2, 5(2) = 4, 17(2) = 8$.

From Theorem 5.6 we find

$$f\,(1) \equiv 2, \qquad f\,(2) \equiv 3 \ (\mod 5)\,;$$

and from the previous example we have

$$\begin{aligned} &\varphi\,(1) \equiv 2, \quad \varphi\,(2) \equiv 8, \quad \varphi\,(3) \equiv 15, \quad \varphi\,(4) \equiv 9, \\ & && (\mod 17) \\ &\varphi\,(5) \equiv 6, \quad \varphi\,(6) \equiv 7, \quad \varphi\,(7) \equiv 11, \quad \varphi\,(8) \equiv 10. \end{aligned}$$

Then the following numbers will be the coefficients in congruence (5.3.15):

34, 51, 2, 15, 9, 6, 7, 11, 10, 19, 25, 32, 26, 23, 24, 28, 27, 36, 42, 49, 43, 40, 41, 45, 44, 53, 59, 66, 60, 57, 58, 62, 61, 70, 76, 83, 77, 74, 75, 79, 78.

Ph-Channel

Theorem 5.8 *If for all prime divisors p of number q $2 \mid p\,(2)$, then there exists a check matrix of size $r \times (q^r - 1)/2$*

$$\mathbf{H} = \left(\overline{h}_1, \overline{h}_2, \ldots, \overline{h}\,(q^r - 1)/2\right) \tag{5.3.18}$$

which null space $\mathbf{V}_{Ph} = \left\{\overline{x} \mid \overline{x}\,H^T = \overline{0}\right\}$ is a linear $((q^r - 1)/2, (q^r - 1)/2 - r)$ perfect code capable of correcting errors of the type $\{1, 2\}$ or $\{-1, -2\}$ on **Ph** *channel.*

Proof Let $q = p_1, p_2, \ldots, p_s$ and for all $1 \leq i \leq s\ 2 \mid p_i\,(2)$. Then by Theorems 5.6 and 5.7 one can find numbers $f\,(1), f\,(2)\,, \ldots, f\,((q - 1)/2)$ such that all $(q - 1)$

numbers $\varepsilon f(i)$ $(1 \le i \le (q-1)/2, \varepsilon \in \{1, 2\})$ differ by modulo q. Taking all the q-ary vectors which first nonzero components are the numbers $f(1), f(2), \ldots, f((q-1)/2)$ as the columns of the matrix (5.3.18), we obtain the matrix of the form

$$\mathbf{H} = \left[H_1 \; H_2 \ldots H_{(q-1)/2} \right],$$

where $\mathbf{H}_k$ $(1 \le k \le (q-1/2))$ is a matrix of size $r \times (q^r-1)/(q-1)$.

$$\mathbf{H}_k = \begin{bmatrix} f(k) & f(k) & \cdots & 0 & 0 & \cdots & 0 \\ \cdot & \cdot & \cdots & f(k) & f(k) & \cdots & 0 \\ \cdot & \cdot & \cdots & \cdot & \cdot & \cdots & \cdot \\ \cdot & \cdot & \cdots & \cdot & \cdot & \cdots & \cdot \\ \cdot & \cdot & \cdots & \cdot & \cdot & \cdots & f(k) \end{bmatrix}.$$

It will be sufficient to prove, using the u.d. criterion, that all q^r-1 error syndromes $\varepsilon_s \overline{h}_l$ $(\varepsilon_s \in \{1, 2\}, 1 \le l \le (q^r-1)/2)$ are distinct vectors of the length r over the ring $\mathbf{Z}_q$.

Actually, all error syndromes in l- and l'th positions corresponding to the columns of different submatrices $\mathbf{H}_k$ and $\mathbf{H}_m$ $k \ne m$, $k(q^r-1)/(q-1) < l \le (k+1)/(q^r-1)$, $m(q^r-1)/(q-1) < l' \le (m+1)(q^r-1)/(q-1)$ differ just by the first component. By the same component of the error syndrome the quantity ε is uniquely defined. The syndromes $\varepsilon \overline{h}_l$ and $\varepsilon \overline{h}_{l_1}$ corresponding to errors in l and $l_1$positions $(l \ne l_1)$, $k(q^r-1)/(q-1) < l, l_1 < (k+1)(q^r-1)/(q-1)$ differ by the remaining components.[4] □

It is easy to prove that the condition of Theorem 5.6 is also a necessary condition on the existence of perfect codes. Give an example of Theorem 5.8 application.

Example $q = 5$, $r = 2$. From the previous example we have $f(1) \equiv 2$, $f(2) \equiv 3 \pmod 5$. Wherefrom by Theorem 5.8 the null space of matrix

$$\mathbf{H} = \begin{bmatrix} 2\,2\,2\,2\,2\,0\,3\,3\,3\,3\,3\,0 \\ 0\,1\,2\,3\,4\,2\,0\,1\,2\,3\,4\,3 \end{bmatrix}$$

over $\mathbf{Z}_5$ is a linear (12, 10) perfect code which corrects errors of the type $\{1, 2\}$ or $\{-1, -2\}$ on **Ph** channel.

Compare the powers of codes constructed in 1, 2 in this section. In case of existence of the codes for **Ph** channel on the same bases of q, we get

$$\mathbf{M}_A \ge q^{\lceil \log_q(2n+1) \rceil - \log_q(2n+1)} \mathbf{M}_{Ph}.$$

[4]Because unlike Hamming codes, the basis q of the given codes is not a power of a prime number, zero divisors may appear. But for the error types under consideration this cannot occur, since $(r, q) = 1$.

5.3.5 $\{+1, -1, +2, -2\}$-*Type Error-Correcting Codes*

A-Channel

First consider the case for prime module $4n + 1 = p$.

Theorem 5.9 *In order that there should exist a function $f(i)$ such that the set of every possible solutions for congruence*

$$\sum_{i=1}^{n} f(i)\, x_i \equiv j \ mod\ 4n+1 \tag{5.3.19}$$

be capable of correcting errors of the type $\{+1, -1, +2, -2\}$, it is necessary and sufficient that $4 \mid p(2)$.

Proof Sufficiency. Let $4 \mid p(2)$. then the function $f(i)$ in congruence (5.3.19) is defined as follows:

$$f(i) = a_{s(i)} 2^{\left[i - s(i)\frac{p(2)}{4}\right] - 1},$$

where $s(i) = \left\lceil \frac{4i}{p(2)} \right\rceil - 1$, $a_0 = 1$, and $a_{s(i)}$ for $s(i) > 0$ is any integer satisfying the condition

$$a_{s(i)} \not\equiv a_t 2^r \mod p \tag{5.3.20}$$

for all $t < s(i)$ and $r \ (1 \le r \le p(2).)$

For $i \le p(2)/4$, $s(i) = 0$ and for the $p(2)$ error syndromes $f(i), 2f(i), -f(i), -2f(i)$ in the first k positions $\left(k = \frac{p(2)}{4}\right)$, and with regard for $2^{\frac{p(2)}{2}} \equiv -1 \mod p$, we get the following values, respectively:

$$\begin{array}{llll} f(i) & 2, & 2^3, & \dots, 2^{\frac{p(2)}{2}-1} \\ 2f(i) & 2^2, & 2^4, & \dots, 2^{\frac{p(2)}{2}} \\ -f(i) & 2^{\frac{p(2)}{2}+1}, & 2^{\frac{p(2)}{2}+3}, & \dots, 2^{p(2)-1} \\ -2f(i) & 2^{\frac{p(2)}{2}+2}, & 2^{\frac{p(2)}{2}+4}, & \dots, 2^{p(2)} \end{array} \tag{5.3.21}$$

All the numbers in matrix (5.3.21) form a subgroup of the multiplicative group of the field $\mathbf{Z}_p^*$ of residue classes of integers modulo p. If take the leaders of the cosets decomposition of the group $\mathbf{Z}_p^*$ with respect to that group as $a_1, a_2, \dots, a_{(p-1)/p(2)-1}$ then all the $4n$ syndromes will correspond to the elements in $\mathbf{Z}_p^*$, and combined with null syndrome to the elements of the field $\mathbf{Z}_p$. Thus by the u.d. criterion the set of all possible solutions for the congruence

$$\sum_{s=0}^{\frac{p-1}{p(2)}-1} \sum_{t=1}^{\frac{p(2)}{4}} a_s\, 2^{2t-1} x_{t+\frac{p(2)s}{2}} \equiv j \mod p,$$

where a_s is defined by the condition (5.3.21), and is the code capable of correcting errors of the type $\{1, -1, 2, -2\}$ on **A** channel.

Necessity. Without loose of generality, we'll seek the values of $f(i)$ among the numbers less than $p-1$. Form the following matrix of size $4 \times (p-1)$

$$\begin{array}{c} i \\ 2i \bmod p \\ p-i \\ p-2i \bmod p \end{array} \begin{bmatrix} 1 & 2 & \cdots \frac{p-1}{2} & \frac{p+1}{2} & \cdots p-1 \\ 2 & 4 & \cdots p-1 & 1 & \cdots p-2 \\ p-1 & p-2 & \cdots \frac{p+1}{2} & \frac{p-1}{2} & \cdots 1 \\ p-2 & p-4 & \cdots 1 & p-1 & \cdots 2 \end{bmatrix}. \tag{5.3.22}$$

In this matrix each number $< p-1$ appears exactly 4 times, once in each row. Thus the problem of finding the values of $f(i)$ satisfying the u.d. criterion is reduced to the one of choosing $(p-1)/4$ columns of the matrix such that all elements in which must be different . Prove that for $4 \dagger p(2)$ this choice is impossible.

Case 1. Let $p(2)$ be an odd number. Consider that columns of matrix (5.3.22) first rows of which include the numbers

$$2 \bmod p,\ 2^2 \bmod p,\ \ldots,\ 2^{p(2)} \bmod p.$$

Then, the second rows of these columns will also contain the same numbers. Besides, the mentioned numbers appear twice in that columns, which first rows include the numbers $p-2 \bmod p,\ p-2^2 \bmod p$. Thus, no other columns in the matrix may include these numbers. And in order to include all these numbers into the chosen columns at least $\frac{p(2)+1}{2}$ columns of the matrix should be taken. Then some of them must be taken twice.

Case 2. Let $p(2) = 2t$, t-an odd number. In this case, since $2^t \equiv -1 \bmod p$, then any column including at least one of the following numbers

$$2 \bmod p,\ 2^2 \bmod p,\ \ldots,\ 2^{p(2)} \bmod p,$$

is wholly consisted of these numbers. Thus, to include all these numbers into the chosen $(p-1)/4$ columns at least $(p(2)+2)/2$ columns must be taken. Then some of these numbers would be repeated. □

Example $p = 17, n = (17-1)/4 = 4, 17(2) = 8$.

$f(1) \equiv 2 \bmod 17$, $f(2) \equiv 8 \bmod 17$. Since for an arbitrary r $3 \not\equiv 2^r \bmod 17$, then taking $a_1 = 3$, we have $f(4) \equiv 3 \cdot 2 \equiv 6 \bmod 17$, $f(5) \equiv 3 \cdot 2^3 \equiv 7 \bmod 17$. Thus the set of all solutions for congruence

$$2x_1 + 8x_2 + 6x_3 + 7x_4 \equiv j \bmod 17$$

is the code of length 4 capable of correcting errors of the type $\{+1, +2, -1, -2\}$ on **A** channel and the power of the best code among them will be

$$\mathbf{M}_A \geq \left\lceil \frac{4i}{p(2)} \right\rceil.$$

For $q = 5$, $j = 1$ we obtain the following 37 code words:

3200	3112	1223
0300	0212	0043
2020	2013	4204
2101	3340	1304
1002	0440	3024
2410	3421	0124
4130	2241	4432
1230	2322	3333
4211	4042	0433
1311	1142	3414
3031	2403	2234
0131	4123	4344
		1444

Lemma 5.8 *For primes of the form* $p = 8k + 5$ $4 \mid p(2)$.

Proof For $4 \dagger p(2)$ represent $p - 1$ in the form

$$(p-1) = 2p(2)\,l \quad \text{or} \quad \frac{p-1}{2} = p(2)\,l.$$

Rasing into lth power the congruence

$$2^{p(2)} \equiv 1 \mod p,$$

we get

$$2^{p(2)l} \equiv 2^{\frac{p-1}{2}} \equiv 1 \mod p. \tag{5.3.23}$$

On the other hand, by the theory of quadratic residues for the prime numbers of the form $p = 8k + 5$, we have

$$2^{\frac{p-1}{2}} \equiv \left(\frac{2}{p}\right) \equiv -1 \mod p$$

which contradictes to (5.3.23) and thus proves the lemma. □

By the lemma we have that the number of primes satisfying the condition is infinite.

Remark The lemma shows that the set of prime numbers which satisfy the conditions of Theorem 5.9 is infinite. However, this set is not restricted only by the prime numbers of the form $p = 8k + 5$. Primes of other forms also satisfy the property of $4 \mid p(2)$, say the numbers of the form $p = 8k + 1 = a2^k + 1$, where a is an odd number, $k \geq 3$ and $2^{2a} < p \quad 4 \mid p\,(2)$, since from $p\,(2) \mid a2^k$ and $2^{p(2)} \equiv 1 \mod p$ follows that $p\,(2) > 4a2^s \quad (s \geq 0)$. For example, from $17 = 1 \cdot 2^4 + 1$, $2^2 < 17$ follows that $4 \mid 17\,(2) \quad (17\,(2) = 8)$ or from $97 = 3 \cdot 2^5 + 1$ and $2^6 < 97$ follows that $4 \mid 97\,(2) \quad (97\,(2) = 48)$.

The following theorem allows, using Theorem 5.9, to find recurrently the coefficients of the congruence for composite module $4n + 1$, which defines the code capable of correcting errors of the type $\{+1, +2, -1, -2\}$ on **A** channel.

Theorem 5.10 *Let $m = 4n + 1 = m_1 m_2$ and for $m_1 = 4r + 1$, $m_2 = 4s + 1$ the coefficients $f\,(1), \ldots, f\,(r)$, $\varphi\,(1), \ldots \varphi\,(s)$, respectively, satisfying the u.d. criterion have been found earlier. Then the set of all possible solutions for congruence*

$$\sum_{i=1}^{r} m_2\, f\,(i)\, x_i + \sum_{k=0}^{m_1-1} \sum_{l=0}^{s} (\varphi\,(l) + k m_2)\, x_{r+ks+l} \equiv j \ mod\ 4n + 1 \qquad (5.3.24)$$

is a close-packed code capable of correcting single errors of the type $\{+1, +2, -1, -2\}$ on **A** *channel.*

We omit the proof of Theorem 5.10, since it is analogous to that given for Theorem 5.7.

Example $m = m_1 m_2 = 13 \cdot 17 = 221$, $n = (m - 1)/4 = 55$, $r = 3$, $k = 4$.

From Theorem 5.9 we have

$$f\,(1) \equiv 2,\ f\,(2) \equiv 8,\ f\,(3) \equiv 6 \pmod{13}$$

and from the previous example

$$\varphi\,(1) \equiv 2,\ \varphi\,(2) \equiv 6,\ \varphi\,(3) \equiv 7\,,\ \varphi\,(4) \equiv 8 \pmod{17}.$$

From Theorem 5.10 we have the set of all possible solutions for the congruence

$$\sum_{i=1}^{55} f\,(i)\, x_i \equiv j \mod 221,$$

where $f\,(i)$ takes the values

34, 112, 129, 2, 6, 7, 8, 19, 23, 24, 25, 36, 40, 41, 42, 53, 57, 58, 59, 70, 74,
75, 76, 87, 91, 92, 93, 104, 108, 109, 110, 121, 125, 126, 127, 138, 142, 143, 144, 155, 159, 160, 161, 172, 176, 177, 178, 189, 193, 194, 195, 206, 210, 211, 212

over an arbitrary basis of q is the code of length 55 which corrects errors of the type $\{+1, -1, +2, -2\}$ with code power $M \geq q^n/221$.

Thus, close-packed codes capable of correcting errors of the type $\{+1, -1, +2, -2\}$ on **A** channel exist on an arbitrary basis of q and of such lengths n that for all primes $p \mid 4n+1 \quad 4 \mid p\,(2)$. For $q=3$ this code corrects symmetrical single errors in Hamming metrics (since any transitions between symbols are possible) which allows to compare it with well-known ternary codes.

The code powers of the codes presented here and the Hamming ternary codes are connected by the congruence

$$\mathbf{M}_A \geq 3^{\lceil \log_3(2n+1) \rceil - \log_3(4n+1)}\,\mathbf{M}_H, \tag{5.3.25}$$

where $\mathbf{M}_H$ is the code power of Hamming ternary codes. It follows from (5.3.25) that $\mathbf{M}_A$ is greater than the power of Hamming ternary codes over a large range of code lengths and exceeds it up to 1.5 times.

Ph-Channel

Theorem 5.11 *If for all prime divisors p of number q,* $4 \mid p\,(2)$, *then there exists a check-matrix of the size* $r \times (q^r-1)/4$

$$\mathbf{H} = \left(\overline{h}_1, \overline{h}_2, \ldots, \overline{h}_{(q^r-1)/4}\right) \tag{5.3.26}$$

which null space $\mathbf{V}_{Ph} = \left\{\overline{x} \mid \overline{x}H^T = 0 \quad over \quad \mathbf{Z}_q\right\}$ *is a linear* $((q^r-1)/4, (q^r-1)/4-r)$ *perfect code capable of correcting errors of the type* $\{+1, -1, +2, -2\}$ *on* **Ph** *channel.*

Proof Let $q = p_1, p_2, \ldots, p_s$ and $4 \mid p_i\,(2) \quad 1 \leq i \leq s$. Then by Theorems 5.9 and 5.10 one can find numbers $f\,(1), f\,(2), \ldots, f\,((q-1)/4)$ such that all $q-1$ numbers $\varepsilon f\,(i)$ $\left(\varepsilon \in \{1,-1,2,-2\},\ 1 \leq i \leq \frac{q-1}{4}\right)$ be different by modulo q. If take all q-nary vectors of length r which first nonzero elements are $f\,(1), f\,(2), \ldots,$ $f((q-1)/4)$ as the columns of matrix (5.3.26), we obtain the matrix of the form

$$\mathbf{H} = \left(H_1, H_2, \ldots, H_{(q-1)/4}\right),$$

where $\mathbf{H}_k \quad 1 \leq k \leq (q-1)/4$ is a matrix of size $r \times (q^r-1)/(q-1)$

$$\mathbf{H}_k = \begin{bmatrix} f\,(k) & f\,(k) & \cdots & 0 & 0 & \cdots & 0 \\ \cdot & \cdot & \cdots & f\,(k) & f\,(k) & \cdots & 0 \\ \cdot & \cdot & \cdots & \cdot & \cdot & \cdots & 0 \\ \cdot & \cdot & \cdots & \cdot & \cdot & \cdots & f\,(k) \end{bmatrix}.$$

Prove that all q^r-1 error syndromes $\varepsilon_s \overline{h}_l, \quad \varepsilon_s \in \{1,-1,2,-2\},\ 1 \leq l \leq \frac{q^r-1}{4}$ are distinct vectors over the field $\mathbf{Z}_q$. In fact, error syndromes in lth and l'th positions corresponding to columns of matrices $\mathbf{H}_k$ and $\mathbf{H}_m$

$$k\left(q^r-1\right)/(q-1) < l \leq (k+1)\left(q^r-1\right)(q-1)$$

and

$$m\left(q^r-1\right)/(q-1) < l' \leq (m+1)\left(q^r-1\right)/(q-1)$$

differ just by the first component. Exactly by the same component is estimated the error value ε_S. $4\,(q^r-1)\,/\,(q-1)$ error syndromes $\varepsilon_S\overline{h}_l$ and $\varepsilon_t\overline{h}_{l_1}$ in lth and l_1th positions $(l \neq l_1)$ corresponding to the columns of matrix $\mathbf{H}_k$; $1 \leq k \leq (q-1)\,/4$; $k\,(q^r-1)\,/\,(q-1) < l_1 l_1' \leq (k+1)\,(q^r-1)\,(q-1)$ differ by the remaining components.

Thus we obtain that the perfect codes capable of correcting errors of the type $\{1-1, 2, -2\}$ on **Ph** channel exist on such basis of q that for primes $p \mid q \quad 4 \mid p\,(2)$. It is easy to prove that this is also the necessary condition.

For $q=5$ the code given by Theorem 5.11 corresponds to the Hamming quinary perfect code which corrects single symmetric errors on **Ph** channel, since for $q=5$ every transitions as a result of errors of the type $\{+1, -1, +2, -2\}$ between the symbols on this channel are possible. In general case, if there exist the both perfect codes on bases of q, the codes constructed in this section, for the same number of check symbols have code lengths $(q-1)\,/4$ times greater than the symmetric single error-correcting Hamming codes, or

$$\mathbf{M}_{ph} = q^{(q^r-1)(q-5)/4(q-1)}\mathbf{M}_H. \qquad \square$$

Example $q=13$, $r=2$. From the previous example we have

$$f\,(1) \equiv 2,\;\; f\,(2) \equiv 8,\;\; f\,(3) \equiv 6 \;\; (\bmod\;\; 13)\,.$$

Hence, from Theorem 5.10 we have that the null space of matrix

$$\mathbf{H} = \begin{bmatrix} 222 \cdots & 20888 & \cdots & 80666 & \cdots & 60 \\ 012 \cdots & 122012 & \cdots & 128012 & \cdots & 120 \end{bmatrix}$$

over $\mathbf{Z}_{13}$ is a 13-nary linear (52, 50) code capable of correcting errors of the type $\{1, -1, 2, -2\}$ on **Ph** channel.

Compare the powers of the codes constructed in 1, 2 of this section. Considering the shortened codes for **Ph** channel on the bases of q for which there exist codes of lengths $\left(q^{r-1}\right)/4 < n < (q^r-1)\,/4$ on both channels, we have

$$\mathbf{M}_A \geq q^{\lceil \log_q(4n+1) \rceil - \log_q(4n+1)}\mathbf{M}_{Ph}. \tag{5.3.27}$$

For $4n+1 = q^k$ from (5.3.27) we have $\mathbf{M}_A = \mathbf{M}_{Ph}$ and in the remaining cases $\mathbf{M}_A > \mathbf{M}_{Ph}$ and may exceed it up to q times.

Problem 5.1 The problem is to find the necessary and sufficient conditions on existence of infinite classes of q-nary close-packed or perfect codes capable of correcting single errors of the type $\{\pm 1, \pm 2, \ldots, \pm t\}$ or $\{1, 2, \ldots, t\}$ for $3 \leq t \leq q-1$.

Problem 5.2 The problem is to find the necessary and sufficient conditions on existence of q-nary close-packed or perfect codes capable of correcting more than single errors of the type $\{\pm 1,\;\; \pm 2, \ldots, \pm t\}$ or $\{1,\;\; 2, \ldots, t\}$.

5.3.6 A Formula for Computing Powers of Codes Defined by Congruences

The idea of using the congruences of the form

$$\sum_{i=1}^{n} f(i)\, x_i \equiv j \mod m \tag{5.3.28}$$

where $f(i)$ is a numerical function $x_i \in \{0, 1, \ldots, q-1\}$, and n, j, m are natural numbers, was first suggested by Varshamov and Tenengol'ts [45] in which the codes capable of correcting single asymmetrical errors of the type $0 \to 1$ $(1 \to 0)$ are defined by the set of all possible solutions for the congruence $\sum_{i=1}^{n} i x_i \equiv j \mod n+1$. It was proved in [41] that this congruence has the maximum number of solutions for $j = 0$, and the exact formula was derived later in [17]. Furthermore, using congruencies of the form (5.3.28) or the sets of such congruencies some good codes were constructed. However, in most cases they do not succeed to estimate the exact value of powers of these codes and as a rule, the following average value is used

$$\mathbf{M}_A = \max_j t_j \geq \frac{q^n}{m}.$$

For sake of convenience we denote the set of solutions of congruence (5.3.28) by t_j here and hereafter.

In this section we derive the formula to compute the number of t_j which depend on the primitive root of 1 for this congruence of the most general form (5.3.28), as well as the exact formulas for two definite congruencies used in code construction (we preserve the designations used in [17] for deriving these formulas).

Consider the generating function.

$$P(z) = \prod_{k=1}^{n} \left(1 + z^{f(k)} + z^{2f(k)} + \cdots + z^{(q-1)f(k)}\right) = \prod_{k=1}^{n} \frac{z^{f(k)q} - 1}{z^{f(k)} - 1}.$$

The number of solutions of equation

$$\sum_{i=1}^{n} f(i)\, x_i = a \tag{5.3.29}$$

equals to the coefficient for z^a in the polynomial $P(z) = \sum_{s=0}^{\infty} c_s z^s$, since there is a one-to- one correspondence between any of the solutions $\overline{\alpha} = (\alpha_1, \alpha_2, \ldots, \alpha_n)$ for this equation and the product

$$z^a = z^{f(1)\alpha_1} z^{f(2)\alpha_n} \cdots z^{f(n)\alpha_n} = z^{\sum_{i=1}^{n} f(i)\alpha_i}.$$

The number of solutions for congruence (5.3.1) is $\sum_{r=0}^{\infty} c_{j+rm}$, since this number equals to the sum of the numbers of solutions for the following equations

$$\sum_{i=1}^{n} f(i)\, x_i = j + rm \qquad (r = 0, 1, 2, \ldots).$$

Hence, t_j is equal to the coefficient at z^j in the remainder $T(z)$ obtained by dividing $P(z)$ by $z^m - 1$. Thus we get

$$P(z) = \left(z^m - 1\right) Q(z) + T(z),$$

where

$$T(z) = \sum_{j=0}^{m-1} t_j\, z^j.$$

Let $\varepsilon = e^{\frac{2\pi}{m} i}$ be mth primitive root of 1. Since for an arbitrary l $(l = 0, 1, \ldots, m-1)$

$$\left(\varepsilon^l\right)^m - 1 = 1^l - 1 = 0,$$

then

$$P\left(\varepsilon^l\right) = T\left(\varepsilon^l\right).$$

Thus, we can express t_j in terms of $P\left(\varepsilon^l\right)$ and ε from the following m equations linear about t_j $(j = 0, 1, \ldots, m-1)$

$$P\left(\varepsilon^l\right) = T\left(\varepsilon^l\right) = \sum_{j=0}^{m-1} t_j \left(\varepsilon^l\right)^j \quad l = 0, 1, \ldots, m-1.$$

Multiplying the lth equation by ε^{-jl} for an arbitrary j and summing up all these equations term by term, we obtain

$$t_0 \sum_{l=0}^{m-1} \varepsilon^{-jl} + \cdots + t_j \sum_{l=0}^{m-1} \varepsilon^{jl-jl} + \cdots + t_{m-1} \sum_{l=0}^{m-1} \varepsilon^{(m-1)l-jl} = \sum_{l=0}^{m-1} P\left(\varepsilon^l\right) \varepsilon^{-jl}.$$

In this expression all the coefficients at t_s for $s \neq j$ equal to 0

$$\sum_{l=0}^{m-1} \varepsilon^{l(s-j)} = \frac{\varepsilon^{m(s-j)} - 1}{\varepsilon^{s-j} - 1} = 0,$$

since $\varepsilon^{m(s-j)} - 1 = 0 \quad \left(\varepsilon^{s-j} - 1 \neq 0, \text{ if s} \neq j\right)$ and the coefficient at t_j equals to m

$$\sum_{l=0}^{m-1} \varepsilon^{l(j-j)} = \sum_{l=0}^{m-1} 1 = m.$$

Finally, we get

$$t_j = \frac{1}{m} \sum_{l=0}^{m-1} P\left(\varepsilon^l\right) \varepsilon^{-jl} = \frac{1}{m} \sum_{l=1}^{m} P\left(\varepsilon^l\right) \varepsilon^{-jl}. \quad (5.3.30)$$

In case when $f(i) = i, \ m = 2n + 1$ the Eq. (5.3.28) has the form

$$\sum_{i=1}^{n} i x_i \equiv j \mod 2n + 1. \quad (5.3.31)$$

As it was proved in Sect. 6.2.5 of this work that the set of all possible solutions for the Eq. (5.3.31) is the code over an arbitrary basis of q capable of correcting errors of the type $\{1, -1\}$ on **A** channel. For this case the expression (5.3.30) has the form

$$t_j = \frac{1}{2n+1} \sum_{l=1}^{2n+1} P\left(\varepsilon^l\right) \varepsilon^{-jl}, \quad (5.3.32)$$

where

$$P(z) = \prod_{k=1}^{n} \sum_{l=0}^{q-1} z^{kl} = \prod_{k=1}^{n} \frac{z^{kq} - 1}{z^k - 1}.$$

Lemma 5.9 *Let the greatest common divisor (GCD) of numbers l and $2n+1$ be $(l, 2n+1) = d$, and $u = \frac{2n+1}{d}$, then*

$$P\left(\varepsilon^l\right) = (-1)^{L(u,q)} q^{\frac{2n+1-u}{2u}} \varepsilon^{lK(u,q)} \delta_l,$$

where $L(u, q)$ is number of elements in the set $q, 2q, \ldots, \frac{u-1}{2} q$, which least positive residues modulo u are greater than $\frac{u-1}{2}$; $K(u, q)$ is any solution for congruence

$$16y \equiv -(q-1) \ mod \ u,$$

and

$$\delta_l = \begin{cases} 0, \ if \ (q, u) \neq 1 \\ 1, \ if \ (q, u) = 1 \end{cases}.$$

Proof Represent n as $n = u\frac{d-1}{2} + \frac{u-1}{2}$. Since in terms of Lemma 5.9 $\varepsilon^l = \gamma$ is the uth primitive root of 1 and taking into account that $\gamma^{ut+p} = \gamma^p$ for any p and $\frac{\gamma^{uq}-1}{\gamma^u-1} = q$, we have

$$p\left(\varepsilon^l\right) = \prod_{k=1}^{n} \frac{\varepsilon^{klq}-1}{\varepsilon^{kl}-1} = \prod_{k=1}^{u(d-1)/2+(u-1)/2} \frac{\varepsilon^{klq}-1}{\varepsilon^{kl}-1} = \prod_{k=1}^{u} \frac{\varepsilon^{klq}-1}{\varepsilon^{kl}-1} \prod_{k=u+1}^{2u} \frac{\varepsilon^{klq}-1}{\varepsilon^{kl}-1} \cdots$$
$$\cdots \prod_{k=u(d-3)/2+1}^{u(d-1)/2} \frac{\varepsilon^{klq}-1}{\varepsilon^{kl}-1} \prod_{k=u(d-1)/2+1}^{u(d-1)/2+(u-1)/2} \frac{\varepsilon^{klq}-1}{\varepsilon^{kl}-1} = \prod_{k=1}^{u} \frac{\gamma^{kq}-1}{\gamma^k-1} \prod_{k=1}^{u} \frac{\gamma^{kq}-1}{\gamma^k-1} \cdots$$
$$\cdots \prod_{k=1}^{u} \frac{\gamma^{kq}-1}{\gamma^k-1} \prod_{k=1}^{(u-1)/2} \frac{\gamma^{kq}-1}{\gamma^k-1} = \left(\prod_{k=1}^{u} \frac{\gamma^{kq}-1}{\gamma^k-1}\right)^{\frac{d-1}{2}} \prod_{k=1}^{(u-1)/2} \frac{\gamma^{kq}-1}{\gamma^k-1}$$
$$= q^{\frac{d-1}{2}} \left(\prod_{k=1}^{u-1} \frac{\gamma^{kq}-1}{\gamma^k-1}\right)^{\frac{d-1}{2}} \times \prod_{k=1}^{(u-1)/2} \frac{\gamma^{kq}-1}{\gamma^k-1}.$$

If $(q, u) = 1$ then all numbers kq $(k = 1, 2, \ldots, u-1)$ form a complete system of residues modulo u, and

$$\prod_{k=1}^{u-1} \frac{\gamma^{kq}-1}{\gamma^k-1} = 1,$$

and

$$\prod_{k=1}^{(u-1)/2} \frac{\gamma^{kq}-1}{\gamma^k-1} \neq 0.$$

If $(u, q) = \tau \neq 1$ $(\tau > 2)$ the factor $\gamma^{k_0 q} - 1 = 0$ appears in numerator, where $k_0 = \frac{u}{\tau}$ $\left(< \frac{u-1}{2}\right)$ and since denominator is not 0, then

$$\prod_{k=1}^{(u-1)/2} \frac{\gamma^{kq}-1}{\gamma^k-1} = 0.$$

Hence, for such l $\mathrm{P}\left(\varepsilon^l\right) = 0$.

Thus, we have

$$P\left(\varepsilon^l\right) = q^{\frac{d-1}{2}} \prod_{k=1}^{(u-1)/2} \frac{\gamma^{kq}-1}{\gamma^k-1} \delta_l. \tag{5.3.33}$$

Consider the factor in (5.3.33)

$$\prod_{k=1}^{(u-1)/2} \frac{\gamma^{kq}-1}{\gamma^k-1} = \frac{(\gamma^q-1)\left(\gamma^{2q}-1\right) \cdots \left(\gamma^{\frac{u-1}{2}q}-1\right)}{(\gamma-1)\left(\gamma^2-1\right) \cdots \left(\gamma^{\frac{u-1}{2}}-1\right)}. \tag{5.3.34}$$

In this expression the same factors $\gamma^{kq} - 1$ in numerator for which $kq \equiv \beta \mod u$ (β is the least positive residue modulo u) $\beta \leq \frac{u-1}{2}$ are cancelled by the corresponding factors $\gamma^\beta - 1$ in denominator.

The number of remaining factors is just the quantity $L(u, q)$ itself.

For each of the remaining factors $\gamma^{k'q} - 1$ in numerator there exists a single factor $\gamma^{k''} - 1$ in denominator such that

$$k'q + k'' \equiv 0 \mod u.$$

Indeed, since the factor $\gamma^{k'q} - 1$ have not been consulted, it means that $k'q \equiv \beta' \mod u$ (β is the least positive residue modulo u) $\beta' > \frac{u-1}{2}$.

Then denoting $k'' = u - \beta'$, which is $< \frac{u-1}{2}$ and as there exists the factor $\gamma^{k''} - 1$ in denominator that is not being cancelled, otherwise in case if its cancellation by a factor $\gamma^{k'''q} - 1$

$$k'''q \equiv k'' \mod u,$$

which with regard for

$$k'q \equiv -k'' \mod u$$

gives

$$\left(k''' + k'\right) q \equiv 0 \mod u,$$

which is impossible for $(u, q) = 1$, because of

$$k''' + k' \leq u - 1 < u.$$

After performing these cancellations rewrite the expression (5.3.34) in the form

$$\prod_{k=1}^{(u-1)/2} \frac{\gamma^{kq} - 1}{\gamma^k - 1} = \frac{\gamma^{\nu_1} - 1}{\gamma^{\nu_1'} - 1} \cdot \frac{\gamma^{\nu_2} - 1}{\gamma^{\nu_2'} - 1} \cdots \frac{\gamma^{\nu L(u,q)} - 1}{\gamma^{\nu' L(u,q)} - 1},$$

where

$$\nu_i + \nu_i' \equiv 0 \mod u.$$

Now change each of the fractions $\frac{\gamma^{\nu_i} - 1}{\gamma^{\nu_i'} - 1}$ $(i = 0, 1, \ldots, L(u, q))$ as follows:

$$\frac{\gamma^{\nu_i} - 1}{\gamma^{\nu_i'} - 1} = \frac{\gamma^{\nu_i}\left(\gamma^{\nu_i} - 1\right)}{\gamma^{\nu_i}\left(\gamma^{\nu_i'} - 1\right)} = \frac{\gamma^{\nu_i}\left(\gamma^{\nu_i} - 1\right)}{\gamma^{\nu_i' + \nu_i} - \gamma^{\nu_i}} = \frac{\gamma^{\nu_i}\left(\gamma^{\nu_i} - 1\right)}{1 - \gamma^{\nu_i}} = -\gamma^{\nu_i}.$$

Finally we have

$$\prod_{k=1}^{(u-1)/2} \frac{\gamma^{kq} - 1}{\gamma^k - 1} = (-1)^{L(u,q)} \gamma^{\sum_{i=1}^{L(u,q)} \nu_i}. \tag{5.3.35}$$

The residue classes modulo u to which the exponent $\sum_{i=1}^{L(u,q)} \nu_i$ of γ belongs can be found using the following reasonings.

Since $\gamma^{-kq} = \gamma^{(u-k)q}$ and $(u, q) = 1$ then we can write

$$1 = \prod_{k=1}^{u-1} \frac{\gamma^{kq} - 1}{\gamma^k - 1} = \prod_{k=1}^{(u-1)/2} \frac{\gamma^{kq} - 1}{\gamma^k - 1} \cdot \prod_{k=(u+1)/2}^{(u-1)} \frac{\gamma^{kq} - 1}{\gamma^k - 1}$$

$$= \left(\prod_{k=1}^{(u-1)/2} \frac{\gamma^{kq} - 1}{\gamma^k - 1} \right)^2 \prod_{k=1}^{(u-1)/2} \frac{1}{\gamma^{k(q-1)}}$$

$$= \left(\prod_{k=1}^{(u-1)/2} \frac{\gamma^{kq} - 1}{\gamma^k - 1} \right)^2 \frac{1}{\gamma^{(q-1)\frac{u^2-1}{8}}}$$

wherefrom

$$\left((-1)^{L(u,q)} \gamma^{\sum_{i=1}^{L(u,q)} \nu_i} \right)^2 = \gamma^{(q-1)\frac{u^2-1}{8}} \quad \text{or} \quad \gamma^{2 \sum_{i=1}^{L(u,q)} \nu_i} = \gamma^{(q-1)\frac{u^2-1}{8}}$$

and the number $\sum_{i=1}^{L(u,q)} \nu_i$ should satisfy the congruence

$$2y \equiv (q-1) \frac{u^2 - 1}{8} \mod u \tag{5.3.36}$$

or

$$16y \equiv -(q-1) \mod u.$$

Hence any solution of congruence (5.3.36) could be taken as the exponent of γ in congruence (5.3.35).

Proceeding from γ to ε we'll get to the statement of Lemma 5.9.

Substituting the obtained expression for $\mathrm{P}(\varepsilon^l)$ into (5.3.5) and grouping the components by that l's that correspond to the same u's, and denoting $\frac{l}{d} = t$, with regard for $(t, u) = 1$, we get

$$t_j = \frac{1}{2n+1} \sum_{l=1}^{2n+1} (-1)^{L(u,q)} q^{\frac{2n+1-u}{2u}} \varepsilon^{l[K(u,q)-j]} \delta_l =$$

$$= \frac{1}{2n+1} \sum_{\substack{u/2n+1 \\ (q,u)=1}} (-1)^{L(u,q)} q^{\frac{2n+1-u}{2u}} \sum_{\substack{t \\ (t,u)=1}} \varepsilon^{\frac{2n+1}{u}[K(u,q)-j]t}. \tag{5.3.37}$$

Denote by $f(u)$ the quantity

$$f(u) = \sum_{\substack{t \\ (t,u)=1}} \varepsilon^{\frac{2n+1}{u}[K(u,q)-j]t}$$

and by $g(u)$

$$g(u) = \sum_{t=1}^{u} \varepsilon^{\frac{2n+1}{u}[K(u,q)-j]t} = \frac{\varepsilon^{\frac{2n+1}{u}[K(u,q)-j]u} - 1}{\varepsilon^{\frac{2n+1}{u}[K(u,q)-j]} - 1}.$$

Then

$$g(u) = \begin{cases} 0, & \text{if } u \dagger K(u,q) - j \\ u, & \text{if } u \mid K(u,q) - j \end{cases}$$

By which, using Möbius transformation formula, we write

$$f(u) = \sum_{t|u} \mu\left(\frac{u}{t}\right) g(t) = \sum_{t|u,\ t|(K(u,q)-j)} \mu\left(\frac{u}{t}\right) \cdot t$$

Substituting this expression for $f(u)$ in (5.3.37) we have

Theorem 5.12 *The number of solutions t_j of congruence (5.3.5) is expressed by the formula*

$$t_j = \frac{1}{2n+1} \sum_{u|2n+1,\ (u,q)=1} (-1)^{L(u,q)} q^{\frac{2n+1-u}{2u}} \sum_{t|u,\ t|[K(u,q)-j]} \mu\left(\frac{u}{t}\right) t \qquad (5.3.38)$$

We'll slightly modify the formula to make it more convenient and simple in application.

In the first place, since any solution of the congruence

$$16y \equiv -(q-1) \mod 2n+1$$

is also the solution of

$$16y \equiv -(q-1) \mod u$$

where $u \mid 2n+1$, then we may replace $K(u,q)$ by $K(2n+1,q) = k$ for all u in (5.3.38).

And in the second place, by Gauss criterion for prime u

$$(-1)^{L(u,q)} = \left(\frac{q}{u}\right),$$

where $\left(\frac{q}{u}\right)$ is Legendre's symbol and as it will be shown below for any odd u, under the condition $(u,q) = 1$

$$(-1)^{L(u,q)} = \left(\frac{q}{u}\right),$$

where $\left(\frac{q}{u}\right)$ is Jacobi's symbol.

We assume that the condition $(u, q) \neq 1$ holds in Lemmas 5.10–5.13, not to repeat this each time.

Lemma 5.10 *If q is odd, then*

$$(-1)^{L(u,q)} = (-1)^{\frac{u-1}{2}\cdot\frac{q-1}{2}+L(q,u)}.$$

Proof We prove the lemma using a method similar to the one used to prove the quadratic reciprocity law for Legendre symbol. □

Lemma 5.11 *If $q \equiv q' \mod u$, then*

$$(-1)^{L(u,q)} = (-1)^{L(u,q')}.$$

The proof of the lemma follows directly from definition of $L(u, q)$. □

Lemma 5.12

$$(-1)^{L(u,q)} = (-1)^{\frac{u-1}{2}-L(u,u-q)} = (-1)^{\frac{u-1}{2}+L(u,u-q)}.$$

Proof It follows from definition of $L(u, q)$ that if for an arbitrary k ($k = 1, 2, \ldots,$) the least positive residue of number kq modulo u is greater than $\frac{u-1}{2}$, then the least positive residue of number $(u - q)\,k$ modulo u does not exceed $\frac{u-1}{2}$ and vice versa. So for the following two sets

$$q, 2q, \ldots, \frac{u-1}{2}q$$

and

$$(u-q), 2(u-q), \ldots, \frac{u-1}{2}(u-q)$$

the number of that elements for which the least positive residue modulo u is greater than $\frac{u-1}{2}$, equals to $\frac{u-1}{2}$, that is

$$L(u, q) + L(u, u-q) = \frac{u-1}{2}$$

from which the statement of the lemma follows. □

Lemma 5.13

$$(-1)^{L(u,q)} = \left(\frac{q}{u}\right),$$

where $\left(\frac{q}{u}\right)$ is Jacobi's symbol.

Proof Using Lemmas 5.10 and 5.11 we reduce the procedure of computing the value of $(-1)^{L(u,q)}$ to the one of computing the value for smaller values of parameters $u^{'}$, $q^{'}$. In case of even q Lemma 5.12 is applied.

On the other hand, when computing Jacobi's symbol $\left(\frac{q}{u}\right)$ by a common algorithm, if not separate the factor 2 in numerator, and replace the numerator by $u-q$ and separate the factor -1 only for even q, we have

$$\left(\frac{q}{u}\right)=\left(\frac{-1}{u}\right)\left(\frac{u-q}{u}\right)=(-1)^{\frac{u-1}{2}}\left(\frac{u-q}{u}\right),$$

where $u-q$ is already odd, which provides a possibility to apply quadratic reciprocity law and reduce parameters values.

In both cases at the kth step of computation we get, respectively

$$(-1)^{L(u,q)}=(-1)^{\alpha_k+L\left(u^{'},q^{'}\right)}$$

$$\left(\frac{q}{u}\right)=(-1)^{\alpha_k}\left(\frac{q^{'}}{u^{'}}\right).$$

And at the end of computing algorithm we have, respectively

$$(-1)^{L(u,q)}=(-1)^{\alpha+L\left(u^{''},1\right)}$$

$$\left(\frac{q}{u}\right)=(-1)^{\alpha}\left(\frac{1}{u^{''}}\right).$$

Besides taking into account also the fact that $L\left(u^{''},\ 1\right)$ for any $u^{''}$ and $\left(\frac{1}{u^{''}}\right)=1$, we obtain

$$(-1)^{L(u,q)}=\left(\frac{q}{u}\right).$$

In fact, Lemma 5.13 for Jacobi's symbol is analogous to the Gauss criterion for the Legendre symbol.

If add also the condition $\left(\frac{q}{u}\right)=0$ for $(q,u)\neq 1$, then formula (5.3.38) could be written in the form

$$t_j=\frac{1}{2n+1}\sum_{u\mid 2n+1}\left(\frac{q}{u}\right)q^{\frac{2n+1-u}{2u}}\sum_{t\mid (u,k-j)}\mu\left(\frac{u}{t}\right)t, \tag{5.3.39}$$

where k is any solution of congruence

$$16y\equiv-(q-1)\mod 2n+1.$$

In case of $q = 2$ this formula has the form

$$t_j = \frac{1}{2n+1} \sum_{u\,|\,2n+1} (-1)^{\frac{u^2-1}{8}}\, 2^{\frac{2n+1-u}{2u}} \sum_{t\,|\,(u,k-j)} \mu\left(\frac{u}{t}\right) t,$$

since $\left(\frac{2}{u}\right) = \frac{u^2-1}{8}$ and k is any solution of congruence

$$16y \equiv -1 \mod 2n+1.$$

Consider another example of a code defined by congruencies of the form (5.3.28).

The congruence

$$\sum_{i=1}^{n} (2i-1)\, x_i \equiv j \mod 2n+1$$

is a subject of interest from point of view of coding theory since it defines a class of symmetric single- error-correcting ternary codes which power considerable exceed the power of analogous Hamming codes over some range of code lengths. The formula for computing the number of its solutions is derived from (5.3.30) in the same way as for the congruence (5.3.31), and is of the form

$$t_j = \frac{1}{2n+1} \sum_{u\,|\,2n+1} (-1)^{L(u,q)}\, 2^{\frac{2n+1-u}{2u}} \sum_{t\,|\,(u,k-j)} \mu\left(\frac{u}{t}\right) t,$$

where $L(u, q)$ is the number of elements in the set $q, 3q, \ldots, (u-2)\,q$ which least positive residues modulo u are even, and k is any solution of the congruence

$$8y \equiv (q-1) \mod 2n+1.$$

5.4 Constructing Defect-Correcting Codes

Let us consider the problem of encoding and decoding of stored information oriented on a given memory containing defective cells in such a way that some cells always read out a 0, and others always a 1, regardless of the binary symbols actually stored in them [22]. The positions or even the errors themselves of such defective cells can usually be determined through special tests, but it is frequently impossible to repair or replace them. This is the case, for instance, when memory unit is an integrated circuit of some kind or when the repair can damage the good cells.

The problem can be viewed as information transmission over a channel with defects given in Fig. 5.1. A message $\mathbf{u}$ is a binary vector of length k which is transformed into a binary vector $\mathbf{x}$ of length n. The parameters E_0 and E_1 that affect the encoding and transmission are non-intersecting subsets of the set $[n]$. They can be

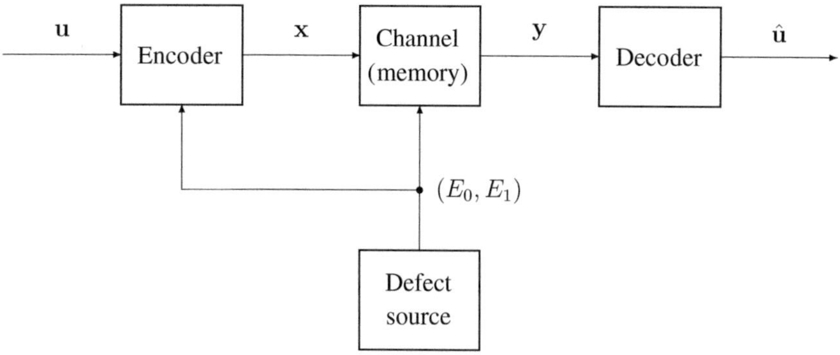

Fig. 5.1 A model of a system of information transmission over a channel with defects

represented as outputs of a 'defective source' since the binary vector $\mathbf{y}$ is formed in accordance with the rule:

$$y_i = \begin{cases} 0, & \text{if } i \in E_0, \\ 1, & \text{if } i \in E_1, \\ x_i, & \text{if } i \notin E_0 \cup E_1. \end{cases}$$

The encoder can use this fact and form a codeword in such a way that the decoder corrects the defects, i.e., exactly recovers $\mathbf{u}$.

We will consider specific constructions of codes correcting all 1- and 2-defects, i.e., $|E_0| + |E_1| = 1$ and $|E_0| + |E_1| = 2$. In our considerations we will say that a vector $\mathbf{v} \in \{0, 1\}^n$ is *compatible* with the defect (E_0, E_1) if

$$v_i = \begin{cases} 0, & \text{if } i \in E_0, \\ 1, & \text{if } i \in E_1. \end{cases}$$

We first construct a code with $M = 2^{n-1}$ codewords correcting all 1-defects. Let

$$\mathbf{v} = (0, u_1, ..., u_{n-1})$$

be a binary vector of length n, where $(u_1, ..., u_{n-1})$ is the message, and let

$$\mathbf{x} = \begin{cases} (0, ..., 0) + \mathbf{v}, & \text{if } \mathbf{v} \text{ is compatible with the defect}, \\ (1, ..., 1) + \mathbf{v}, & \text{otherwise}. \end{cases}$$

The decoder constructs the estimate:

$$\hat{\mathbf{u}} = \begin{cases} (y_2, ..., y_n), & \text{if } y_1 = 0, \\ (y_2 + 1, ..., y_n + 1), & \text{if } y_1 = 1. \end{cases}$$

and, as it is easy to check, $\hat{\mathbf{u}} = \mathbf{u}$ for all 1-defects.

Now let us construct a code of length $n = k + r + 1$ correcting all 1- and 2-defects. Let $q = \log k$ and let C_k be a binary $(2q + 4) \times n$ matrix, which is constructed using the matrices B_q, B'_q, and A of dimensions $q \times k$, $q \times k$, and $(q - r) \times k$, respectively. Let c_{ij}, b_{ij}, b'_{ij}, and a_{ij} denote the elements of C_k, B_q, B'_q, and A (the indices take values over corresponding ranges). We set

$$c_{ij} = \begin{cases} 0, & \text{if } i = 1,\ j = 1, ..., q \\ & \text{if } j = 2q + 1, 2q + 3, \\ & \text{if } i = 2, ..., r + 1, j \neq i - 1, \\ 1, & \text{if } i = 1, j = q + 1, ..., 2q, \\ & \text{if } j = 2q + 2, 2q + 4, \\ & \text{if } i = 2, ..., r + 1, j = i - 1, \\ a_{i-r,j-1}, & \text{if } i = r + 1, ..., r + q,\ j = 2, ..., q + 1, \\ a_{i-r-q,j-1}, & \text{if } i = r + q + 1, ..., r + 2q,\ j = 2, ..., q + 1, \\ b_{i,j-r-1}, & \text{if } i = 1, ..., q,\ j = r + 1, ..., r + q, \\ b'_{i-q,j-r-1}, & \text{if } i = r + q + 1, ..., r + 2q,\ j = r + 1, ..., r + q. \end{cases}$$

The columns of B_q are distinct binary vectors of length q situated in such a way that the i-th column is the binary representation of its index i. The matrix B'_q is obtained from B_q by replacing all elements by their opposites. Each row of A must be distinct and have a weight different from 0, 1, and r. The matrix C_{16} is given in Table 5.1.

Let us assign a codeword $\mathbf{x}$ to the message $\mathbf{u}$ and the defect (E_0, E_1) in such a way that

$$\mathbf{x} = \mathbf{v} + \mathbf{c}_\nu, \tag{5.4.1}$$

Table 5.1 The matrix C_{16} $(q = 4)$

0	1	0	0	0	0	0	0	0	0	0	0	1	1	1	1	1	1	1	1
0	0	1	0	0	0	0	0	1	1	1	1	0	0	0	0	1	1	1	1
0	0	0	1	0	0	1	1	0	0	1	1	0	0	1	1	0	0	1	1
0	0	1	1	0	1	0	1	0	1	0	1	0	1	0	1	0	1	0	1
1	1	0	0	1	1	1	1	1	1	1	1	0	0	0	0	0	0	0	0
1	0	1	0	1	1	1	1	0	0	0	0	1	1	1	1	0	0	0	0
1	0	0	1	1	1	0	0	1	1	0	0	1	1	0	0	1	1	0	0
1	0	1	1	1	0	1	0	1	0	1	0	1	0	1	0	1	0	1	0
0	0	0	0	0	0	0	0	0	0	0	0	0	0	0	0	0	0	0	0
1	1	1	1	1	1	1	1	1	1	1	1	1	1	1	1	1	1	1	1
1	0	0	0	0	0	0	0	0	0	0	0	0	0	0	0	0	0	0	0
0	1	1	1	1	1	1	1	1	1	1	1	1	1	1	1	1	1	1	1

Table 5.2 The rules of selection of the row of the matrix C_k given message and defect

	ν	Condition
1	$2q+1$	$f=g=0,\ 1\le i,j\le n$
2	$2q+2$	$f=g=1,\ 1\le i,j\le n$ $f=0, g=1$
3	$2q+4$	$i=1,\ 2\le j\le n$
4	$j-1$	$2\le i,\ j\le r+1$
5	2	$i=2,\ r+1<j\le n,\ b_{2,j-r-1}=1$
6	$2+q$	$i=2,\ r+1<j\le n,\ b_{2,j-r-1}=0$
7	1	$3\le i\le r+1,\ r+1<j\le n,\ b_{2,j-r-1}=1$
8	$1+q$	$3\le i\le r+1,\ r+1<j\le n,\ b_{2,j-r-1}=0$
9	ν_0	ν_0 is the first (leading) position in which the binary representations of the numbers $i-r-2$ and $j-r-2$ differ; $r+1<i,j\le n$
10	$2q+3$	$i=1,\ 2\le j\le n$
11	$i-1$	$2\le i,\ j\le r+1$
12	$i-1$	$2\le i\le r+1,\ r+1<j\le n,\ b_{i-1,j-r-1}=0$
13	$i-1+q$	$2\le i\le r+1,\ r+1<j\le n,\ b_{i-1,j-r-1}=1$
14	$q+\nu_0$	$r+1<i,j\le n$

where

$$\mathbf{v}=(0,...,0,u_1,...,u_k)\in\{0,1\}^n, \tag{5.4.2}$$

$\mathbf{c}_\nu$ denotes the ν-th row of C_k and the value of ν is defined by $\mathbf{u}$ and (E_0, E_1) in accordance with the rules given in Table 5.2. The notations accepted in Table 5.2 are as follows. We assume that

$$E_0\bigcup E_1=\{i,j\}$$

and denote by $\mathbf{f}$ a tag, which is equal to 0 if the i-th component of $\mathbf{v}$ is compatible with the defect, and to 1 otherwise; $\mathbf{g}$ is a tag, which is equal to 0 if the j-th component of $\mathbf{v}$ is compatible with the defect, and to 1 otherwise.

The rules considered above allow us to construct a class of *additive codes* when the transmitted codeword is defined as a sum modulo 2 of the message shifted by $r=n-k$ positions to the right and a binary vector $\mathbf{c}$ assigned in a special way as function of the message and defect. (see (5.4.1) and (5.4.2)). We present the result that states an asymptotic optimality of additive codes [22].

Theorem 5.13 *Let $M(n,t)$ be the size of an additive code correcting t defects. Then there exist codes such that*

$$M(n,t)\ge n-t-\lceil\log\ln 2^t\binom{n}{t}\rceil. \tag{5.4.3}$$

Proof Let us consider the $2^r\times n$ random binary matrix

$$\theta = \begin{bmatrix} 0 \ldots 0 \; \xi_{01} & \xi_{02} & \ldots \; \xi_{0k} \\ 0 \ldots 1 \; \xi_{11} & \xi_{12} & \ldots \; \xi_{1k} \\ \cdots\cdots & \cdot & \cdots \\ \cdots\cdots & \cdot & \cdots \\ 1 \ldots 1 \; \xi_{2^r-1,1} & \xi_{2^r-1,2} & \ldots \; \xi_{2^r-1,k} \end{bmatrix},$$

where the first r components run over all binary representations of the integers $0, \ldots, 2^r - 1$, and ξ_{ij} are independent binary variables taking values 0 and 1 with probability 1/2. Let θ_i denote the i-th row of the matrix θ.

For a given t-defect d, let $P_i(d)$ be the probability the i-th row of the matrix θ is compatible with the defect d, and let $P_i(d)$ be the probability that none of the rows of the matrix is compatible with d. We denote by a the cardinality of the intersection of $S_0(d) \cup S_1(d)$ with the set of numbers $\{1, \ldots, r\}$ and set $b = t - a$.

If all the first r components of θ_i are compatible with d, then

$$P_i(d) = 1 - 2^{-b}.$$

It is readily verified that the rows θ_i for a given d include exactly 2^{r-a} rows whose first r components are compatible with d.

Using the independence of the rows for different i we have

$$P(d) = \prod_{i=1}^{2^r-1} P_i(d) = (1 - 2^{-b})^{2^{r-a}}.$$

Therefore,

$$\ln P(d) < -2^{r-a-b} = -2^{r-t}.$$

Denote by P the probability that at least one t-defect will be compatible with d. Using the additive bound, we write

$$P \leq 2^t \binom{n}{t} \max P(d),$$

where the maximum is taken with respect to all t-defects. Hence,

$$\ln P \leq -2^{r-t} + \ln 2^t \binom{n}{t}.$$

Direct calculations show that if

$$r \geq t + \log \ln 2^t \binom{n}{t}$$

then $P < 1$ and the existence of additive codes satisfying (5.4.3) follows. □

5.5 Results for the Z-Channel

5.5.1 *Introduction*

An extensive theory of error control coding has been developed (cf. [27, 28, 34]) under the assumption of symmetric errors in the data bits; i.e. errors of type $0 \to 1$ and $1 \to 0$ can occur simultaneously in a codeword.

However in many digital systems such as fiber optical communications and optical disks the ratio between probability of errors of type $1 \to 0$ and $0 \to 1$ can be large. Practically we can assume that only one type of errors can occur in those systems. These errors are called asymmetric. Thus the binary asymmetric channel, also called Z-channel (shown in Fig. 5.2),
has the property that a transmitted 1 is always received correctly but a transmitted 0 may be received as a 0 or 1.

It seems that on the Z-channel the most comprehensive survey (until 1995) was given by T. Kløve [21]. We report here only basic results without proofs.

A code $\mathcal{U}$ is a t-code (i.e., asymmetric error-correcting code) if it can correct up to t errors, that is, there is a decoder such that if $x \in \mathcal{U}$ and v is obtained from x by changing at most t 1's in x into 0's, then the decoder will recover x from v.

Please note that a code correcting t errors for the BSC is also a t-code.

The maximal size of a code in $\mathcal{A}(n, t)$, where $\mathcal{A}(n, t)$ is the set of all t-codes of length n, will be denoted by $A(n, t)$.

5.5.2 *Upper Bounds*

The Varshamov Bound

Obviously, if $\mathcal{U} \in \mathcal{A}(n, t)$, then $\overline{\mathcal{U}} = \{\overline{x} : x \in \mathcal{U}\} \in \mathcal{A}(n, t)$.

Theorem 5.14 *For* $n, t \geq 1$

$$A(n, t) \leq \frac{2^{n+1}}{\sum_{i=1}^{t} \binom{\lfloor n/2 \rfloor}{i} + \binom{\lceil n/2 \rceil}{i}}$$

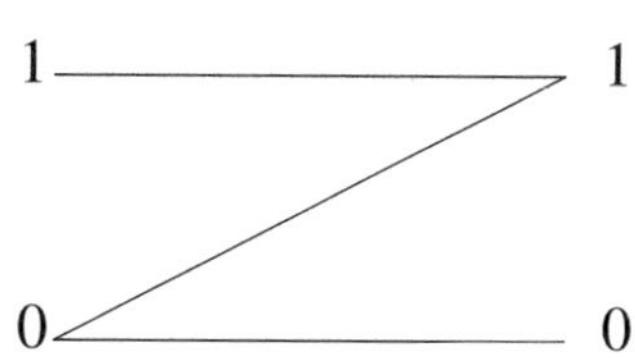

Fig. 5.2 The Z-channel

The Programming Bound

Lemma 5.14 *For $n > t \geq 1$ $\mathcal{U} \in \mathcal{A}(n,t)$ implies the existence of a code $\mathcal{U}' \in \mathcal{A}(n,t)$ with $\underline{0}, \underline{1} \in \mathcal{U}'$ and $\#\mathcal{U}' \geq \#\mathcal{U}$.*

The best upper bound known for $A(n,t)$ is not explicit, but is given as the solution of an integer programming problem involving $M(n,d,w)$, the maximal number of vectors of weight w in $\{0,1\}^n$ of Hamming distance at least d.

Theorem 5.15 *For $n \geq 2$, $t \geq 2$ let*

$$B(n,t) = \max \sum_{i=0}^{n} b_i,$$

where the maximum goes over all $(b_0, b_1, \dots, b_n)$ meeting the following constraints

(i) b_i *are non-negative integers,*
(ii) $b_0 = b_n = 1$, $b_i = b_{n-i} = 0$ *for* $1 \leq i \leq t$,
(iii) $\sum_{j=0}^{s} \binom{i+j}{i} b_{i+j} + \sum_{k=1}^{t-s} \binom{n-i+1}{n-i} z_{i-k} \leq \binom{n}{i}$ *for* $0 \leq s \leq t$, $0 \leq i \leq n$,
(iv) $\sum_{j=s}^{i} M(i-s, 2t+2, i-j) b_j \leq M(n+i-s, 2t+2, i)$ *for* $0 \leq s \leq i$,
(v) $\sum_{j=s}^{i} M(i-s, 2t+2, i-j) b_{n-j} \leq M(n+i-s, 2t+2, i)$ *for* $0 \leq s \leq i$.

Then $A(n,t) \leq B(n,t)$.

An Almost Explicit Bound

By relaxing some constraints in Theorem 5.15 a solvable linear programming problem can be obtained.

Theorem 5.16 *For $n > 2t \geq 2$ let $a_0, a_1, \dots, a_n$ be defined by*

$$a_0 = 1$$

$$a_i = 0 \quad \text{for } 1 \leq i \leq t$$

$$a_{t+i} = \frac{1}{\binom{t+i}{t}} \left(\binom{n}{i} - \sum_{j=0}^{t-1} a_{i+j} \binom{i+j}{j} \right) \quad \text{for } 1 \leq i \leq \frac{n}{2} - t$$

$$a_{n-i} = a_i \quad \text{for } 0 \leq i \leq \frac{n}{2},$$

then

$$A(n,t) \leq \sum_{i=0}^{n} a_i.$$

The bound is weaker than $B(n,t)$. However, it is quite simple to compute. Furthermore, there is a more explicit expression for a_i.

Theorem 5.17 *Let $c_t(k)$ be defined by*

$$c_t(k) = 0 \quad for\ k < 0$$
$$c_t(0) = 1$$
$$c_t(k) = -\sum_{j=0}^{t} c_t(k+j-t)\frac{t!}{j!} \quad for\ k > 0$$

then

$$a_{t+i} = \sum_{k=1}^{i} c_t(i-k)\frac{t!k!}{(t+i)!}\binom{n}{k} \quad for\ 0 \le i \le \frac{n}{2} - t.$$

Also, the $c_t(k)$ can be calculated by linear recursion methods.

The Borden Bounds

Theorem 5.18 *For $n \ge t$*

$$A(n,t) \le M(n+t, 2t+1)$$

Theorem 5.19 *For $n \ge t$*

$$A(n,t) \le (t+1)M(n, 2t+1)$$

Corollary 5.1 *For $n \ge t$ we have*

$$A(n,t) \le \frac{(t+1)2^n}{\sum_{j=0}^{t}\binom{n}{j}} = \frac{(t+1)!2^n}{n^t}(1+o(n))$$

Proof From Theorem 5.19 and the Hamming bound $M(n, 2t+1) \le \frac{2^n}{\sum_{j=0}^{t}\binom{n}{j}}$ the result follows. □

The Constant Weight Code Bound

Theorem 5.20 *For $n > 2t \ge 2$ let $B_t, B_{t+1}, \ldots, B_{n-t-1}$ be defined by*

$$B_t = 2$$
$$B_r = \min_{t \le j < r}\{B_j + M(n+r-j-1, 2t+2, r) \quad for\ r > t,$$

then

$$A(n,t) \le B_{n-t-1}.$$

5.5.3 Single Error-Correcting Codes

We do not discuss decoding algorithms here and refer to Kløve [21], where they are presented in a Pascal-like language.

Kim-Freiman Codes

Let K_m be a code of length $m \geq 1$, which is able to correct one symmetric error. F_n codes constructed as follows.

If $n = 2m$, then define via concatenations

$$F_n = \{x(x \oplus y) : x \in \{0,1\}^m, w(x) \text{ even }, y \in K_m \setminus \{\underline{0}\}\} \cup \{xx : x \in \{0,1\}^m\}.$$

If $n = 2m + 1$, then

$$F_n = \{x(x0 \oplus y) : x \in \{0,1\}^m, w(x) \text{ even }, y \in K_{m+1} \setminus \{\underline{0}\}\} \cup \{xx0 : x \in \{0,1\}^m\}.$$

For the size of the codes we have

$$\#F_n = 2^{m-1}(1 + \#K_m) \qquad \text{if } n = 2m$$

$$\#F_n = 2^{m-1}(1 + \#K_{m+1}) \qquad \text{if } n = 2m + 1$$

Note that for $n = 2^{r-1}$ the Kim-Freiman code of length n is smaller than the Hamming code of the same length, for all other values of n it is larger, if K_m is chosen optimally. Actually, the authors originally used Hamming codes as K_m in the construction.

Stanley-Yoder Codes

Let G be a group of order $n + 1$ such that every element commutes with its conjugates, i.e., $abab^{-1} = bab^{-1}a$ for all $a, b \in G$. Let $g_1, g_2, \dots, g_n, g_{n+1}$ be an ordering of the elements of G such that every conjugacy class appears as a set of consecutive elements $g_m, g_{m+1}, \dots, g_{m+k}$, in the ordering, and $g_{n+1} = e$, the identity. For every $g \in G$ let

$$S_g = \{x_1 x_2 \dots x_n \in \{0,1\}^n : \prod_{i=1}^{n} g_i^{x_i} = g\}.$$

Since $\{S_g : g \in G\}$ is a partition of $\{0,1\}^n$ into $n + 1$ parts

$$\max_{g \in G} \#S_g \geq \frac{2^n}{n+1}.$$

Determination of $\#S_g$ has been done only for Abelian groups G.

Constantin–Rao Codes

These codes are the Stanley-Yoder codes based on an Abelian group G. Writing the group operation as $+$ we get

$$S_g = \{(x_1, x_2, \ldots, x_n) : \sum_{i=1}^{n} x_i g_i = g\},$$

where $g_1, g_2, \ldots, g_n \neq g_0$, the identity element.

Ananiashvili Codes

Let $m = \lceil \log_2(k+1) \rceil$ and define $\varphi : \{0,1\}^k \to \{0,1\}^m$ as follows: for $(x_1, x_2, \ldots, x_k) \in \{0,1\}^k$ define s by

$$s \equiv \sum_{i=1}^{k} x_i i \mod k+1, \quad 0 \leq s \leq k$$

and let $\sum_{i=1}^{m-1} s_i 2^{i-1}$ be the binary expansion of s. Finally, let

$$s_m \equiv \sum_{i=1}^{m-1} \mod 2 \quad \in \{0,1\},$$

set $\varphi(x) = (s_1, \ldots, s_m)$, and define

$$\mathcal{U} = \{x\varphi(x) : x \in \{0,1\}^k\}.$$

This code has length $n = k + \lceil \log_2(k+1) \rceil$ and $\#\mathcal{U} = 2^k$.

Delsarte–Piret Codes

The main idea of the constructions is to look for codes $\mathcal{U}$ without words of weights $w_1, w_2, \ldots, w_s$ and use some known combinatorial construction to get codewords of weights $w_i + 1, w_i + 2, \ldots, w_{i+1} - 1$, $i = 1, 2, \ldots$. The point is that if $w(u) < w_i$ and $w(v) > w_i$, then $\Delta(u, v) = \max(\sum_{i=1}^{n} \max\{v_i - u_i, 0\}, \sum_{i=1}^{n} \max\{u_i - v_i, 0\} \geq 2$, hence the various constructions may be done independently. For all the constructions $\underline{0}, \underline{1} \in \mathcal{U}$ and so there are no words of weight 1 and $n-1$.

Remarks

Many codes, in addition to correcting one single error, are also able to detect many combinations of multiple errors. In many cases the algorithms can be rewritten accordingly without too much effort.

It was shown by Varshamov [39] that most linear codes correcting t asymmetric errors also correct t symmetric errors. Therefore non-linear constructions of asymmetric error-correcting codes are needed to go beyond t-symmetric error-correcting codes.

The main idea underlying the constructions by Stanley-Yoder and Constantin–Rao is due to Varshamov and Tenengol'ts [46]. They used a cyclic group G. The general construction is due to Stanley and Yoder [37]. It was rediscovered by Constantin and Rao [10], who used Abelian groups.

For most code lengths Ananiashvili [4] codes are smaller than Hamming codes. However, they are able to detect a large fraction of double errors.

Varshamov gave several classes of codes to correct multiple errors (see [43, 44, 47–50]) generalizing his original ideas. For these contributions and those by many others see Chaps. 6 and 7 of Kløve [21].

5.5.4 *Error Burst Correction*

If $x \in \{0,1\}^n$ is transmitted and errors occur in positions $i_1, i_2, \ldots, i_r$, where $i_1 < i_2 < \cdots < i_r$, then an (error) burst of length $i_r - i_1 + 1$ has occurred. The codes described below are able to correct a burst of length less than or equal to some specified bound, i.e., the length of the burst, not the number of errors is the focus of attention.

Generalized Oganesyan–Yagdzhyan Codes

$b \in \mathbb{N}$ stands for the maximal burst length, $c = 2b - 1$, $m \in \mathbb{N}$ satisfies $\gcd(m, c) = 1$ and that all its prime factors exceed b, and $n = cm$.

For $a_0 \in \mathbb{Z}_m$ and $a_j \in \mathbb{Z}_2$ for $1 \leq j \leq c$ let

$$\mathcal{U}_{a_0,a_1,\ldots,a_c} = \{x \in \{0,1\}^n : \sum_{i=1}^{n} i x_i \equiv a_0 \mod m, \sum_{k=0}^{m-1} x_{j+kc} \equiv a_j \mod 2 \text{ for } 1 \leq j \leq c\}.$$

These codes correct a burst of length b or less.

Davydov–Dzodzuashvili–Tenengol'ts Codes

In this construction $k, b \in \mathbb{N}$, $\kappa = \lceil k/b \rceil$, and $m = \lceil \log_2 k \rceil$.

For $x \in \{0,1\}^k$ let

$$x^{(i)} = (x_{k-ib+1}, x_{k-ib+2}, \ldots, x_{k-ib+b})$$

for $i = 1, 2, \ldots, \kappa$ (where $x_j = 0$ for $j \leq 0$) and

$$x^{(0)} = \bigoplus_{i=1}^{n} x^{(i)}.$$

Let

$$s(x) \equiv \sum_{i=1}^{\kappa} i w(x^{(i)}) \mod 2^{m+1}, 0 \leq s(x) < 2^{m+1}$$

and let

$$s(x) = \sum_{j=0}^{m} s_j 2^j$$

be the binary expansion of $s(x)$ and define $u(x) = (\overline{s}_m, s_0, s_1, \ldots, s_m)$.

The code is $\mathcal{U} = \{xx^{(0)}u(x) : x \in \{0,1\}^k\}$. It corrects a burst of length $\leq b$. The length of the code is $n = k + b + \lceil \log_2 k \rceil$ and its size is $\#\mathcal{U} = 2^k$.

5.6 On q-Ary Codes Correcting All Unidirectional Errors of a Limited Magnitude

We consider codes over the alphabet $\mathcal{X}_q = \{0, 1, \ldots, q-1\}$ intended for the control of unidirectional errors of level ℓ. That is, the transmission channel is such that the received word cannot contain both a component larger than the transmitted one and a component smaller than the transmitted one. Moreover, the absolute value of the difference between a transmitted component and its received version is at most ℓ.

We introduce and study q-ary codes capable of correcting all unidirectional errors of level ℓ. Lower and upper bounds for the maximal size of those codes are presented.

We also study codes for this aim that are defined by a single equation on the codeword coordinates (similar to the Varshamov–Tennengol'ts codes for correcting binary asymmetric errors). We finally consider the problem of detecting all unidirectional errors of level ℓ.

5.6.1 Introduction

Unidirectional errors slightly differ from asymmetric type of errors: both $1 \to 0$ and $0 \to 1$ type of errors are possible, but in any particular word all the errors are of the same type. The statistics shows that in some of LSI/VLSI ROM and RAM memories the most likely faults are of the unidirectional type. The problem of protection against unidirectional errors arises also in designing of fault-tolerant sequential machines, in write-once memory system, in asynchronous systems et al.

Clearly any code capable of correcting (detecting) t-symmetric errors can be also used to correct (to detect) t-unidirectional or t-asymmetric errors. Obviously also any t-unidirectional error correcting (detecting) code is capable of correcting (detecting) t-asymmetric errors. Note that there are t-asymmetric error correcting codes with higher information rate than that of t-symmetric error correcting codes ([11, 19, 44]). For constructions of codes correcting unidirectional errors see [15, 51]. Note also (as can be easily seen) that the detection problems for asymmetric and unidirectional errors are equivalent (see [7]) i.e. any t-error detecting asymmetric code is also a t-error detecting unidirectional code.

First results on asymmetric error correcting codes are due to Kim and Freiman [20], and Varshamov [39, 40]. In [40] Varshamov introduced an asymmetric metric and obtained bounds for codes correcting asymmetric errors. In [39] Varshamov (and later Weber et al. [51]) proved that linear codes capable of correcting t-asymmetric errors are also capable of correcting t-symmetric errors. Thus only non-linear constructions may go beyond symmetric error correcting codes.

In 1965 Varshamov and Tenengol'ts gave the first construction of nonlinear codes correcting asymmetric errors [47].

The idea behind these (VT-codes) codes is surprisingly simple. Given $n \in \mathbb{N}$ and an integer a the VT-code $\mathcal{C}(n, a)$ is defined by

$$\mathcal{C}(n,a) = \left\{ (x_1, \ldots, x_n) \in \{0,1\}^n : \sum_{i=1}^{n} i x_i \equiv a \pmod{m} \right\} \quad (5.6.1)$$

where $m \geq n + 1$ is an integer.

Varshamov and Tenengol'ts showed that the code $\mathcal{C}(n, a)$ is capable of correcting any single asymmetric error. Moreover taking $m = n + 1$ there exists an $a \in \{0, \ldots, n\}$ so that

$$|\mathcal{C}(n,a)| \geq \frac{2^n}{n+1}. \quad (5.6.2)$$

Recall that for the maximum size of binary single symmetric error correcting codes we have

$$A(n,1) \leq \frac{2^n}{n+1}. \quad (5.6.3)$$

Varshamov [42] showed that $|\mathcal{C}(n, 0)| \geq |\mathcal{C}(n, a)|$.

A number theoretical result due to von Sterneck (1902) [13, p. 87] allows to determine the weight distribution of VT-codes. This result and its special cases were rediscovered many times (see [17, 29, 30, 37]). From a practical point of view VT-codes have the advantage of a very simple decoding algorithm. For systematic encoding of VT-codes see [1, 9].

In general we call a code of length n, correcting t-asymmetric errors a VT-code if it is given by the set of solutions $(x_1, \ldots, x_n) \in \{0, 1\}^n$ of a congruence (or several congruences) of the type

$$\sum_{i=1}^{n} f(i) x_i \equiv a \pmod{M} \quad (5.6.4)$$

where $f : [n] \to \mathbb{Z}$ is an injection, a and M are integers.

We note that there are deep relationships between VT-codes and some difficult problems in Additive Number Theory [14, 44].

The idea of VT-codes was further developed by Constantin and Rao [11], (see also Helleseth and Kløve [19]) by constructing group-theoretical codes based on Abelian Groups.

Levenshtein noticed that VT-codes can also be used to correct single insertion/deletion errors [25].

Modifications of VT-codes where used to construct new codes correcting t-asymmetric errors [8, 16, 32, 44] and bursts of errors [33, 49] (see also [9, 12, 15] for other constructions). For an excellent survey on the results in this direction see Kløve [21].

Very few constructions are known for codes correcting unidirectional errors (for more information see [6]). Note that VT-codes (1.1) and its known modifications are not capable of correcting unidirectional errors.

In 1973 Varshamov introduced a q-ary asymmetric channel [44].

The inputs and outputs of the channel are n-sequences over the q-ary alphabet $\mathcal{X}_q = \{0, 1, \ldots, q-1\}$. If the symbol i is transmitted then the only symbols which the receiver can get are $\{i, i+1, \ldots, q-1\}$. Thus for any transmitted vector $(x_1, \ldots, x_n)$ the received vector is of the form $(x_1 + e_1, \ldots, x_n + e_n)$ where $e_i \in \mathcal{X}_q$ and

$$x_i + e_i \leq q - 1, \; i = 1, \ldots, n. \tag{5.6.5}$$

Then it is said that t-errors have occurred if $e_1 + \cdots + e_n = t$. Generalizing the idea of VT-codes, Varshamov [44] presented several constructions of t-error correcting codes for the defined channel. These codes have been shown in [31] to be superior to BCH codes correcting t errors for $q \geq 2$ and for large n.

We continue here the work started in [2]. We consider *a special type of asymmetric errors in a q-ary channel*, where the magnitude of each component of $\mathbf{e}$ satisfies $0 \leq e_i \leq \ell$ for $i = 1, \ldots, n$. We refer to ℓ as level.

Correspondingly we say that *an unidirectional error of level ℓ has occurred*, if the output is either $\mathbf{x} + \mathbf{e}$ or $\mathbf{x} - \mathbf{e}$ (in the latter case, it is of course required that $x_i \geq e_i$ for all i).

If the error vector $\mathbf{e}$ has Hamming weight $d_H(\mathbf{e}) = t$, then we say that t errors of level ℓ have occurred.

Thus the general problem is the following.

Given n, ℓ, t, q construct q-ary codes of length n capable of correcting t errors of level ℓ. Of course we wish the size of a code to be as big as possible.

Note the difference between the channel described above and Varshamov's channel when $q > 2$. This is shown for $q = 3, l = 1, t \geq 2$ in Fig. 5.3.

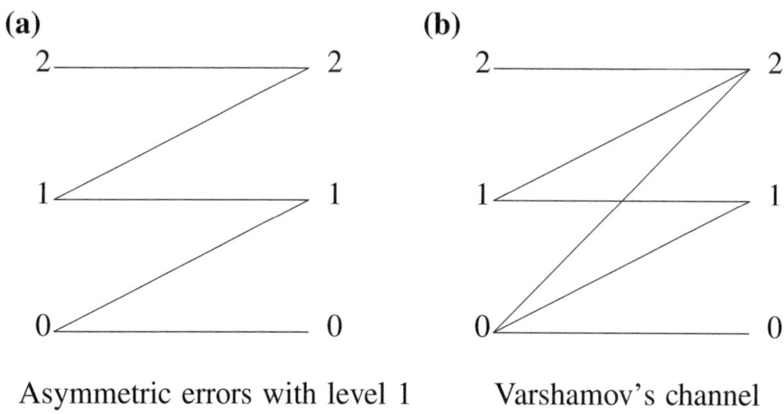

Fig. 5.3 **a** Asymmetric errors with level 1, **b** Varshamov's channel

In this section we consider q-ary codes correcting all asymmetric errors of given level ℓ, (that is $t = n$) for which we use the abbreviation ℓ-AEC code, and ℓ-UEC codes that correct all unidirectional errors of level ℓ. As above our alphabet is $\mathcal{X}_q \triangleq \{0, 1, \dots, q-1\}$.

In Sect. 5.6.2 we define distances that capture the capabilities of a code to correct all asymmetric or unidirectional errors of level ℓ.

For given ℓ, let $A_a(n,\ell)_q$ and $A_u(n,\ell)_q$ denote the maximum number of words in a q-ary AEC code, or UEC code respectively, of length n. Clearly $A_u(n,\ell)_q \leq A_a(n,\ell)_q$.

In Sect. 5.6.3 we determine $A_a(n,\ell)_q$ exactly for all n, ℓ and q.

In Sect. 5.6.4 we give upper and lower bounds on $A_u(n,\ell)_q$, which imply that for fixed q and ℓ the asymptotic growth rate for $A_u(n,\ell)_q$ equals that of $A_a(n,\ell)$.

In Sect. 5.6.5 we study ℓ-AEC and ℓ-UEC codes of VT-type. It is shown that any ℓ-AEC code of VT-type can be transformed into an ℓ-UEC code of VT-type of equal length and cardinality. Upper and lower bounds on the maximum number of codewords in a q-ary ℓ-UEC code of length n of VT-type are derived. For certain pairs (ℓ, q) we give a construction of maximal ℓ-UEC codes.

In Sect. 5.6.9 we consider the problem of detecting all errors of level ℓ.

5.6.2 Distances and Error-Correcting Capabilities

In this section we introduce two distances that capture the capabilities of a code for correcting all symmetrical and unidirectional errors of a certain level. Throughout this section we write L for $[0,\ell]$ (where for integers $a < b$ we use the abbreviation $[a,b] \triangleq \{a, a+1, \dots, b\}$).

Definition 5.2 For $\mathbf{x} = (x_1, x_2, \dots, x_n) \in \mathcal{X}_q^n$ and $\mathbf{y} = (y_1, y_2, \dots, y_n) \in \mathcal{X}_q^n$,

$$d_{\max}(\mathbf{x},\mathbf{y}) = \max\{|x_i - y_i| : i = 1, 2, \dots, n\}$$

$$d_u(\mathbf{x},\mathbf{y}) = \begin{cases} d_{\max}(\mathbf{x},\mathbf{y}) & if\ \mathbf{x} \geq \mathbf{y}\ or\ \mathbf{y} \geq \mathbf{x}, \\ 2d_{\max}(\mathbf{x},\mathbf{y}) & if\ \mathbf{x}\ and\ \mathbf{y}\ are\ incomparable, \end{cases}$$

where $\mathbf{x} \geq \mathbf{y}$ means that $x_i \geq y_i$ for all i.

Later on for short we will write $d(\mathbf{x},\mathbf{y})$ for $d_{\max}(\mathbf{x},\mathbf{y})$.

Note that d_u does not define a metric: take $\mathbf{x}$=(0,2), $\mathbf{y}$=(1,0) and $\mathbf{z}$=(1,2). Then $d_u(\mathbf{x},\mathbf{y}) = 4 > 1 + 2 = d_u(\mathbf{x},\mathbf{z}) + d_u(\mathbf{z},\mathbf{y})$.

Lemma 5.15 *Let* $\mathbf{x}, \mathbf{y} \in \mathcal{X}_q^n$. *The two following assertions are equivalent:*

(i) $d(\mathbf{x},\mathbf{y}) \leq \ell$

(ii) there exist $\mathbf{e} \in L^n$, $\mathbf{f} \in L^n$ *such that* $\mathbf{x} + \mathbf{e} = \mathbf{y} + \mathbf{f} \in \mathcal{X}_q^n$.

Proof Suppose that (i) holds. We define **e** and **f** as

$$e_i = \max(0, y_i - x_i) \text{ and } f_i = \max(0, x_i - y_i), \quad i = 1, 2, \ldots, n.$$

As $d(\mathbf{x}, \mathbf{y}) \le \ell$, the vectors **e** and **f** are in L^n, and for each i, we have that $x_i + e_i = y_i + f_i = \max(x_i, y_i) \in \mathcal{X}_q$. That is (ii) holds.

Conversely, suppose that (ii) holds, then for each i we have that $|x_i - y_i| = |f_i - e_i| \le \max(f_i, e_i) \le \ell$, where the first inequality holds since e_i and f_i both are non-negative. □

The following proposition readily follows from Lemma 5.15.

Proposition 5.1 *A code $C \subset \mathcal{X}_q^n$ is an ℓ-AEC code if and only if $d(\mathbf{x}, \mathbf{y}) \ge \ell + 1$ for all distinct* **x**,**y** *in C.*

Note that Proposition 5.1 and the definition of $d(\mathbf{x}, \mathbf{y})$ imply that for $\ell \ge q - 1$, an ℓ-AEC code (and therefore also an ℓ-UEC code) contains at most a single codeword. *For this reason, we assume in the remainder of the section that $\ell \le q - 2$.*

Lemma 5.16 *Let* $\mathbf{x}, \mathbf{y} \in \mathcal{X}_q^n$. *The two following assertions are equivalent.*

(i) $\mathbf{y} \ge \mathbf{x}$ *and* $d(\mathbf{x}, \mathbf{y}) \le 2\ell$,
(ii) there exist $\mathbf{e} \in L^n, \mathbf{f} \in L^n$ *such that* $\mathbf{x} + \mathbf{e} = \mathbf{y} - \mathbf{f} \in \mathcal{X}_q^n$.

Proof Suppose that (i) holds. We define **e** and **f** as

$$e_i = \lceil \frac{1}{2}(y_i - x_i) \rceil \text{ and } f_i = \lfloor \frac{1}{2}(y_i - x_i) \rfloor, \quad i = 1, 2, \ldots, n.$$

As $\mathbf{y} \ge \mathbf{x}$, both **e** and **f** have only non-negative components and for each i, we have that $f_i \le e_i \le \lceil \frac{1}{2}(2l) \rceil = \ell$; moreover, we obviously have that $\mathbf{e} + \mathbf{f} = \mathbf{y} - \mathbf{x}$. Finally, for each i we have that $x_i + e_i = y_i - f_i \le y_i \le q - 1$, so $\mathbf{x} + \mathbf{e} = \mathbf{y} - \mathbf{f} \in \mathcal{X}_q^n$. We conclude that (ii) holds.

Conversely suppose that (ii) holds. Then $\mathbf{y} - \mathbf{x} = \mathbf{e} + \mathbf{f}$ and so $\mathbf{y} \ge \mathbf{x}$, and for each i we have that $|y_i - x_i| = y_i - x_i = e_i + f_i \le \ell + \ell = 2\ell$. That is (i) holds. □

Combination of Lemmas 5.15 and 5.16 yields the following

Proposition 5.2 *A code $C \subset \mathcal{X}_q^n$ is an ℓ-UEC code if and only if $d_u(\mathbf{x}, \mathbf{y}) \ge 2\ell + 1$ for all distinct* **x**, **y** *in C.*

5.6.3 ℓ-AEC Codes

It turns out that $A_a(n, \ell)_q$ can be determined exactly for all integers n and each $\ell \in \mathcal{X}_q$.

Theorem 5.21 (Ahlswede, Aydinian, Khachatrian, and Tolhuizen 2006 [3]) *For all integers n and each* $\ell \in \mathcal{X}_q$, $A_a(n, \ell)_q = \lceil \frac{q}{\ell+1} \rceil^n$.

Proof Let $\mathcal{C} \subset \mathcal{X}_q^n$ be an ℓ-AEC-code. Let $\varphi : \mathcal{X}_q \to \left\{0, 1, \dots \left\lfloor \frac{q-1}{\ell+1} \right\rfloor\right\}$, be defined as

$$\varphi(j) = \left\lfloor \frac{j}{\ell+1} \right\rfloor, \quad j = 0, \dots, q-1.$$

For any codeword $\mathbf{x} = (x_1, \dots, x_n) \in \mathcal{C}$ define $\varphi^n(\mathbf{x}) = \big(\varphi(x_1), \dots, \varphi(x_n)\big)$. Clearly φ^n is injective: if $\mathbf{x}, \mathbf{y} \in \mathcal{C}$ are such that $\varphi^n(\mathbf{x}) = \varphi^n(\mathbf{y})$, then $|x_i - y_i| \le \ell$, $(i = 1, \dots, n)$, that is, $d(\mathbf{x}, \mathbf{y}) \le \ell$ and so $\mathbf{x} = \mathbf{y}$. This implies that $|\varphi^n(\mathcal{C})| = |\mathcal{C}|$ and since $\left\lfloor \frac{q-1}{\ell+1} \right\rfloor + 1 = \lceil \frac{q}{\ell+1} \rceil$ we get

$$|\mathcal{C}| \le \left\lceil \frac{q}{\ell+1} \right\rceil^n. \tag{5.6.6}$$

The code $\mathcal{C}$ defined as

$$\mathcal{C} = \left\{(x_1, x_2, \dots, x_n) \in \mathcal{X}_q^n : x_i \equiv 0 \mod (\ell+1) \text{ for } i = 1, 2, \dots, n\right\}$$

obviously is an ℓ-AEC code that achieves equality in (5.6.6). A received vector can be decoded by component-wise rounding downwards to the nearest multiple of ℓ+1. □

5.6.4 ℓ-UEC Codes

In this section, we study $A_u(n, \ell)_q$, the maximum number of words in a q-ary ℓ-UEC code of length n. As any ℓ-UEC code is an ℓ-AEC code, Theorem 5.21 implies that

$$A_u(n, \ell)_q \le A_a(n, \ell)_q = \left\lceil \frac{q}{\ell+1} \right\rceil^n. \tag{5.6.7}$$

In some special cases the upper bound (5.6.7) is met with equality.

Proposition 5.3 *For all n and* ℓ, $A_u(n, \ell)_{2\ell+2} = 2^n$.

Proof By Proposition 5.2 the code $\{0, 2\ell+1\}^n$ meeting 2^n has the desired property and $A_u(n, \ell)_{2\ell+2} \le 2^n$ by (5.6.7). □

In Sect. 5.6.5 we will construct q-ary ℓ-UEC codes of VT type. For various classes of pairs (q, ℓ), (for example, if $\ell+1$ divides q), these codes have cardinality $\lceil \frac{q}{\ell+1} \rceil^{n-1}$ and thus they are below the upperbound (5.6.8) only by a multiplicative factor.

We continue the present section with two constructions for q-ary ℓ-UEC codes valid for all pairs (q, ℓ). We denote by $\mathcal{X}_{q,\ell+1}$ all integers in $\mathcal{X}_q = [0, q-1]$ that are multiples of $\ell + 1$, that is

$$\mathcal{X}_{q,\ell+1} = \{m \in \{0, 1, \dots, q-1\} : m \equiv 0 \pmod{\ell+1}\} = \{a(\ell+1) : 0 \leq a \leq b-1\}, \tag{5.6.8}$$

where

$$b = |\mathcal{X}_{q,\ell+1}| = \left\lceil \frac{q}{\ell+1} \right\rceil.$$

It is clear that $d(\mathbf{x}, \mathbf{y}) \geq \ell + 1$ for any two distinct words $\mathbf{x}, \mathbf{y}$ in $\mathcal{X}^n_{q,\ell+1}$. In the subsequent two subsections we use $\mathcal{X}^n_{q,\ell+1}$ to construct a code with minimum asymmetric distance ℓ+1 for which any two codewords are incomparable. Thus we have created a code with undirectional distance at least $2\ell + 2$.

Construction 1: Taking a Subset of $\mathcal{X}^{\mathbf{n}}_{\mathbf{q},\ell+1}$

For each j let

$$C(j) = \{(x_1, x_2, \dots, x_n) \in \mathcal{X}^n_{q,\ell+1} : \sum_{i=1}^{n} \frac{x_i}{\ell+1} = j\}.$$

Any two distinct words from $C(j)$ clearly are incomparable and so $C(j)$ is an ℓ-UEC code. It is clear that

$$|C(j)| = |\{(y_1, y_2, \dots, y_n) \in \{0, 1, \dots, b-1\}^n : \sum_{i=1}^{n} y_i = j\}|.$$

It is known [5, Theorem 4.1.1] that $|C(j)|$ is maximized for $j = j^* \triangleq \lfloor \frac{1}{2} n(b-1) \rfloor$. Moreover, according to [5, Theorem 4.3.6], the following bounds are valid.

Proposition 5.4 *There exist positive constants c_1 and c_2 (depending on $b = \lceil \frac{q}{\ell+1} \rceil$) such that*

$$c_1 \frac{1}{\sqrt{n}} b^n \leq |C(j^*)| \leq c_2 \frac{1}{\sqrt{n}} b^n.$$

Proposition 5.4 implies the following theorem.

Theorem 5.22 (Ahlswede, Aydinian, Khachatrian, and Tolhuizen 2006 [3]) *For each integer q and $\ell \in \mathcal{X}_q$, there is a constant $c > 0$ such that for each n,*

$$A_u(n, \ell)_q \geq c \frac{1}{\sqrt{n}} \lceil \frac{q}{\ell+1} \rceil^n .$$

Clearly, (5.6.8) and Theorem 5.22 imply that for fixed q and ℓ the asymptotic growth rate of $A_u(n, \ell)_q$ is known.

Corollary 5.2 *For each q and each $\ell \in [0, q-1]$* $\lim_{n\to\infty} \sqrt[n]{A_u(n,\ell)_q} = \lceil \frac{q}{\ell+1} \rceil$.

Construction 2: Adding Tails to Words from $\mathcal{X}^{\mathbf{n}}_{\mathbf{q},\ell+1}$

In order to formulate our second construction clearly, we cast it in the form of a proposition. Later we take appropriate values for certain parameters in this construction to obtain a lower bound on $A_u(n,\ell)_q$.

Proposition 5.5 *Let $X \subset \mathcal{X}_q^n$ be an ℓ-AEC code. For $\mathbf{x} \in X$, let $S(\mathbf{x})$ denote the sum of its entries, and let s_1, s_2 be such that for each $\mathbf{x} \in X$, $s_1 \leq S(x) \leq s_2$. Let $\phi : [s_1, s_2] \to \mathcal{X}_q^m$ be such that for all $a, b \in [s_1, s_2]$ with $a > b$, there is an $i \in \{1, 2, \ldots, m\}$ such that $(\phi(a))_i < (\phi(b))_i$. Then $\mathcal{C} = \{(\mathbf{x}, \phi(S(x)) : \mathbf{x} \in X\} \subset \mathcal{X}_q^{n+m}$ is an ℓ-UEC code.*

Proof Let $\mathbf{u} = (\mathbf{x}, \phi(S(\mathbf{x})))$ and $\mathbf{v} = (\mathbf{y}, \phi(S(\mathbf{y})))$ be two distinct words in $\mathcal{C}$. As $d(\mathbf{x}, \mathbf{y}) \geq \ell + 1$, all we have to show is that $\mathbf{u}$ and $\mathbf{v}$ are incomparable. This is clear if $\mathbf{x}$ and $\mathbf{y}$ are incomparable. Now suppose that $\mathbf{x}$ and $\mathbf{y}$ are comparable, say $\mathbf{x} \geq \mathbf{y}$. Then $S(\mathbf{x}) > S(\mathbf{y})$ and hence, by the property imposed on ϕ, $u_j < v_j$ for some $j \in [n+1, n+m]$. □

We now apply the construction from Proposition 5.5. Given s_1 and s_2, we take $m \triangleq \lceil \log_q(s_2 - s_1 + 1) \rceil$, and define $\phi(s)$ as the m-symbols q-ary representation of $s_2 - s$. We choose for X a large subset of $\mathcal{X}^n_{q,\ell+1}$ such that $s_2 - s_1 + 1$ is small, so that m can be small. As shown below we can invoke Chebyshev's inequality to show the existence of a set X such that $|X| > \frac{3}{4}b^n$, while $s_2 - s_1 + 1 < K_1\sqrt{n}$ for some constant K_1. As a consequence, m can be as small as $\frac{1}{2}\log_q n + K_2$ for some constant K_2.

Theorem 5.23 (Ahlswede, Aydinian, Khachatrian, and Tolhuizen 2006 [3]) *For each q and ℓ, there exists a positive constant K such that for each n,*

$$A_u(n,\ell)_q \geq K b^n n^{-\frac{1}{2}\log_q b}, \textit{where } b = \lceil \frac{q}{\ell+1} \rceil .$$

Proof We start with the well-known Chebyshev inequality.

Proposition 5.6 *Let $Y_1, Y_2, \ldots, Y_n$ be independent, identically distributed random variables, each with average μ and variance σ^2. For each $\epsilon > 0$, we have that*

$$prob(|\sum_{i=1}^{n} Y_i \quad - n\mu| > \epsilon \cdot n) \leq \frac{\sigma^2}{n\epsilon^2} \quad .$$

We choose now $\epsilon = \frac{2\sigma}{\sqrt{n}}$ and get

$$\text{Prob}(|\sum_{i=1}^{n} Y_i - n\mu| \leq 2\sigma\sqrt{n}) \geq \frac{3}{4}. \tag{5.6.9}$$

In the above, we take each Y_i uniformly distributed in $\mathcal{X}_{q,\ell+1} = \{a(\ell+1) : 0 \le a \le b-1\}$. It follows from (5.6.9) that the set X defined as

$$X = \{x \in \mathcal{X}_{q,\ell+1}^n : n\mu - 2\sigma\sqrt{n} \le \sum_{i=1}^n x_i \le n\mu + 2\sigma\sqrt{n}\}$$

has cardinality at least $\frac{3}{4}b^n$.

As a consequence of this and Proposition 5.5, there exists a constant K_2 such that for each n, there is an ℓ-AUEC code of length at most $n + \frac{1}{2}\log_q n + K_2$.

Now let n be a positive integer. Choose n_0 such that

$$n_0 + \frac{1}{2}\log_q n_0 + K_2 \le n \text{ and } (n_0+1) + \frac{1}{2}\log_q(n_0+1) + K_2 \ge n.$$

Our construction shows the existence of an ℓ-AUEC code of length n with at least $\frac{3}{4}b^{n_0}$ words. The definition of n_0 implies that

$$\log_q(n_0+1) \le \log_q(n+1-\frac{1}{2}\log_q n_0 - K_2) \le \log_q(n+1-K_2), \text{ and so}$$

$$n_0 \ge n-1-K_2-\frac{1}{2}\log_q(n_0+1) \ge n-1-K_2-\frac{1}{2}\log_q(n+1-K_2).$$

From the final inequality, it follows that there exists a constant K_3 such that $n_0 \ge n - \frac{1}{2}\log_q n - K_3$. We conclude that

$$\frac{3}{4}b^{n_0} \ge \frac{3}{4}b^n n^{-\frac{1}{2}\log_q b} b^{-K_3}.$$

□

5.6.5 ℓ-UEC Codes of Varshamov–Tennengol'ts Type

In this section we study VT-type ℓ-UEC codes. Note however that unlike the VT-codes, the codes we introduce here are defined by means of some linear equation (rather than a congruence) over the real field. Namely given $\mathcal{X}_q = [0, q-1] \subset \mathbb{R}$ and $a_0, \dots, a_{n-1}, a \in \mathbb{Z}$ let

$$X = \{(x_0, \dots, x_{n-1}) \in \mathcal{X}_q^n : \sum_{i=0}^{n-1} a_i x_i = a\}. \tag{5.6.10}$$

Note that X defines an ℓ-UEC code if and only if for each distinct $\mathbf{x}, \mathbf{y} \in X$ holds $\mathbf{x} - \mathbf{y} \notin [-\ell, \ell]^n$ and $\mathbf{x} - \mathbf{y} \notin [0, 2\ell]^n$.

Thus an obvious sufficient condition for the set of vectors $X \subset \mathcal{X}_q^n$ to be an ℓ-UEC code is that the hyperplane H defined by

$$H = \left\{ (x_0, \ldots, x_{n-1}) \in \mathbb{R}^n : \sum_{i=0}^{n-1} a_i x_i = 0 \right\}$$

does not contain vectors from $[-\ell, \ell]^n \cup [0, 2\ell]^n$, except for the zero vector.

An ℓ-UEC code of VT type may have the advantage of a simple encoding and decoding procedure. In particular, let $\mathcal{C}$ be a code given by (5.6.10) where for $i = 0, 1, \ldots, n-1$, $a_i = (\ell+1)^i$. Suppose for the received vector $\mathbf{y} = (y_0, \ldots, y_{n-1})$ we have

$$\sum_{i=0}^{n-1} (\ell+1)^i y_i = a'$$

with $a' \geq a$. Then the transmitted vector $(x_0, \ldots, x_{n-1}) = (y_0 - e_0, \ldots, y_{n-1} - e_{n-1})$, where the error vector $(e_0, \ldots, e_{n-1})$ is just the $(\ell+1)$-ary representation of the number $a' - a$.

For given ℓ, q and n, we define $LA_u(n, \ell)_q =$the maximum size of an ℓ-UEC code, over the alphabet $[0, q-1]$, defined by a linear Eq. (5.6.10).

Correspondingly we use $LA_a(n, \ell)_q$ for ℓ-AEC codes.

Theorem 5.24 (Ahlswede, Aydinian, Khachatrian, and Tolhuizen 2006 [3]) *For all n, q and ℓ, $LA_a(n, \ell)_q = LA_u(n, \ell)_q$.*

Proof Suppose an ℓ-AEC code $\mathcal{C}$ is defined by (5.6.10), that is $\mathcal{C} = X$. Suppose also w.l.o.g. that $a_0, \ldots, a_k < 0$ $(k < n-1)$, $a_{k+1}, a_{k+1}, \ldots a_n \geq 0$, and $s \triangleq a_0 + \cdots + a_k$. Let $\mathcal{C}'$ be the code defined by the equation

$$-\sum_{i=0}^{k} a_i y_i + \sum_{j=k+1}^{n-1} a_j y_j = a - s(q-1) \tag{5.6.11}$$

Note that for each $\mathbf{c} = (c_o, \ldots, c_{n-1}) \in \mathcal{C}$ the vector $\mathbf{c}' = (q-1-c_0, \ldots, q-1-c_k, c_{k+1}, \ldots, c_{n-1}) \in \mathcal{X}_q^n$ is a solution of (5.6.11), that is $\mathbf{c}' \in \mathcal{C}'$. The opposite is also true. Hence we have $|\mathcal{C}| = |\mathcal{C}'|$. Note further that the condition $\mathbf{c} - \mathbf{b} \notin [-\ell, \ell]^n$ for each distinct $\mathbf{c}, \mathbf{b} \in \mathcal{C}$ (this we have since $\mathcal{C}$ is an ℓ-AEC code) implies that for the corresponding $\mathbf{c}', \mathbf{b}' \in \mathcal{C}'$ we also have $\mathbf{c}' - \mathbf{b}' \notin [-\ell, \ell]^n$. Moreover since $-a_0, \ldots, -a_k, a_{k+1}, \ldots, a_{n-1} > 0$ we have $\mathbf{c}' - \mathbf{b}' \notin \mathcal{X}_q^n$, which implies that $\mathcal{C}'$ is an ℓ-UEC code. Thus we have

$$LA_a(n, \ell)_q \leq LA_u(n, \ell)_q.$$

This completes the proof since we also have the inverse inequality. □

For future reference, we note the obvious fact that for all n, ℓ, q and q', we have

$$LA_u(n,\ell)_q \geq LA_u(n,\ell)_{q'} \text{ if } q \geq q'. \tag{5.6.12}$$

Remark Given ℓ and q let $a_0, a_1, \ldots, a_n$ be nonzero integers such that the code $\mathcal{C} = X$ defined by (5.6.10) is an ℓ-UEC code over the alphabet $\mathcal{X}_q = [0, q-1]$. Then the following is true.

Proposition 5.7 *The code $\mathcal{C}^*$ defined by*

$$\mathcal{C}^* = \left\{ (z_0, \ldots, z_{n-1}) \in \mathcal{X}_q^n : \sum_{i=0}^{n-1} a_i z_i \equiv a \pmod{2\ell S + 1} \right\},$$

where $S \triangleq a_0 + \cdots + a_{n-1}$ is an ℓ-UEC code.

Proof If for two distinct $\mathbf{z}, \mathbf{z}' \in \mathcal{C}^*$ holds $\sum_{i=0}^{n-1} a_i(z_i - z_i') = 0$ then $\mathbf{z}, \mathbf{z}'$ belong to some translate of code $\mathcal{C}$ and hence $d_u(\mathbf{z}, \mathbf{z}') \geq 2\ell + 1$. Conversely if $\sum_{i=0}^{n-1} a_i(\mathbf{z} - \mathbf{z}_\mathbf{i}') \neq 0$ then there exists j (by the pigeonhole principle) such that $|z_j - z_j'| \geq 2\ell + 1$. Therefore in both cases $d_u(\mathbf{z}, \mathbf{z}') \geq 2\ell + 1$. □

Thus we have $|\mathcal{C}^*| \geq |\mathcal{C}|$ which shows that in general the codes given by some congruence could have better performance. Note however that by construction given above we cannot have much gain as compared to the code given by (5.6.10). This is clear since $|\mathcal{C}| \geq c|\mathcal{C}^*|$ for some constant $c \leq \frac{(q-1)S}{2S\ell+1} < \frac{q-1}{2\ell}$.

5.6.6 *Lower and Upper Bounds for* $\mathbf{LA_u(n,\ell)_q}$

Theorem 5.25 (Ahlswede, Aydinian, Khachatrian, and Tolhuizen 2006 [3]) *For all integers q, n and ℓ satisfying $q > \ell + 1$ we have*

$$\frac{\ell}{q-1}\left(\frac{q}{\ell+1}\right)^n \leq LA_u(n,\ell)_q \leq \lceil \frac{q}{\ell+1} \rceil^{n-1}.$$

Proof Consider the equation

$$\sum_{i=0}^{n-1} (\ell+1)^i x_i = a, \tag{5.6.13}$$

and let X be the set of vectors $\mathbf{x} \in \mathcal{X}_q^n$ satisfying (5.6.13). The equation $\sum_{i=0}^{n-1}(\ell + 1)^i x_i = 0$ has no non-zero solutions $\mathbf{x} \in [-\ell, \ell]^n \cup [0, 2\ell]^n$. Thus X is a q-ary ℓ-

UEC code. Note also that $X = \varnothing$ if $a \notin I \triangleq [0, (q-1)\frac{(\ell+1)^n-1}{\ell}]$. Hence we infer that there exists an $a \in I$ such that

$$|X| \geq \frac{|\mathcal{X}_q^n|}{|I|} = q^n / \left((q-1)\frac{(\ell+1)^n - 1}{\ell} + 1\right) \geq \left(\frac{q}{\ell+1}\right)^n \cdot \frac{\ell}{q-1}.$$

This gives the lower bound for $LA_u(n,\ell)_q$.

Let now X be a q-ary ℓ-UEC code defined by (5.6.10).

To prove the upper bound we consider the mapping $\psi : \mathcal{X}_q \to \mathbb{Z}_b$, where $b \triangleq \lceil \frac{q}{\ell+1} \rceil$, defined by

$$\psi(j) \equiv j \pmod{b};\ j = 0, \ldots, q-1.$$

Correspondingly for a codeword $\mathbf{x} = (x_0, \ldots, x_{n-1}) \in X$ we define $\psi^n(\mathbf{x}) = (\psi(x_0), \ldots, \psi(x_{n-1}))$. Let us show that ψ^n is an injection on X. Suppose $\psi^n(\mathbf{x}) = \psi^n(\mathbf{x}')$ for two codewords $\mathbf{x}, \mathbf{x}' \in X$. By definition of ψ we have $\mathbf{x} - \mathbf{x}' = b\mathbf{e}$, where $\mathbf{e} \in [-\ell, \ell]^n$. As $\mathbf{x}$ and $\mathbf{x}'$ both are in X we have

$$\sum_{i=0}^{n-1} a_i e_i = 0. \tag{5.6.14}$$

We define $\mathbf{x}^* = \mathbf{x}' + (b-1)\mathbf{e}$ and claim that $\mathbf{x}^*$ is in X. In view of (5.6.14), it is sufficient to show that $\mathbf{x}^* \in \mathcal{X}_q^n$. For $1 \leq i \leq n$ let now $e_i \geq 0$. Then $x_i^* = x_i' + (b-1)e_i \geq x_i' \geq 0$ and $x_i^* = x_i - e_i \leq x_i \leq q-1$, so $x_i^* \in \mathcal{X}_q$. In a similar way it is proved that $x_i^* \in \mathcal{X}_q$ if $e_i \leq 0$. Since $\mathbf{x} - \mathbf{x}^* = \mathbf{e} = [-\ell, \ell]^n$, and $\mathbf{x}$ and $\mathbf{x}^*$ both are in X, we conclude that $\mathbf{e} = \mathbf{0}$, so $\mathbf{x} = \mathbf{x}'$. Thus ψ^n is an injection, which implies that $|X| = |\psi^n(X)|$.

Define now

$$H' = \{(y_0, \ldots, y_{n-1}) \in \mathbb{Z}_b^n : \sum_{i=0}^{n-1} a_i y_i \equiv a (\text{mod } b)\}.$$

It is easy to see that $\psi^n(X) \subset H'$. We can assume without loss of generality that $g.c.d.(a_0, \ldots, a_{n-1}) = 1$, so $(a_0(\text{ mod } b), \ldots, a_{n-1}(\text{ mod } b)) \neq (0, \ldots, 0)$.

Thus $H' \subset \mathbb{Z}_b^n$ is a hyperplane over $\mathbb{Z}_b$ and hence

$$|X| = |\psi^n(X)| \leq |H'| = b^{n-1}.$$

□

5.6.7 Construction of Optimal Codes

We call a VT-type ℓ-UEC code VT-type optimal or shortly optimal if it attains the upper bound in Theorem 5.25. In this section we construct, for various classes of pairs (ℓ, q), maximal q-ary ℓ-UEC codes for each length n.

Given integers $\ell \in [1, q-1]$, n, r we define

$$\mathcal{C}_n(r) = \left\{ (x_0, \dots, x_{n-1}) \in \mathcal{X}_q^n : \sum_{i=0}^{n-1} (\ell+1)^i x_i = \alpha S_n + r \right\}, \tag{5.6.15}$$

$$\text{where } S_n \triangleq \sum_{i=0}^{n-1} (\ell+1)^i = \frac{(\ell+1)^n - 1}{\ell}, \quad \text{and } \alpha \triangleq \lfloor \frac{q-1}{2} \rfloor. \tag{5.6.16}$$

As we have seen in the proof of Theorem 5.25, $\mathcal{C}_n(r)$ is an ℓ-UEC code for all n and r.

For notational convenience, we denote the cardinality of $\mathcal{C}_n(r)$ by $\gamma_n(r)$, that is,

$$\gamma_n(r) = |\mathcal{C}_n(r)| \ . \tag{5.6.17}$$

Proposition 5.8 *For each $n \geq 2$ and each r,*

$$\gamma_n(r) = \sum_{x_0} \gamma_{n-1}\left((\alpha + r - x_0)/(\ell+1) \right) \ ,$$

where the sum extends over all $x_0 \in \mathcal{X}_q$ satisfying $x_0 \equiv \alpha + r \pmod{\ell+1}$.

Proof By definition $\mathbf{x} = (x_0, x_1, \dots, x_{n-1})$ is in $\mathcal{C}_n(r)$ if and only if $\sum_{i=0}^{n-1} (\ell+1)^i x_i - \alpha S_n = r$. Using that $S_n = (\ell+1)S_{n-1} + 1$, the latter equality can also be written as $\sum_{i=1}^{n-1} (\ell+1)^i x_i - \alpha S_{n-1} = r - x_0 + \alpha$. In other words $\mathbf{x}$ is in $\mathcal{C}_n(r)$ if and only if $x_0 \equiv r + \alpha \ (\text{mod}\ \ell+1)$ and $(x_1, \dots, x_{n-1})$ is in $\mathcal{C}_{n-1}(r')$, where $r' = (r - x_0 + \alpha)/(\ell+1)$. □

In the remainder of this section, we use the notation $\langle x \rangle_y$ to denote the integer in $[0, y-1]$ that is equivalent to x modulo y. In other words, $\langle x \rangle_y = x - \lfloor \frac{x}{y} \rfloor \cdot y$.

Lemma 5.17 *Let e and f be integers such that $0 \leq e \leq f-1$. We have that*

$$|\{x \in \mathcal{X}_q : x \equiv e \ (\text{mod}\ f)\}| = \begin{cases} \lceil \frac{q}{f} \rceil & if\ e < \langle q \rangle_f \\ \lfloor \frac{q}{f} \rfloor & if\ e \geq \langle q \rangle_f \end{cases}$$

Proof We obviously have that

$$\{x \in \mathcal{X}_q : x \equiv e \ (\text{mod}\ f)\} = \{e + f, e + 2f, \dots, e + mf\},$$

where m is such that $e + mf \leq q - 1$ and $e + (m+1)f \geq q$. In other words $m = \lfloor \frac{q-1-e}{f} \rfloor$. Writing $q = \lambda f + \langle q \rangle_f$, we have $m - \lambda = \lfloor \frac{\langle q \rangle_f - 1 - e}{f} \rfloor$, which equals 0 if $\langle q \rangle_f \geq e + 1$, and -1 otherwise. This proves the lemma. □

Theorem 5.26 (Ahlswede, Aydinian, Khachatrian, and Tolhuizen 2006 [3]) *Let $u_1, u_2, \ldots$ and $v_1, v_2, \ldots$ be sequences of integers such that:*

(i) $0 \leq u_1 + \alpha \leq v_1 + \alpha \leq q - 1$,

and for each $n \geq 2$

(ii) $\lceil \frac{1}{\ell+1}(u_n + \alpha - (q-1)) \rceil \geq u_{n-1}$,
(iii) $\lfloor \frac{1}{\ell+1}(v_n + \alpha) \rfloor \leq v_{n-1}$, *and*
(iv) $\ell + 1$ *divides* q, *or for each* $r \in [u_n, v_n]$, $\langle \alpha + r \rangle_{\ell+1} < \langle q \rangle_{\ell+1}$.

Then for each $n \geq 1$ and $r \in [u_n, v_n]$ we have $\gamma_n(r) = \lceil \frac{q}{\ell+1} \rceil^{n-1}$.

Proof We proceed by induction on n.

For $n = 1$ the assertion is true because of condition (i).

Now let $n \geq 2$, and suppose the assertion is true for $n - 1$. Let $r \in [u_n, v_n]$. According to Proposition 5.8, we have that

$$\gamma_n(r) = \sum_{x_0} \gamma_{n-1} \left(\frac{r + \alpha - x_0}{\ell + 1} \right) . \tag{5.6.18}$$

According to condition (iv), either $\ell + 1$ divides q, or $\langle \alpha + r \rangle_{\ell+1} < \langle q \rangle_{\ell+1}$. In both cases Lemma 5.17 implies that the sum in (5.6.18) has $\lceil \frac{q}{\ell+1} \rceil$ terms.

For each $x_0 \in \mathcal{X}_q$ we have that $r + \alpha - x_0 \leq r + \alpha \leq v_n + \alpha$ and $r + \alpha - x_0 \geq r + \alpha - (q-1) \geq u_n + \alpha - (q-1)$. That is, for each $x_0 \in \mathcal{X}_q$

$$u_n + \alpha - (q-1) \leq r + \alpha - x_0 \leq v_n + \alpha. \tag{5.6.19}$$

Combining (5.6.19) with conditions (ii) and (iii) we find that for each x_0 in $\mathcal{X}_q$, such that $r + \alpha - x_0$ is a multiple of $\ell + 1$, we have

$$\frac{r + \alpha - x_0}{\ell + 1} \in [u_{n-1}, v_{n-1}].$$

The induction hypothesis implies that each term in the sum in (5.6.18) equals $\lceil \frac{q}{\ell+1} \rceil^{n-2}$. □

Theorem 5.27 (Ahlswede, Aydinian, Khachatrian, and Tolhuizen 2006 [3]) *Let ℓ and q be such that $\ell + 1$ divides q. Let $u_1 = -\alpha$, $v_1 = \alpha$, and for $n \geq 2$, $u_n = (\ell+1)u_{n-1} + \alpha$ and $v_n = (\ell+1)v_{n-1} - \alpha$. In other words, for $n \geq 1$, $v_n = -u_n = \frac{\alpha}{\ell}\left[(\ell-1)(\ell+1)^{n-1} + 1\right]$.*

Then for each $n \geq 1$ and $r \in [u_n, v_n]$, we have

$$\gamma_n(r) = LA_u(n, \ell)_q = \left(\frac{q}{\ell + 1} \right)^{n-1} .$$

Proof We apply Theorem 5.26. It is immediately clear that conditions (i), (iii) and (iv) are satisfied. Moreover, for each $n \geq 2$, $u_n + \alpha - (q-1) = (\ell+1)u_{n-1} + 2\alpha - (q-1) \geq (\ell+1)u_{n-1} - 1$, so condition (iii) is satisfied as well. □

Theorem 5.28 (Ahlswede, Aydinian, Khachatrian, and Tolhuizen 2006 [3]) *Let $c \in [0, \ell]$, $\delta \in \{0, 1\}$, and m be such that*

$$q = 2m(\ell+1) + 2c + 1 + \delta \text{ and } 2c + \delta \neq \ell.$$

We define $\lambda_1 = 0$, and for $n \geq 2$,

$$\lambda_n = (\ell+1)\lambda_{n-1} - \eta, \text{ where } \eta = \begin{cases} 0 & \text{if } 2c+\delta \leq \ell - 1, \\ \lceil \frac{1}{2}(\ell-\delta)\rceil & \text{if } 2c+\delta \geq \ell+1. \end{cases}$$

Moreover, for $n \geq 1$, we define

$$u_n = -c + \lambda_n(\ell+1) \text{ and } v_n = -c + \lambda_n(\ell+1) + \langle q\rangle_{\ell+1} - 1.$$

If $m \leq c - 1 - \lceil\frac{1}{2}(\ell-\delta)\rceil$ or $2c+\delta \leq \ell$ and $m \leq c$, then for each integer n and $r \in [u_n, v_n]$,

$$\gamma_n(r) = LA_u(n,\ell)_q = \lceil \frac{q}{\ell+1} \rceil^{n-1}.$$

Proof We apply Theorem 5.26. Note that

$$\alpha = \lfloor \frac{q-1}{2} \rfloor = m(\ell+1) + c.$$

We first check condition (i): $u_1 + \alpha = -c + \alpha = m(\ell+1) \geq 0$ and $u_1 + \alpha \leq v_1 + \alpha = m(\ell+1) + \langle q\rangle_{\ell+1} - 1 \leq q - 1$.

The definition of u_n and v_n implies that for each n and each $r \in [u_n, v_n]$ we have that

$$r + \alpha \in [u_n+\alpha, v_n+\alpha] = [(\lambda_n + m)(\ell+1), (\lambda_n+m)(\ell+1) + \langle q\rangle_{\ell+1} - 1],$$

so condition (iv) is satisfied.

For verifying Condition (ii), we note that

$$\lceil \frac{1}{\ell+1}(u_n + \alpha - (q-1))\rceil = \lceil \frac{1}{\ell+1}(u_n - \alpha - \delta)\rceil = (\lambda_n - m) + \lceil \frac{-\delta-2c}{\ell+1}\rceil.$$

As $\lambda_n = \lambda_{n-1}(\ell+1) - \eta = u_{n-1} + c - \eta$ condition (ii) is satisfied if and only if

$$m \leq c - \eta - \lfloor \frac{\delta+2c}{\ell+1} \rfloor. \tag{5.6.20}$$

For verifying condition (iii) we note that

$$\lfloor\frac{1}{\ell+1}(v_n+\alpha)\rfloor = \lfloor\frac{1}{\ell+1}((\lambda_n+m)(\ell+1)+\langle q\rangle_{\ell+1})\rfloor = \lambda_n+m.$$

As $\lambda_n = (\ell+1)\lambda_{n-1} - \eta = v_{n-1} + c - \langle q\rangle_{\ell+1} + 1 - \eta$, condition (iii) is satisfied if and only if

$$m \le \langle q\rangle_{\ell+1} - 1 - c + \eta \tag{5.6.21}$$

We distinguish between two cases.

Case 1. $2c+\delta \le \ell - 1$.
Then $\langle q\rangle_{\ell+1} = 2c+\delta+1$, and $\lfloor\frac{\delta+2c}{\ell+1}\rfloor = 0$. That is, (5.6.20) reduces to the inequality $m \le c-\eta$ and (5.6.21) reduces to $m \le c+\delta+\eta$. As $\eta = 0$, we see that (5.6.20) and (5.6.21) both are satisfied if $m \le c$.
Case 2. $2c+\delta \ge \ell+1$.
Then $\langle q\rangle_{\ell+1} = 2c+\delta-\ell$, and $\lfloor\frac{\delta+2c}{\ell+1}\rfloor = 1$. Consequently, (5.6.20) reduces to the inequality $m \le c-\eta-1$, and (5.6.21) reduces to $m \le c+\delta-\ell-1+\eta$. With our choice for η, we see that (5.6.20) and (5.6.21) both are satisfied if $m \le c-\eta-1 = c-1-\lceil\frac{1}{2}(\ell-\delta)\rceil$. □

Corollary 5.3 *Let $q = (b-1)(\ell+1)+d$ for integers $1 \le b-1 < d \le \ell$. Then for each n*

$$LA_u(n,\ell)_q = b^{n-1} = \left\lceil\frac{q}{\ell+1}\right\rceil^{n-1}.$$

Proof Suppose $b-1$ is even. Then we can write

$$q = 2m(\ell+1)+d = 2m(\ell+1)+2c+1+\delta,$$

where $c = (d-1-\delta)/2$ and $m = (b-1)/2$. The condition $b-1 < d \le \ell$ implies that $2c+\delta \le \ell-1$ and $m \le c$. Therefore by Theorem 5.28 we have

$$\gamma_n(r) = b^{n-1}, \text{ where } r \in [-c,c].$$

Suppose now $b-1$ is odd. Then

$$q = (2m+1)(\ell+1)+d = 2m(\ell+1)+d+\ell+1 = 2m(\ell+1)+2c+1+\delta,$$

where $c = (d+\ell-\delta)/2$ and $m = (b-2)/2$.

Now the condition $b-1 < d$ implies $m \le c-1-\lceil\frac{1}{2}(\ell-\delta)\rceil$ and hence by Theorem 5.28 we have

$$\gamma_n(r) = b^{n-1}, \text{ where } r \in [u_n, v_n].$$

□

In conclusion of this section let us note that the determination of $LA_u(n,\ell)_q$ in general seems to be a difficult problem. As was shown above codes defined by (5.6.15) are best possible for certain parameters q and ℓ, mentioned in Theorems 5.26 and 5.27. However we do not know how good these codes are for other parameters.

An interesting open problem is to decide what is the $\max_r |\mathcal{C}_n(r)|$ for given ℓ and q. Note that for some cases the code $\mathcal{C}_n(0)$ has the size bigger than the lower bound in Theorem 5.25. Let for example $\ell = 2,\ q = 7$. Then it is not hard to observe that the number of solutions c_n of (5.6.15) satisfies the recurrence $c_n = 2c_{n-1} + c_{n-2}$. This gives the bound $|\mathcal{C}_n(r)| \geq K(2, 41)^n$, where $2, 41 \approx 1 + \sqrt{2}$ is the largest root of the characteristic equation $x^2 - 2x - 1 = 0,\ K$ is a constant. The same recurrence we obtain for any $q = 2\ell + 3$, which implies that for $q = 2\ell + 3$ and $\ell \geq 2$ one has $|\mathcal{C}_n(r)| \geq K(2, 41)^n > \frac{\ell}{q-1}\left(\frac{q}{\ell+1}\right)^n$ (the lower bound in Theorem 5.25). Note however that this is not the case for $\ell = 1,\ q = 5$.

One can also observe that for $q = 7,\ \ell = 1$ we have $|\mathcal{C}_n(r)| \geq K(3, 51)^n$. Without going into detail we note that this can be derived from the recurrence $c_n = 4c_{n-1} - 2c_{n-2} + c_{n-3}$ for the number of solutions c_n of (5.6.15) (with $r = 0,\ q = 7,\ \ell = 1$).

One may use a generating functions approach to analize the problem.

Let $f(x)$=$1 + x + x^2 + \ldots + x^{q-1}$. We are interested in the largest coefficient of the polynomial $f(x)f(x^{\ell+1})f(x^{(\ell+1)^2})f(x^{(\ell+1)^3})\cdots f(x^{(\ell+1)^{n-1}})$. If, for example, we take $q = 5, \ell = 1$ and $n = 4$, the largest coefficient equals 20 (attained with x^{24}, x^{28}, x^{32} and x^{36}), while the coefficient of x^a for $a = \lfloor\frac{q-1}{2}\rfloor\frac{(\ell+1)^n-1}{\ell} = 30$ only equals 17.

5.6.8 Asymptotic Growth Rate of ℓ-UEC Codes of VT Type

In the previous section we explicitly constructed maximal q-ary ℓ-UEC codes of VT type of arbitrary length for some classes of pairs (ℓ, q) – but not for all.

In this section we state a less ambitious goal, namely, given ℓ and q, to determine the asymptotic behaviour of $\sqrt[n]{LA_u(n, \ell)_q}$. We will show that this quantity converges if $n \to \infty$. As a preparation we need the following

Lemma 5.18 *Let $a, b, a_0, a_1, \ldots, a_{m-1}, b_0, b_1, \ldots, b_{n-1}$ be integers such that the codes $\mathcal{A}$ and $\mathcal{B}$, defined as*

$$\mathcal{A} = \{(x_0, x_1, \ldots, x_{m-1}) \in \mathcal{X}_q^m : \sum_{i=0}^{m-1} a_i x_i = a\}$$

$$and \mathcal{B} = \{(y_0, y_1, \ldots, y_{n-1}) \in \mathcal{X}_q^n : \sum_{j=0}^{n-1} b_j y_j = b\}$$

both are non-empty ℓ-UEC codes. Let $\mathcal{A} \times \mathcal{B} \subset \mathcal{X}_q^{m+n}$ be the direct product of $\mathcal{A}$ and $\mathcal{B}$:

$$\mathcal{A} \times \mathcal{B} = \{(\mathbf{x}; \mathbf{y}) : \mathbf{x} \in \mathcal{A}, \mathbf{y} \in \mathcal{B}\}.$$

Let M be an integer such that $\sum_{i=0}^{n-1} |a_i|(q-1) < M$, and define $\mathcal{C}$ as

$$\mathcal{C} = \{(z_0, z_1, \ldots, z_{n+m-1}) \in \mathcal{X}_q^{n+m} : \sum_{i=0}^{n-1} a_i z_i + \sum_{i=n}^{n+m-1} M b_{i-n} z_i = a + Mb\}.$$

Then $\mathcal{C} = \mathcal{A} \times \mathcal{B}$, and $\mathcal{A} \times \mathcal{B}$ is a q-ary ℓ-AUEC code.

Proof It is clear that $\mathcal{A} \times \mathcal{B} \subset \mathcal{C}$. Moreover, $\mathcal{A} \times \mathcal{B}$ is an ℓ-UEC code: a received word can be decoded by decoding its m leftmost and n rightmost symbols to $\mathcal{A}$ and $\mathcal{B}$, respectively. All we are left with to show is that $\mathcal{C} \subset \mathcal{A} \times \mathcal{B}$. Therefore, let $(z_0, z_1, \ldots, z_{n+m-1})$ be in $\mathcal{C}$. By definition, we have that

$$a + Mb = \sum_{i=0}^{m-1} a_i z_i + M \cdot \sum_{i=m}^{m+n-1} b_{i-m} z_i, \tag{5.6.22}$$

and so

$$a - \sum_{i=0}^{m-1} a_i z_i \equiv 0 \bmod M. \tag{5.6.23}$$

As $\mathcal{A} \neq \emptyset$, there is an $\mathbf{x} \in \mathcal{X}_q^m$ such that $a = \sum_{i=0}^{m-1} a_i x_i$, and whence

$$|a - \sum_{i=0}^{m-1} a_i z_i| = |\sum_{i=0}^{m-1} a_i (x_i - z_i)| \leq \sum_{i=0}^{m-1} |a_i||x_i - z_i| \leq \sum_{i=0}^{m-1} |a_i|(q-1) < M. \tag{5.6.24}$$

From (5.6.23) and (5.6.24) we conclude that $a = \sum_{i=0}^{m-1} a_i z_i$ and so $(z_0, z_1, \ldots, z_{m-1}) \in \mathcal{A}$. Furthermore using (5.6.22) we find that $(z_m, z_{m+1}, \ldots, z_{m+n-1})$ is in $\mathcal{B}$. □

Lemma 5.18 immediately implies that

$$LA_u(\ell, m+n)_q \geq LA_u(\ell, m)_q \cdot LA_u(\ell, n)_q. \tag{5.6.25}$$

As $LA_u(\ell, n)_q \leq \lceil \frac{q}{\ell+1} \rceil^{n-1}$ we can invoke Fekete's lemma to derive the following result from (5.6.25):

Proposition 5.9 *For each q and $\ell \in \mathcal{X}_q$, there exists a constant $\beta(\ell, q) \leq \lceil \frac{q}{\ell+1} \rceil$ such that*

$$\lim_{n \to \infty} \sqrt[n]{LA_u(n, \ell)_q} = \beta(\ell, q).$$

Theorem 5.25 implies that for all ℓ and q,

$$\frac{q}{\ell+1} \leq \beta(\ell, q) \leq \lceil \frac{q}{\ell+1} \rceil.$$

In particular, $\beta(\ell, q) = \frac{q}{\ell+1}$ if $\ell + 1$ divides q (of course, this is also implied by the much stronger Theorem 5.27). Note also that for pairs (ℓ, q) for which the conditions from Theorem 5.28 applies, we have $\beta(\ell, q) = \lceil \frac{q}{\ell+1} \rceil$.

Inequality (5.6.25) implies that for each n, $\beta(\ell, q) \geq \sqrt[n]{LA_u(n, \ell)_q}$. For example, consider the case that $q = \ell + 2$. The code

$$\{(x_0, x_1, x_2, x_3) \in \mathcal{X}_q^4 : \sum_{i=0}^{3} (\ell + 1)^i x_i = \ell + 1 + (\ell + 1)^3\}$$

has five words, *viz.* $(1 + \ell, 1 + \ell, \ell, 0), (1 + \ell, 0, 1 + \ell, 0), (1 + \ell, 0, 0, 1), (0, 1, 1 + \ell, 0)$, and $(0, 1, 0, 1)$. That is, $\beta(\ell, \ell + 2) \geq \sqrt[4]{5} \approx 1.495$. Note that Theorem 5.25 only allows us to deduce that $\beta(\ell, \ell + 2) \geq \frac{\ell+2}{\ell+1}$.

Also note that Corollary 5.3 with $b = 2$ states that for $\ell \geq 2 \quad \beta(\ell, \ell + 3) = 2$.

5.6.9 The Error Detection Problem

We find it interesting to consider also the error detection problem, i.e. codes detecting unconventional errors of a certain level. It is easy to see that codes detecting asymmetric errors of level ℓ can be also used to detect unidirectional errors of level ℓ. For codes detecting all asymmetric (unidirectional) errors of level ℓ we use the abbreviation ℓ-AED codes (or ℓ-UED codes).

For integers ℓ, q, n satisfying $1 \leq \ell < q$ and $n \geq 1$, we define

$$P_i = \{(a_1, \dots, a_n) \in \mathcal{X}_q^n : \sum_{j=1}^{n} a_j = i\}.$$

It is clear that P_i detect each unidirectional error pattern. Note that $|P_i|$ is maximal for $i = i^* = \lfloor \frac{1}{2} n(q - 1) \rfloor$, see [5, Theorem 4.1.1]. For $a \in [0, \ell n]$, let $\mathcal{C}_a \subset \mathcal{X}_q^n$ be defined as

$$\mathcal{C}_a = \bigcup_{i : i \equiv a (\bmod \ \ell n+1)} P_i \tag{5.6.26}$$

Proposition 5.10 *$\mathcal{C}_a$ is an ℓ-UED-code over the alphabet $\mathcal{X}_q$.*

Proof Clearly $\mathcal{C}_a$ is an ℓ-UED code iff for each $\mathbf{x}, \mathbf{y} \in \mathcal{C}_a$ either $\mathbf{x}$ and $\mathbf{y}$ are incomparable or $d(\mathbf{x}, \mathbf{y}) \geq \ell + 1$. Suppose that for some $\mathbf{x} = (x_1, \dots, x_n)$ and $\mathbf{y} = (y_1, \dots, y_n)$ we have $\mathbf{x} > \mathbf{y}$. Then clearly by definition of $\mathcal{C}$ there exists a coordinate $i \in [1, n]$ such that $x_i - y_i \geq \ell + 1$, i.e. $d(\mathbf{x}, \mathbf{y}) \geq \ell + 1$. □

This simple construction gives us a lower bound for the maximum size of an ℓ-UED code over alphabet $\mathcal{X}_q$. However we don't know whether it is possible to improve this bound, even for the case $\ell = 1$.

Remark Asymptotically, taking the union of several P_i's does not really help as the largest P_i contains $c\frac{1}{\sqrt{n}}q^n$ words, while nearly all words in $\mathcal{X}_q^n$ are in the union of about $\sqrt{n}$ sets P_i with consecutive i's.

Remark The construction is not optimal in general. For example take $\ell = 1$ and $q = n = 3$. It can easily be checked that $(|P_0|, |P_1|, \ldots, |P_6|) = (1, 3, 6, 7, 6, 3, 1)$. Therefore for each $a \in [0, \ell n] = [0, 3]$, $|C_a| \leq 7$. The code consisting of (0,0,0), (2,2,2) and the six permutations of (0,1,2) has eight words and is a 1-UED code.

Consider also two other small cases.

For $\ell = 1, q = 4$ and $n = 3$ one easily checks that $(|P_0|, |P_1|, \ldots, |P_9|) = (1, 3, 6, 10, 12, 10, 6, 3, 1)$ and so $|C_a| = 16$ for all $a \in [0, \ell n] = [0, 3]$.

Similarly for $\ell = 1, q = 5$ and $n = 3$ one easily checks that $(|P_0|, |P_1|, \ldots, |P_{12}|) = (1, 3, 6, 10, 15, 18, 19, 18, 15, 10, 6, 3, 1)$. It follows that $|C_0| = 32$ and $|C_1| = |C_2| = |C_3| = 31$. Note that C_0, the largest of the four codes, does not contain P_6, the largest P_i.

References

1. K.A.S. Abdel-Ghaffar, H. Fereira, Systematic encoding of the Varshamov-Tennengolts codes and the Constantin-Rao codes. IEEE Trans. Inf. Theory **44**(1), 340–345 (1998)
2. R. Ahlswede, H. Aydinian, L.H. Khachatrian, Undirectional error control codes and related combinatorial problems, in *Proceedings of Eight International Workshop on Algebraic and Combinatorial Coding Theory, 8–14 September, Tsarskoe Selo, Russia* (2002), pp. 6–9
3. R. Ahlswede, H. Aydinian, L. Khachatrian, L. Tolhuizen, On q-ary codes correcting all unidirectional errors of a limited magnitude (2006)
4. G.G. Ananiashvili, On a class of asymmetric single-error correcting non-linear codes. Doklady Akad. Nauk. Georgian SSR **53**(3), 549–552 (1969)
5. I. Anderson, *Combinatorics of Finite Sets* (Clarendon Press, Oxford, 1987)
6. M. Blaum (ed.), *Codes for Detecting and Correcting Unidirectional Errors* (IEEE Computer Society Press Reprint Collections, IEEE Computer Society Press, Los Alamitos, 1993)
7. J.M. Borden, Optimal asymmetric error detecting codes. Inf. Control **53**(1–2), 66–73 (1982)
8. B. Bose, S. Cunningham, *Asymmetric Error Correcting Codes, Sequences, II (Positiano 1991)* (Springer, New York, 1993), pp. 24–35
9. B. Bose, S.A. Al-Bassam, On systematic single asymmetric error correcting codes. IEEE Trans. Inf. Theory **46**(2), 669–672 (2000)
10. S.D. Constantin, T.R.N. Rao, Concatenated group theoretic codes for binary asymmetric channels. AFIPS Conf. Proc. **46**, 837–842 (1979)
11. S.D. Constantin, T.R.N. Rao, On the theory of binary asymmetric error correcting codes. Inf. Control **40**(1), 20–36 (1979)
12. Ph Ph Delsarte, Piret, Bounds and constructions for binary asymmetric error correcting codes. IEEE Trans. Inf. Theory **27**(1), 125–128 (1981)
13. L.E. Dickson, *History of the Theory of Numbers*, vol. 2 (Chelsea, New York, 1952)
14. P. Erdős, Problems and results from additive number theory. Colloq. Theoretic des Nombres, Bruxelles, 1955, Liege&Paris, 1956
15. G. Fang, H.C.A. van Tilborg, Bound and constructions of asymmetric or unidirectional error-correcting codes. Appl. Algebra Eng. Commun. Eng. **3**(4), 269–300 (1992)
16. D. Gevorkian, A.G. Mhitarian, Classes of codes that correct single asymmetric errors (in Russian). Dokl. Akad. Nauk Armyan. SSR **70**(4), 216–218 (1980)

17. B.D. Ginzburg, A number-theoretic function with an application in the theory of coding. Probl. Kybern. **19**, 249–252 (1967)
18. R.W. Hamming, Bell Syst. Tech. J. **29**, 147 (1950)
19. T. Helleseth, T. Kløve, On group-theoretic codes for asymmetric channels. Inf. Control **49**(1), 1–9 (1981)
20. W.H. Kim, C.V. Freiman, Single error-correcting-codes for asymmetric binary channels. IRE Trans. Inf. Theory **IT–5**, 62–66 (1959)
21. T. Kløve, Error correcting codes for the asymmetric channel. Report, Department of Mathematics, University of Bergen, 1981 (with updated bibliography in 1995)
22. A.V. Kuznetsov, B.S. Tsybakov, Coding in a memory with defective cells. Problemy Peredachi Informatsii **10**(2), 52–60 (1974)
23. V.I. Levenshtein, A class of systematic codes. Sov. Math.-Dokl. **1**, 368–371 (1960)
24. V.I. Levenshtein, Binary codes with correction for bit losses, gains, and substitutions. Dokl. Akad. Nauk SSSR **163**(4), 845–848 (1965)
25. V.I. Levenshtein, Binary codes capable of correcting deletions and insertions, and reversals. Sov. Phys. Dokl. **10**, 707–710 (1966)
26. V.I. Levenshtein, Asymptotically optimum binary code with correction for losses of one or two adjacent bits. Probl. Cybern. **19**, 298–304 (1967)
27. S. Lin, D.J. Costello Jr., *Error Control Coding: Fundamentals and Applications* (Prentice-Hall Inc, Englewood Cliffs, 1983)
28. F.J. MacWilliams, N.J.A. Sloane, *The Theory of Error-Correcting Codes* (North-Holland, Amsterdam, 1977)
29. S.S. Martirossian, Single-error correcting close-packed and perfect codes, in *Proceedings of First INTAS International Seminar on Coding Theory and Combinatorics, (Tsahkadzor Armenia)* (1996), 90–115
30. L.E. Mazur, Certain codes that correct non-symmetric errors. Probl. Inf. Transm. **10**(4), 308–312 (1976)
31. R.J. McEliece, Comments on class of codes for asymmetric channels and a problem from additive theory of numbers. IEEE Trans. Inf. Theory **19**(1), 137 (1973)
32. M.N. Nalbandjan, A class of codes that correct multiple asymmetric errors (in Russian). Dokl. Acad. Nauk Georgian SSR **77**, 405–408 (1975)
33. O.S. Oganesyan, V.G. Yagdzhyan, Classes of codes correcting bursts of errors in an asymmetric channel. Problemy Peredachi Informatsii **6**(4), 27–34 (1970)
34. V. Pless, W.C. Huffman, R.A. Brualdi (eds.), *Handbook of Coding Theory*, vol. I (II (North-Holland, Amsterdam, 1998)
35. F.F. Sellers Jr., Bit loss and gain correction code. IRE Trans. Inf. Theory **IT–8**(1), 35–38 (1962)
36. V.I. Siforov, Radiotechn. i. Elektron. **1**, 131 (1956)
37. R.P. Stanley, M.F. Yoder, A study of Varshamov codes for asymmetric channels. Jet Propulsion Laboratory, Technical report, 32-1526, vol. 14 (1982), pp. 117–122
38. R.R. Varshamov, Dokl. Akad. Nauk SSSR **117** (1957)
39. R.R. Varshamov, On some features of asymmetric error-correcting linear codes (in Russian). Rep. Acad. Sci. USSR **157**(3), 546–548 (1964) (transl: Sov. Phys.-Dokl. **9**, 538–540, 1964)
40. R.R. Varshamov, Estimates of the number of signals in codes with correction of nonsymmetric errors (in Russian). Avtomatika i Telemekhanika 25(11), 1628–1629 (1964) (transl. Autom. Remote Control **25**, 1468–1469 (1965)
41. R.R. Varshamov, On an arithmetical function applied in coding theory. DAN USSR, Moscow **161**(3), 540–542 (1965)
42. R.R. Varshamov, On the theory of asymmetric codes (in Russian). Dokl. Akademii Nauk USSR **164**, 757–760 (1965) (transl: Sov. Phys.-Dokl. **10**, 185–187, 1965)
43. R.R. Varshamov, A general method of constructing asymmetric coding systems, related to the solution of a combinatorial problem proposed by Dixon. Dokl. Akad. Nauk. SSSR **194**(2), 284–287 (1970)
44. R.R. Varshamov, A class of codes for asymmetric channels and a problem from the additive theory of numbers. IEEE Trans. Inf. Theory **19**(1), 92–95 (1973)

45. R.R. Varshamov, G.M. Tenengol'ts, Asymmetrical single error-correcting code. Autom. Telem. **26**(2), 288–292 (1965)
46. R.R. Varshamov, G.M. Tenengol'ts, A code that corrects single unsymmetric errors. Avtomatika i Telemekhanika **26**(2), 288–292 (1965)
47. R.R. Varshamov, G.M. Tennengolts, A code which corrects single asymmetric errors (in Russian) Avtomat. Telemeh. **26**, 282–292 (1965) (transl: Autom. Remote Control 286–290, 1965)
48. R.R. Varshamov, E.P. Zograbjan, A class of codes correcting two asymmetric errors. Trudy Vychisl. Centra Akad. Nauk. Armjan. SSR i Erevan **6**, 54–58 (1970)
49. R.R. Varshamov, E.P. Zograbian, Codes correcting packets of non-symmetricerrors (in Russian), in *Proceedings of the 4'th Symposium on Problems in Information Systems*, vol. 1 (1970), 87–96 (Review in RZM No. 2, V448, 1970)
50. R.R. Varshamov, S.S. Oganesyan, V.G. Yagdzhyan, Non-linear binary codes which correct one and two adjacent errors for asymmetric channels, in *Proceedings of the First Conference of Young Specialists at Computer Centers, Erevan*, vol. 2 (1969)
51. J.H. Weber, C. de Vroedt, D.E. Boekee, Bounds and constructions for codes correcting unidirectional errors. IEEE Trans. Inf. Theory **35**(4), 797–810 (1989)

Further Readings

52. M.J. Aaltonen, Linear programming bounds for tree codes. IEEE Trans. Inf. Theory **25**, 85–90 (1977)
53. M.J. Aaltonen, A new bound on nonbinary block codes. Discret. Math. **83**, 139–160 (1990)
54. N. Alon, O. Goldreich, J. Hastad, R. Peralta, Simple construction of almost k-wise independent random variables. Random Struct. Algorithms **3**(3), 289–304 (1992)
55. H. Batemann, A. Erdelyi, *Higher Transcendental Functions*, vol. 2 (McGraw-Hill, New York, 1953)
56. E. Bannai, T. Ito, *Algebraic Combinatorics. 1. Association Schemes* (Benjamin/Cummings, London, 1984)
57. R.C. Bose, Mathematical theory of the symmetrical factorial design. Sankhya **8**, 107–166 (1947)
58. A.E. Brouwer, A.M. Cohen, A. Neumaier, *Distance-Regular Graphs* (Springer, Berlin, 1989)
59. R. Calderbank, On uniformly packed $[n, n-k-4]$ codes over $GF(q)$ and a class of caps in $PG(k-1, q)$. J. Lond. Math. Soc. **26**, 365–384 (1982)
60. R. Calderbank, W.M. Kantor, The geometry of two-weight codes. Bull. Lond. Math. Soc. **18**, 97–122 (1986)
61. J.H. Conway, N.J.A. Sloane, A new upper bound on the minimal distance of self-dual codes. IEEE Trans. Inf. Theory **36**, 1319–1333 (1990)
62. Ph Delsarte, Four fundamental parameters of a code and their combinatorial significance. Inf. Control **23**, 407–438 (1973)
63. Ph. Delsarte, An algebraic approach to the association schemes of coding theory. Philips Res. Rep. Suppl. **10** (1973)
64. R.H.F. Denniston, Some maximal arcs in finite projective planes. J. Comb. Theory **6**, 317–319 (1969)
65. C.F. Dunkl, Discrete quadrature and bounds on t-design. Mich. Math. J. **26**, 81–102 (1979)
66. E.N. Gilbert, F.J. MacWilliams, N.J.A. Sloane, Codes with detect deception. Bell Syst. Tech. J. **53**, 405–424 (1974)
67. M.J.E. Golay, Notes on digital coding. Proc. IRE **37**, 657 (1949)
68. R.W. Hamming, Error detecting and error correcting codes. Bell Syst. Tech. J. **29**, 147–160 (1950)
69. R. Hill, On the largest size cap in $S_{5,3}$. Rend. Acad. Naz. Lincei **54**(8), 378–384 (1973)

70. R. Hill, Caps and groups, in Atti dei Covegni Lincei. Colloquio Intern. sulle Theorie Combinatorie (Roma 1973), vol. 17 (Acad. Naz. Lincei) (1976), pp. 384–394
71. G.A. Kabatiansky, V.I. Levenshtein, Bounds for packings on a sphere and in space. Probl. Inf. Transm. **14**(1), 1–17 (1978)
72. M. Krawtchouk, Sur une géneralisation des polynómes d'Hermite. Compt. Rend. **189**, 620–622 (1929)
73. C.W.M. Lam, V. Pless, There is no (24, 12, 10) self-dual quaternary codes. IEEE Trans. Inf. Theory **36**, 1153–1156 (1990)
74. V.I. Levenshtein, On choosing polynomials to obtain bounds in packing problem, in *Proceedings of the 7th All-Union Conference on Coding Theory and Information Transmission, pt.2, Moscow-Vilnus, USSR* (1978), pp. 103–108
75. V.I. Levenshtein, Bounds on the maximal cardinality of a code with bounded modulus of the inner product. Sov. Math.-Dokl. **25**(2), 526–531 (1982)
76. V.I. Levenshtein, Bounds for packings of metric spaces and some their applications. Problemy Cybernetiki 40, 43–110, *(Moscow* (USSR, Nauke), 1983)
77. V.I. Levenshtein, Designs as maximum codes in polynomial metric spaces. Act. Applicandae Mathematicae **29**, 1–82 (1992)
78. V.I. Levenshtein, Bounds for self-complementary codes and their applications, *Eurocode-92*, vol. 339, CISM Courses and Lectures (Springer, Wien, 1993), pp. 159–171
79. V.I. Levenshtein, *Split orthogonal arrays and maximum resilient systems of functions* (Codes, and Cryptography, subm, Designs, 1994)
80. V.I. Levenshtein, Krawtchouk polynomials and universal bounds for codes and designs in Hamming spaces. IEEE Trans. Inf. Theory **41**(5), 1303–1321 (1995)
81. V.I. Levenshtein, Universal bounds for codes and designs, in *Handbook of Coding Theory*, ed. by V.S. Pless, W.C. Huffman (Elsevier Science, Amsterdam, 1998)
82. V.I. Levenshtein, Efficient reconstruction of sequences. IEEE Trans. Inf. Theory **47**(1), 2–22 (2001)
83. C.L. Mallows, N.J.A. Sloane, An upper bounds for self-dual codes. Inf. Control **22**, 188–200 (1973)
84. R.J. McEliece, E.R. Rodemich, H. Rumsey Jr., L.R. Welch, New upper bounds on the rate of a code via the Delsarte-MacWilliams inequalities. IEEE Trans. Inf. Theory **23**, 157–166 (1977)
85. A. Neumaier, Combinatorial configurations in terms of distances, Eindhoven University of Technology, Eindhoven, The Netherlands, Memo, 81-00 (Wiskunde) (1981)
86. V. Pless, *Introduction to the Theory of Error-Correcting Codes*, 2nd edn. (Wiley, New York, 1989)
87. B. Quist, Some remarks concerning curves of the second degree in finite plane. Ann. Acad. Fenn. Sci. Ser. A **134** (1952)
88. C.R. Rao, Factorial experiments derivable from combinatorial arrangement of arrays. J. R. Stat. Soc. **89**, 128–139 (1947)
89. I. Schoenberg, G. Szegø, An extremum problem for polynomials. Composito Math. **14**, 260–268 (1960)
90. N.V. Semakov, V.A. Zinoviev, Equidistant q-ary codes and resolved balanced incomplete designs. Probl. Inf. Transm. **4**(2), 1–7 (1968)
91. N.V. Semakov, V.A. Zinoviev, G.V. Zaitsev, Class of maximal equidistant codes. Probl. Inf. Transm. **5**(2), 65–69 (1969)
92. V.M. Sidelnikov, On mutual correlation of sequences. Sov. Math.-Dokl. **12**(1), 197–201 (1971)
93. V.M. Sidelnikov, On extremal polynomials used to estimate the size of code. Probl. Inf. Transm. **16**(3), 174–186 (1980)
94. R.C. Singleton, Maximum distance q-ary codes. IEEE Trans. Inf. Theory **10**, 116–118 (1964)
95. G. Szegø, *Orthogonal Polynomials*, vol. 23 (AMS Publications, Providence, 1979)
96. H.N. Ward, A bound for divisible codes. IEEE Trans. Inf. Theory **38**, 191–194 (1992)
97. L.R. Welch, Lower bounds on the maximum correlation of signals. IEEE Trans. Inf. Theory **20**, 397–399 (1974)

Chapter 6
Orthogonal Polynomials in Information Theory

The following lectures are based on the works [115–118, 120–125] of Tamm.

6.1 Introduction

6.1.1 Orthogonal Polynomials

Let $(t_j(x))_{j=0,1,2...}$ be a sequence of polynomials, where $t_j(x)$ is of degree j for all j. These polynomials are orthogonal with respect to some linear operator T if

$$T(t_j(x) \cdot t_m(x)) = 0 \qquad \text{for all } j \neq m$$

Usually, the linear operator is an integral. If the polynomials are orthogonal with respect to a weighted sum, we speak of discrete orthogonal polynomials. For instance, for the Krawtchouk polynomials it is $\sum_{i=0}^{n} \binom{n}{i}(q-1)^i K_j(i) K_m(i) = q^n(q-1)^j \binom{n}{j} \delta_{jm}$ with $\delta_{jm} = 1$ if $j = m$ and 0 else. In the proof of the refined alternating sign matrix conjecture, Zeilberger used a discrete integral describing the orthogonality relation of a discrete version of the Legendre polynomials.

An important property of orthogonal polynomials is that they obey a *three – term recurrence*, i. e.

$$t_j(x) = (\xi_j x - \alpha_j) \cdot t_{j-1}(x) - \beta_{j-1} \cdot t_{j-2}(x), \quad t_0(x) = 1, \quad , t_1(x) = \xi_1 x - \alpha_1$$

with real coefficients ξ_j, α_j, β_j for all $j \geq 1$. The polynomials are orthogonal exactly if $\xi_j \xi_{j-1} \beta_j > 0$ for all $j \geq 1$.

A. Ahlswede et al. (eds.), *Combinatorial Methods and Models*,
Foundations in Signal Processing, Communications and Networking 13,
DOI 10.1007/978-3-319-53139-7_6

Examples of orthogonal polynomials are Chebyshev polynomials of the first and second kind, Hermite polynomials, Jacobi polynomials, Laguerre polynomials, Legendre polynomials, etc.

Most of these polynomials had been known before, but a general theory of orthogonal polynomials is usually attributed to the work of Chebyshev, Markov, and Stieltjes, who worked out the close connection between orthogonal polynomials and continued fractions.

This relation will be sketched now. Let

$$F(x) = c_0 + c_1 x + c_2 x^2 + \dots$$

be the (formal) power series expression of a function $F(x)$. We denote by

$$d_n^{(k)} = \det(A_n^{(k)})$$

the determinant of a Hankel matrix $A_n^{(k)}$ of size n with the consecutive coefficients $c_m, m = k, \dots, k+2n-2$ as above.

If all determinants $d_n^{(0)}$ and $d_n^{(1)}$ are different from 0 the series $F(x)$ can be expressed as the continued fraction

$$F(x) = \cfrac{c_0}{1 - \cfrac{q_1 x}{1 - \cfrac{e_1 x}{1 - \cfrac{q_2 x}{1 - \cfrac{e_2 x}{1 - \dots}}}}}.$$

whose coefficients can be expressed in terms of Hankel determinants, namely

$$q_n = \frac{d_n^{(1)} \cdot d_{n-1}^{(0)}}{d_{n-1}^{(1)} \cdot d_n^{(0)}}, \qquad e_n = \frac{d_{n+1}^{(0)} \cdot d_{n-1}^{(1)}}{d_n^{(0)} \cdot d_n^{(1)}}.$$

The above S – fraction can be transformed to the J – fraction of the function $\frac{1}{x}F(\frac{1}{x})$

$$\cfrac{c_0}{x - \alpha_1 - \cfrac{\beta_1}{x - \alpha_2 - \cfrac{\beta_2}{x - \alpha_3 - \cfrac{\beta_3}{x - \alpha_4 - \dots}}}}$$

where

$$\alpha_1 = q_1, \quad , \beta_j = q_j e_j, \quad \alpha_{j+1} = q_{j+1} + e_j \text{ for } j \geq 1$$

Now the j-th convergent $\frac{p_j(x)}{t_j(x)}$ to $\frac{1}{x}F(\frac{1}{x})$ is obtained by polynomials $p_j(x)$ and $t_j(x)$ defined by the three – term recurrence

$$p_0(x) = 0, \quad p_1(x) = c_0, p_j(x) = (x - \alpha_j)p_{j-1}(x) - \beta_{j-1}p_{j-2}(x) \text{ for } j \geq 2,$$

$$t_0(x) = 1, \quad t_1(x) = x - \alpha_1, t_j(x) = (x - \alpha_j)t_{j-1}(x) - \beta_{j-1}t_{j-2}(x) \text{ for } j \geq 2.$$

Chebyshev applied the methods he derived on continued fractions to the problem of finding an approximation to a function u given its values $u_1, u_2, \dots, u_n$ at positions $x_1, x_2, \dots, x_n$ by the method of least squares. One forms the successive sums (for $m = 0, 1, 2, \dots$)

$$K_0 t_0(x) + \cdots + K_m t_m(x)$$

until the quality criterion via least squares is achieved, where $K_0, K_1, K_2, \dots$ are constant coefficients and the polynomials $t_j(x)$ are just the denominators of the convergents to the function $\sum_{i=1}^{n} \frac{1}{x - x_i}$.

Markov and Stieltjes considered moment problems, for instance, Stieltjes asked for a given infinite sequence $c_0, c_1, c_2, \dots$ to find a measure μ on $[0, \infty)$ such that $c_l = \int_0^\infty x^l d\mu(x)$ for all $l = 0, 1, 2, \dots$. He could show that if the Hankel determinants $\det(A_n^{(0)})$ and $\det(A_n^{(1)})$ are both greater than 0 then there exists a solution to this Stieltjes' moment problem. Continued fractions come in by the formal expansion $\int \frac{d\mu(t)}{x+t} = \sum_{l=0}^{\infty}(-1)^l \frac{c_l}{x^{l+1}}$. Further moment problems had been studied by Hamburger, Nevanlinna, Hausdorff, et al. .

For a thorough treatment of the topic and further applications and results on orthogonal polynomials see e.g. the standard textbooks by Perron or Wall on continued fractions and by Chihara, Freud, or Szegö on orthogonal polynomials.

Orthogonal polynomials are an important tool in Algebraic Combinatorics. We concentrate here mainly on their applications in Information Theory. Delsarte recognized the importance of association schemes – formerly studied as a tool in the design of experiments in Statistics – in Coding Theory. The eigenvalues of the matrices in the association scheme form a family of discrete orthogonal polynomials. Especially for the Hamming association scheme the Krawtchouk polynomials arise. Their analysis allowed, for instance, to obtain the best known asymptotic upper bounds on the code size due to McEliece, Rodemich, Rumsey, and Welch. Further, Zinoviev/Leontiev and Tietäväinen could characterize all parameters for which perfect codes in the Hamming metric over an alphabet of size being a prime power exist, exploiting the fact that all zeros of a so-called Lloyd polynomial, which for the Hamming distance is a special Krawtchouk polynomial, must be integers in order to guarantee the existence of a perfect code.

6.2 Splittings of Cyclic Groups and Perfect Shift Codes

6.2.1 Introduction

In algebraic and combinatorial coding, the errors are usually such that single components are distorted by adding the noise e_i to the original value x_i, i.e. the received i-th component is $x_i + e_i \mod q$ when x_i had been sent.

A code should be able to correct all errors within a tolerated minimum distance d, which means that the decoder decides in favor of a message M if the error vector $(e_1, \ldots, e_n)$ is within a distance less or equal to $\lfloor \frac{d-1}{2} \rfloor$ to the codeword $(x_1, \ldots, x_n)$ corresponding to M.

The *distance function*, like Hamming distance or Lee distance, is usually of *sum-type*, i.e., the distance $d((x_1, \ldots, x_n), (y_1, \ldots, y_n))$ is the sum $\sum_{i=1}^{n} d(x_i, y_i)$ of the componentwise distances.

In this case a code can be regarded as a packing of the space $\{0, \ldots, q-1\}^n$ with spheres of the same type (just the error spheres around the codewords, which should not overlap in order to be able to uniquely conclude to the correct message). If the packing is as well a covering of the space, i.e., each possible word in $\{0, \ldots, q-1\}^n$ is in a sphere around exactly one codeword, the code is said to be perfect. A *perfect code* hence corresponds to a *tiling* (or partition) of the space.

Further error types are, for instance, deletion or insertion of components or permutations of several components. When timing or synchronization problems arise, as in coding for digital storage media as CD's or harddiscs, usually *run-length limited* sequences are used as codewords, i.e., the number of 0's (a run) between two consecutive 1's is limited to be between a minimal value d and a maximal value k. The errors to be corrected here are *peak-shifts*, i.e., a 1 (or peak), which is originally in position i, is shifted by t positions and can hence be found in position $i - t$ or $i + t$ in the received word.

Such distance measures are usually hard to analyze. However, for single errors, combinatorial methods exists as shown in Sect. 5.3. If the codes turn out to be perfect algebra comes into play.

Splitting of Groups and Perfect Shift Codes

Let $(\mathcal{G}, +)$ be an additive Abelian group. For any element $g \in \mathcal{G}$ and any positive integer m we define $m \cdot g = \underbrace{g + \cdots + g}_{m}$ and for a negative integer m it is $m \cdot g = -((-m) \cdot g)$. A *splitting* of an additive Abelian group $\mathcal{G}$ is a pair $(\mathcal{M}, \mathcal{S})$, where $\mathcal{M}$ is a set of integers and $\mathcal{S}$ is a subset of the group $\mathcal{G}$ such that every nonzero element $g \in \mathcal{G}$ can be uniquely written as $m \cdot h$ for some $m \in \mathcal{M}$ and $h \in \mathcal{S}$. It is also said that $\mathcal{M}$ *splits* $\mathcal{G}$ with *splitting set* $\mathcal{S}$. The notation here is taken from [105] which may also serve as an excellent survey on splittings of groups, see also [61, 103, 108].

In [68] Levenshtein and Vinck investigated perfect run-length limited codes which are capable of correcting single peak shifts. As a basic combinatorial tool for the construction of such codes they introduced the concept of a *k-shift code*, which is defined to be a subset $\mathcal{H}$ of a finite additive Abelian group $\mathcal{G}$, with the property that for any $m = 1, \dots, k$ and any $h \in \mathcal{H}$ all elements $\pm m \cdot h$ are different and not equal to zero. Such a code is said to be *perfect* if for every nonzero element $g \in \mathcal{G}$ there are exactly one $h \in \mathcal{H}$ and $m \in \{1, \dots, k\}$ such that $g = m \cdot h$ or $g = -m \cdot h$. Hence a perfect shift code corresponds to a splitting of a group $\mathcal{G}$ by the set

$$F(k) = \{\pm 1, \pm 2, \dots, \pm k\}.$$

Levenshtein and Vinck [68] also gave explicit constructions of perfect shift codes for the special values $k = 1, 2$ and for the case $k = \frac{p-1}{2}$, where p is a prime number. Later Munemasa [79] gave necessary and sufficient conditions for the existence of perfect shift codes for the parameters $k = 3$ and $k = 4$. Munemasa also introduced the notion shift code (originally in [68] it was called shift design). One may think of the elements $h \in \mathcal{H}$ as codewords and the set $\{\pm h, \pm 2h, \dots, \pm kh\}$ as the sphere around the codeword h. Implicitly, this code concept is already contained in [45]. Here Golomb refers to Stein's paper [102], in which had been introduced splittings by $F(k)$ and by the set

$$S(k) = \{1, 2, \dots, k\}.$$

Tilings by the Cross and Perfect Lee Codes

Originally, in [102] splittings by the sets $F(k)$ and $S(k)$ had been introduced to study the equivalent geometric problem of tiling $\mathbb{R}^n$ by certain star bodies, the cross and the semicross (see Sect. 6.2.4).

A (k, n)-*semicross* is a translate of the cluster consisting of the $kn + 1$ unit cubes in $\mathbb{R}^n$ with edges parallel to the coordinate axes and with centers

$$(0, 0, \dots, 0), (j, 0, \dots, 0), (0, j, \dots, 0), \dots, (0, 0, \dots, j)$$

$j = 1, 2 \dots, k$. Accordingly, a (k, n)-*cross* (or *full cross*) is a translate of the cluster consisting of the $2kn + 1$ n-dimensional unit cubes with centers (for $j = 1, 2 \dots, k$)

$$(0, 0, \dots, 0), (\pm j, 0, \dots, 0), (0, \pm j, \dots, 0), \dots, (0, 0, \dots, \pm j).$$

These star bodies correspond to the error spheres discussed in [45, 104] as Stein sphere and Stein corner, respectively. We shall also use the notion *sphere* around h for the set $h \cdot \mathcal{M}, h \in \mathcal{S}$ for any splitting $(\mathcal{M}, \mathcal{S})$.

It turned out in [102] that a lattice tiling of the Euclidean space $\mathbb{R}^n$ by the (k, n)-cross exists exactly if $F(k)$ splits some Abelian group of order $2kn + 1$. This result hence links the two code concepts introduced which at first glance do not seem to be too closely related.

The *Stein sphere* induced by the cross arises if an error in the transmission of a word $(x_1, \dots, x_n)$ results in the distortion of a single component x_i such that after the transmission the received letter in component i is from the set $x_i + F(k)$. For $k = 1$ this is just the pattern caused by a single error in the Lee metric. Golomb and Welch [48] demonstrated that a tiling by the $(1, n)$-cross exists in $\mathbb{R}^n$ for all

n. They used a Varshamov/Tenengol'ts construction [127] from which they derived perfect single-error correcting codes in the Lee metric. Martirossian [71] studied the case $k = 2$ and gave also results for further errors spheres, e.g. the *Stein corner*, which arises, if after transmission the letter in component i is from the set $x_i + S(k)$ (see Sect. 6.2.5).

The Main Result

We shall investigate splittings of cyclic groups $\mathbb{Z}_p$ of prime order p by sets of the form

$$\mathcal{M}_1 = \{1, a, a^2, \ldots, a^r, b, b^2, \ldots, b^s\}, \tag{1}$$

and

$$\mathcal{M}_2 = \{\pm 1, \pm a, \pm a^2, \ldots, \pm a^r, \pm b, \pm b^2, \ldots, \pm b^s\},$$

where $r \geq 1$ and $s \geq 1$ are nonnegative integers.

Observe that for the special choice $a = 2, b = 3$ and $r = 1, s = 1$ (or $r = 2, s = 1$, respectively) the sets $S(3)$ and $F(3)$ (or $S(4)$ and $F(4)$) arise. It was shown by Galovich and Stein [40] and by Munemasa [79], respectively, that the analysis of splittings of finite Abelian groups by $S(k)$ and $F(k)$, respectively, for the parameters $k = 3, 4$ can essentially be reduced to the analysis of splittings by the same sets in cyclic groups $\mathbb{Z}_p$ of prime order (the exact results are presented in Sect. 6.2.4).

In this case a splitting of the additive group $(\mathbb{Z}_p, +)$ corresponds to a *factorization* of the multiplicative group $(\mathbb{Z}_p^* = \mathbb{Z}_p \setminus \{0\}, \cdot)$. A factorization of a (multiplicative) group $\mathcal{G}$ is a representation $\mathcal{G} = \mathcal{A} \cdot \mathcal{B}$, where $\mathcal{A}$ and $\mathcal{B}$ are subsets of $\mathcal{G}$ such that every element $g \in \mathcal{G}$ can be uniquely written as a product $a \cdot b$ for some $a \in \mathcal{A}$ and $b \in \mathcal{B}$. So for p prime a splitting $(\mathcal{M}, \mathcal{S})$ in $\mathbb{Z}_p$ yields the factorization $\mathbb{Z}_p^* = \mathcal{M} \cdot \mathcal{S}$ and vice versa, since now $\mathcal{M}$ is also a subset of $\mathbb{Z}_p^*$.

Further, the splittings $(\mathcal{M}_2, \mathcal{S})$ of the cyclic group $\mathbb{Z}_p$ by the set $\mathcal{M}_2$ correspond to factorizations of the group $\mathbb{Z}_p^*/\{1, -1\} = \mathcal{M}_1 \cdot \mathcal{S}$. Since only one of the elements g and $-g$ can be contained in the splitting set $\mathcal{S}$ and since both elements must be simultaneously contained in the same sphere around an element of the splitting set $\mathcal{S}$, we can identify them by considering $\mathbb{Z}_p^*$ modulo the units $\{1, -1\}$. On the other hand, every factorization $\mathbb{Z}_p^*/\{1, -1\} = \mathcal{M}_1 \cdot \mathcal{S}$ yields a factorization $\mathbb{Z}_p^* = \mathcal{M}_2 \cdot \mathcal{S}'$ with the elements of the factor S' in $\mathbb{Z}_p^*$ regarded as the coset representatives for the cosets belonging to S in $\mathbb{Z}_p^*/\{1, -1\}$. Hence, in the search for splittings of $\mathbb{Z}_p$ by the sets $\mathcal{M}_1$ and $\mathcal{M}_2$ we can concentrate on finding factorizations $\mathcal{G} = \mathcal{M}_1 \cdot \mathcal{S}$, where $\mathcal{G} = \mathbb{Z}_p^*$ or $\mathcal{G} = \mathbb{Z}_p^*/\{1, -1\}$, respectively.

In order to characterize the structure of a splitting set $\mathcal{S}$ for splittings by $\mathcal{M}_1$ or $\mathcal{M}_2$ we first need some further notation. As usual, we denote by

$$< a, b >= \{a^i \cdot b^j, i = 1, \ldots, \text{ord}(a), j = 1, \ldots, \text{ord}(b)\} \tag{2}$$

the subgroup of $\mathcal{G} = \mathbb{Z}_p^*$ or $\mathcal{G} = \mathbb{Z}_p^*/\{1, -1\}$, respectively, generated by the elements a and b, whose orders are $\text{ord}(a)$ and $\text{ord}(b)$. Furthermore let

$$\mathcal{F} = \{a^i \cdot b^j, i - j \equiv 0 \bmod (r + s + 1)\}. \tag{3}$$

Observe that $\mathcal{F}$ is the subgroup in $\mathcal{G}$ generated by the elements a^{r+s+1}, b^{r+s+1}, and ab. It can be shown that a splitting of $\mathbb{Z}_p$ by the set $\mathcal{M}_1$ (by the set $\mathcal{M}_2$) exists exactly if for each element $f \in \mathcal{F}$, where $\mathcal{F}$ is a subgroup of $\mathcal{G} = \mathbb{Z}_p^*$ (a subgroup of $\mathcal{G} = \mathbb{Z}_p^*/\{1, -1\}$ for splittings by $\mathcal{M}_2$), all possible representations $f = a^i \cdot b^j$ as a product of powers of a and b are such that $i - j \equiv 0 \bmod (r + s + 1)$. Necessary and sufficient conditions are given in the following theorem.

Theorem 6.1 *Let $\mathcal{M}_1 = \{1, a, a^2, \dots, a^r, b, b^2, \dots, b^s\}$. A factorization $\mathcal{G} = \mathcal{M}_1 \cdot \mathcal{S}$ of the group $\mathcal{G} = \mathbb{Z}_p^*$ ($\mathcal{G} = \mathbb{Z}_p^*/\{1, -1\}$) by the set $\mathcal{M}_1$ and hence a splitting of $\mathbb{Z}_p$ by $\mathcal{M}_1 (\mathcal{M}_2 = \{\pm 1, \pm a, \dots, \pm a^r, \pm b, \dots, \pm b^s\})$ exists if and only if the splitting set $\mathcal{S}$ is of the form*

$$\mathcal{S} = x_0 \cdot \mathcal{F} \dot{\cup} x_1 \cdot \mathcal{F} \dot{\cup} \dots \dot{\cup} x_{\rho-1} \cdot \mathcal{F}, \tag{4}$$

where ρ is the number of cosets of the subgroup $< a, b >$ of $\mathcal{G}$ and the $x_i, i = 0, \dots, \rho - 1$ are representatives of each of these cosets and where

$$ord(a) \text{ and } ord(b) \text{ are divisible by } r + s + 1. \tag{5}$$

Further, if $b^{l_1} = a^{l_2}$ for some integers l_1 and l_2, then

$$l_1 + l_2 \equiv 0 \; mod \; (r + s + 1). \tag{6}$$

Observe that condition (5) is a special case of (6). However, in the proof of Theorem 6.1, which will be carried out in Sect. 6.2.2, we shall first derive (5). Further (5) is also important in order to find splittings by computer research (see Sect. 6.2.3).

Further Results and Discussion

There have been derived necessary and sufficient conditions on the existence of splittings of $\mathbb{Z}_p$ (and derived from them also conditions for arbitrary finite Abelian groups) by $\mathcal{M}_1$ and $\mathcal{M}_2$ for the special cases $\mathcal{M} = \{1, a, b\}$ by Galovich and Stein [40], for $\mathcal{M} = \{1, 2, 3\}$ by Stein [106], and for $\mathcal{M} = \{\pm 1, \pm 2, \pm 3\}$ and $\mathcal{M} = \{\pm 1, \pm 2, \pm 3, \pm 4\}$ by Munemasa [79]. We shall discuss these conditions in Sect. 6.2.2 and compare them with those of Theorem 6.1.

Further in Sect. 6.2.2, as a consequence of Theorem 6.1, it will be derived that if $r + s$ is an even number, then the set of prime numbers p, for which a splitting of $\mathbb{Z}_p$ by $\mathcal{M}_1$ exists, is the same as the set of prime numbers p for which a splitting of $\mathbb{Z}_p$ by $\mathcal{M}_2$ exists. Especially, this holds for the sets $\mathcal{M}_1 = \{1, 2, 3\}$ and $\mathcal{M}_2 = \{\pm 1, \pm 2, \pm 3\}$, which answers a general respective question by Galovich and Stein [40] for this special case. Hence it is possible to treat splittings by $\{1, 2, 3\}$ and $\{\pm 1, \pm 2, \pm 3\}$ simultaneously, to show the equivalence of several of the above – mentioned conditions, and to apply results on splittings by $\{1, 2, 3\}$ in the analysis of perfect 3-shift codes (which are just splittings by $\{\pm 1, \pm 2, \pm 3\}$).

In principle, splittings by sets $\mathcal{M}_1$ and $\mathcal{M}_2$ are completely characterized by Theorem 6.1. However, in order to find such splittings, several conditions imposed on the orbit of the elements a and b by (5) and (6) have to be verified. This will be done in Sect. 6.2.3 for sets of the form $\mathcal{M} = \{1, a, b\}$ and $\mathcal{M} = \{1, a, a^2, b\}$, from which the perfect 3– and 4-shift codes can be obtained by the special choice of the parameters $a = 2$ and $b = 3$.

In Sect. 6.2.4 we shall discuss the relation between splittings by $F(k)$ and $S(k)$ and tiling of the Euclidean space $\mathbb{R}^n$ by certain star bodies, the cross and the semicross. Further, we shall present the above – mentioned results from [40, 79] on splittings of groups of composite order.

Finally, in Sect. 6.2.5, the application of shift codes in peak – shift correction of run-length limited codes is briefly discussed and further connections between splittings of groups and Coding Theory are pointed out.

6.2.2 Factorizations of $\mathbb{Z}_p^*$ and $\mathbb{Z}_p^*/\{1,-1\}$ with the Set $\{1, a, \ldots, a^r, b, \ldots, b^s\}$

We mentioned already that $\mathcal{F}$ is generated by the elements a^{r+s+1}, b^{r+s+1}, and ab in $\mathcal{G} = \mathbb{Z}_p^*$ or $\mathcal{G} = \mathbb{Z}_p^*/\{1,-1\}$. The next lemma shows that with $x \in \mathcal{S}$ these three elements multiplied by x necessarily are also contained in $\mathcal{S}$.

Lemma 6.1 *In a factorization $\mathcal{G} = \mathcal{M}_1 \cdot \mathcal{S}$ with every element $x \in \mathcal{S}$, also $(ab)x \in \mathcal{S}$ and $a^{r+s+1}x$ and $b^{r+s+1}x$ must be the next powers of a and b, respectively, which multiplied by x are contained in $\mathcal{S}$.*

Proof Let $x \in \mathcal{S}$ be an element of the splitting set $\mathcal{S}$. Then $(ab)x$ must also be contained in $\mathcal{S}$. If this would not be the case, then $(ab)x = a^k y$ for some $k \in \{1, \ldots, r\}$ and $y \in \mathcal{S}$ or $(ab)x = b^l y'$ for some $l \in \{1, \ldots, s\}$ and $y' \in \mathcal{S}$. If $(ab)x = a^k y$ then $x' = bx = a^{k-1}y$ and x' would have two different representations $mh, m'h'$ with $m, m' \in \mathcal{M}_1$ and $h, h' \in \mathcal{S}$, which contradicts the definition of a splitting. Analogously, if $(ab)x = b^l y'$ then $x' = ax = b^{l-1}y'$ would have two different representations, which is not possible.

Now let k^* be the minimum power k such that $a^k x \in \mathcal{S}$ for some $x \in \mathcal{S}$. We have to show that $k^* = r + s + 1$.

First observe that obviously $k^* \geq r + 1$, since otherwise $x' = a^{k^*} (= 1 \cdot x')$ would occur in two different ways as product of elements of $\mathcal{M}_1$ and $\mathcal{S}$.

In order to see that $k^* \notin \{r + 1, \ldots, r + s\}$, we shall prove by induction the stronger statement

$$\begin{aligned} &\text{If } x \in \mathcal{S} \text{ then for all } i = 1, \ldots, s + 1 \text{ it is } a^{r+i}x = b^{s+1-i}y_i \\ &\text{with } y_i = (ab)^{i-1}y \text{ for some } y \in \mathcal{S} \end{aligned} \tag{7}$$

The statement (7) holds for $i = 1$, since $a^{r+1}x = b^j y$ for some $y \in \mathcal{S}$ and $j \in \{0, \dots, s-1\}$ is not possible. Otherwise, $a^{r+1}bx = a^r(abx) = b^{j+1}y$ could be written in two different ways as product of members of the splitting set (abx and y) and elements of $\mathcal{M}_1$ (a^r and b^{j+1}).

So we proved that for $x \in \mathcal{S}$ it is $a^{r+1}x = b^s y$ for some $y \in \mathcal{S}$. Hence $a^{r+2}x = b^{s-1} \cdot aby = b^{s-1}y_2$ with $y_2 = aby \in \mathcal{S}$ and further $a^{r+3}x = b^{s-2}y_3$ with $y_3 = (ab)^2 y \in \mathcal{S}$, ..., $a^{r+s}x = by_s$ with $y_s = (ab)^{s-1}y \in \mathcal{S}$, and $a^{r+s+1}x = (ab)^s y \in \mathcal{S}$.

So $a^{r+s+1}x \in \mathcal{S}$ and $a^{r+i}x \notin \mathcal{S}$ for $i \in \{1, \dots, s\}$, since otherwise the element $a^{r+i+1}b^{i+1}x = a^r((ab)^{i+1}x) = b^s y_i$ could be represented in two different ways as a product of a member of the splitting set ($(ab)^{i+1}x$ or y_i) and an element of $\mathcal{M}_1$ (a^r or b^s).

Analogously, it can be shown that $r + s + 1$ is also the minimum power l^* such that $b^{l^*}x \in \mathcal{S}$ when $x \in \mathcal{S}$ (obviously $l^* > s$ and $l^* \notin \{s+1, \dots, s+r\}$ by an argument as (7)). □

Proof of Theorem 6.1. First we shall demonstrate that a set $\mathcal{S}$ with the properties (4), (5), and (6) from Theorem 6.1 is a splitting set as required. It suffices to show that every element $z = a^i b^j$ of the subgroup $< a, b >$ can uniquely be obtained from one element $a^{i'}b^{j'}$, $i' - j' \equiv 0 \bmod (r+s+1)$ in $\mathcal{F}$ by multiplication with a power of a or b from the set $\mathcal{M}_1 = \{1, a, a^2, \dots a^r, b, b^2, \dots b^s\}$.

To see this let $z = a^i b^j$ with $i - j \equiv k \bmod (r+s+1)$. If $k = 0$, then by (5) and (6) z must be contained in $\mathcal{S}$. If $k \in \{1, \dots, r\}$, then $z = a^k h = a^k(a^{i-k}b^j)$ is the only possibility to write z as a product of an element (namely a^k) from $\mathcal{M}_1$ and a member h of $\mathcal{S}$. Finally, if $k \in \{r+1, \dots, r+s\}$, then again by (5) and (6) $z = b^{r+s+1-k}h' = b^{r+s+1-k}(a^i b^{j-(r+s+1-k)})$ is the unique way of representing z as product of an element ($b^{r+s+1-k}$) of $\mathcal{M}_1$ and one (h') of $\mathcal{S}$.

In order to show that a splitting set $\mathcal{S}$ must have a structure as in (4), let now $x \in \mathcal{S}$. Lemma 6.1 then implies that all powers of a^{r+s+1}, b^{r+s+1} and ab and all combinations of them, i.e., all elements of the form

$$h = (a^{r+s+1})^{k_1}(b^{r+s+1})^{k_2}(ab)^{k_3} = a^{(r+s+1)\cdot k_1 + k_3}b^{(r+s+1)\cdot k_2 + k_3} \tag{8}$$

multiplied by x must also be contained in $\mathcal{S}$, which just yields that $hx \in \mathcal{S}$ for all

$$h = a^i \cdot b^j \text{ with } i - j \equiv 0 \bmod r + s + 1. \tag{9}$$

It is also clear that every element of this form (9) can occur as a combination (8). So all elements of $\mathcal{F}$ as defined under (3) multiplied by x must be contained in $\mathcal{S}$, if $x \in \mathcal{S}$.

Further observe that with $x \in \mathcal{S}$ an element of the form $a^i b^j x$ with $i - j$ not divisible by $r + s + 1$ cannot be contained in a splitting set $\mathcal{S}$, since in this case the unique representability would be violated (see above).

Since the elements of a proper coset $\mathcal{N}$ in $\mathcal{G}$ of $< a, b >$ cannot be obtained from elements of another such coset by multiplication with powers of a or b, for every such coset $\mathcal{N}$ we can choose a representative $x_{\mathcal{N}}$ and the elements $x_{\mathcal{N}} \cdot \mathcal{F}$ can be included in the splitting set in order to assure that every element from $\mathcal{N}$ can be uniquely written as a product $m \cdot h, m \in \mathcal{M}_1, h \in \mathcal{S}$.

The conditions (5) and (6) are necessary conditions on the orbits of a and b that must be fulfilled if a perfect shift code should exist in $\mathcal{G}$.

It is easy to verify that the elements a and b each must have an order divisible by $r+s+1$ in order to assure the existence of a factorization $\mathcal{G} = \mathcal{M}_1 \cdot \mathcal{S}, \mathcal{G} = \mathbb{Z}_p^*, \mathbb{Z}_p^*/\{1,-1\}$. Assume that this would not be the case, e.g.,

$$\text{ord}(a) = k_1 \cdot (r+s+1) + k_2, 0 < k_2 < r+s+1.$$

By Lemma 6.1 with $x \in \mathcal{S}$ all elements $(a^{r+s+1})^k x$ must also be contained in the set $\mathcal{S}$. However, for $k = k_1 + 1$ this yields

$$\mathcal{S} \ni (a^{r+s+1})^{k_1+1} x = a^{(r+s+1)k_1+k_2+((r+s+1)-k_2)} x = a^{r+s+1-k_2} x$$

which is not possible by Lemma 6.1.

Further, from Lemma 6.1 we can conclude that with $x \in \mathcal{S}$ also $(ab)^{l_1} x$ and $(ab)^{l_2} x$ for all l_1 and l_2 are contained in the splitting set $\mathcal{S}$. Now if $a^{l_1} = b^{l_2}$ then $(ab)^{l_1} = a^{l_1} b^{l_1} = b^{l_1+l_2}$, $(ab)^{l_2} = a^{l_2} b^{l_2} = a^{l_1+l_2}$ and with Lemma 6.1 and the preceding considerations $l_1 + l_2$ must be divisible by $r+s+1$. □

Remarks

1. Observe that (for a and b fulfilling (5) and (6)) all the sets $\mathcal{M}_1 = \{1, a, \dots, a^r, b, \dots, b^s\}$ ($\mathcal{M}_2 = \{\pm 1, \pm a, \dots, \pm a^r, \pm b, \dots, \pm b^s\}$) with the same sum $r+s$ yield the same splitting set.

2. Obviously, the group $\mathcal{F}$ fulfilling conditions (5) and (6) also is defined for $i - j \equiv 0 \mod 2$, although this case for our considerations is not of importance since then $r = 0$ or $s = 0$ and one of the parameters a or b will not occur.

It can be shown (following the considerations in [68] about 2-shift codes) that a splitting by $\{1, a, \dots, a^r\}$ ($\{\pm 1, \pm a, \dots, \pm a^r\}$) in $\mathbb{Z}_p$ exists exactly if the order of a in $\mathbb{Z}_p^*$ ($\mathbb{Z}_p^*/\{1,-1\}$) is divisible by $r+1$. This is clear, since with $x \in \mathcal{S}$ also $a^{r+1}x$ must be a member of the splitting set $\mathcal{S}$ in this case.

Galovich and Stein [40] already considered splittings of Abelian groups by the set $\{1, a, b\}$. They could derive necessary and sufficient conditions on the existence of splittings. Especially for $a = 2, b = 3$, they found for cyclic groups of prime order $p \equiv 1 \mod 3$:

(i) Let g be a generator of $< 2, 3 >$ and let $2 = g^u, 3 = g^v, d = \gcd(u, v), u' = \frac{u}{d}, v' = \frac{v}{d}$, and $d_1 = \gcd(d, p-1)$. Then $\{1, 2, 3\}$ splits $\mathbb{Z}_p$ if and only if 3 divides $\frac{p-1}{d_1}$ and $u' \cdot v' \equiv 2 \mod 3$.

Later Stein [106] obtained further necessary and sufficient conditions for splittings by $\{1, 2, 3\}$ using number theoretic methods involving Newton sums (see also Sect. 6.2.4).

(ii) The set $\{1, 2, 3\}$ splits $\mathbb{Z}_p$ if and only if for some positive integer u dividing $p-1$ it is $1^{(p-1)/3u} + 2^{(p-1)/3u} + 3^{(p-1)/3u} \equiv 0 \bmod p$ and $(1^{(p-1)/3u})^2 + (2^{(p-1)/3u})^2 + (3^{(p-1)/3u})^2 \equiv 0 \bmod p$.

Munemasa considered splittings by $\{\pm 1, \pm 2, \pm 3\}$ and by $\{\pm 1, \pm 2, \pm 3, \pm 4\}$. He found necessary and sufficient conditions based on the behaviour of the subgroups $< -1, 2, 3 >$ and $< -1, 6 >$ of $\mathbb{Z}_p^*$ generated by the elements $-1, 2, 3$ and $-1, 6$ respectively.

(iii) A splitting of $\mathbb{Z}_p$, where p is a prime number, by $\{\pm 1, \pm 2, \pm 3\}$ exists if and only if
$| < -1, 2, 3 >:< -1, 6 > | \equiv 0 \bmod 3$.
(iv) A splitting of $\mathbb{Z}_p$, where p is a prime number, by $\{\pm 1, \pm 2, \pm 3, \pm 4\}$ exists if and only if $| < -1, 2, 3 >:< -1, 6 > | \equiv 0 \bmod 4$.

Whereas in order to check condition (i) one has to find a generator of the subgroup $< 2, 3 >$, the approach in [79] is very similar to the one in this paper.

Munemasa [79] proved (for $a = 2, b = 3, r = 1, 2, s = 1$) that with $a, b \in \mathcal{S}$ also the elements ab and a^{r+s+1} must be contained in the splitting set $\mathcal{S}$. His proof then follows a different line compared to the proof of Theorem 6.1, where it is used next that also b^{r+s+1} must be a member of $\mathcal{S}$. Theorem 6.1 describes the structure of the splitting set such that only the conditions (5) and (6) have to be checked, which can be done faster than checking (iii) or (iv). However, the aim in [40, 79] was to characterize splittings of arbitrary Abelian groups and not only of $\mathbb{Z}_p$. In order to do so, a general approach does not work. Such conditions have to be verified for each set $\mathcal{M}$ individually, if the group orders are composite numbers (see Sect. 6.2.4).

Further, Saidi [93] obtained necessary conditions on the existence of splittings by $\{1, 2, 3\}$ and by $\{\pm 1, \pm 2, \pm 3\}$ based on cubic residues.

(v) Assume that 2 and 3 are cubic nonresidues mod p and $L \equiv \pm M \bmod 12$, where $4p = L^2 + 27M^2$, $L \equiv 1 \bmod 3$. Then $\{1, 2, 3\}$ splits $\mathbb{Z}_p$.

The primes fulfilling the above conditions are also characterized in [93], they are of the form $p = 7M^2 \pm 6MN + 36N^2$, where $L = M \pm 12N$. Since any primitive quadratic form represents infinitely many primes [131], Saidi further concludes that the set $\{1, 2, 3\}$ splits $\mathbb{Z}_p$ for infinitely many primes p. It can be shown that this last condition (v) is not sufficient, since there is a splitting in $\mathbb{Z}_p$ for $p = 919$ but 919 is not of the form as required in (v) (cf. also Sect. 6.2.3). In [93] similar conditions (with $L \equiv \pm M \bmod 24$) as in (v) are derived for the set $\{\pm 1, \pm 2, \pm 3\}$, from which by the same argumentation as above follows that there are infinitely many primes p for which a splitting of $\mathbb{Z}_p$ by $\{\pm 1, \pm 2, \pm 3\}$ and hence a perfect 3-shift code exists.

Observe, that in [93] splittings by the sets $\{1, 2, 3\}$ and $\{\pm 1, \pm 2, \pm 3\}$ are treated separately. As a consequence of Theorem 6.1, we shall now show that they can be analyzed simultaneously. This follows from a more general result.

Theorem 6.2 *Let r, s be positive integers such that $r + s$ is an even number. Then a splitting of the group $\mathbb{Z}_p$, p prime, by the set $\mathcal{M}_1 = \{1, a, \dots, a^r, b, \dots, b^s\}$ exists if and only if there also is a splitting of $\mathbb{Z}_p$ by the set $\mathcal{M}_2 = \{\pm 1, \pm a, \dots, \pm a^r, \pm b, \dots, \pm b^s\}$*

Proof It is easy to see that from every splitting $(\mathcal{M}_2, \mathcal{S})$ of $\mathbb{Z}_p$ by the set $\mathcal{M}_2$ we obtain a splitting $(\mathcal{M}_1, \mathcal{S} \dot{\cup} -\mathcal{S})$ of $\mathbb{Z}_p$ by the set $\mathcal{M}_1$. This holds because for every $h \in \mathcal{S}$ by definition of a splitting the sphere $\{\pm h, \pm ha, \dots, \pm ha^r, \pm hb, \dots, \pm hb^s\} = \{h, ha, \dots, ha^r, hb, \dots, hb^s\} \dot{\cup} \{-h, -ha, \dots, -ha^r, -hb, \dots, -hb^s\}$ and hence h and $-h$ can be chosen as members of the splitting set in a splitting of $\mathbb{Z}_p$ by $\mathcal{M}_1$.

In order to show the converse direction, we shall prove that whenever there exists a splitting of $\mathbb{Z}_p$ by $\mathcal{M}_1$, it is also possible to find a splitting $(\mathcal{M}_1, \mathcal{S})$ of $\mathbb{Z}_p$ such that with every element $h \in \mathcal{S}$ also its additive inverse $-h$ is contained in the splitting set. in this case the two spheres $\{h, ha, \dots, ha^r, hb, \dots, hb^s\}$ and $\{-h, -ha, \dots, -ha^r, -hb, \dots, -hb^s\}$ are disjoint by definition of a splitting such that their union $\{\pm h, \pm ha, \dots, \pm ha^r, \pm hb, \dots, \pm hb^s\}$ is a sphere in a splitting by $\mathcal{M}_2$.

By Theorem 6.1 the splitting set $\mathcal{S}$ is essentially determined by the subgroup $\mathcal{F} = \{a^i b^j : i - j \equiv 0 \bmod (r + s + 1) \subset < a, b >$ (with the conditions (5) and (6) fulfilled). Obviously, the element $1 \in \mathcal{F}$. Now there are two possible structures of $\mathcal{F}$ depending on the behaviour of -1:

Case 1: $-1 \in < a, b >$. Then also $-1 \in \mathcal{F}$, since otherwise $-1 = a^i b^j$ for some $i - j \not\equiv 0 \bmod (r + s + 1)$ and hence $1 = (-1)^2 = a^{2i} b^{2j} \notin \mathcal{F}$ (since $2(i - j) \not\equiv 0 \bmod (r + s + 1)$) which is not possible. Since $\mathcal{F}$ is a group, with every element $h \in \mathcal{F}$ hence also $-h \in \mathcal{F}$. Also, for every coset $x_i < a, b >, i = 0, \dots, \rho - 1$, obviously, with $x_i h, h \in \mathcal{F}$ also $-x_i h$ must be contained in the splitting set $\mathcal{S}$.

Case 2: $-1 \notin < a, b >$. Then -1 must be contained in the coset $- < a, b >$ and one can choose $x_1 = -x_0$ as representative of this coset and include $x_1 \mathcal{F} = -x_0 \mathcal{F}$ in the splitting set $\mathcal{S}$, if x_0 is the representative from $< a, b >$ such that $x_0 \mathcal{F} \in \mathcal{S}$.

From the next coset (if there are still some cosets left not used so far for the splitting set) we include some representative x_2 and hence also $x_2 \mathcal{F}$ into the splitting set. Now the element $-x_2$ cannot be contained in any coset from which already elements are included into the splitting set so far. Obviously $-x_2$ is not contained in $x_2 < a, b >$ (since $-1 \notin < a, b >$) and if it were contained in $< a, b >$, then x_2 would be an element of $- < a, b >$ and vice versa, which is not possible by construction.

In the same way we can continue to include pairs $h, -h$ from the cosets of $< a, b >$ not used so far and with them the sets $h\mathcal{F}$ and $-h\mathcal{F}$ into the splitting set $\mathcal{S}$ until there is no further coset left. □

Remarks

1. For $r + s$ odd a similar result does not hold, since then it is possible that $-1 \in < a, b >$ but $-1 \notin \mathcal{F}$, since now $2(i - j)$ may be divisible by $r + s + 1$ although $i - j \not\equiv 0 \bmod (r + s + 1)$. For instance, there exist splittings of $\mathbb{Z}_p$ by $\mathcal{M}_1 = \{1, 2, 3, 4\}$ for $p = 409, 1201, 2617, 3433$, but there do not exist splittings by $\{\pm 1, \pm 2, \pm 3, \pm 4\}$ in the same groups.

2. For $r + s$ even there are two possible structures for a splitting set $\mathcal{S}$. Either $-1 \in < a, b >$, then automatically $-1 \in \mathcal{F}$ and hence with every element $h \in \mathcal{S}$ also $-h$ is forced to be in the splitting set $\mathcal{S}$. If $-1 \notin < a, b >$, then there also exist splittings for which there are elements h in the splitting set $\mathcal{S}$ such that $-h \notin \mathcal{S}$ – depending on the choice of the representatives of the cosets.

By the special choice of the parameters $a = 2, b = 3, r = s = 1$ the following corollary is immediate.

Corollary 6.1 *A splitting of the group $\mathbb{Z}_p$, p prime, by the set $\{1, 2, 3\}$ exists if and only if there also is a splitting of $\mathbb{Z}_p$ by the set $\{\pm 1, \pm 2, \pm 3\}$.*

Obviously, by the first argument in the proof of Theorem 6.2 (derive a splitting $(\mathcal{S} \dot{\cup} -\mathcal{S}, \mathcal{M}_1)$ from a splitting $(\mathcal{S}, \mathcal{M}_2)$) it holds that for every positive integer k the group $\mathbb{Z}_p$ is split by $S(k) = \{1, \dots, k\}$ if it is split by $F(k) = \{\pm 1, \dots, \pm k\}$. In [40] it is asked for which parameters k the converse holds (for arbitrary finite Abelian groups). Hickerson ([61], p. 168) demonstrated by the example $p = 281$ that the converse does not hold for $k = 2$. Corollary 6.1 now demonstrates that it holds for $k = 3$ and with Remark 1 it is clear that the converse does not hold for $k = 4$. However, for arbitrary Abelian groups splittings by $\{1, 2, 3\}$ and $\{\pm 1, \pm 2, \pm 3\}$ are not equivalent, since there is the trivial splitting $(\{1, 2, 3\}, \{1\})$ in $\mathbb{Z}_4$ and obviously in $\mathbb{Z}_4$ a splitting by $\{\pm 1, \pm 2, \pm 3\}$ does not exist. In Sect. 6.2.4 we shall see that this is essentially the only exception. Further from Corollary 6.1, it is immediate that the conditions (i), (ii), and also (iii) now characterize splittings of $\mathbb{Z}_p$ by $\{1, 2, 3\}$ as well as by $\{\pm 1, \pm 2, \pm 3\}$.

6.2.3 Computational Results on Splittings and Perfect 3- and 4-Shift Codes

Theorem 6.1 in principle completely characterizes splittings of $\mathbb{Z}_p$ by $\mathcal{M}_1 = \{1, a, \dots, a^r, b, \dots, b^s\}$ and $\mathcal{M}_2 = \{\pm 1, \pm a, \dots, \pm a^r, \pm b, \dots, \pm b^s\}$. The conditions on the existence of such splittings, however, have to be verified for each $\mathbb{Z}_p^*$ or $\mathbb{Z}_p^*/\{1, -1\}$ individually. Especially, condition (6) requires that many products $a^i b^j$ have to be calculated. Often it is enough to check the orbits of the elements a and b, since a splitting cannot exist if (5) is violated. The complexity can also be reduced when a generator of the subgroup $< a, b >$ is known (cf. the Galovich/Stein condition (i) from the previous section). A good candidate is the element ba^{-1}, since if a is contained on its orbit, so is $b = a(ba^{-1})$ and hence all products $a^i b^j$.

Corollary 6.2 *If a is on the orbit of ba^{-1} in $\mathcal{G} = \mathbb{Z}_p^*$ or $\mathcal{G} = \mathbb{Z}_p^*/\{1, -1\}$, i.e., $a = (ba^{-1})^l$ for some l, then (with $\mathcal{F}$ fulfilling the conditions (5) and (6))*

(i) $< a, b > = \{(ba^{-1})^m : m = 0, \dots, ord(ba^{-1}) - 1\}$ *is generated by the element* ba^{-1},

(ii) *For* $r = s = 1$, *i.e.*, $r + s + 1 = 3$, $\mathcal{F}$ *is generated by* $(ba^{-1})^3$, *hence*

$$\mathcal{F} = \{(ba^{-1})^{3m} : m = 0, \ldots, \frac{1}{3} ord(ba^{-1}) - 1\},$$

(iii) *A factorization* $\mathcal{G} = \{1, a, b\} \cdot \mathcal{F}$ *exists, exactly if*

$$l \equiv 1 \; mod \; 3.$$

Proof (i) is clear from the preceding discussion. In order to prove (ii) and (iii), observe that obviously by (6) the order of ba^{-1} must be divisible by 3. Hence, $(ba^{-1})^m = b^m a^{-m} \in \mathcal{F}$ exactly if m is divisible by 3. Further observe that $\mathcal{F}$ is generated by $a^3 = (ba^{-1})^{3l} \in \mathcal{F}, b^3 = (ba^{-1})^{3(l+1)} \in \mathcal{F}$ and $ab = (ba^{-1})^{2l+1}$ which is contained in $\mathcal{F}$ exactly if $l \equiv 1$ mod 3. □

For $\mathcal{G} = \mathbb{Z}_p^*$ the results in Corollary 6.2 already can be derived from the considerations in [40]. Following the same line of proof, Corollary 6.2 can be extended to the case $r + s + 1$ odd, where then (iii) reads $l \equiv \frac{r+s}{2}$ mod $(r + s + 1)$.

With the set $\mathcal{M}_1 = \{1, a, b\}$ the perfect 3-shift codes (splittings by $\{\pm 1, \pm 2, \pm 3\}$ or with Corollary 6.1 even by $\{1, 2, 3\}$ of $\mathbb{Z}_p$) arise for the special choice of the parameters $a = 2$ and $b = 3$. It is possible to formulate necessary and sufficient conditions on the existence of perfect 3-shift codes depending only on the behaviour of the element $3 \cdot 2^{-1}$, even if this does not generate the subgroup $< 2, 3 >$[119].

As mentioned before, perfect shift codes are much faster to find if the subgroup $\mathcal{F}$ is generated by one element and one might first check the orbit of 2, 3, or $3 \cdot 2^{-1}$ by Theorem 6.1 and Corollary 6.2. This way, it was calculated that the first perfect 3-shift codes for primes up to 1000 exist in $\mathbb{Z}_p$ for $p = 7$, 37, 139, 163, 181, 241, 313, 337, 349, 379, 409, 421, 541, 571, 607, 631, 751, 859, 877, 919, 937.

Saidi [93] computed a list of all primes $p < 1000$ such that a splitting of $\mathbb{Z}_p$ by $\{1, 2, 3\}$ fulfilling condition (v) of the previous section exists. It was mentioned before that condition (v) is not sufficient, since $p = 919$ is not of the required form. However, for all other primes, the list of [93] coincides with our list above.

4-shift codes are just splitting sets obtained from splittings of $\mathbb{Z}_p$ by the $\{\pm 1, \pm a, \pm a^2, \pm b\}$ for the special choice of the parameters $a = 2$ and $b = 3$. Again it might be useful to consider the orbit of further elements besides a and b to speed up the computation of an algorithm which finds perfect shift codes. However, the element ba^{-1} now does not generate $< a, b >$, but also the orbit of the element ba^{-2} may be checked, since $\mathcal{F}$ is the union of the cosets of the subgroup $< (ba^{-2})^4 >$ generated by the element $(ba^{-2})^4$.

Corollary 6.3 *If a factorization* $\mathcal{G} = \{1, a, a^2, b\} \cdot \mathcal{S}$ *of a group* $\mathcal{G} = \mathbb{Z}_p^*$ *or* $\mathcal{G} = \mathbb{Z}_p^*/\{1, -1\}$ *exists, then*

(i) ba^{-1} *has even order in* $\mathcal{G}$,
(ii) a *is not on the orbit of* ba^{-1},

(iii) if $a^{4l_1} = (ba^{-1})^{l_2}$ *for some positive integers* l_1 *and* l_2*, then* $l_2 \equiv 0 \bmod 2$.
(iv) the order of the element $(ba^{-2})^4$ *is divisible by* 4.

Proof (i) is immediate from conditions (5) and (6), since if $1 = (ba^{-1})^m = (ba)^m \cdot a^{-2m}$ then $2m$ must be divisible by 4.

(ii) If $(ba^{-1})^m = a$ for some m, then $b^m = a^{m+1}$, which cannot occur since $2m+1$ is not divisible by 4 (condition (6)).

(iii) follows from condition (6).

(iv) For every t with $(ba^{-2})^t = 1 \in \mathcal{F}$ it is $b^t = a^{2t}$. If a factorization of the required form exists, then by (6) the order of ba^{-2} must be divisible by 4. □

There are only 21 prime numbers $p = 8N + 1 < 25000$ for which a perfect 4-shift code exists in $\mathbb{Z}_p$, namely $p = 97, 1873, 2161, 3457, 6577, 6673, 6961, 7297, 7873, 10273, 12721, 13537, 13681, 13729, 15601, 15649, 16033, 16561, 16657, 21121, 22129$.

Observe that $p = 2161$ is the first number for which as well a perfect 3-shift as a perfect 4-shift code in $\mathbb{Z}_p$ exists.

For $q = r + s + 1 > 3$ it is not known if there are infinitely many primes for which a splitting by the sets $\mathcal{M}_1$ or $\mathcal{M}_2$ in $\mathbb{Z}_p$ exists. For $q = 3$ this follows from Saidi's investigations (cf. the considerations after condition (v) in Sect. 6.2.2).

Stein [102] could demonstrate that the set of positive integers N for which a splitting by $\{\pm1, \pm2, \pm3\}$ exists in $\mathbb{Z}_N$ has density 0.

The distribution of 3– and 4-shift codes among the first 500 primes of the required form $p = 2qN + 1$ can be seen in the table below. Here are listed the numbers of primes $p = 2 \cdot |\mathcal{M}| \cdot N + 1$, $N = 1, \dots, 500$, for which a factorization $< a, b >= \mathcal{M} \cdot \mathcal{F}$ in $\mathbb{Z}_p^*/\{1, -1\}$ with $\mathcal{F} = \{a^i b^j : i - j \equiv 0 \bmod q\}$ exists. Here for $q = r+s+1 \geq 3$ just the splittings of $\mathbb{Z}_p$ by $\mathcal{M} = \{\pm1, \pm a, \dots, \pm a^r, \pm b \dots, \pm b^s\}$ (and especially for $(a, b) = (2, 3)$, $q = 3$ and 4 the perfect 3– and 4-shift codes) are counted.

$(a, b)\backslash q$	2	3	4	5	6	7	8	9	10
(2, 3)	48	46	4	23	6	12	1	13	8
(2, 5)	50	48	5	22	17	18	1	13	1
(2, 7)	46	48	5	23	9	21	1	17	1
(3, 4)	3	50	1	28	2	19	1	17	1
(3, 5)	41	51	19	32	2	18	11	15	2
(3, 7)	44	46	20	29	3	24	9	16	7
(4, 5)	4	50	2	31	0	17	0	17	1
(4, 7)	3	51	0	28	0	13	0	14	0
(5, 7)	43	49	18	19	15	18	9	15	1

Observe that there are usually more splittings when q is odd. However in this case by Theorem 6.2 splittings by $\mathcal{M}_1$ and $\mathcal{M}_2$ are equivalent. For even q this is not the case and there may also exist primes of the form $p = qn + 1$ yielding a splitting in $\mathbb{Z}_p$ by $\mathcal{M}_1$. The following table contains the number of such primes $p = |\mathcal{M}| \cdot N + 1$, $N = 1, \ldots, 1000$, for which a factorization $< a, b >= \mathcal{M} \cdot \mathcal{F}$ in $\mathbb{Z}_p^*$ exists. Observe that for $q = r+s+1 \geq 3$ just the splittings of $\mathbb{Z}_p$ by $\mathcal{M} = \{1, a, \ldots, a^r, b \ldots, b^s\}$ are counted. These numbers have been obtained by checking the conditions in Theorem 6.1 or Corollaries 2.1 and 3.1, respectively, for each group $\mathbb{Z}_p$.

$(a, b)\backslash q$	2	3	4	5	6	7	8	9	10
(2, 3)	85	46	41	23	35	12	4	13	11
(2, 5)	88	48	43	22	28	18	2	13	1
(2, 7)	82	48	40	23	18	21	2	17	16
(3, 4)	23	50	1	28	17	19	3	17	6
(3, 5)	84	51	40	32	34	18	17	15	5
(3, 7)	86	46	35	29	31	24	17	16	11
(4, 5)	26	50	5	31	8	17	1	17	8
(4, 7)	25	51	2	28	6	13	1	14	4
(5, 7)	88	49	33	19	27	18	16	15	3

6.2.4 *Tilings by the Cross and Semicross and Splittings of Groups of Composite Order*

We considered splittings of Abelian groups by the sets $S(k)$ and $F(k)$ in order to analyze perfect shift codes. Such splittings have been studied in literature for another reason. They are closely related to tilings (partitioning into translates of a certain cluster) of the n-dimensional Euclidean space $\mathbb{R}^n$ by the (k, n)-cross and the (k, n)-semicross, respectively. Recall that a (k, n)-*semicross* is a translate of the cluster consisting of the $kn + 1$ unit cubes in $\mathbb{R}^n$ with edges parallel to the coordinate axes and with centers

$$(0, 0, \ldots, 0), (j, 0, \ldots, 0), (0, j, \ldots, 0), \ldots, (0, 0, \ldots, j)$$

$j = 1, 2 \ldots, k$ and that a (k, n)-*cross* (or *full cross*) is a translate of the cluster consisting of the $2kn + 1$ n-dimensional unit cubes with centers (for $j = 1, 2 \ldots, k$)

$$(0, 0, \ldots, 0), (\pm j, 0, \ldots, 0), (0, \pm j, \ldots, 0), \ldots, (0, 0, \ldots, \pm j).$$

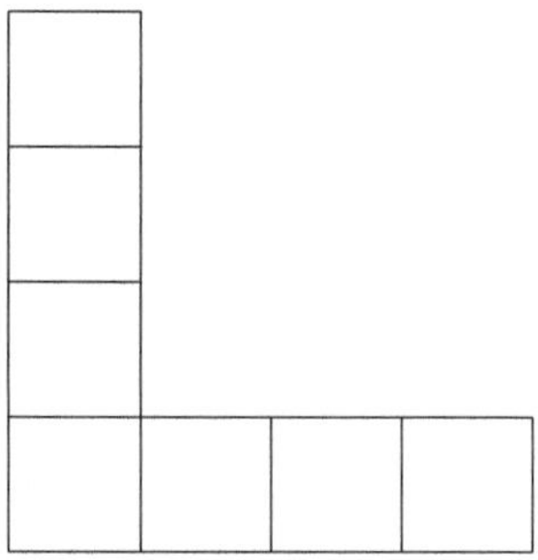

the $(3, 2)$ – semicross

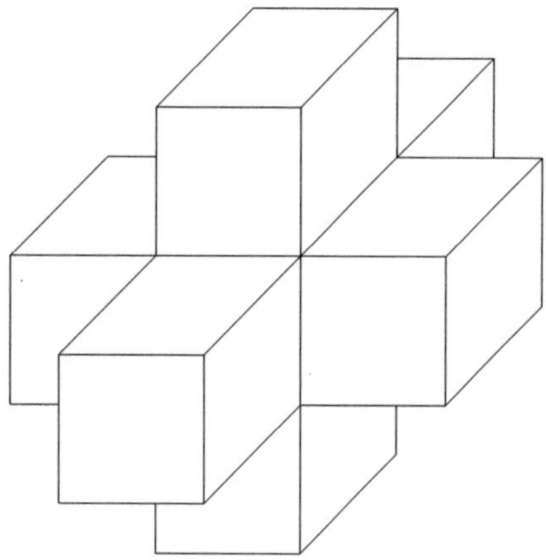

the $(1, 3)$ – cross

The following results concerning lattice tilings are proved, for instance, in [108]. A lattice tiling is a tiling, where the translates of any fixed point of a cluster (e.g. the center of the cross) form a lattice. A lattice tiling by the cross (semicross) corresponds to a splitting of some Abelian group by the set $F(k)$ ($S(k)$). The analysis can further be reduced to cyclic groups $\mathbb{Z}_N = \mathbb{Z}/\mathbb{Z}_N$.

Fact 6.1 ([102]) A lattice tiling of the n-dimensional Euclidean space $\mathbb{R}^n$ by the (k, n)-semicross (by the (k, n)-cross) exists, if and only if the set $\{1, 2, \ldots, k\}$ (the set $\{\pm 1, \pm 2, \ldots, \pm k\}$) splits an Abelian group of order $kn + 1$ ($2kn + 1$).

Fact 6.2 ([61]) If $S(k)$ ($F(k)$) splits an Abelian group of order N, then it also splits the cyclic group $\mathbb{Z}_N$ of the same order.

It should be mentioned that Fact 6.2 does not hold for arbitrary sets $\mathcal{M}$. For small parameters $k = 3, 4$ one can concentrate on cyclic groups $\mathbb{Z}_p$ of prime order p. It was shown by Galovich and Stein [40] for $S(3)$ and $S(4)$ and by Munemasa [79] for $F(3)$ and $F(4)$, respectively, that there exists a splitting of $\mathbb{Z}_N$, where N is a composite number, by the set $S(k)$ ($F(k)$) if and only if there exists a splitting of $\mathbb{Z}_p$ by $S(k)$ (or $F(k)$, respectively) for every prime factor p of N (with the exception of the primes 2 and 3, which can easily be handled separately). Hence the analysis of splittings by those sets for Abelian groups $\mathcal{G}$, where $|\mathcal{G}|$ is a composite number, can easily be done with the following results.

Fact 6.3 ([40]) The set $\{1, 2, 3\}$ splits the finite Abelian group $\mathcal{G}$ if and only if it splits $\mathbb{Z}_p$ for every odd prime p dividing $|\mathcal{G}|$ and the 2– Sylow subgroup of $\mathcal{G}$ is either trivial or isomorphic to $\mathbb{Z}_4$.

The set $\{1, 2, 3, 4\}$ splits the finite Abelian group $\mathcal{G}$ if and only if it splits $\mathbb{Z}_p$ for every odd prime $p \neq 3$ dividing $|\mathcal{G}|$ and the 3– Sylow subgroup of $\mathcal{G}$ is either trivial or isomorphic to $\mathbb{Z}_9$.

Fact 6.4 ([79]) For $k = 1, 2$, and 3 the set $F(k) = \{\pm 1, \pm 2, \ldots, \pm k\}$ splits the finite Abelian group $\mathcal{G}$ if and only if it splits $\mathbb{Z}_p$ for every prime p dividing $|\mathcal{G}|$. The set $\{\pm 1, \pm 2, \pm 3, \pm 4\}$ splits the finite Abelian group $\mathcal{G}$ if and only if it splits $\mathbb{Z}_p$ for every odd prime $p \neq 3$ dividing $|\mathcal{G}|$ and the 3– Sylow subgroup of $\mathcal{G}$ is either trivial or isomorphic to $\mathbb{Z}_9$.

The trivial splittings $(\{1, 2, 3\}, \{1\})$ in $\mathbb{Z}_4$ and $(\{1, 2, 3, 4\}, \{1, -1\})$ as well as $(\{\pm 1, \pm 2, \pm 3, \pm 4\}, \{1\})$ in $\mathbb{Z}_9$ are responsible for the exceptional behaviour of the primes 2 and 3.

Results similar to Facts 6.3 and 6.4 are derived for $S(5)$ and $S(6)$ in [105]. More results on splittings of Abelian groups and further conditions under which a splitting by $S(k)$ or $F(k)$ exists are presented e.g. in [105] or [108].

6.2.5 Concluding Remarks

1. Most of the results in Sect. 6.2.4 are recalled from [103, 108], where the interplay between algebra and tiling is investigated. Starting point of this direction of research was a problem due to Minkowski [76] from 1907. Originally motivated by a problem on diophantine approximation, Minkowski conjectured the following statement: In a lattice tiling of $\mathbb{R}^n$ by unit cubes there must be a pair of cubes which share a complete $(n-1)$-dimensional face.

This problem (for general n) remained open for 35 years and was finally settled by Hajós [60] in 1942 using factorizations of finite Abelian groups by cyclic subsets, which are of the form $\{1, a, a^2, \ldots, a^r\}$ for some r less than the order of a Hajós proved that in a factorization of a finite Abelian group by cyclic subsets one of the factors is a subgroup.

Hajós' work motivated research on the structure of factorizations of finite Abelian groups (cf. also [39], Chap. XV), for instance, by de Bruijn [27, 28] and Sands [94, 95]. The most far–reaching result, generalizing Hajós' original theorem, in this direction is due to Redei [89].

2. Stein in [106] used results on Newton sums (for the set $\mathcal{M}$ the j-th Newton sum is $\sum_{m \in \mathcal{M}} m^j$) in order to compute all splittings of $\mathbb{Z}_p$, p prime, by sets $S(k)$ for $k = 5, \ldots, 12$ up to quite large prime numbers. The smallest primes p for which a splitting of $\mathbb{Z}_p$ by $S(k)$ (besides the trivial splittings with splitting set $\{1\}$ or $\{1, -1\}$) exists are

k	5	6	7	8	9	10	11	12
p	421	103	659	3617	27127	3181	56431	21061

It is easy to see that, if there is no splitting by $S(k)$, then there also does not exist a splitting by $F(k)$ and hence no perfect k – shift code. Hence Stein's results also suggest that perfect shift codes seem to be quite sparsely distributed for $k > 3$ (for $k = 4$ cf. Sect. 6.2.3). Especially, for the application in run–length limited coding, groups of small order, in which a perfect shift code exists, are of interest. The reason is that simultaneously $|\mathcal{G}|$ perfect run-length limited codes correcting single peak shifts are obtained (one for each $g \in \mathcal{G}$) by the construction

$$\mathcal{C}(g) = \{(x_1, \ldots, x_n) : \sum_{i=1}^{n} f(i) \cdot x_i = g\}, \tag{10}$$

where x_i is the length of the i-th run (the number of consecutive 0's between the $(i-1)$-th and the i-th 1) and the $f(i)$'s are obtained from the members of a perfect shift code $\mathcal{S} = \{h_1, \ldots, h_n\}$ by $f(n) = h_n$, $f(i) - f(i+1) = h_i$ for $i = 1, \ldots, n-1$. Observe that the same shift code may yield several perfect run-length limited codes depending on the order of the h_i's. This is intensively discussed in [68].

So the size of the best such code will be about $\frac{1}{|\mathcal{G}|}$ times the number of all possible codes $(x_1, \ldots, x_n)$ with n peaks (=ones). The above table suggests that for $k \geq 4$ groups in which a splitting by $F(k)$ exists are hard to find.

3. One might relax the conditions and no longer require perfectness but a good packing, cf. also [68]. Packings of $\mathbb{R}^n$ by the cross or the semicross and packings of $\mathbb{Z}_n$ by the sets $F(k)$ or $S(k)$ have been considered e.g. in [62, 107], for further results also on coverings see [35, 114]. Some applications to Information Theory have been discussed in [58, 104].

In [105] several results on packings of $\mathbb{Z}_n$ by the cross $F(k)$ are presented. For instance, an almost close packing by $F(p-1)$ exists if $n = 2p^2$ for an odd prime number p.

We say that $F(k)$ packs $\mathbb{Z}_n$ with packing set $\mathcal{S}$ if all products $m \cdot h$ with $m \in F(k), h \in \mathcal{S} \subset \mathbb{Z}_n$ are different.

The following construction may yield good packings for parameters n divisible by 4 and such that the order of the element 3 in $\mathbb{Z}_n^*/\{1, -1\}$ is divisible by 2: Let $\mathcal{F} = \{3^{2l} : l = 0, \ldots, \frac{1}{2}\text{ord}(3)\}$ denote the subgroup of even powers of 3 in $\mathbb{Z}_n^*/\{1, -1\}$ and include in the splitting set $\mathcal{S}$ as many sets of the form $a \cdot \mathcal{F}$ as possible.

For instance, the packing of $\mathbb{Z}_{40}$ by $F(3)$ with packing set $\{1, 4, 5, 7, 9, 17\}$ improves the value for $k = 3$ in Table V-4 on p. 316 in [105], where only an example of a packing of $\mathbb{Z}_{43}$ by $F(3)$ was given (however, of course, there also exists the splitting of $\mathbb{Z}_{37}$ by $F(3)$).

4. Tilings of metric spaces are intimately related to perfect codes (see [24], Chaps. 11 and 16 or [98]). For recent results on binary perfect codes and tilings of binary spaces see e.g. [25, 33, 34]. From tilings of the Euclidean space $\mathbb{R}^n$ by the $(1, n)$-cross one can obtain perfect nonbinary single-error correcting codes in the Lee - metric, since the sphere around the codeword of such a code in the Lee metric corresponds to a full $(1, n)$-cross. Whereas Fact 6.1 just guarantees the existence of a tiling, Golomb and Welch [47] could demonstrate by the construction (10) (where now the x_i's are the components of a codeword $(x_1, \ldots, x_n)$ and $f(i) = i$ for all $i = 1, \ldots, n$) that the $(1, n)$-cross always tiles $\mathbb{R}^n$, from which they could derive perfect single-error correcting codes in the Lee metric.

The Lee metric is a special case of an error measure for codes over an alphabet $\{0, \ldots, q-1\}$, $q \geq 3$ for which a single error distorting coordinate x_i in a codeword $(x_1, \ldots, x_n)$ results in one of the letters $x_i + j \bmod q$, $j \in \{\pm 1, \ldots, \pm k\}$ (the Lee

metric arises for $k = 1$). Relations between such nonbinary single-error correcting codes and splittings of groups can already be found in [126] (cf. also [108], p. 80).
5. Martirossian [71] considers the case $k = 2$, which is closely related to perfect 2-shift codes. Again in [71] the construction (10) is used by choosing the $f(i)$'s appropriately. Construction (10) had been introduced by Varshamov/Tenengol'ts [127] for $\mathcal{G} = \mathbb{Z}_n$ and extended by Levenshtein [67] (in a more general setting) and Constantin/Rao [26] for arbitrary Abelian groups (cf. also [1]). Martirossian [71] also derives a formula for the size of the set $\mathcal{C}(g)$.

Perfect 2-shift codes or splittings by the set $\{\pm 1, \pm 2\}$ have been studied e.g. in [68, 102]. The (necessary and sufficient) conditions on the prime p for the existence of such a 2-shift code in the group $\mathbb{Z}_p$ is that the element 2 has order divisible by 4 in $\mathbb{Z}_p^*$. In [71] it is further analyzed for which primes this condition is fulfilled. Especially, this holds for primes of the form $p \equiv 5 \bmod 8$, hence there are infinitely many perfect 2-shift codes.
6. Saidi [92] gave conditions in terms of a kind of Lloyd's polynomial for the existence of perfect codes correcting more than one error of type "Stein sphere" and "Stein corner".

Shift codes correcting more than one error have also been discussed by Vinck and Morita [59] as a special case of codes over the the ring of integers modulo m, which also comprise the codes for the amplitude and phase modulation channel studied by Martirossian in [71].
7. The semicross and the cross are special polyominoes as studied by Golomb in [46], e.g., a right trominoe just corresponds to the $(1, 2)$-semicross. As a further application in Information Theory, tilings of a bounded region by the cross and similar clusters have also been considered in [7, 97] in the study of memory with defects.
8. Let us finally mention a relation between splittings of groups and dominating sets in graphs. Namely, in [55] results on the existence of perfect Lee codes were used to deduce the asymptotic values of the domination numbers in Cartesian products of paths and cycles, cf. also [66].

6.3 Some Aspects of Hankel Matrices in Coding Theory and Combinatorics

6.3.1 *Introduction*

A Hankel matrix (or persymmetric matrix)

$$A_n = \begin{pmatrix} c_0 & c_1 & c_2 & \dots & c_{n-1} \\ c_1 & c_2 & c_3 & \dots & c_n \\ c_2 & c_3 & c_4 & \dots & c_{n+1} \\ \vdots & \vdots & \vdots & & \vdots \\ c_{n-1} & c_n & c_{n+1} & \dots & c_{2n-2} \end{pmatrix}. \tag{1}$$

is a matrix (a_{ij}) in which for every r the entries on the diagonal $i + j = r$ are the same, i.e., $a_{i,r-i} = c_r$ for some c_r.

For a sequence $c_0, c_1, c_2, \ldots$ of real numbers we also consider the collection of Hankel matrices $A_n^{(k)}$, $k = 0, 1, \ldots, n = 1, 2, \ldots$, where

$$A_n^{(k)} = \begin{pmatrix} c_k & c_{k+1} & c_{k+2} & \cdots & c_{k+n-1} \\ c_{k+1} & c_{k+2} & c_{k+3} & \cdots & c_{k+n} \\ c_{k+2} & c_{k+3} & c_{k+4} & \cdots & c_{k+n+1} \\ \vdots & \vdots & \vdots & & \vdots \\ c_{k+n-1} & c_{k+n} & c_{k+n+1} & \cdots & c_{k+2n-2} \end{pmatrix}. \tag{2}$$

So the parameter n denotes the size of the matrix and the $2n - 1$ successive elements $c_k, c_{k+1}, \ldots, c_{k+2n-2}$ occur in the diagonals of the Hankel matrix.

We shall further denote the determinant of a Hankel matrix (2) by

$$d_n^{(k)} = \det(A_n^{(k)}). \tag{3}$$

Hankel matrices have important applications, for instance, in the theory of moments, and in Padé approximation. In Coding Theory, they occur in the Berlekamp - Massey algorithm for the decoding of BCH - codes. Their connection to orthogonal polynomials often yields useful applications in Combinatorics: as shown by Viennot [128] Hankel determinants enumerate certain families of weighted paths, Catalan-like numbers as defined by Aigner [2] via Hankel determinants often yield sequences important in combinatorial enumeration, and as a recent application, they turned out to be an important tool in the proof of the refined alternating sign matrix conjecture.

The framework for studying combinatorial applications of Hankel matrices and further aspects of orthogonal polynomials was set up by Viennot [128]. Of special interest are determinants of Hankel matrices consisting of Catalan numbers $\frac{1}{2m+1}\binom{2m+1}{m}$. Desainte-Catherine and Viennot [31] provided a formula for $\det(A_n^{(k)})$ and all $n \geq 1, k \geq 0$ in case that the entries c_m are Catalan numbers, namely:

For the sequence $c_m = \frac{1}{2m+1}\binom{2m+1}{m}$, $m = 0, 1, \ldots$ of Catalan numbers it is

$$d_n^{(0)} = d_n^{(1)} = 1, \qquad d_n^{(k)} = \prod_{1 \leq i \leq j \leq k-1} \frac{i + j + 2n}{i + j} \qquad \text{for } k \geq 2, n \geq 1. \tag{4}$$

Desainte-Catherine and Viennot [31] also gave a combinatorial interpretation of this determinant in terms of special disjoint lattice paths and applications to the enumeration of Young tableaux, matchings, etc.

They studied (4) as a companion formula for $\prod_{1 \leq i \leq j \leq k} \frac{i+j-1+c}{i+j-1}$, which for integer c was shown by Gordon (cf. [99]) to be the generating function for certain Young tableaux.

For even $c = 2n$ this latter formula also can be expressed as a Hankel determinant formed of successive binomial coefficients $\binom{2m+1}{m}$.

For the binomial coefficients $c_m = \binom{2m+1}{m}$, $m = 0, 1, \ldots$

$$d_n^{(0)} = 1, \qquad d_n^{(k)} = \prod_{1 \le i \le j \le k} \frac{i + j - 1 + 2n}{i + j - 1} \qquad \text{for } k, n \ge 1. \tag{5}$$

We are going to derive the identities (4) and (5) simultaneously in the next section.

Our main interest, however, concerns a further generalization of the Catalan numbers and their combinatorial interpretations.

In Sect. 6.3.3 we shall study Hankel matrices whose entries are defined as generalized Catalan numbers $c_m = \frac{1}{3m+1}\binom{3m+1}{m}$. In this case we could show that

$$d_n^{(0)} = \prod_{j=0}^{n-1} \frac{(3j+1)(6j)!(2j)!}{(4j+1)!(4j)!}, \qquad d_n^{(1)} = \prod_{j=1}^{n} \frac{\binom{6j-2}{2j}}{2\binom{4j-1}{2j}}. \tag{6}$$

These numbers are of special interest, since they coincide with two Mills-Robbins-Rumsey determinants, which occur in the enumeration of cyclically symmetric plane partitions and alternating sign matrices which are invariant under a reflection about a vertical axis. The relation between Hankel matrices and alternating sign matrices will be discussed in Sect. 6.3.4.

Let us recall some properties of Hankel matrices. Of special importance is the equation

$$\begin{pmatrix} c_0 & c_1 & c_2 & \ldots & c_{n-1} \\ c_1 & c_2 & c_3 & \ldots & c_n \\ c_2 & c_3 & c_4 & \ldots & c_{n+1} \\ \vdots & \vdots & \vdots & & \vdots \\ c_{n-1} & c_n & c_{n+1} & \ldots & c_{2n-2} \end{pmatrix} \cdot \begin{pmatrix} a_{n,0} \\ a_{n,1} \\ a_{n,2} \\ \vdots \\ a_{n,n-1} \end{pmatrix} = \begin{pmatrix} -c_n \\ -c_{n+1} \\ -c_{n+2} \\ \vdots \\ -c_{2n-1} \end{pmatrix}. \tag{7}$$

It is known (cf. [17], p. 246) that, if the matrices $A_n^{(0)}$ are nonsingular for all n, then the polynomials

$$t_j(x) := x^j + a_{j,j-1}x^{j-1} + a_{j,j-2}x^{j-2} + \ldots a_{j,1}x + a_{j,0} \tag{8}$$

form a sequence of monic orthogonal polynomials with respect to the linear operator T mapping x^l to its moment $T(x^l) = c_l$ for all l, i.e.

$$T(t_j(x) \cdot t_m(x)) = 0 \text{ for } j \ne m. \tag{9}$$

and that

$$T(x^m \cdot t_j(x)) = 0 \text{ for } m = 0, \ldots, j-1. \tag{10}$$

In Sect. 6.3.5 we shall study matrices $L_n = (l(m, j))_{m,j=0,1,\ldots,n-1}$ defined by

$$l(m, j) = T(x^m \cdot t_j(x)) \tag{11}$$

By (10) these matrices are lower triangular. The recursion for Catalan-like numbers, as defined by Aigner [2] yielding another generalization of Catalan numbers, can be derived via matrices L_n with determinant 1. Further, the Lanczos algorithm as discussed in [14] yields a factorization $L_n = A_n \cdot U_n^t$, where A_n is a nonsingular Hankel matrix as in (1), L_n is defined by (11) and

$$U_n = \begin{pmatrix} 1 & 0 & 0 & \dots & 0 & 0 \\ a_{1,0} & 1 & 0 & \dots & 0 & 0 \\ a_{2,0} & a_{2,1} & 1 & \dots & 0 & 0 \\ \vdots & \vdots & \vdots & & \vdots & \vdots \\ a_{n-1,0} & a_{n-1,1} & a_{n-2,2} & \dots & a_{n-1,n-2} & 1 \end{pmatrix}. \tag{12}$$

is the triangular matrix whose entries are the coefficients of the polynomials $t_j(x)$, $j = 0, \dots, n-1$.

In Sect. 6.3.5 we further shall discuss the Berlekamp-Massey algorithm for the decoding of BCH-codes, where Hankel matrices of syndromes resulting after the transmission of a code word over a noisy channel have to be studied. Via the matrix L_n defined by (11) it will be shown that the Berlekamp-Massey algorithm applied to Hankel matrices with real entries can be used to compute the coefficients in the corresponding orthogonal polynomials and the three-term recurrence defining these polynomials.

Several methods to find Hankel determinants are presented in [87]. We shall mainly concentrate on their occurrence in the theory of continued fractions and orthogonal polynomials. If not mentioned otherwise, we shall always assume that all Hankel matrices A_n under consideration are nonsingular.

Hankel matrices come into play when the power series

$$F(x) = c_0 + c_1 x + c_2 x^2 + \dots \tag{13}$$

is expressed as a continued fraction. If the Hankel determinants $d_n^{(0)}$ and $d_n^{(1)}$ are different from 0 for all n the so-called S-fraction expansion of $1 - xF(x)$ has the form

$$1 - xF(x) = 1 - \cfrac{c_0 x}{1 - \cfrac{q_1 x}{1 - \cfrac{e_1 x}{1 - \cfrac{q_2 x}{1 - \cfrac{e_2 x}{1 - \dots}}}}}. \tag{14}$$

Namely, then (cf. [82], p. 304 or [130], p. 200) for $n \geq 1$ and with the convention $d_0^{(k)} = 1$ for all k it is

$$q_n = \frac{d_n^{(1)} \cdot d_{n-1}^{(0)}}{d_{n-1}^{(1)} \cdot d_n^{(0)}}, \qquad e_n = \frac{d_{n+1}^{(0)} \cdot d_{n-1}^{(1)}}{d_n^{(0)} \cdot d_n^{(1)}}. \tag{15}$$

For the notion of S- and J- fraction (S stands for Stieltjes, J for Jacobi) we refer to the standard books by Perron [82] and Wall [130]. We follow here mainly the (q_n, e_n)-notation of Rutishauser [91].

For many purposes it is more convenient to consider the variable $\frac{1}{x}$ in (13) and study power series of the form

$$\frac{1}{x}F(\frac{1}{x}) = \frac{c_0}{x} + \frac{c_1}{x^2} + \frac{c_2}{x^3} + \dots \tag{16}$$

and its continued S-fraction expansion

$$\cfrac{c_0}{x - \cfrac{q_1}{1 - \cfrac{e_1}{x - \cfrac{q_2}{1 - \cfrac{e_2}{x - \dots}}}}}$$

which can be transformed to the J-fraction

$$\cfrac{c_0}{x - \alpha_1 - \cfrac{\beta_1}{x - \alpha_2 - \cfrac{\beta_2}{x - \alpha_3 - \cfrac{\beta_3}{x - \alpha_4 - \dots}}}} \tag{17}$$

with $\alpha_1 = q_1$, and $\alpha_{j+1} = q_{j+1} + e_j, \qquad \beta_j = q_j e_j$ for $j \geq 1$. (cf. [82], p.375 or [91], pp. 13).

The J-fraction corresponding to (14) was used by Flajolet [36, 37] to study combinatorial aspects of continued fractions, especially he gave an interpretation of the coefficients in the continued fractions expansion in terms of weighted lattice paths. This interpretation extends to parameters of the corresponding orthogonal polynomials as studied by Viennot [128]. For further combinatorial aspects of orthogonal polynomials see e.g. [38, 111].

Hankel determinants occur in Padé approximation and the determination of the eigenvalues of a matrix using their Schwarz constants, cf. [91]. Especially, they have been studied by Stieltjes in the theory of moments [109, 110]. He stated the problem to find out if a measure μ exists such that

$$\int_0^\infty x^l d\mu(x) = c_l \quad \text{for all } l = 0, 1, \dots \tag{18}$$

for a given sequence $c_0, c_1, c_2, \dots$ by the approach $\int \frac{d\mu(t)}{x+t} = \sum_{l=0}^\infty (-1)^l \frac{c_l}{x^{l+1}}$.

Stieltjes could show that such a measure exists if the determinants of the Hankel matrices $A_n^{(0)}$ and $A_n^{(1)}$ are positive for all n. Indeed, then (9) results from the quality of the approximation to (16) by quotients of polynomials $\frac{p_j(x)}{t_j(x)}$ where $t_j(x)$ are just the polynomials (8). They hence obey the three-term recurrence

$$t_j(x) = (x - \alpha_j)t_{j-1}(x) - \beta_{j-1} \cdot t_{j-2}(x), \qquad t_0(x) = 1, \qquad t_1(x) = x - \alpha_1, \tag{19}$$

where

$$\alpha_1 = q_1, \text{ and } \alpha_{j+1} = q_{j+1} + e_j, \qquad \beta_j = q_j e_j \text{ for } j \geq 1. \tag{20}$$

In case that we consider Hankel matrices of the form (2) and hence the corresponding power series $c_k + c_{k+1}x + c_{k+2}x^2 + \ldots$, we introduce a superscript (k) to the parameters in question.

Hence, $q_n^{(k)}$ and $e_n^{(k)}$ denote the coefficients in the continued fractions expansions

$$\cfrac{c_k}{1 - \cfrac{q_1^{(k)}x}{1 - \cfrac{e_1^{(k)}x}{1 - \cfrac{q_2^{(k)}x}{1 - \ldots}}}}, \qquad \cfrac{c_k}{x - q_1^{(k)} - \cfrac{e_1^{(k)}q_1^{(k)}}{x - q_2^{(k)} - e_1^{(k)} - \cfrac{e_2^{(k)}q_2^{(k)}}{x - q_3^{(k)} - e_2^{(k)} - \ldots}}}$$

and

$$t_j^{(k)}(x) = x^j + a_{j,j-1}^{(k)}x^{j-1} + a_{j,j-2}^{(k)}x^{j-2} + \ldots a_{j,1}^{(k)}x + a_{j,0}^{(k)}$$

are the corresponding polynomials obeying the three-term recurrence

$$t_j^{(k)}(x) = (x - \alpha_j^{(k)})t_{j-1}^{(k)}(x) - \beta_{j-1}^{(k)}t_{j-2}^{(k)}(x).$$

Several algorithms are known to determine this recursion. We mentioned already the Berlekamp-Massey algorithm and the Lanczos algorithm. In the quotient-difference algorithm due to Rutishauser [91] the parameters $q_n^{(k)}$ and $e_n^{(k)}$ are obtained via the so-called rhombic rule

$$e_n^{(k)} = e_{n-1}^{(k)} + q_n^{(k+1)} - q_n^{(k)}, \qquad e_0^{(k)} = 0 \text{ for all } k, \tag{21}$$

$$q_{n+1}^{(k)} = q_n^{(k+1)} \cdot \frac{e_n^{(k+1)}}{e_n^{(k)}}, \qquad q_1^{(k)} = \frac{c_{k+1}}{c_k} \text{ for all } k. \tag{22}$$

As will be seen in Sect. 6.3.3, the Hankel matrices consisting of generalized Catalan numbers have an application in the enumeration of tuples of disjoint lattice paths, where the single paths are not allowed to go above the diagonal $(p-1)x = y$. This result can be generalized to Hankel matrices consisting of numbers counting paths which never touch or cross the diagonal $cx = dy$, for coprime integers $c, d > 1$. However, a closed expression for these numbers is not known. We shall conclude with Sect. 6.3.6, where this enumeration problem is analyzed. A probabilistic approach due to Gessel allows to study the case $d = 2$. We shall derive the generating function for the number of paths starting in the origin and then not touching or crossing the diagonal $cx = 2y$ before they terminate in $(2n, cn)$. Further, a combinatorial interpretation of the numbers $2\binom{n+k}{k} - c\binom{n+k}{k-1}$ will be supplied, which can be regarded as a generalization of the ballot numbers $\binom{n+k}{k} - \binom{n+k}{k-1}$.

6.3.2 Hankel Matrices and Chebyshev Polynomials

Let us illustrate the methods introduced by computing determinants of Hankel matrices whose entries are successive Catalan numbers. In several recent papers (e.g. [2, 72, 81, 88]) these determinants have been studied under various aspects and formulae were given for special parameters. Desainte-Catherine and Viennot in [31] provided the general solution $d_n^{(k)} = \prod_{1 \le i \le j \le k-1} \frac{i+j+2n}{i+j}$ for all n and k. This was derived as a companion formula (yielding a "90 % bijective proof" for tableaux whose columns consist of an even number of elements and are bounded by height $2n$) to Gordon's result [52] in the proof of the Bender-Knuth conjecture [8]. Gordon proved that $\prod_{1 \le i \le j \le k} \frac{c+i+j-1}{i+j-1}$ is the generating function for Young tableaux with entries from $\{1, \ldots, n\}$ strictly increasing in rows and not decreasing in columns consisting of $\le c$ columns and largest part $\le k$. Actually, this follows from the more general formula in the Bender-Knuth conjecture by letting $q \to 1$, see also [99], p. 265.

By refining the methods of [31], Choi and Gouyou-Beauchamps [22] could also derive Gordon's formula for even $c = 2n$. In the following proposition we shall apply a well - known recursion for Hankel determinants allowing to see that in this case also Gordon's formula can be expressed as a Hankel determinant, namely the matrices then consist of consecutive binomial coefficients of the form $\binom{2m+1}{m}$. Simultaneously, this yields another proof of the result of Desainte-Catherine and Viennot, which was originally obtained by application of the quotient-difference algorithm [129].

Proposition 6.1 *(i) For the sequence* $c_m = \frac{1}{2m+1}\binom{2m+1}{m}$, $m = 0, 1, \ldots$ *of Catalan numbers it is*

$$d_n^{(0)} = d_n^{(1)} = 1, \qquad d_n^{(k)} = \prod_{1 \le i \le j \le k-1} \frac{i+j+2n}{i+j} \qquad \text{for } k \ge 2, n \ge 1. \quad (23)$$

(ii) For the binomial coefficients $c_m = \binom{2m+1}{m}$, $m = 0, 1, \ldots$

$$d_n^{(0)} = 1, \qquad d_n^{(k)} = \prod_{1 \le i \le j \le k} \frac{i+j-1+2n}{i+j-1} \qquad \text{for } k, n \ge 1. \quad (24)$$

Proof The proof is based on the following identity for Hankel determinants.

$$d_n^{(k+1)} \cdot d_n^{(k-1)} - d_{n-1}^{(k+1)} \cdot d_{n+1}^{(k-1)} - [d_n^{(k)}]^2 = 0. \tag{25}$$

This identity can for instance be found in the book by Polya and Szegö [86], Ex. 19, p. 102. It is also an immediate consequence of Dodgson's algorithm for the evaluation of determinants (e.g. [135]).

We shall derive both results simultaneously. The proof will proceed by induction on $n + k$.

It is well known, e.g. [101], that for the Hankel matrices $A_n^{(k)}$ with Catalan numbers as entries it is $d_n^{(0)} = d_n^{(1)} = 1$. For the induction beginning it must also be verified that $d_n^{(2)} = n + 1$ and that $d_n^{(3)} = \frac{(n+1)(n+2)(2n+3)}{6}$ is the sum of squares, cf. [72], which can also be easily seen by application of recursion (25).

Furthermore, for the matrix $A_n^{(k)}$ whose entries are the binomial coefficients $\binom{2k+1}{k}$, $\binom{2k+3}{k+1}$, ... it was shown in [2] that $d_n^{(0)} = 1$ and $d_n^{(1)} = 2n + 1$. Application of (25) shows that $d_n^{(2)} = \frac{(n+1)(2n+1)(2n+3)}{3}$, i.e., the sum of squares of the odd positive integers.

Also, it is easily seen by comparing successive quotients $\frac{c_{k+1}}{c_k}$ that for $n = 1$ the product in (23) yields the Catalan numbers and the product in (24) yields the binomial coefficients $\binom{2k+1}{k+1}$, cf. also [31].

Now it remains to be verified that (23) and (24) hold for all n and k, which will be done by checking recursion (25). The sum in (25) is of the form (with either $d = 0$ for (23) or $d = 1$ for (24) and shifting k to $k + 1$ in (23))

$$\prod_{i,j=1}^{k} \frac{i+j-d+2n}{i+j-d} \cdot \prod_{i,j=1}^{k-2} \frac{i+j-d+2n}{i+j-d} - \prod_{i,j=1}^{k} \frac{i+j-d+2(n+1)}{i+j-d} \cdot \prod_{i,j=1}^{k-2} \frac{i+j-d+2(n-1)}{i+j-d} -$$

$$- \left[\prod_{i,j=1}^{k-1} \frac{i+j-d+2n}{i+j-d} \right]^2$$

$$= \left[\prod_{i,j=1}^{k-1} \frac{i+j-d+2n}{i+j-d} \right]^2 \cdot$$

$$\cdot \left(\frac{\prod_{j=1}^{k}(k+j-d+2n)}{\prod_{j=1}^{k}(k+j-d)} \cdot \frac{\prod_{j=1}^{k-1}(k-1+j-d)}{\prod_{j=1}^{k-1}(k-1+j-d+2n)} - \frac{\prod_{j=0}^{k-1}(j-d+2n)}{\prod_{j=1}^{k}(k+j-d)} \cdot \frac{\prod_{j=1}^{k-1}(k-1+j-d)}{\prod_{j=1}^{k-1}(1+j-d+2n)} - 1 \right)$$

$$= \left[\prod_{i,j=1}^{k-1} \frac{i+j-d+2n}{i+j-d} \right]^2 \cdot$$

$$\cdot \left(\frac{(2n+2k-d)(2n+2k-1-d)(k-d)}{(2n+k-d)(2k-d)(2k-1-d)} - \frac{(2n-d)(2n+1-d)(k-d)}{(2n+k-d)(2k-d)(2k-1-d)} - 1 \right).$$

This expression is 0 exactly if

$$(2n+2k-d)(2n+2k-1-d)(k-d)-(2n-d)(2n+1-d)(k-d)-(2n+k-d)(2k-d)(2k-1-d)=0. \tag{26}$$

In order to show (23), now observe that here $d=0$ and then it is easily verified that

$$(n+k)(2n+2k-1)-n(2n+1)-(2n+k)(2k-1)=0.$$

In order to show (24), we have to set $d=1$ and again the analysis simplifies to verifying

$$(2n+2k-1)(n+k-1)-(2n-1)n-(2n+k-1)(2k-1)=0.$$

□

Remarks

1. As pointed out in the introduction, Desainte-Catherine and Viennot [31] derived identity (23) and recursion (25) simultaneously proves (24). The identity $\det(A_n^{(0)})=1$, when the c_m's are Catalan numbers or binomial coefficients $\binom{2m+1}{m}$ can already be found in [78], pp. 435–436. $d_n^{(1)}$, $d_n^{(2)}$, and $d_n^{(3)}$ for this case were already mentioned in the proof of Proposition 6.1. The next determinant in this series is obtained via $\frac{d_n^{(4)}}{d_{n-1}^{(4)}}=\frac{d_{n+1}^{(3)}}{d_{n-1}^{(3)}}$. For the Catalan numbers then $d_n^{(4)}=\frac{d_{n+1}^{(3)}\cdot d_n^{(3)}}{5}=\frac{n(n+1)^2(n+2)(2n+1)(2n+3)}{180}$.
2. Formula (23) was also studied by Desainte-Catherine and Viennot [31] in the analysis of disjoint paths in a bounded area of the integer lattice and perfect matchings in a certain graph as a special Pfaffian. An interpretation of the determinant $d_n^{(k)}$ in (23) as the number of k-tuples of disjoint positive lattice paths (see the next section) was used to construct bijections to further combinatorial configurations. Applications of (23) in Physics have been discussed by Guttmann, Owczarek, and Viennot [56].
3. The central argument in the proof of Proposition 6.1 was the application of recursion (25). Let us demonstrate the use of this recursion with another example. Aigner [3] could show that the Bell numbers are the unique sequence $(c_m)_{m=0,1,2,...}$ such that

$$\det(A_n^{(0)})=\det(A_n^{(1)})=\prod_{k=0}^{n}k!,\qquad \det(A_n^{(2)})=r_{n+1}\prod_{k=0}^{n}k!, \tag{27}$$

where $r_n=1+\sum_{l=1}^{n}n(n-1)\cdots(n-l+1)$ is the total number of permutations of n things (for $\det(A_n^{(0)})$ and $\det(A_n^{(1)})$ see [30, 37]). In [3] an approach via generating functions was used in order to derive $d_n^{(2)}=\det(A_n^{(2)})$ in (27). Setting $d_n^{(2)}=r_{n+1}\cdot\prod_{k=0}^{n}k!$ in (27), with (25) one obtains the recurrence $r_{n+1}=(n+1)\cdot r_n+1$, $r_2=5$, which just characterizes the total number of permutations of n things, cf. [90], p. 16, and hence can derive $\det(A_n^{(2)})$ from $\det(A_n^{(0)})$ and $\det(A_n^{(1)})$ also this way.
4. From the proof of Proposition 6.1 it is also clear that $\prod_{1\le i,j\le k}\frac{i+j-d+2n}{i+j-d}$ yields a sequence of Hankel determinants $d_n^{(k)}$ only for $d=0,1$, since otherwise recursion (25) is not fulfilled.

As pointed out, in [31] formula (23) was derived by application of the quotient-difference algorithm, cf. also [22] for a more general result. The parameters $q_n^{(k)}$ and $e_n^{(k)}$ also can be obtained from Proposition 6.1.

Corollary 6.4 *For the Catalan numbers the coefficients $q_n^{(k)}$ and $e_n^{(k)}$ in the continued fractions expansion of $\sum_{m=0}^{\infty} \frac{1}{2(k+m)+1}\binom{2(k+m)+1}{k+m}x^m$ as in (14) are given as*

$$q_n^{(k)} = \frac{(2n+2k-1)(2n+2k)}{(2n+k-1)(2n+k)}, \qquad e_n^{(k)} = \frac{(2n)(2n+1)}{(2n+k)(2n+k+1)}. \tag{28}$$

For the binomial coefficients $\binom{2m+1}{m}$ the corresponding coefficients in the expansion of $\sum_{m=0}^{\infty} \binom{2(k+m)+1}{k+m}x^m$ are

$$q_n^{(k)} = \frac{(2n+2k)(2n+2k+1)}{(2n+k-1)(2n+k)}, \qquad e_n^{(k)} = \frac{(2n-1)(2n)}{(2n+k)(2n+k+1)}. \tag{29}$$

Proof Equations (28) and (29) can be derived by application of the rhombic rule (21) and (22). They are also immediate from the previous Proposition 6.1 by application of (15), which for $k > 0$ generalizes to the following formulae from [91], p. 15, where the $d_n^{(k)}$'s are Hankel determinants as (3).

$$q_n^{(k)} = \frac{d_n^{(k+1)} d_{n-1}^{(k)}}{d_n^{(k)} d_{n-1}^{(k+1)}}, \qquad e_n^{(k)} = \frac{d_{n+1}^{(k)} d_{n-1}^{(k)}}{d_n^{(k)} d_n^{(k+1)}}.$$

□

Corollary 6.5 *The orthogonal polynomials associated to the Hankel matrices $A_n^{(k)}$ of Catalan numbers $c_m = \frac{1}{2m+1}\binom{2m+1}{m}$ are*

$$t_n^{(k)}(x) = (x - \alpha_n^{(k)})t_{n-1}^{(k)} - \beta_{n-1}^{(k)} t_{n-2}^{(k)}(x), \quad t_0^{(k)}(x) = 1, \quad t_1^{(k)}(x) = x - \frac{4k+2}{k+2}$$

where

$$\alpha_{n+1}^{(k)} = 2 - \frac{2k(k-1)}{(2n+k+2)(2n+k)}, \qquad \beta_n^{(k)} = \frac{(2n+2k-1)(2n+2k)(2n)(2n+1)}{(2n+k-1)(2n+k)^2(2n+k+1)}.$$

Proof By (20), $\beta_n^{(k)} = q_n^{(k)} \cdot e_n^{(k)}$ as in the previous corollary and

$$\alpha_{n+1}^{(k)} = q_{n+1}^{(k)} + e_n^{(k)} = \frac{(2n+2k+1)(2n+2k+2)((2n+k) + (2n)(2n+1)(2n+k+2)}{(2n+k+1)(2n+k+2)(2n+k)}$$

$$= \frac{8n^2 + 8nk + 8n + 2k + 4k^2}{(2n+k+2)(2n+k)} = 2 - \frac{2k(k-1)}{(2n+k+2)(2n+k)}.$$

□

Especially for small parameters k the following families of orthogonal polynomials arise here.

$$t_n^{(0)}(x) = (x-2)\cdot t_{n-1}^{(0)}(x) - t_{n-2}^{(0)}(x), \qquad t_0^{(0)}(x) = 1, \qquad t_1^{(0)}(x) = x-1,$$

$$t_n^{(1)}(x) = (x-2)\cdot t_{n-1}^{(1)}(x) - t_{n-2}^{(1)}(x), \qquad t_0^{(1)}(x) = 1, \qquad t_1^{(1)}(x) = x-2,$$

$$t_n^{(2)}(x) = \left(x - \frac{(n+1)^2+n^2}{n(n+1)}\right)\cdot t_{n-1}^{(2)}(x) - \frac{n^2-1}{n^2} t_{n-2}^{(2)}(x), \quad t_0^{(2)}(x) = 1, t_1^{(2)}(x) = x - \frac{5}{2}.$$

It is well - known that the Chebyshev-polynomials of the second kind

$$u_n(x) = \sum_{i=0}^{\lfloor \frac{n}{2} \rfloor} (-1)^i \binom{n-i}{i} (2x)^{n-2i}$$

with recursion

$$u_n(x) = 2x\cdot u_{n-1}(x) - u_{n-2}(x), \qquad u_0(x) = 1, \qquad u_1(x) = 2x$$

come in for Hankel matrices with Catalan numbers as entries. For instance, in this case the first orthogonal polynomials in Corollary 6.5 are

$$t_n^{(0)}(x^2) = \frac{1}{x} u_{2n}\left(\frac{x}{2}\right), \qquad t_n^{(1)}(x^2) = \frac{1}{x} u_{2n+1}\left(\frac{x}{2}\right).$$

Corollary 6.6 *The orthogonal polynomials associated to the Hankel matrices* $A_n^{(k)}$ *of binomial coefficients* $c_m = \binom{2m+1}{m}$ *are*

$$t_n^{(k)}(x) = (x - \alpha_n^{(k)})t_{n-1}^{(k)} - \beta_{n-1}^{(k)} t_{n-2}^{(k)}(x), \quad t_0^{(k)}(x) = 1, \quad t_1^{(k)}(x) = x - \frac{4k+6}{k+2}$$

where

$$\alpha_{n+1}^{(k)} = 2 - \frac{2k(k+1)}{(2n+k+2)(2n+k)}, \qquad \beta_{n+1}^{(k)} = \frac{(2n+2k)(2n+2k+1)(2n-1)(2n)}{(2n+k-1)(2n+k)^2(2n+k+1)}.$$

Proof Again, $\beta_n^{(k)} = q_n^{(k)} \cdot e_n^{(k)}$ as in the previous corollary and

$$\alpha_{n+1}^{(k)} = q_{n+1}^{(k)} + e_n^{(k)} = \frac{(2n+2k+2)(2n+2k+3)((2n+k)+(2n-1)(2n)(2n+k+2)}{(2n+k)(2n+k+1)(2n+k+2)}$$

$$= \frac{8n^2+8nk+8n+2k^2+4k}{(2n+k+2)(2n+k)} = 2 - \frac{2k(k+1)}{(2n+k+2)(2n+k)}.$$

□

6.3.3 Generalized Catalan Numbers and Hankel Determinants

For an integer $p \geq 2$ we shall denote the numbers $\frac{1}{pm+1}\binom{pm+1}{m}$ as generalized Catalan numbers. The Catalan numbers occur for $p = 2$. (The notion "generalized Catalan numbers" as in [63] is not standard, for instance, in [54], pp. 344–350 it is suggested to denote them "Fuss numbers").

Their generating function

$$C_p(x) = \sum_{m=0}^{\infty} \frac{1}{pm+1}\binom{pm+1}{m} x^m \tag{30}$$

fulfills the functional equation

$$C_p(x) = 1 + x \cdot C_p(x)^p,$$

from which immediately follows that

$$\frac{1}{C_p(x)} = 1 - x \cdot C_p(x)^{p-1}. \tag{31}$$

Further, it is

$$C_p(x)^{p-1} = \sum_{m=0}^{\infty} \frac{1}{pm+p-1}\binom{pm+p-1}{m+1} x^m. \tag{32}$$

It is well known that the generalized Catalan numbers $\frac{1}{pm+1}\binom{pm+1}{m}$ count the number of paths in the integer lattice $\mathbb{Z}\times\mathbb{Z}$ (with directed vertices from (i, j) to either $(i, j+1)$ or to $(i+1, j)$) from the origin $(0, 0)$ to $(m, (p-1)m)$ which never go above the diagonal $(p-1)x = y$. Equivalently, they count the number of paths in $\mathbb{Z}\times\mathbb{Z}$ starting in the origin $(0, 0)$ and then first touching the boundary $\{(l+1, (p-1)l+1) : l = 0, 1, 2, \dots\}$ in $(m, (p-1)m+1)$ (cf. Sect. 6.3.6).

Viennot [128] gave a combinatorial interpretation of Hankel determinants in terms of disjoint Dyck paths. In case that the entries of the Hankel matrix are consecutive Catalan numbers this just yields an equivalent enumeration problem analyzed by Mays and Wojciechowski [72]. The method of proof from [72] extends to Hankel matrices consisting of generalized Catalan numbers as will be seen in the following proposition.

Proposition 6.2 *If the c_m's in (2) are generalized Catalan numbers, $c_m = \frac{1}{pm+1}\binom{pm+1}{m}$, $p \geq 2$ a positive integer, then $det(A_n^{(k)})$ is the number of n-tuples $(\gamma_0, \dots, \gamma_{n-1})$ of vertex-disjoint paths in the integer lattice $\mathbb{Z}\times\mathbb{Z}$ (with directed vertices from (i, j) to either $(i, j+1)$ or to $(i+1, j)$) never crossing the diagonal $(p-1)x = y$, where the path γ_r is from $(-r, -(p-1)r)$ to $(k+r, (p-1)(k+r))$.*

Proof The proof follows the same lines as the one in [44], which was carried out only for the case $p = 2$ and is based on a result in [70] on disjoint path systems in directed graphs. We follow here the presentation in [72].

Namely, let $\mathcal{G}$ be an acyclic directed graph and let $\mathcal{A} = \{a_0, \ldots, a_{n-1}\}$, $\mathcal{B} = \{b_0, \ldots, b_{n-1}\}$ be two sets of vertices in $\mathcal{G}$ of the same size n. A disjoint path system in $(\mathcal{G}, \mathcal{A}, \mathcal{B})$ is a system of vertex disjoint paths $(\gamma_0, \ldots, \gamma_{n-1})$, where for every $i = 0, \ldots, n-1$ the path γ_i leads from a_i to $b_{\sigma(i)}$ for some permutation σ on $\{0, \ldots, n-1\}$. Now let p_{ij} denote the number of paths leading from a_i to b_j in $\mathcal{G}$, let p^+ be the number of disjoint path systems for which σ is an even permutation and let p^- be the number of disjoint path systems for which σ is an odd permutation. Then $\det((p_{ij})_{i,j=0,\ldots,n-1}) = p^+ - p^-$ (Theorem 3 in [72]).

Now consider the special graph $\mathcal{G}'$ with vertex set

$$\mathcal{V} = \{(u, v) \in \mathbb{Z} \times \mathbb{Z} : (p-1)u \leq v\},$$

i.e. the part of the integer lattice on and above the diagonal $(p-1)x = y$, and directed edges connecting (u, v) to $(u, v+1)$ and to $(u+1, v)$ (if this is in $\mathcal{V}$, of course).

Further let $\mathcal{A} = \{a_0, \ldots a_{n-1}\}$ and $\mathcal{B} = \{b_0, \ldots b_{n-1}\}$ be two sets disjoint to each other and to $\mathcal{V}$. Then we connect $\mathcal{A}$ and $\mathcal{B}$ to $\mathcal{G}'$ by introducing directed edges as follows

$$a_i \to (-i, -(p-1)i), \qquad (k+i, (p-1)(k+i)) \to b_i, \qquad i = 0, \ldots, n-1. \quad (33)$$

Now denote by $\mathcal{G}''$ the graph with vertex set $\mathcal{V} \cup \mathcal{A} \cup \mathcal{B}$ whose edges are those from $\mathcal{G}'$ and the additional edges connecting $\mathcal{A}$ and $\mathcal{B}$ to $\mathcal{G}'$ as described in (33).

Observe that any permutation σ on $\{0, \ldots, n-1\}$ besides the identity would yield some j and l with $\sigma(j) > j$ and $\sigma(l) < l$. But then the two paths γ_j from a_j to $b_{\sigma(j)}$ and γ_l from a_l to $b_{\sigma(l)}$ must cross and hence share a vertex. So the only permutation yielding a disjoint path system for $\mathcal{G}'$ is the identity. The number of paths p_{ij} from a_i to b_j is the generalized Catalan number $\frac{1}{p(k+i+j)+1}\binom{p(k+i+j)+1}{(k+i+j)}$. So the matrix (p_{ij}) is of Hankel type as required and its determinant gives the number of n-tuples of disjoint paths as described in Proposition 6.2. □

Remarks

1. The use of determinants in the enumeration of disjoint path systems is well known, e.g. [43]. In a similar way as in Proposition 6.2 we can derive an analogous result for the number of tuples of vertex-disjoint lattice paths, with the difference that the paths now are not allowed to touch the diagonal $(p-1)x = y$ before they terminate in $(m, (p-1)m)$. Since the number of such paths from $(0, 0)$ to $(m, (p-1)m)$ is $\frac{1}{pm+p-1}\binom{pm+p-1}{m+1}$ (cf. e.g. the appendix), this yields a combinatorial interpretation of Hankel matrices $A_n^{(k)}$ with these numbers as entries as in (2).

2. For the Catalan numbers, i.e. $p = 2$, lattice paths are studied which never cross the diagonal $x = y$. Viennot provided a combinatorial interpretation of orthogonal polynomials by assigning weights to the steps in such a path, which are obtained from the coefficients in the three-term recurrence of the orthogonal polynomials ([128], cf also. [36]). In the case that all coefficients α_j are 0, a Dyck path arises with vertical steps having all weight 1 and horizontal steps having weight β_j for some j. For the Catalan numbers as entries in the Hankel matrix all β_j's are 1, since the Chebyshev polynomials of second kind arise. So the total number of all such paths is counted. Observe that Proposition 6.2 extends the path model for the Catalan numbers in another direction, namely the weights of the single steps are still all 1, but the paths now are not allowed to cross a different boundary.

In order to evaluate the Hankel determinants we further need the following identity.

Lemma 6.2 *Let $p \geq 2$ be an integer. Then*

$$\left(\sum_{m=0}^{\infty}\binom{pm}{m}x^m\right)\cdot\left(\sum_{m=0}^{\infty}\frac{1}{pm+1}\binom{pm+1}{m}x^m\right)=\sum_{m=0}^{\infty}\binom{pm+1}{m}x^m. \tag{34}$$

Proof We are obviously done if we could show that for all $m = 0, 1, 2, \dots$

$$\binom{pm+1}{m}=\sum_{l=0}^{m}\frac{1}{pl+1}\binom{pl+1}{l}\cdot\binom{p(m-l)}{m-l}.$$

In order to do so, we count the number $\binom{pm+1}{m}$ of lattice paths (where possible steps are from (i, j) to either $(i, j+1)$ or to $(i+1, j)$) from $(0, 0)$ to $(m, (p-1)m+1)$ in a second way. Namely each such path must go through at least one of the points $(l, (p-1)l+1), l = 0, 1, \dots, m$. Now we divide the path into two subpaths, the first subpath leading from the origin $(0, 0)$ to the first point of the form $(l, (p-1)l+1)$ and the second subpath from $(l, (p-1)l+1)$ to $(m, (p-1)m+1)$. Recall that there are $\frac{1}{pl+1}\binom{pl+1}{l}$ possible choices for the first subpath and obviously there exist $\binom{p(m-l)}{m-l}$ possibilities for the choice of the second subpath. □

Theorem 6.3 *For $m = 0, 1, 2\dots$ let denote $c_m = \frac{1}{3m+1}\binom{3m+1}{m}$ and $b_m = \frac{1}{3m+2}\binom{3m+2}{m+1}$. Then*

$$\begin{pmatrix} c_0 & c_1 & c_2 & \dots & c_{n-1} \\ c_1 & c_2 & c_3 & \dots & c_n \\ c_2 & c_3 & c_4 & \dots & c_{n+1} \\ \vdots & \vdots & \vdots & & \vdots \\ c_{n-1} & c_n & c_{n+1} & \dots & c_{2n-2} \end{pmatrix} = \prod_{j=0}^{n-1}\frac{(3j+1)(6j)!(2j)!}{(4j+1)!(4j)!},$$

$$\begin{pmatrix} c_1 & c_2 & c_3 & \dots & c_n \\ c_2 & c_3 & c_4 & \dots & c_{n+1} \\ c_3 & c_4 & c_5 & \dots & c_{n+2} \\ \vdots & \vdots & \vdots & & \vdots \\ c_n & c_{n+1} & c_{n+2} & \dots & c_{2n-1} \end{pmatrix} = \prod_{j=1}^{n}\frac{\binom{6j-2}{2j}}{2\binom{4j-1}{2j}} \tag{35}$$

and

$$\begin{pmatrix} b_0 & b_1 & b_2 & \dots & b_{n-1} \\ b_1 & b_2 & b_3 & \dots & b_n \\ b_2 & b_3 & b_4 & \dots & b_{n+1} \\ \vdots & \vdots & \vdots & & \vdots \\ b_{n-1} & b_n & b_{n+1} & \dots & b_{2n-2} \end{pmatrix} = \prod_{j=1}^{n} \frac{\binom{6j-2}{2j}}{2\binom{4j-1}{2j}},$$

$$\begin{pmatrix} b_1 & b_2 & b_3 & \dots & b_n \\ b_2 & b_3 & b_4 & \dots & b_{n+1} \\ b_3 & b_4 & b_5 & \dots & b_{n+2} \\ \vdots & \vdots & \vdots & & \vdots \\ b_n & b_{n+1} & b_{n+2} & \dots & b_{2n-1} \end{pmatrix} = \prod_{j=0}^{n} \frac{(3j+1)(6j)!(2j)!}{(4j+1)!(4j)!}. \tag{36}$$

Proof Observe that

$$\binom{3m}{m} = \frac{\prod_{j=1}^{m}(3j)\prod_{j=0}^{m-1}(3j+1)\prod_{j=0}^{m-1}(3j+2)}{m!\prod_{j=1}^{m}(2j)\prod_{j=0}^{m-1}(2j+1)} = (\frac{27}{4})^m \frac{\prod_{j=0}^{m-1}(\frac{2}{3}+j)\prod_{j=0}^{m-1}(\frac{1}{3}+j)}{m!\prod_{j=0}^{m-1}(\frac{1}{2}+j)}$$

and accordingly

$$\binom{3m+1}{m} = \frac{\prod_{j=1}^{m}(3j)\prod_{j=0}^{m-1}(3j+4)\prod_{j=0}^{m-1}(3j+2)}{m!\prod_{j=1}^{m}(2j)\prod_{j=0}^{m-1}(2j+3)} = (\frac{27}{4})^m \frac{\prod_{j=0}^{m-1}(\frac{2}{3}+j)\prod_{j=0}^{m-1}(\frac{4}{3}+j)}{m!\prod_{j=0}^{m-1}(\frac{3}{2}+j)}.$$

Then with (31) and 34) we have the representation

$$D(x) := 1 - x \cdot C_3(x)^2 = \frac{\sum_{m=0}^{\infty}\binom{3m}{m}x^m}{\sum_{m=0}^{\infty}\binom{3m+1}{m}x^m} = \frac{F(\alpha,\beta,\gamma,y)}{F(\alpha,\beta+1,\gamma+1,y)},$$

which is the quotient of two hypergeometric series, where

$$F(\alpha,\beta,\gamma,y) = 1 + \frac{\alpha\beta}{\gamma}y + \frac{\alpha(\alpha+1)\beta(\beta+1)}{2!\cdot\gamma(\gamma+1)}y^2 + \frac{\alpha(\alpha+1)(\alpha+2)\beta(\beta+1)(\beta+2)}{3!\cdot\gamma(\gamma+1)(\gamma+2)}y^2 + \dots$$

with the parameter choice

$$\alpha = \frac{2}{3},\quad \beta = \frac{1}{3},\quad \gamma = \frac{1}{2},\quad y = \frac{27}{4}x. \tag{37}$$

For quotients of such hypergeometric series the continued fractions expansion as in (14) was found by Gauss (see [82], p. 311 or [130], p. 337). Namely for $n = 1, 2, \dots$ it is

$$e_n = \frac{(\alpha+n)(\gamma-\beta+n)}{(\gamma+2n)(\gamma+2n+1)},\qquad q_n = \frac{(\beta+n)(\gamma-\alpha+n)}{(\gamma+2n-1)(\gamma+2n)}.$$

Now denoting by $q_n^{(D)}$ and $e_n^{(D)}$ the coefficients in the continued fractions expansion of the power series $D(x) = 1 - xC_3(x)^2$ under consideration, then taking into account that $y = \frac{27}{4}x$ we obtain with the parameters in (37) that

$$e_n^{(D)} = \frac{3}{2}\frac{(6n+1)(3n+2)}{(4n+1)(4n+3)}, \qquad q_n^{(D)} = \frac{3}{2}\frac{(6n-1)(3n+1)}{(4n-1)(4n+1)}. \tag{38}$$

The continued fractions expansion of $1 + xC_3(x)^2$ differs from that of $1 - xC_3(x)^2$ only by changing the sign of c_0 in (14).

So, by application of (15) the identity (3) for the determinants $d_n^{(0)}$ and $d_n^{(1)}$ of Hankel matrices with the numbers $\frac{1}{3m+2}\binom{3m+2}{m+1}$ as entries is easily verified by induction. Namely, observe that

$$\frac{3}{2}\frac{(6n-1)(3n+1)}{(4n-1)(4n+1)} = \frac{2(6n)(6n-1)(2n)(3n+1)}{(4n+1)(4n)^2(4n-1)}$$

$$= \frac{(3n+1)(6n)!(2n)!}{(4n+1)!(4n)!} \cdot \frac{2\binom{4n-1}{2n}}{\binom{6n-2}{2n}} = \frac{d_n^{(1)}}{d_{n-1}^{(1)}} \cdot \frac{d_{n-1}^{(0)}}{d_n^{(0)}}$$

and that

$$\frac{3}{2}\frac{(6n+1)(3n+2)}{(4n+1)(4n+3)} = \frac{(6n+4)(6n+3)(6n+2)(6n+1)(2n+1)}{2(4n+3)(4n+2)^2(4n+1)(3n+1)}$$

$$= \frac{\binom{6n+4}{2n+2}}{2\binom{4n+3}{2n+1}} \cdot \frac{(4n+1)!(4n)!}{(3n+1)(6n)!(2n)!} = \frac{d_{n+1}^{(0)}}{d_n^{(0)}} \cdot \frac{d_{n-1}^{(1)}}{d_n^{(1)}},$$

where $d_{n-1}^{(0)}, d_{n-1}^{(1)}, d_n^{(0)}, d_n^{(1)}, d_{n+1}^{(0)}$ are the determinants for the Hankel matrices in (3).

In order to find the determinants for the Hankel matrices in (3) with generalized Catalan numbers $\frac{1}{3m+1}\binom{3m+1}{m}$ as entries, just recall that $D(x) = 1 - xC_3(x)^2 = \frac{1}{C_3(x)}$. So the continued fractions expansion of

$$1 + xC_3(x) = 1 - \frac{-x}{1 - xC_3(x)^2} = 1 - \cfrac{-x}{1 - \cfrac{q_1^{(C)}x}{1 - \cfrac{e_1^{(C)}x}{1 - \cfrac{q_2^{(C)}x}{1 - \dots}}}}$$

is obtained by setting $q_1^{(C)} = 1$, $e_n^{(C)} = q_n^{(D)}$ for $n \geq 1$ and $q_n^{(C)} = e_{n-1}^{(D)}$ for $n \geq 2$. □

Research Problem In the last section we were able to derive all Hankel determinants $d_n^{(k)}$ with Catalan numbers as entries. So the case $p = 2$ for Hankel determinants (2) consisting of numbers $\frac{1}{pm+1}\binom{pm+1}{m}$ is completely settled. For $p = 3$, the above theorem yields $d_n^{(0)}$ and $d_n^{(1)}$. However the methods do not work in order to determine

$d_n^{(k)}$ for $k \geq 2$. Also they do not allow to find determinants of Hankel matrices consisting of generalized Catalan numbers when $p \geq 4$. What can be said about these cases?

Let us finally discuss the connection to the Mills-Robbins-Rumsey determinants

$$T_n(x,\mu) = \det\left(\sum_{t=0}^{2n-2} \binom{i+\mu}{t-i}\binom{j}{2j-t} x^{2j-t}\right)_{i,j=0,\ldots,n-1}, \tag{39}$$

where μ is a nonnegative integer (discussed e.g. in [5, 6, 23, 75, 84]). For $\mu = 0, 1$ it is $T_n(1,\mu) = d_n^{(\mu)}$ - the Hankel determinants in (6.3). This coincidence does not continue for $\mu \geq 2$.

Using former results by Andrews [4], Mills, Robbins, and Rumsey [75] could derive that

$$T_n(1,\mu) = \det\left(\binom{\mu+i+j}{2j-i}\right)_{i,j=0,\ldots,n-1} = \frac{1}{2^n}\prod_{k=0}^{n-1} \Delta_{2k}(2\mu) \tag{40}$$

where $\Delta_0(\mu) = 2$ and with $(x)_j = x(x+1)(x+2)\cdots(x+j-1)$

$$\Delta_{2k}(\mu) = \frac{(\mu+2k+2)_k(\frac{1}{2}\mu+2k+\frac{3}{2})_{k-1}}{(k)_k(\frac{1}{2}\mu+k+\frac{3}{2})_{k-1}}, \quad k > 0.$$

They also state that the proof of formula (40) is quite complicated and that it would be interesting to find a simpler one. One might look for an approach via continued fractions for further parameters μ, however, application of Gauss's theorem only works for $\mu = 0, 1$, where (38) also follows from (40).

Robbins, and Rumsey [75] found the number of cyclically symmetric plane partitions of size n, which are equal to its transpose-complement to be the determinant $T_n(1,0)$. They also conjectured $T_n(x,1)$ to be the generating function for alternating sign matrices invariant under a reflection about a vertical axis, especially $T_n(1,1)$ should then be the total number of such alternating sign matrices as stated by Stanley [100]. We shall further discuss this conjecture in Sect. 6.3.4.

The determinant $T_n(1,\mu) = \det\left(\sum_{t=0}^{2n-2}\binom{i+\mu}{t-i}\binom{j}{t-j}\right)_{i,j=0,\ldots,n-1}$, comes in as counting function for another class of vertex-disjoint path families in the integer lattice. Namely, for such a such a tuple $(\gamma_0,\ldots,\gamma_{n-1})$ of disjoint paths, path γ_i leads from $(i, 2i+\mu)$ to $(2i, i)$.

By a bijection to such disjoint path families for $\mu = 0$ the enumeration problem for the above-mentioned family of plane partitions was finally settled in [75].

6.3.4 Alternating Sign Matrices

An *alternating sign matrix* is a square matrix with entries from $\{0, 1, -1\}$ such that i) the entries in each row and column sum up to 1, ii) the nonzero entries in each row and column alternate in sign. An example is

$$\begin{pmatrix} 0&0&0&1&0&0&0 \\ 1&0&0&-1&0&0&1 \\ 0&0&0&1&0&0&0 \\ 0&1&0&-1&0&1&0 \\ 0&0&0&1&0&0&0 \\ 0&0&1&-1&1&0&0 \\ 0&0&0&1&0&0&0 \end{pmatrix} \tag{41}$$

Robbins and Rumsey discovered the alternating sign matrices in the analysis of Dodgson's algorithm in order to evaluate the determinant of an $n \times n$-matrix. Reverend Charles Lutwidge Dodgson, who worked as a mathematician at the Christ College at the University of Oxford is much wider known as Lewis Carroll, the author of [19]. His algorithm, which is presented in [17], pp. 113–115, is based on the following identity for any matrix ([32], for a combinatorial proof see [135]).

$$\det\left((a_{i,j})_{i,j=1,\dots,n}\right) \cdot \det\left((a_{i,j})_{i,j=2,\dots,n-1}\right) =$$
$$\det\left((a_{i,j})_{i,j=1,\dots,n-1}\right) \cdot \det\left((a_{i,j})_{i,j=2,\dots,n}\right) - \det\left((a_{i,j})_{i=1,\dots,n-1,j=2,\dots,n}\right) \cdot \det\left((a_{i,j})_{i=2,\dots,n,j=1,\dots,n-1}\right). \tag{42}$$

If $(a_{i,j})_{i,j=1,\dots,n}$ in (42) is a Hankel matrix, then all the other matrices in (42) are Hankel matrices, too. Hence recursion (25) from the introduction is an immediate consequence of Dodgson's result.

In the course of Dodgson's algorithm only 2×2 determinants have to be calculated. Robbins asked what would happen, if in the algorithm we would replace the determinant evaluation $a_{ij}a_{i+1,j+1} - a_{i,j+1}a_{i+1,j}$ by the prescription $a_{ij}a_{i+1,j+1} + xa_{i,j+1}a_{i+1,j}$, where x is some variable.

It turned out that this yields a sum of monomials in the a_{ij} and their inverses, each monomial multiplied by a polynomial in x. The monomials are of the form $\prod_{i,j=1}^{n} a_{ij}^{b_{ij}}$ where the b_{ij}'s are the entries in an alternating sign matrix. The exact formula can be found in Theorem 3.13 in the book "Proofs and Confirmations: The Story of The Alternating Sign Matrix Conjecture" by David Bressoud [17].

The alternating sign matrix conjecture concerns the total number of $n \times n$ alternating sign matrices, which was conjectured by Mills, Robbins, and Rumsey to be $\prod_{j=0}^{n-1} \frac{(3j+1)!}{(n+j)!}$.

The problem was open for fifteen years until it was finally settled by Zeilberger [133]. The development of ideas is described in the book by Bressoud. There are deep relations to various parts of Algebraic Combinatorics, especially to plane partitions, where the same counting function occurred, and also to Statistical Mechanics, where the configuration of water molecules in "square ice" can be described by an alternating sign matrix.

As an important step in the derivation of the refined alternating sign matrix conjecture [134], a Hankel matrix comes in, whose entries are $c_m = \frac{1-q^{m+1}}{1-q^{3(m+1)}}$. The relevant orthogonal polynomials in this case are a discrete version of the Legendre polynomials.

Many problems concerning the enumeration of special types of alternating sign matrices are still unsolved, cf. [17], pp. 201. Some of these problems have been presented by Stanley in [100], where it is also conjectured that the number $V(2n+1)$ of alternating sign matrices of odd order $2n+1$ invariant under a reflection about a vertical axis is

$$V(2n+1) = \prod_{j=1}^{n} \frac{\binom{6j-2}{2j}}{2\binom{4j-1}{2j}}$$

A more refined conjecture is presented by Mills, Robbins, and Rumsey [75] relating this type of alternating sign matrices to the determinant $T_n(x, 1)$ in (39). Especially, $T_n(1, 1) = \prod_{j=1}^{n} \frac{\binom{6j-2}{2j}}{2\binom{4j-1}{2j}}$ is conjectured to be the total number $V(2n+1)$. As we saw in Sect. 6.3.3, the same formula comes in as the special Hankel determinant $d_n^{(1)}$, where in (2) we choose generalized Catalan numbers $\frac{1}{3m+1}\binom{3m+1}{m}$ as entries.

Let us consider this conjecture a little closer. If an alternating sign matrix (short: ASM) is invariant under a reflection about a vertical axis, it must obviously be of odd order $2n+1$, since otherwise there would be a row containing two successive nonzero entries with the same sign. For the same reason, such a matrix cannot contain any 0 in its central column as seen in the example (41).

In [16], cf. also [17], Ch. 7.1, an equivalent counting problem via a bijection to families of disjoint paths in a square lattice is presented. Denote the vertices corresponding to the entry a_{ij} in the ASM by (i, j), $i, j = 0, \ldots, n-1$. Then following the outmost path from $(n-1, 0)$ to $(0, n-1)$, the outmost path in the remaining graph from $(0, n-2)$ to $(n-2, 0)$, and so on until the path from $(0, 1)$ to $(1, 0)$ one obtains a collection of lattice paths, which are edge-disjoint but may share vertices.

Since there can be no entry 0 in the central column of the ASM invariant under a reflection about a vertical axis, the entries $a_{0,n}, a_{2,n}, a_{4,n}, \ldots, a_{2n,n}$ must be 1 and $a_{1,n} = a_{3,n} = a_{5,n} = \ldots a_{2n,n} = -1$. This means that for $i = 0, \ldots n-1$ the path from $(2n-i, 0)$ to $(0, 2n-i)$ must go through $(2n-i, n)$ where it changes direction from East to North and after that in $(2n-i-1, n)$ it again changes direction to East and continues in $(2n-i-1, n+1)$.

Because of the reflection-invariance about the central column the matrix of size $(2n+1) \times (2n+1)$ is determined by its columns nos. $n+1, n+2, \ldots 2n$. So, by the above considerations the matrix can be reconstructed from the collection of subpaths $(\mu_0, \mu_1, \ldots, \mu_{n-1})$ where μ_i leads from $(2n-i-1, n+1)$ to $(0, 2n-i)$.

By a reflection about the horizontal and a 90 degree turn to the left, we now map the collection of these paths to a collection of paths $(\nu_0, \nu_1, \ldots, \nu_{n-1})$ the integer lattice $\mathbb{Z} \times \mathbb{Z}$, such that the inner most subpath in the collection leads from $(-1, 0)$ to $(0, 0)$ and path ν_i leads from $(-2i - 1, 0)$ to $(0, i)$.

Denoting by $v_{i,s}$ the y-coordinate of the sth vertical step (where the path is followed from the right to the left) in path number i, $i = 1, \dots, n-1$ - path ν_0 does not contain vertical steps- the collection of paths $(\nu_0, \nu_1, \dots, \nu_{n-1})$ can be represented by a two-dimensional array (plane partition) of positive integers

$$\begin{array}{llllll} v_{n-1,1} & v_{n-1,2} & v_{n-1,2} & \dots & v_{n-1,n-2} & v_{n-1,n-1} \\ v_{n-2,1} & v_{n-2,2} & \dots & v_{n-2,n-2} & & \\ \vdots & \vdots & & & & \\ v_{2,1} & v_{2,2} & & & & \\ v_{1,1} & & & & & \end{array} \tag{43}$$

with weakly decreasing rows, i.e. $v_{i,1} \geq v_{i,2} \geq \dots \geq v_{i,i}$ for all i, and the following restrictions:

(i) $2i - 1 \leq v_{i,1} \leq 2i + 1$ for all $i = 1, \dots, n-1$,
(ii) $v_{i,s} - v_{i,s-1} \leq 1$ for all i, s with $s > i$.
(iii) $v_{i+1,i+1} \geq v_{i,i}$ for all $1 \leq i \leq n-1$.

So for $n = 1$ there is only the empty array and for $n = 2$ there are the three possibilities $v_{1,1} = 1$, $v_{1,1} = 2$, or $v_{1,1} = 3$. For $n = 3$ the following 26 arrays obeying the above restrictions exist:

$$\begin{array}{ccccccccccc} 3\,1 & 3\,2 & 3\,3 & 4\,1 & 4\,2 & 4\,3 & 4\,4 & 5\,1 & 5\,2 & 5\,3 & 5\,4 \\ 1 & 1 & 1 & 1 & 1 & 1 & 1 & 1 & 1 & 1 & 1 \end{array}$$

$$\begin{array}{ccccccccccc} 5\,5 & 4\,2 & 4\,3 & 4\,4 & 5\,2 & 5\,3 & 5\,4 & 5\,5 & 5\,3 & 5\,4 & 5\,5 \\ 1 & 2 & 2 & 2 & 2 & 2 & 2 & 2 & 3 & 3 & 3 \end{array}$$

$$\begin{array}{cccc} 3\,2 & 3\,3 & 4\,3 & 4\,4 \\ 2 & 2 & 3 & 3 \end{array}$$

Now consider a collection $(\gamma_0, \gamma_1, \dots, \gamma_{n-1})$ of vertex disjoint paths in the integer lattice as required in Theorem 6.3, where the single paths are not allowed to cross the diagonal $2x = y$ and path γ_i leads from $(-i, -2i)$ to $(i + 1, 2i + 2)$. Obviously, the initial segment of path γ_i must be the line connecting $(-i, -2i)$ and $(-i, i + 2)$. Since no variation is possible in this part, we can remove these initial segments and obtain a collection $(\eta_0, \dots, \eta_{n-1})$ of vertex-disjoint paths, where now η_i leads from $(-i, i + 2)$ to $(i + 1, 2i + 2)$.

We now denote by $v_{i,s}$ the position of the sth vertical step (i.e. the number of the horizontal step before the sth vertical step in the path counted from right to left) in path η_i, $i = 1, \dots, n-1$ and obtain as a representation of the collection $(\eta_0, \dots, \eta_{n-1})$ a two-dimensional array of positive integers with weakly decreasing rows as in (43), where the restrictions now are:

(i) $2i - 1 \leq v_{i,1} \leq 2i + 1$ for all $i = 1, \dots, n$,
(ii') $v_{i,s} - v_{i,s-1} \leq 2$ for all i, s with $s > i$.

Again, for $n = 1$ there is only the empty array and for $n = 2$ there are the three choices $v_{1,1} = 1$, $v_{1,1} = 2$, or $v_{1,1} = 3$ as above. For $n = 3$ the first 22 arrays above also fulfill the conditions (ii'), whereas the four arrays in the last row do not. However, they can be replaced by

$$\begin{matrix} 4\,1 \\ 2 \end{matrix} \quad \begin{matrix} 5\,1 \\ 2 \end{matrix} \quad \begin{matrix} 5\,1 \\ 3 \end{matrix} \quad \begin{matrix} 5\,2 \\ 3 \end{matrix}$$

in order to obtain a total number of 26 as above. Unfortunately, we did not find a bijection between these two types of arrays or the corresponding collections of paths yet.

6.3.5 Catalan-Like Numbers and the Berlekamp-Massey Algorithm

In this section we shall study two-dimensional arrays $l(m, j)$, $m, j = 0, 1, 2, \dots$ and the matrices $L_n = (l(m, j))_{m,j=0,1,\dots,n-1}$ defined by

$$l(m, j) = T(x^m \cdot t_j(x)), \tag{44}$$

where T is the linear operator defined under (9). Application of the three-term-recurrence (19)

$$t_j(x) = (x - \alpha_j)t_{j-1}(x) - \beta_{j-1}t_{j-2}(x)$$

and the linearity of T gives the recursion

$$l(m, j) = l(m - 1, j + 1) + \alpha_{j+1}l(m - 1, j) + \beta_j l(m - 1, j - 1) \tag{45}$$

with initial values $l(m, 0) = c_m$, $l(0, j) = 0$ for $j \neq 0$ (and $\beta_0 = 0$, of course).

Especially, cf. also [130], p. 195,

$$l(m, m) = c_0\beta_1\beta_2\cdots\beta_m, \quad l(m + 1, m) = c_0\beta_1\beta_2\cdots\beta_m(\alpha_1 + \alpha_2 + \cdots + \alpha_{m+1}) \tag{46}$$

We shall point out two connections of the matrices L_n to Combinatorics and Coding Theory. Namely, for the case that $\beta_j = 1$ for all j the matrices L_n occur in the derivation of Catalan-like numbers as defined by Aigner in [2]. They also can be determined in order to find the factorization $L_n = A_n \cdot U_n^t$, where A_n is a nonsingular Hankel matrix of the form (1) and U_n is the matrix (12) with the coefficients of the orthogonal polynomials in (8). Via formula (46) the Berlekamp-Massey algorithm can be applied to find the parameters α_j and β_j in the three-term recurrence of the orthogonal polynomials (8).

Aigner in [2] introduced Catalan-like numbers and considered Hankel determinants consisting of these numbers. For positive reals $a, s_1, s_2, s_3, \ldots$ Catalan-like numbers $C_m^{(a,\vec{s})}$, $\vec{s} = (s_1, s_2, s_3, \ldots)$ can be defined as entries $b(m, 0)$ in a two-dimensional array $b(m, j), m = 0, 1, 2, \ldots, j = 0, 1, \ldots, m$, with initial conditions $b(m, m) = 1$ for all $m = 0, 1, 2, \ldots, b(0, j) = 0$ for $j > 0$, and recursion

$$\begin{aligned} b(m,0) &= a \cdot b(m-1,0) + b(m-1,1), \\ b(m,j) &= b(m-1,j-1) + s_j \cdot b(m-1,j) + b(m-1,j+1) \text{ for } j = 1, \ldots, m. \end{aligned} \tag{47}$$

The matrices $B_n = (b(m, j))_{m,j=0,\ldots,n-1}$, obtained from this array, have the property that $B_n \cdot B_n^t$ is a Hankel matrix, which has, of course, determinant 1, see also [96] for the Catalan numbers.

The matrices B_n can be generalized in several ways. For instance, with $\beta_j = 1$ for all $j \geq 2$, $\alpha_1 = a$ and $\alpha_{j+1} = s_j$ for $j \geq 2$ the recursion (45) now yields the matrix $L_n = (l(m, j)_{m,j=0,\ldots,n-1})$. Another generalization of the matrices B_n will be mentioned below.

Aigner [2] was especially interested in Catalan-like numbers with $s_j = s$ for all j and some fixed s denoted here by $C_m^{(a,s)}$. In the example below the binomial coefficients $\binom{2m+1}{m}$ arise as $C_m^{(3,2)}$.

$$\begin{array}{l} 1 \\ 3 \;\; 1 \\ 10 \;\; 5 \;\; 1 \\ 35 \;\; 21 \;\; 7 \;\; 1 \\ 126 \;\; 84 \;\; 36 \;\; 9 \;\; 1 \end{array}$$

So, by the previous considerations, choosing $c_m = C_m^{(a,\vec{s})}$ we have that the determinant $d_n^{(0)} = 1$ for all n. In [2] it is also computed the determinant $d_n^{(1)}$ via the recurrence

$$d_n^{(1)} = s_{n-1} \cdot d_{n-1}^{(1)} - d_{n-2}^{(1)}.$$

with initial values $d_0^{(1)} = 1, d_1^{(1)} = a$.

Remarks

1. One might introduce a new leading element c_{-1} to the sequence $c_0, c_1, c_2, \ldots$ and define the $n \times n$ Hankel matrix $A_n^{(-1)}$ and its determinant $d_n^{(-1)}$ for this new sequence. Let $(c_m = C_m^{(s,s)})_{m=0,1,\ldots}$ be the sequence of Catalan-like numbers with parameters (s, s), $s > 1$ and let $c_{-1} = 1$. Let $A_n^{(k)}$ be the Hankel matrix of size $n \times n$ as under (2) and let $d_n^{(k)}$ denote its determinant. Then

$$d_n^{(-1)} = (s-1)(n-1)+1, \qquad d_n^{(0)} = 1, \qquad d_n^{(1)} = sn+1, \qquad d_n^{(2)} = \sum_{j=1}^{n+1}(sj+1)^2.$$

This result follows, since $d_n^{(0)}$ and $d_n^{(1)}$ are known from Propositions 6 and 7 in [2]. So the sequences $d_n^{(k)}$ are known for two successive k's, such that the formulae for $d_n^{(-1)}$ and $d_n^{(2)}$ are easily found using recursion (25).
2. In [2] it is shown that $C_m^{(1,1)}$ are the Motzkin numbers, $C_m^{(2,2)}$ are the Catalan numbers and $C_m^{(3,3)}$ are restricted hexagonal numbers. Guy [57] gave an interpretation of the numbers $C_m^{(4,4)}$ starting with $1, 4, 17, 76, 354, \ldots$. They come into play when determining the number of walks in the three-dimensional integer lattice from $(0, 0, 0)$ to (i, j, k) terminating at height k, which never go below the (i, j)-plane. With the results of [2] their generating function is $\frac{1-4x-\sqrt{1-8x+12x^2}}{2x^2}$.

Lower triangular matrices L_n as defined by (44) are also closely related to the Lanczos algorithm. Observe that with (46) we obtain the parameters in the three-term recursion in a form which was already known to Chebyshev in his algorithm in [20], p. 482, namely

$$\alpha_1 = \frac{l(1,0)}{l(0,0)} \text{ and } \alpha_{j+1} = \frac{l(j+1,j)}{l(j,j)} - \frac{l(j,j-1)}{l(j-1,j-1)}, \quad \beta_j = \frac{l(j,j)}{l(j-1,j-1)} \text{ for } j \geq 1. \tag{48}$$

Since further $l(m, 0) = c_m$ for all $m \geq 0$ by (46) it is $l(m-1, 1) = l(m, 0) - \alpha_1 l(m-1, 0)$ and

$$l(m-1, j+1) = l(m, j) - \alpha_{j+1} l(m-1, j) - \beta_j l(m-1, j-1)$$

for $j > 0$, from which the following recursive algorithm is immediate.

Starting with $\vec{l}_1 = \begin{pmatrix} c_0 \\ c_1 \\ \vdots \\ c_{2n-2} \end{pmatrix}$ and defining $Z = \begin{pmatrix} 0\,0 \ldots 0\,0 \\ 1\,0 \ldots 0\,0 \\ 0\,1 \ldots 0\,0 \\ \vdots\ \vdots \quad \vdots\ \vdots \\ 0\,0 \ldots 1\,0 \end{pmatrix}$ of size $(2n-1) \times (2n-1)$ and Z^t its transpose, we obtain recursively

$$\vec{l}_1 = Z^t \cdot \vec{l}_0 - \alpha_1 \vec{l}_0, \quad \vec{l}_{j+1} = Z^t \cdot \vec{l}_j - \alpha_{j+1} \cdot \vec{l}_j - \beta_j \cdot \vec{l}_{j-1} \quad \text{for } j > 0$$

The subvectors of the initial n elements of $\vec{l}_{j+1}$ then form the $(j+1)$th column $(j = 1, \ldots, n-2)$ of L_n.

In a similar way the matrix U_n^t, the transpose of the matrix (12) consisting of the coefficients of the orthogonal polynomials, can be constructed. Here $\vec{u}_0 = \begin{pmatrix} 1 \\ 0 \\ \vdots \\ 0 \end{pmatrix}$ is the first unit column vector of size $2n - 1$ and then the further columns are obtained via

$$\vec{u}_1 = Z \cdot \vec{u}_0 - \alpha_1 \cdot \vec{u}_0, \quad \vec{u}_{j+1} = Z \cdot \vec{u}_j - \alpha_{j+1} \cdot \vec{u}_j - \beta_j \cdot \vec{u}_{j-1}$$

Again the first n elements of $\vec{u}_j$ form the jth column of U_n^t.

This is the asymmetric Lanczos algorithm yielding the factorization $A_n \cdot U_n^t = L_n$ as studied by Boley, Lee, and Luk [14], where A_n is an $n \times n$ Hankel matrix as in (1.1). Their work is based on a former paper by Phillips [85]. The algorithm is $O(n^2)$ due to the fact that the columns in L_n and U_n^t are obtained only using the entries in the previous two columns.

The symmetric Lanczos algorithm in [14] yields the factorization $A_n = M_n \cdot D_n \cdot M_n^t$. Here, cf. [14], p. 120, $L_n = M_n \cdot D_n$ where $M_n = U_n^{-1}$ is the inverse of U_n and D_n is the diagonal matrix with the eigenvalues of A_n. A combinatorial interpretation of the matrix M_n was given by Viennot [128].

When D_n is the identity matrix, then $L_n = M_n$ and the matrix M_n was used in [81] to derive combinatorial identities as for Catalan-like numbers. Namely, in [81], the Stieltjes matrix $S_n = M_n^{-1} \cdot \overline{M}_n$ was applied, where $\overline{M}_n = (m_{n+1,j})_{m,j=0,\ldots,n-1}$ for $M_n = (m_{n,j})_{m,j=0,\ldots,n-1}$. Then

$$S_n = \begin{pmatrix} \alpha_0 & 1 & 0 & 0 & \ldots & 0 \\ \beta_0 & \alpha_1 & 1 & 0 & \ldots & 0 \\ 0 & \beta_1 & \alpha_2 & 1 & \ldots & 0 \\ \vdots & \vdots & \vdots & \vdots & & \vdots \\ 0 & 0 & 0 & 0 & \ldots & \alpha_{n-1} \end{pmatrix}$$

is tridiagonal with the parameters of the three-term recurrence on the diagonals.

Important for the decoding of BCH codes, studied in the following, is also a decomposition of the Hankel matrix $A_n = V_n D_n V_n^t$ as a product of a Vandermonde matrix V_n, its transpose V_n^t and the diagonal matrix D_n. Here the parameters in the Vandermonde matrix are essentially the roots of the polynomial $t_n(x)$. This decomposition was known already to Baron Gaspard Riche de Prony [29] (rather known as the leading engineer in the construction of the Pont de la Concorde in Paris and as project head of the group producing the logarithmic and trigonometric tables from 1792–801), cf. also [15].

Let us now discuss the relation of the Berlekamp-Massey algorithm to orthogonal polynomials. Via (46) the parameters r_j in the Berlekamp-Massey algorithm presented below will be explained in terms of the three-term recurrence of the orthogonal polynomials related to A_n.

Peterson [83] and Gorenstein and Zierler [53] presented an algorithm for the decoding of BCH codes. The most time-consuming task is the inversion of a Hankel matrix A_n as in (1), in which the entries c_i now are syndromes resulting after the transmission of a codeword over a noisy channel. Matrix inversion, which takes $O(n^3)$ steps was proposed to solve equation (7).

Berlekamp found a way to determine the $a_{n,j}$ in (7) in $O(n^2)$ steps. His approach was to determine them as coefficients of a polynomial $u(x)$ which is found as appropriate solution of the "key equation"

$$F(x)u(x) = q(x) \bmod x^{2t+1}.$$

Here the coefficients $c_0, \dots, c_{2t}$ up to degree $2t$ of $F(x)$ can be calculated from the received word. Further, the roots of $u(x)$ yield the locations of the errors (and also determine $q(x)$). By the application in Coding Theory one is interested in finding polynomials of minimum possible degree fulfilling the key equation. This key equation is solved by iteratively calculating solutions $(q_k(x), u_k(x))$ to $F(x)u_k(x) = q_k(x) \bmod z^{k+1}$, $k = 0, \dots, 2t$.

Massey [73] gave a variation of Berlekamp's algorithm in terms of a linear feedback shift register. The algorithm is presented by Berlekamp in [10]. We follow here Blahut's book [12], p. 180.

The algorithm consist in constructing a sequence of shift registers $(\ell_j, u_j(x))$, $j = 1, \dots, 2n-2$, where ℓ_j denotes the length (the degree of u_j) and

$$u_j(x) = b_{j,j}x^j + b_{j,j-1}x^{j-1} + \dots + b_{j,1}x + 1.$$

the feedback-connection polynomial of the jth shift register. For an introduction to shift registers see, e.g., [12], pp. 131, The Berlekamp-Massey algorithm works over any field and will iteratively compute the polynomials $u_j(x)$ as follows using a second sequence of polynomials $v_j(x)$.

Berlekamp-Massey Algorithm (as in [12], p. 180): Let $u_0(x) = 1$, $v_0(x) = 1$ and $\ell_0 = 0$. Then for $j = 1, \dots, 2n-2$ set

$$r_j = \sum_{t=0}^{\ell_j} b_{j-1,t} c_{j-1-t}, \tag{49}$$

$$\ell_j = \delta_j (j - \ell_{j-1}) + (1 - \delta_j)\ell_{j-1}, \tag{50}$$

$$\begin{pmatrix} u_j(x) \\ v_j(x) \end{pmatrix} = \begin{pmatrix} 1 & -r_j x \\ \delta_j \cdot 1/r_j & (1-\delta_j)x \end{pmatrix} \cdot \begin{pmatrix} u_{j-1}(x) \\ v_{j-1}(x) \end{pmatrix}, \tag{51}$$

where

$$\delta_j = \begin{cases} 1 & \text{if } r_j \neq 0 \text{ and } 2\ell_{j-1} \leq j-1 \\ 0 & \text{otherwise} \end{cases}. \tag{52}$$

Goppa [49] introduced a more general class of codes (containing the BCH-codes as special case) for which decoding is based on the solution of the key equation $F(x)u(x) = q(x) \bmod G(x)$ for some polynomial $G(x)$. Berlekamp's iterative algorithm does not work for arbitrary polynomial $G(x)$ (cf. [11]). Sugiyama et al. [112] suggested to solve this new key equation by application of the Euclidean algorithm for the determination of the greatest common divisor of $F(x)$ and $G(x)$, where the algorithm stops, when the polynomials $u(x)$ and $q(x)$ of appropriate degree are found. They also showed that for BCH codes the Berlekamp algorithm usually has a better performance than the Euclidean algorithm. A decoding procedure based on continued fractions for separable Goppa codes was presented by Goppa in [50] and later for general Goppa codes in [51]. The relation of Berlekamp's algorithm to continued fraction techniques was pointed out by Mills [74] and thoroughly studied by Welch and Scholtz [132].

Cheng [21] analysed that the sequence ℓ_j provides the information when Berlekamp's algorithm completes one iterative step of the continued fraction, which happens when $\ell_j < j+\frac{1}{2}$ and when $\ell_j \neq \ell_{j+1}$. This means that if this latter condition is fulfilled, the polynomials $q_j(x)$ and $u_j(x)$ computed so far give the approximation $\frac{q_j(x)}{u_j(x)}$ to $F(x)$, which would also be obtained as convergent from the continued fractions expansion of $F(x)$.

Indeed, the speed of the Berlekamp-Massey algorithm is due to the fact that it constructs the polynomials $u_j(x)$ in the denominator of the convergent to $F(x)$ via the three-term recursion

$$u_j(x) = u_{j-1}(x) - \frac{r_j}{r_m} x^{j-m} u_{m-1}(x).$$

Here r_m and r_j are different from 0 and $r_{m+1} = ... = r_{j-1} = 0$, which means that in (50) $\delta_{m+1} = \cdots = \delta_{j-1} = 0$ and $\delta_j = 1$, such that at time j for the first time after m a new shift register must be designed. This fact can be proved inductively as implicit in [13], p. 374. An approach reflecting the mathematical background of these "jumps" via the Iohvidov index of the Hankel matrix or the block structure of the Padé table is carried out by Jonckheere and Ma [65].

Several authors (e.g. [69], p. 156, [14, 64, 65]) point out that the proof of the above recurrence is quite complicated or that there is need for a transparent explanation. We shall see now that the analysis is much simpler for the case that all principle submatrices of the Hankel matrix A_n are nonsingular. As a useful application, then the r_j's yield the parameters from the three-term recurrence of the underlying polynomials. Via (48) the three-term recurrence can also be transferred to the case that calculations are carried out over finite fields.

So, let us assume from now on that all principal submatrices A_i, $i \leq n$ of the Hankel matrix A_n are nonsingular. For this case, Imamura and Yoshida [64] demonstrated that $\ell_j = \ell_{j-1} = \frac{j}{2}$ for even j and $\ell_j = j - \ell_{j-1} = \frac{j+1}{2}$ for odd j such that δ_j is 1 if j is odd and 0 if j is even ($\frac{q_{2j}(x)}{u_{2j}(x)}$ then are the convergents to $F(x)$).

This means that there are only two possible recursions for $u_j(x)$ depending on the parity of j, namely

$$u_{2j}(x) = u_{2j-1}(x) - \frac{r_{2j}}{r_{2j-1}} x u_{2j-2}(x), \quad u_{2j-1}(x) = u_{2j-2}(x) - \frac{r_{2j-1}}{r_{2j-3}} x^2 u_{2j-4}(x).$$

So the algorithm is simplified in (50) and we obtain the recursion

$$\begin{pmatrix} u_{2j}(x) \\ v_{2j}(x) \end{pmatrix} = \begin{pmatrix} 1 - \frac{r_{2j}}{r_{2j-1}} x & -r_{2j-1} x \\ \frac{1}{r_{2j-1}} x & 0 \end{pmatrix} \cdot \begin{pmatrix} u_{2j-2}(x) \\ v_{2j-2}(x) \end{pmatrix}. \tag{53}$$

By the above considerations we have the following three-term recurrence for $u_{2j}(x)$ (and also for $q_{2j}(x)$ with different initial values).

$$u_{2j}(x) = (1 - \frac{r_{2j}}{r_{2j-1}}x)u_{2j-2}(x) - \frac{r_{2j-1}}{r_{2j-3}}x^2 u_{2j-4}(x).$$

Since the Berlekamp - Massey algorithm determines the solution of equation (9) it must be

$$x^j \cdot u_{2j}(\frac{1}{x}) = t_j(x).$$

as under (8). This is consistent with (16) where we consider the function $F(\frac{1}{x})$ rather than $F(x)$. By the previous considerations, for $t_j(x)$, we have the recurrence

$$t_j(x) = (x - \frac{r_{2j}}{r_{2j-1}})t_{j-1}(x) - \frac{r_{2j-1}}{r_{2j-3}}t_{j-2}(x) \tag{54}$$

Equation (54) now allows us to give a simple interpretation of the calculations in the single steps carried out in the course of the Berlekamp-Massey algorithm for the special case that all principle submatrices of the Hankel matrix A_n are nonsingular.

Proposition 6.3 *Let A_n be a Hankel matrix with real entries such that all principal submatrices A_i, $i = 1, \dots, n$ are nonsingular and let T be the linear operator mapping $T(x^l) = c_l$ as in (9). Then for the parameters r_j obtained via (49) it is*

$$r_{2j-1} = T(x^{j-1} \cdot t_{j-1}(x)) = c_0\beta_1\beta_2 \cdots \beta_{j-1},$$

$$r_{2j} = \alpha_j T(x^{j-1} \cdot t_{j-1}(x)) = c_0\beta_1\beta_2 \cdots \beta_{j-1}\alpha_j, \tag{55}$$

where α_j and $\beta_1, \dots, \beta_{j-1}$ are the parameters from the three-term recurrence of the orthogonal polynomials $t_i(x)$, $i = 0, \dots, j$.

Proof The proposition, of course, follows directly from (54), since the three-term recurrence immediately yields the formula for the r_j's. Let us also verify the identities directly. From the considerations under (49)–(54) it is clear that the degree of u_{2j-2} is $j-1$. Hence in this case $b_{2j-2,j} = b_{2j-2,j+1} = \dots = b_{2j-2,2j-2} = 0$ in (49) and

$$r_{2j-1} = \sum_{t=0}^{j-1} b_{2j-2,t}c_{2j-2-t} = \sum_{t=0}^{j-1} b_{2j-2,t}T(x^{2j-2-t})$$

$$= T\left(\sum_{t=0}^{j-1} b_{2j-2,t}x^{2j-2-t}\right) = T\left(x^{j-1}\sum_{t=0}^{j-1} b_{2j-2,t}x^{j-1-t}\right) = T\left(x^{j-1}\sum_{t=0}^{j-1} b_{2j-2,j-1-t}x^t\right)$$

$$= T\left(x^{j-1}\sum_{t=0}^{j-1} a_{j-1,t}x^t\right) = T(x^{j-1}t_{j-1}(x)) = c_0\beta_1\beta_2\cdots\beta_{j-1}$$

where the last equation follows by (5.3). A similar calculation shows that

$$r_{2j} = T\left(x^j t_{j-1}(x) - \frac{r_{2j-1}}{r_{2j-3}}x^{j-1}t_{j-2}(x)\right) = T\left(x^j t_{j-1}(x) - \beta_{j-1}x^{j-1}t_{j-2}(x)\right)$$

since by the previous calculation $\frac{r_{2j-1}}{r_{2j-3}} = \beta_{j-1}$. So by (46) further

$$r_{2j} = c_0\beta_1\beta_2\cdots\beta_{j-1}\left[(\alpha_1+\alpha_2+\cdots+\alpha_j) - (\alpha_1+\alpha_2+\cdots+\alpha_{j-1})\right] = c_0\beta_1\beta_2\cdots\beta_{j-1}\alpha_j.$$

□

Remarks

1. Observe that with Proposition 6.3, the Berlekamp-Massey algorithm can be applied to determine the coefficients α_j and β_j from the three-term recurrence of the orthogonal polynomials $t_j(x)$. From the parameters r_{2j-1} obtained by (49) in the odd steps of the iteration $\beta_{j-1} = \frac{r_{2j-1}}{r_{2j-3}}$ can be immediately calculated, and in the even steps $\alpha_j = \frac{r_{2j}}{r_{2j-1}}$ is obtained. By (15) and (20) it is $\beta_{j-1} = \frac{r_{2j-1}}{r_{2j-3}} = \frac{\det(A_j)\det(A_{j-2})}{\det(A_{j-1})^2}$. Hence $r_{2j-1} = \frac{\det(A_j)}{\det(A_{j-1})}$, which means that the Berlekamp-Massey algorithm also yields a fast procedure to compute the determinant of a Hankel matrix.

2. By Proposition 6.3 the identity (49) reduces to

$$\sum_{t=0}^{j} a_{j,t}c_{j+t} = c_0\beta_1\beta_2\cdots\beta_j$$

where the $a_{j,t}$ are the coefficients of the polynomial $t_j(x)$, the β_i's are the coefficients in their three-term recurrence and the c_i's are the corresponding moments. For the classical orthogonal polynomials all these parameters are usually known, such that one might also use (49) in the Berlekamp-Massey algorithm to derive combinatorial identities.

6.3.6 Lattice Paths not Touching a Given Boundary

Introduction

A path starting in the origin of the lattice $\{(x, y) : x, y \text{ integers}\}$ of pairs of integers here is a sequence of pairs (x_i, y_i) of nonnegative integers where $(x_0, y_0) = (0, 0)$ and (x_i, y_i) is either $(x_{i-1}+1, y_{i-1})$ or $(x_{i-1}, y_{i-1}+1)$. So, a particle following such a path can move either one step to the right, i.e. $x_i = x_{i-1} + 1$, or one step upwards, i.e. $y_i = y_{i-1} + 1$ in each time unit i.

Several methods for the enumeration of lattice paths are discussed in the books by Mohanty [77] and Narayana [80]. For the number of paths $N(\underline{u}, n)$ first touching the boundary $(0, u_0), (1, u_1), (2, u_2), \dots$ in $(n-1, u_{n-1})$ (and not touching or crossing this boundary before) characterized by the infinite nondecreasing sequence $\underline{u} = (u_0, u_1, u_2, \dots)$ of nonnegative integers the following recursion is presented in [80], p. 21.

$$N(\underline{u}, n) = \sum_{j=1}^{n} (-1)^{j-1} \binom{u_{n-j}+1}{j} \cdot N(\underline{u}, n-j).$$

One might further be interested in an expression of closed form. For instance, if the boundary is given by the sequence $\underline{u} = (1, 2, 3, \dots)$, then $N(\underline{u}, n)$ is the nth Catalan number $\frac{1}{2n+1}\binom{2n+1}{n}$ and, more generally, for $\underline{u} = (1 + (p-1) \cdot n)_{n=0,1,2,\dots}$ as counting function arise the generalized Catalan numbers $\frac{1}{pn+1}\binom{pn+1}{n}$. (The notion “generalized Catalan numbers” as in [63] is not standard, for instance, in [54], pp. 344–350 it is suggested to denote them “Fuss numbers”).

Note that this describes the case in which the sequence of differences $(u_m - u_{m-1})_{m=1,2,\dots}$ is periodic with period length 1. We shall derive similar identities for period length 2, hereby following a probabilistic method introduced by Gessel [41], which allows to apply Lagrange inversion. For instance, it can be shown that if $\underline{u}^{(1)}$ and $\underline{u}^{(2)}$ are such that $u_{2i}^{(1)} = u_{2i}^{(2)} = s+ci$, $u_{2i+1}^{(1)} = s+\mu+ci$ and $u_{2i+1}^{(2)} = s+(c-\mu)+ci$, then

$$N(\underline{u}^{(1)}, 2n) + N(\underline{u}^{(2)}, 2n) = \frac{2}{(c+2)n+1}\binom{(c+2)n+1}{2n} \tag{56}$$

By the same approach, a new expression for the number of paths not crossing or touching the line $cx = 2y$ for odd c will be obtained.

Further, an application of (56) in the analysis of two-dimensional arrays will be studied. For $i = 1, 2, \dots$ let $\lambda_i^{(\nu)}$ denote the frequency of the number i in the sequences $\underline{u}^{(\nu)}$ describing two boundaries for $\nu = 1, 2$ and let $\lambda^{(\nu)} = (\lambda_1^{(\nu)}, \lambda_2^{(\nu)}, \dots)$. Denoting by $\gamma^{(\nu)}(n, k)$ the number of paths from the origin to (n, k) not touching or crossing the boundary described by $\underline{u}^{(\nu)}$, in the case that $\underline{\lambda}^{(1)} = (\lambda, c-\lambda, \lambda, c-\lambda, \dots)$ and $\underline{\lambda}^{(2)} = (c-\lambda, \lambda, c-\lambda, \lambda, \dots)$ are both periodic with period length 2 it is

$$\gamma^{(1)}(n, k) + \gamma^{(2)}(n, k) = 2 \cdot \binom{n+k}{k} - c \cdot \binom{n+k}{k-1} \tag{57}$$

which can be derived using (56). Further, (57) can be regarded as a generalization of the ballot numbers $\binom{n+k}{k} - \binom{n+k}{k-1}$.

From (57) results are immediate for the numbers $\alpha^{(\nu)}(n, k) = \gamma^{(\nu)}(n+k, k)$. Such a two-dimensional array $\alpha(n, k)$ had been found by Berlekamp in the study of burst-error correcting convolutional codes and thoroughly analyzed by Carlitz, Rosselle, and Scoville [18].

Gessel's Probabilistic Approach

We shall consider paths in an integer lattice from the origin (0, 0) to the point (n, u_n), which never touch any of the points (m, u_m), $m = 0, 1, \dots, n-1$. In [41] Gessel introduced a general probabilistic method to determine the number of such paths, denoted by f_n, which he studied for the case that the subsequence $(u_m)_{m=1,2\dots}$ is periodic.

In this case the elements of the sequence $(u_m)_{m=0,1,2,\dots}$ are on the d lines (for $i = 0, 1, 2, \dots$)

$$u_{di} = \mu_0 + ci \text{ and } u_{di+1} = \mu_0 + \mu_1 + ci, \quad \dots \quad, u_{di+d-1} = \mu_0 + \mu_1 + \dots + \mu_{d-1} + ci \tag{58}$$

so $\mu_0 = u_0 > 0$, and $c = \mu_1 + \mu_2 + \dots + \mu_d$, where $\mu_j = u_j - u_{j-1}$ for $j = 1, \dots, d$.

Gessel's probabilistic method is as follows. A particle starts at the origin (0, 0) and successively moves with probability p one unit to the right and with probability $q = 1 - p$ one unit up. The particle stops if it touches one of the points (i, u_i).

The probability that the particle stops at (n, u_n) is $p^n q^{u_n} \cdot f_n$, which is $p^n q^{\mu_0 + \dots + \mu_j + cn}$ if $n \equiv j \mod d$. Setting

$$f(t) = \sum_{n=0}^{\infty} f_n t^n = \sum_{j=0}^{d-1} t^j f^{(j)}(t^d)$$

(so $f^{(j)}(t) = \sum_{n=0}^{\infty} f_n^{(j)} t^{dn} = \sum_{n=0}^{\infty} f_{dn+j} t^n$ are the generating functions for the f_n's with indices congruent j modulo d), the probability that the particle eventually stops is

$$q^{u_0} f^{(0)}(p^d q^c) + p q^{u_1} f^{(1)}(p^d q^c) + p^2 q^{u_2} f^{(2)}(p^d q^c) + \dots + p^{d-1} q^{u_{d-1}} f^{(d-1)}(p^d q^c) = 1$$

where $u_j = \mu_0 + \dots + \mu_j$.

If p is sufficiently small, the particle will touch the boundary $(m, u_m)_{m=0,1,\dots}$, or equivalently, enter the *forbidden area*, i.e. the lattice points on and behind this boundary, with probability 1. So for small p and with $t = pq^{c/d}$ it is

$$q(t)^{u_0} f^{(0)}(t^d) + p(t) q(t)^{u_1} f^{(1)}(t^d) + \dots + p(t)^{d-1} q(t)^{u_{d-1}} f^{(d-1)}(t^d) = 1$$

For p sufficiently small one may invert $t = p(1-p)^{c/d}$ to express p as a power series in t, namely $p = p(t)$. Then changing t to $\omega^i t$, $i = 1, \dots, d-1$, where ω is a primitive dth root of unity, yields the system of equations

$$A \cdot \begin{pmatrix} f^{(0)}(t^d) \\ f^{(1)}(t^d) \\ \vdots \\ f^{(d-1)}(t^d) \end{pmatrix} = \begin{pmatrix} 1 \\ 1 \\ \vdots \\ 1 \end{pmatrix}. \tag{59}$$

with $A = (p(\omega^i t)^j q(\omega^i t)^{u_j})_{i,j=0,\ldots,d-1}$, from which the functions $f^{(j)}(t^d)$, $j = 0,\ldots,d-1$ might be determined.

For period length $d = 1$ the interpretations for the generalized Catalan numbers $\frac{1}{pn+1}\binom{pn+1}{n}$ and the numbers $\frac{1}{pn+p-1}\binom{pn+p-1}{n+1}$ in terms of lattice paths given in Sect. 6.3.3 can easily be derived by (59).

We shall now take a closer look at the period length $d = 2$.

Let us denote $s = \mu_0$ and $\mu = \mu_1$. Then the boundary $(n, u_n)_{n=0,1,\ldots}$ is characterized by

$$u_{2i} = s + ci \text{ and } u_{2i+1} = s + \mu + ci, \tag{60}$$

Further, denoting $p(-t)$ by $\overline{p}(t)$ and similarly $q(-t)$ by $\overline{q}(t)$ and setting $g(t^2) = f^{(0)}(t^2)$ and $h(t^2) = f^{(1)}(t^2)$ (as in [41]) we obtain the two equations

$$q^s \cdot g(t^2) + p \cdot q^{s+\mu} \cdot h(t^2) = 1,$$

$$\overline{q}^s \cdot g(t^2) + \overline{p} \cdot \overline{q}^{s+\mu} \cdot h(t^2) = 1$$

which for $g(t^2)$ and $h(t^2)$ yield the solutions

$$g(t^2) = \frac{p^{-1}q^{-s-\mu} - \overline{p}^{-1}\overline{q}^{-s-\mu}}{p^{-1}q^{-\mu} - \overline{p}^{-1}\overline{q}^{-\mu}} = \frac{q^{c/2-\mu-s} + \overline{q}^{c/2-\mu-s}}{q^{c/2-\mu} + \overline{q}^{c/2-\mu}} \tag{61}$$

and

$$h(t^2) = \frac{q^{-s} - \overline{q}^{-s}}{t \cdot (q^{\mu-c/2} + \overline{q}^{\mu-c/2})} \tag{62}$$

By Lagrange inversion (cf. e.g. [42], pp. 1032–1034) for any α it is

$$q^{-\alpha} = \sum_{n=0}^{\infty} \frac{\alpha}{(c/2+1)n+\alpha} \binom{(c/2+1)n+\alpha}{n} \cdot t^n \tag{63}$$

Gessel analyzed the case $\mu = \lambda$, $c = 2\lambda + 1$ for a positive integer λ, which arises in the enumeration of paths never touching or crossing the line $y = s - \frac{1}{2} + \frac{c}{2} \cdot x$. For the special case $s = 1$ he derived the following nice identity for the function $h(t^2)$

Proposition 6.4 (Gessel 1986, [41]) *Let c be an odd positive integer, $s = 1$ and $\mu = \frac{c-1}{2}$. Then*

$$h(t^2) = \frac{q^{-1/2} - \overline{q}^{-1/2}}{t} = \sum_{n=0}^{\infty} \frac{1}{(c+2)n+\mu+2} \binom{(c+2)n+\mu+2}{2n+1} t^{2n},$$

So, the coefficients in the expansion of $h(t^2)$ have a similar form as the Catalan numbers. It is also possible to show that for these parameters

$$g(t^2) = \sum_{n=0}^{\infty} \frac{1}{(c+2)n+1} \binom{(c+2)n+1}{2n} t^{2n} - \frac{t^2}{2}[h(t^2)]^2$$

This is a special case of a more general result which we are going to derive now. Since we are going to look at several random walks in parallel, we shall introduce the parameters determining the restrictions as a superscript to the generating functions. So, $g^{(s,c,\mu)}$ and $h^{(s,c,\mu)}$ are the generating functions for even and odd n, respectively, for the random walk of a particle starting in the origin and first touching the boundary $(i, u_i)_{i=0,1,\dots}$ determined by the parameters s, c, and μ as in (60) in (n, u_n).

Proposition 6.5 *Let s, c, μ be the parameters defined above with $0 \le \mu < \frac{c}{2}$.*

(i)

$$g^{(s,c,\mu)}(t^2) + g^{(s,c,c-\mu)}(t^2) = q^{-s} + \overline{q}^{-s} = \sum_{n=0}^{\infty} \frac{2s}{(c+2)n+s} \binom{(c+2)n+s}{2n} t^{2n}$$

(ii)

$$g^{(s,c,c-\mu)}(t^2) - g^{(s,c,\mu)}(t^2) = t^2 \cdot h^{(s,c,\mu)}(t^2) \cdot h^{(c-2\mu,c,\mu)}(t^2)$$

Proof (i) In order to derive the first identity observe that with $a = \frac{c}{2} - \mu$ it is

$$g^{(s,c,\mu)}(t^2) + g^{(s,c,c-\mu)}(t^2) = \frac{q^{a-s} + \overline{q}^{a-s}}{q^a + \overline{q}^a} + \frac{q^{-a-s} + \overline{q}^{-a-s}}{q^{-a} + \overline{q}^{-a}}$$

$$= \frac{(q^{a-s} + \overline{q}^{a-s})(q^{-a} + \overline{q}^{-a}) + (q^{-a-s} + \overline{q}^{-a-s})(q^a + \overline{q}^a)}{(q^a + \overline{q}^a)(q^{-a} + \overline{q}^{-a})}$$

$$= \frac{2q^{-s} + 2\overline{q}^{-s} + \overline{q}^{a-s}q^{-a} + q^{-a-s}\overline{q}^a + q^{a-s}\overline{q}^{-a} + \overline{q}^{-a-s}q^a}{2 + \overline{q}^a q^{-a} + q^a \overline{q}^{-a}} = q^{-s} + \overline{q}^{-s}$$

Since by definition $\overline{q}(t) = q(-t)$, with Lagrange inversion it is

$$q^{-s} + \overline{q}^{-s} =$$

$$= \sum_{n=0}^{\infty} \frac{s}{(c/2+1)n+s} \binom{(c/2+1)n+s}{n} \cdot t^n + \sum_{n=0}^{\infty} \frac{s}{(c/2+1)n+s} \binom{(c/2+1)n+s}{n} \cdot (-t)^n$$

$$= \sum_{n=0}^{\infty} \frac{2s}{(c+2)n+s} \binom{(c+2)n+s}{2n} t^{2n}.$$

(ii) Let again $a = \frac{c}{2} - \mu$. Then

$$g^{(s,c,c-\mu)}(t^2) - t^2 \cdot h^{(s,c,\mu)}(t^2) \cdot h^{(c-2\mu,c,\mu)}(t^2) = \frac{q^{-a-s} + \overline{q}^{-a-s}}{q^{-a} + \overline{q}^{-a}} - t^2 \frac{q^{-s} - \overline{q}^{-s}}{t(q^{-a} + \overline{q}^{-a})} \cdot \frac{q^{-2a} - \overline{q}^{-2a}}{t(q^{-a} + \overline{q}^{-a})}$$

$$= \frac{q^{-a-s} + \overline{q}^{-a-s}}{q^{-a} + \overline{q}^{-a}} - \frac{(q^{-s} - \overline{q}^{-s})(q^{-a} - \overline{q}^{-a})}{(q^{-a} + \overline{q}^{-a})} = \frac{q^{-a}\overline{q}^{-s} + \overline{q}^{-a}q^{-s}}{q^{-a} + \overline{q}^{-a}} = \frac{q^{a-s} + \overline{q}^{a-s}}{q^{a} + \overline{q}^{a}} = g^{(s,c,\mu)}(t^2)$$

□

Similar identities can be derived for the case $s + \mu = c$.

Proposition 6.6 *Let $c > 0$ be a positive integer, and $s + \mu = c$ with $s \geq \mu$. Then*
(i)

$$h^{(s,c,c-s)}(t^2) + h^{(c-s,c,s)}(t^2) = \frac{1}{t^2} \cdot (p + \overline{p}) = \sum_{n=1}^{\infty} \frac{2}{(c+2)n - 1} \binom{(c+2)n - 1}{2n} \cdot t^{2(n-1)}$$

(ii) In the special case c odd, $s = \frac{c+1}{2}$ and $\mu = \frac{c-1}{2}$ it is

$$h^{(\frac{c+1}{2},c,\frac{c-1}{2})}(t^2) - h^{(\frac{c-1}{2},c,\frac{c+1}{2})}(t^2) = \left(g^{(\frac{c+1}{2},c,\frac{c-1}{2})}(t^2)\right)^2$$

where

$$g^{(\frac{c+1}{2},c,\frac{c-1}{2})}(t^2) = \frac{1}{t} \cdot (\overline{q}^{\frac{1}{2}} - q^{\frac{1}{2}}) = \sum_{n=0}^{\infty} \frac{1}{(c+2)n + \frac{c+1}{2}} \binom{(c+2)n + \frac{c+1}{2}}{2n+1} \cdot t^{2n}$$

Proof (i) By (62)

$$h^{(s,c,c-s)}(t^2) + h^{(c-s,c,s)}(t^2) = \frac{q^{-s} - \overline{q}^{-s}}{t \cdot (q^{c/2-s} + \overline{q}^{c/2-s})} + \frac{q^{s-c} - \overline{q}^{s-c}}{t \cdot (q^{s-c/2} + \overline{q}^{s-c/2})}$$

$$= \frac{q^{-s} - \overline{q}^{-s}}{pq^{c-s} - \overline{pq}^{c-s}} + \frac{q^{s-c} - \overline{q}^{s-c}}{pq^{s} - \overline{pq}^{s}} = \frac{2(p + \overline{p}) - q^s\overline{q}^{-s}(p + \overline{pq}^c q^{-c}) - \overline{q}^s q^{-s}(\overline{p} + pq^c\overline{q}^{-c})}{p^2q^c + \overline{p}^2\overline{q}^c - q^s\overline{q}^{-s}(p\overline{pq}^c) - \overline{q}^s q^{-s}(p\overline{pq}^c)}$$

$$= \frac{(p + \overline{p})(2 - q^s\overline{q}^{-s}(p/\overline{p}) - \overline{q}^s q^{-s}(\overline{p}/p))}{t^2(2 - q^s\overline{q}^{-s}(p/\overline{p}) - \overline{q}^s q^{-s}(\overline{p}/p))} = \frac{p + \overline{p}}{t^2}$$

since $p^2q^c = \overline{p}^2\overline{q}^c = t^2$ by definition of t and since $\overline{p}(p + \overline{pq}^c q^{-c}) = p(p + \overline{p})$.
(ii) With $s = \frac{c+1}{2}$ again by (62) as under (i)

$$h^{(s,c,c-s)}(t^2) - h^{(c-s,c,s)}(t^2) = \frac{q^{-s} - \overline{q}^{-s}}{pq^{c-s} - \overline{pq}^{c-s}} - \frac{q^{s-c} - \overline{q}^{s-c}}{pq^s - \overline{pq}^s}$$

$$= \frac{\overline{pq}^{c-s}q^{-(c-s)} + pq^{c-s}\overline{q}^{-(c-s)} - \overline{pq}^s q^{-s} - pq^s\overline{q}^{-s}}{t^2(2 - (pq^s)/(\overline{pq}^s) - (\overline{pq}^s)/(pq^s))}$$

$$= \frac{\overline{pq}^{\frac{c-1}{2}} q^{-\frac{c+1}{2}}(q-\overline{q}) - pq^{\frac{c-1}{2}}\overline{q}^{-\frac{c+1}{2}}(q-\overline{q})}{t^2(2-(tq^{\frac{1}{2}})/(-t\overline{q}^{\frac{1}{2}}) - (-t\overline{q}^{\frac{1}{2}})/(tq^{\frac{1}{2}}))} = \frac{q^{-\frac{1}{2}}\overline{q}^{-\frac{1}{2}}(q-\overline{q})(\overline{pq}^{\frac{c}{2}}q^{-\frac{c}{2}} - pq^{\frac{c}{2}}\overline{q}^{-\frac{c}{2}})}{t^2(2+q^{\frac{1}{2}}/\overline{q}^{\frac{1}{2}} + \overline{q}^{\frac{1}{2}}/q^{\frac{1}{2}})}$$

$$= \frac{q^{-\frac{1}{2}}\overline{q}^{-\frac{1}{2}}(q-\overline{q})(\overline{p}-p)}{t^2 q^{-\frac{1}{2}}\overline{q}^{-\frac{1}{2}}(2q^{\frac{1}{2}}\overline{q}^{\frac{1}{2}} + q + \overline{q})} = \frac{(\overline{p}-p)^2}{t^2(q^{\frac{1}{2}} + \overline{q}^{\frac{1}{2}})^2} = \left(g^{(\frac{c+1}{2},c,\frac{c-1}{2})}(t^2)\right)^2$$

since $t = pq^{\frac{c}{2}} = -\overline{pq}^{\frac{c}{2}}$ and $p = 1-q, \overline{p} = 1-\overline{q}$ and by (61)

$$g^{(\frac{c+1}{2},c,\frac{c-1}{2})}(t^2) = \frac{q^{-c/2} + \overline{q}^{-c/2}}{q^{1/2} + \overline{q}^{1/2}} = \frac{p-\overline{p}}{t\cdot(q^{1/2}+\overline{q}^{1/2})} = \frac{\overline{q}-q}{t\cdot(q^{1/2}+\overline{q}^{1/2})} = \frac{1}{t}(\overline{q}^{1/2} - q^{1/2})$$

□

Further, several convolution identities for the generating functions can be derived. For instance:

Proposition 6.7 *(i)*

$$\left(g^{(s,c,\mu)}(t^2) + g^{(s,c,c-\mu)}(t^2)\right)\cdot h^{(s,c,\mu)}(t^2) = h^{(2s,c,\mu)}(t^2)$$

(ii)

$$g^{(c-2\mu,c,\mu)}(t^2)\cdot g^{(\mu,c,c-\mu)}(t^2) = g^{(c-\mu,c,\mu)}(t^2)$$

(iii) For $s_1 + \mu_1 + \mu_2 = c$ it is

$$g^{(s_1,c,\mu_1)}(t^2)\cdot h^{(s_2,c,\mu_2)}(t^2) = h^{(s_2,c,s_1+\mu_2)}(t^2)$$

Especially, for odd c

$$g^{(1,c,\frac{c-1}{2})}(t^2)\cdot h^{(1,c,\frac{c-1}{2})}(t^2) = h^{(1,c,\frac{c+1}{2})}(t^2)$$

Proof (i) is immediate from the fact that $g^{(s,c,\mu)}(t^2) + g^{(s,c,c-\mu)})(t^2) = q^{-s} + \overline{q}^{-s}$ (Proposition 6.5(i)) and (ii) is immediate, since the nominator of $g^{(c-2\mu,c,\mu)}(t^2)$ in (61) is at the same time denominator of $g^{(\mu,c,c-\mu)}(t^2)$. The nominator of $g^{(s_1,c,\mu_1)}(t^2)$ in (iii) by (61) is $q^{c/2-\mu_1-s_1} + \overline{q}^{c/2-\mu_1-s_1}$ and this is the term in brackets in the denominator in (62) of $h^{(s_2,c,\mu_2)}(t^2) = \frac{q^{-s_2}-\overline{q}^{-s_2}}{t(q^{\mu_2-c/2}+\overline{q}^{-\mu_2-c/2})}$. □

Let us discuss the case $c = 3$ a little closer and hereby illustrate the derived identities. The parameter choices ($s = 1, \mu = 1$), ($s = 1, \mu = 2$), and ($s = 2, \mu = 1$) will be of interest in the combinatorial applications, we shall speak about later on. By application of the previous results, the generating functions for these parameters (after mapping $t^2 \to x$) look as follows. Observe that they all can be expressed in terms of $a(x) := g^{(1,3,1)}(x)$ and $b(x) := g^{(1,3,2)}(x)$.

Corollary 6.7

$$a(x) = g^{(1,3,1)}(x) = \sum_{n=0}^{\infty} \frac{1}{5n+1}\binom{5n+1}{2n} x^n - \frac{x}{2} \cdot [h^{(1,3,1)}(x)]^2 = 1 + 2x + 23x^2 + 377x^3 + \dots$$

$$b(x) = g^{(1,3,2)}(x) = \sum_{n=0}^{\infty} \frac{1}{5n+1}\binom{5n+1}{2n} x^n + \frac{x}{2} \cdot [h^{(1,3,1)}(x)]^2 = 1 + 3x + 37x^2 + 624x^3 + \dots$$

$$g^{(2,3,1)}(x) = \sum_{n=0}^{\infty} \frac{1}{5n+2}\binom{5n+2}{2n+1} x^n = 1 + 5x + 66x^2 + 1156x^3 + \dots = a(x) \cdot b(x)$$

$$h^{(1,3,1)}(x) = \sum_{n=0}^{\infty} \frac{1}{5n+3}\binom{5n+3}{2n+1} x^n = 1 + 7x + 99x^2 + 1768x^3 + \dots = a(x)^2 \cdot b(x)$$

$$h^{(1,3,2)}(x) = \sum_{n=1}^{\infty} \frac{1}{5n-1}\binom{5n-1}{2n} x^{n-1} - \frac{1}{2}[g^{(2,3,1)}(x)]^2 = 1 + 9x + 136x^2 + \dots = a(x)^3 \cdot b(x)$$

$$h^{(2,3,1)}(x) \sum_{n=1}^{\infty} \frac{1}{5n-1}\binom{5n-1}{2n} x^{n-1} + \frac{1}{2} \cdot [g^{(2,3,1)}(x)]^2 = 2 + 19x + 293x^2 + 5332x^3 + \dots$$

$$= h^{(1,3,2)}(x) + [g^{(2,3,1)}(x)]^2 = a(x)^3 \cdot b(x) + a(x)^2 \cdot b(x)^2$$

$$= (a(x) + b(x)) \cdot a(x)^2 \cdot b(x) = (g^{(1,3,1)}(x) + g^{(1,3,2)}(x)) \cdot h^{(1,3,1)}(x)$$

It is also possible to express all six functions in terms of either $a(x)$ or $b(x)$, namely it can be shown that

$$b(x) = \frac{a(x)}{2}\left[(a(x) - 1) + \sqrt{(a(x) - 1)^2 + 4)}\right], \qquad a(x) = \frac{b(x)}{2(b(x) + 1)}\left[1 + \sqrt{4b(x) + 2}\right]$$

As pointed out before, as an example to illustrate his probabilistic approach Gessel in [41] analyzed half-integer slopes for odd c and $d = 2$ hereby counting paths starting in the origin and not touching the line $y = r + \frac{c}{2}x$ before (n, u_n). This line determines a boundary, which is given as in (60) by the parameters $s = r + \frac{1}{2}$, $\mu = \frac{c-1}{2}$ if r is a half-integer and $s = r$, $\mu = \frac{c+1}{2}$ if r is an integer. The number of paths first touching the line $y = r + \frac{c}{2}x$ in $(2n, u_{2n})$ then obviously is the nth coefficient of $g^{(s,c,\mu)}(x)$.

Observe that the original approach only works for $s > 0$, since for $s = 0$ the system of equations $g(t^2) + pq^\mu h(t^2) = 1$, $g(t^2) + \overline{pq}^\mu h(t^2) = 1$ does not yield a solution.

Several authors studied the number of paths starting in the origin and hereafter touching the line $cx = dy$ for the first time in (dn, cn) (the only intersections of the line with the integer lattice when c and d are coprime). In [77] on pp. 12–14 a recursive approach due to Bizley is described. Namely, denoting by f_n the number of such paths to (dn, cn) it is

$$(c+d)f_1 = \binom{c+d}{d}, 2(c+d)f_2 = \binom{2(c+d)}{2d} - \binom{c+d}{d} f_1$$

$$3(c+d)f_3 = \binom{3(c+d)}{3d} - \binom{2(c+d)}{2d} f_1 - \binom{c+d}{d} f_2, \dots$$

As an example, for $c = 3$ and $d = 2$ this recursion yields the numbers $f_1 = 2$, $f_2 = 19$, $f_3 = 293, \dots$. These are just the coefficients in $h^{(2,3,1)}(x)$ studied in Corollary 6.7 and this holds in wider generality.

Let us consider $d = 2$. Assume that the first step from the origin is to the right (by reversing the paths, i.e. mapping the path $(0, 0), \dots, (nd, nc)$ to $(nd, nc), \dots, (0, 0)$ the analysis for a first step upwards is analogous). Then, after this first step, the boundary is given by the parameters $s = \frac{c+1}{2}$ and $\mu = \frac{c-1}{2}$ where contrasting to the original model now $s = u_1$ (and not $s = u_0$). This has the effect that the generating function for the paths to (n, u_n) with even n now is $h^{(\frac{c+1}{2}, c, \frac{c-1}{2})}$. By Proposition 6.6 hence

Theorem 6.4 *The number of paths from the origin first touching the line $cx = 2y$ in $(2n, cn)$, $n \geq 1$ and not crossing or touching this line before is the coefficient of $t^{2(n-1)}$ in*

$$\sum_{n=1}^{\infty} \frac{1}{(c+2)n-1} \binom{(c+2)n-1}{2n} \cdot t^{2(n-1)} + \frac{1}{2} \left(\sum_{n=0}^{\infty} \frac{1}{(c+2)n + \frac{c+1}{2}} \binom{(c+2)n + \frac{c+1}{2}}{2n+1} \cdot t^{2n} \right)^2$$

Two-Dimensional Arrays Generalizing the Ballot Numbers

We saw that we have to enumerate lattice paths not touching a given boundary. This immediately yields a fast algorithm to determine these numbers recursively. Since the lattice paths arriving in (n, k) - by definition of the single steps - must pass either $(n, k-1)$ or $(n-1, k)$, the number $\gamma(n, k)$ of paths from the origin $(0, 0)$ to (n, k) obeys the recursion

$$\gamma(n, k) = \gamma(n, k-1) + \gamma(n-1, k)$$

with initial values

$$\gamma(0, 0) = 1, \gamma(n, u_n) = 0 \text{ for all } n.$$

The initial values just translate the fact that the boundary $(n, u_n), n = 0, 1, 2, \dots$ cannot be touched.

Let

$$\underline{u} = (u_0, u_1, u_2, \dots)$$

be the vector representing the boundary $(m, u_m)_{m=0,1,\dots}$ which is not allowed to be crossed or touched by a path in a lattice and let

$$\underline{\mu} = (\mu_1, \mu_2, \mu_3, \dots)$$

be the sequence of differences $\mu_i = u_i - u_{i-1}$. Let us denote

$$\underline{\lambda} = (\lambda_1, \lambda_2, \lambda_3, \dots) \tag{64}$$

where λ_i counts the frequency of the number i in $\underline{u}$ and let

$$\underline{v} = (v_0, v_1, v_2, \dots) \tag{65}$$

with $v_i = v_0 + \sum_{j=1}^{i} \lambda_j$.

By interchanging the roles of n and k (mapping $(n, k) \to (k, n)$), the pairs $(\underline{u}, \underline{\mu})$ and $(\underline{v}, \underline{\lambda})$ are somehow dual to each other. Namely, consider a path from $(0, 0)$ to (v_k, k) not touching the boundary $(0, u_0), (1, u_1), \dots$. Then the reverse path (just obtained by going backwards from (v_k, k) to $(0, 0)$) corresponds to a path from the origin to (k, v_k) not touching the boundary $(0, v_0), (1, v_0 + \lambda_k), \dots, (k, v_0 + \lambda_k + \lambda_{k-1} + \dots + \lambda_1)$. Hence:

Proposition 6.8 *The number of paths from the origin* $(0, 0)$ *to* (v_k, k)*, where* $v_k = v_0 + \lambda_1 + \lambda_2 + \dots + \lambda_k$ *not touching or crossing the boundary* $(0, u_0), (1, u_1), \dots$ *is the same as the number of paths from the origin to the point* (k, v_k) *which never touch or cross the boundary* $(0, v_0), (1, v_0+\lambda_k), \dots, (k, v_0+\lambda_k+\lambda_{k-1}+\dots+\lambda_1)$.

We shall compare the array γ with a two-dimensional array β with entries $\beta(n, k)$, $n \geq -1, k \geq 0$ defined by

$$\beta(n, k) = \beta(n, k-1) + \beta(n-1, k)$$

with initial values

$$\beta(n, 0) = d \text{ for all } n \geq -1, \beta(-1, k) = -c \text{ for all } k \geq 1.$$

For any d it can be easily verified that

$$\beta(n, k) = d \cdot \binom{n+k}{k} - c \cdot \binom{n+k}{k-1} = \frac{d(n+1) - ck}{n+k+1} \binom{n+k+1}{k}$$

For $d = 1$ this just coincides with the arrays studied by Sulanke [113] defined by $\beta(n, 0) = 1$ for all n and $\beta(ck-1, k) = 0$ for all $k = 1, 2, \dots$. Especially for $c = 2, d = 1$ the positive entries are just the ballot numbers.

When $d \geq 2$, the model studied in [113] is no longer valid, since the arrays contain rows with all entries different from 0. Observe that in each case the entries

$\beta(ck-1, dk) = 0$, when d and c are coprime. However, the results obtained so far now allow us to derive similar identities for the case $d = 2$.

Theorem 6.5 *Let $\gamma^{(1)}(n,k)$ denote the number of paths from the origin to (n,k) not touching or crossing the boundary $(m, u_m^{(1)})_m$ determined as defined above by $\underline{\lambda}^{(1)} = (\lambda_1^{(1)}, \lambda_2^{(1)}, \dots)$ and let $\gamma^{(2)}(n,k)$ denote the number of such paths where the boundary $(m, u_m^{(2)})_m$ is determined by $\underline{\lambda}^{(2)} = (\lambda_1^{(2)}, \lambda_2^{(2)}, \dots)$. If $\underline{\lambda}^{(1)} = (\lambda, c-\lambda, \lambda, c-\lambda, \dots)$ and $\underline{\lambda}^{(2)} = (c-\lambda, \lambda, c-\lambda, \lambda, \dots)$ are periodic with period length 2, then for all k and $n > \max\{\sum_{j=1}^k \lambda_j^{(1)}, \sum_{j=1}^k \lambda_j^{(2)}\}$ it is*

$$\gamma^{(1)}(n,k) + \gamma^{(2)}(n,k) = 2 \cdot \binom{n+k}{k} - c \cdot \binom{n+k}{k-1}.$$

Proof In order to prove the theorem we shall compare the array γ defined by $\gamma(n,k) = \gamma^{(1)}(n,k) + \gamma^{(2)}(n,k)$ with the array β where $\beta(n,k) = 2 \cdot \binom{n+k}{k} - c \cdot \binom{n+k}{k-1}$ and show that $\gamma(n,k) = \beta(n,k)$ for all $n \geq \max\{\sum_{j=1}^k \lambda_j^{(1)}, \sum_{j=1}^k \lambda_j^{(2)}\}$. W. l. o. g. let $\sum_{j=1}^k \lambda_j^{(1)} \geq \sum_{j=1}^k \lambda_j^{(2)}$. Then we are done if we can show that $\gamma(\lambda_1 + \dots + \lambda_k + 1, k) = \beta(\lambda_1 + \dots + \lambda_k + 1, k)$ for all k, since both arrays from then on follow the same recursion. Namely, $\gamma(n,k) = \gamma(n,k-1) + \gamma(n-1,k)$, because $\gamma^{(\nu)}(n,k) = \gamma^{(\nu)}(n,k-1) + \gamma^{(\nu)}(n-1,k)$ for $\nu = 1,2$ and $\beta(n,k) = \beta(n,k-1) + \beta(n-1,k)$ was seen to hold even beyond the boundary.

So let us proceed by induction in k. The induction beginning for $k = 1$ and $k = 2$ is easily verified. Assume that for all $k = 1, 2, \dots, 2K-2$ it is $\gamma(n,k) = \beta(n,k)$ whenever n is big enough as specified in the theorem.

Now observe that since the period length in $\underline{\lambda}^{(1)}$ and $\underline{\lambda}^{(2)}$ is 2, it is

$$\sum_{j=1}^{2K} \lambda_j^{(1)} = \sum_{j=1}^{2K} \lambda_j^{(2)} = cK.$$

This means that for $\nu = 1, 2$ by the Proposition 6.8 $\gamma^{(\nu)}(cK+1, 2K)$ is the number of paths from the origin to $(cK+1, 2K)$ never touching the boundary $(0,1)$, $(1, \lambda_{2K}^{(\nu)} + 1)$, $(2, \lambda_{2K}^{(\nu)} + \lambda_{2K-1}^{(\nu)} + 1), \dots, (2K, \lambda_{2K}^{(\nu)} + \lambda_{2K-1}^{(\nu)} + \dots + \lambda_1^{(\nu)} + 1)$.

These boundaries now are periodic with period length 2 as we studied before. The parameters as in (60) are $s = 1$, c and λ for $\nu = 1$ (or $c - \lambda$ for $\nu = 2$, respectively). The generating functions for the numbers of such paths are $g^{(s,c,\lambda)}(t^2)$ and $g^{(s,c,c-\lambda)}(t^2)$ as studied above and by Proposition 6.5

$$\gamma(cK+1, 2K) = \gamma^{(1)}(cK+1, 2K) + \gamma^{(2)}(cK+1, 2K) = \frac{2}{(c+2)K+1}\binom{(c+2)K+1}{2K}$$

$$= 2 \cdot \binom{(c+2)K}{2K} - c \cdot \binom{(c+2)K}{2K-1}$$

Now observe that also

$$\gamma(cK+1, 2K-1) = \gamma(cK+1, 2K) = \frac{2}{(c+2)K+1}\binom{(c+2)K+1}{2K}$$

because in both arrays $\gamma^{(1)}$ and $\gamma^{(2)}$ all paths from the origin to $(cK+1, 2K)$ must pass through $(cK+1, 2K-1)$. It is also clear that

$$\beta(cK+1, 2K-1) = \beta(cK+1, 2K) = \frac{2}{(c+2)K+1}\binom{(c+2)K+1}{2K}$$

Thus we found that in position $cK+1$ in each of the columns $2K-1$ and $2K$ the two arrays γ and β coincide. Since γ and β obey the same recursion under the boundary $(m, u_m^{(1)})_m$, the theorem is proven. □

Berlekamp at the Waterloo Combinatorics Conference presented an algorithm for computing numbers of the form $\beta(n, k)$, which seemingly arose in the study of burst-error correcting convolutional codes [9]. This algorithm was thoroughly analyzed by Carlitz, Rosselle and Scoville in [18]. The idea is to consider a two-dimensional array with a recursion like in Pascal's triangle. This array can be obtained from β via $\alpha(n, k) = \beta(n+k, k)$, the recursion is hence

$$\alpha(n, k) = \alpha(n-1, k) + \alpha(n-1, k-1)$$

In [18] the part of the array α consisting of positive entries was considered, which are described by the conditions $\alpha(n, 0) = 1$ for all n and $\alpha(\sigma(k)+1, k+1) = \alpha(\sigma(k), k)$. (Indeed, the array $d(k, j)$ in [18] was presented in a slightly different form. With n taking the role of j and by placing the elements of the kth chain in the kth column of our array α, the two arrays d and α are equivalent). With the above discussion, it can now be seen that $\sigma(k) = v_k + k - 1$, where v_k is as in (65).

Observe that we extend the array α by introducing the row $\alpha(-1, k)$. The reason is that in this row the numbers δ_k from [18] are contained. These numbers are defined recursively via

$$\delta_k = -\sum_{r=1}^{k}\binom{v_k+k-1}{r}\delta_{k-r} \tag{66}$$

with initial value $\delta_0 = 1$,

Reading out the numbers δ_k as entries $\alpha(-1, k)$ is a second method to derive the defining recursion. In [18] a different approach was chosen. Also it was derived that

$$\alpha(n, k) = \sum_{r=0}^{k}\binom{n+1}{r}\delta_{k-r}.$$

Corollary 6.8 *Let $\gamma^{(1)}$ and $\gamma^{(2)}$ be defined as in the previous theorem. Arrays $\alpha^{(\nu)}$ for $\nu = 1, 2$ are defined by $\alpha^{(\nu)}(n, k) = \gamma^{(\nu)}(n + k, k)$ for all n, k with $n \geq \upsilon_k + k$. The corresponding parameters $\delta^{(1)}(k)$ and $\delta^{(2)}(k)$ as defined under (A.11) fulfill for all $k \geq 1$.*

$$\delta^{(1)}(k) + \delta^{(2)}(k) = (-1)^k \cdot (c + 2)$$

Proof Extend the array beyond the boundary by the recursion $\alpha(n, k) = \alpha(n - 1, k) + \alpha(n - 1, k - 1)$ if $n + k < u_n$. As mentioned above, the numbers $\delta^{(1)}(k) = \alpha^{(1)}(-1, k)$ and $\delta^{(2)}(k) = \alpha^{(2)}(-1, k)$ can be found as entries of row No. −1. in the arrays $\alpha^{(\nu)}$. □

Example For $d = 2$, $c = 3$ the arrays $\alpha^{(1)}$ and $\alpha^{(2)}$ look as follows.

	0	1	2	3	4 …
−1	1	−2	0	7	−40 …
0	1	−1	−2	7	−33 …
1	1	0	−3	5	−26 …
2	1	1	−3	2	−21 …
3	1	2	−2	−1	−19 …
4	1	3	0	−3	−20 …
5	1	4	3	−3	−23 …
6	1	5	7	0	−26 …
7	1	6	12	7	−26 …
8	1	7	18	19	−19 …
9	1	8	25	37	0 …
⋮	⋮	⋮	⋮	⋮	⋮

	0	1	2	3	4 …
−1	1	−3	5	−12	45 …
0	1	−2	2	−7	33 …
1	1	−1	0	−5	26 …
2	1	0	−1	−5	21 …
3	1	1	−1	−6	16 …
4	1	2	0	−7	10 …
5	1	3	2	−7	3 …
6	1	4	5	−5	−4 …
7	1	5	9	0	−9 …
8	1	6	14	9	−9 …
9	1	7	20	23	0 …
⋮	⋮	⋮	⋮	⋮	⋮

The sum array $\alpha = \alpha^{(1)} + \alpha^{(2)}$ hence is

	0	1	2	3	4 ...
−1	2	−5	−5	−5	−5 ...
0	2	−3	0	0	0 ...
1	2	−1	−3	0	0 ...
2	2	1	−4	−3	0 ...
3	2	3	−3	−7	−3 ...
4	2	5	0	−10	−10 ...
5	2	7	5	−10	−20 ...
6	2	9	12	−5	−30 ...
7	2	11	21	7	−35 ...
8	2	13	32	28	−28 ...
9	2	15	45	60	0 ...
⋮	⋮	⋮	⋮	⋮	⋮

Computer observations strongly suggest that the generalization of the ballot numbers holds for all positive integers d. More exactly, let $\underline{\lambda}^{(\nu)} = (\lambda_1^{(\nu)}, \lambda_2^{(\nu)}, \lambda_3^{(\nu)}, \dots)$, $\nu = 1, \dots, d$ be periodic sequences of period length d, such that the initial segment of length d in $\underline{\lambda}^{(\nu)}$ is a cyclic shift of order $\nu - 1$ of the initial segment of $\underline{\lambda}^{(1)}$, i.e.

$$\underline{\lambda}^{(1)} = (\lambda_1, \lambda_2, \dots, \lambda_{d-1}, \lambda_d, \lambda_1, \lambda_2, \dots, \lambda_{d-1}, \lambda_d, \lambda_1, \dots),$$

$$\underline{\lambda}^{(2)} = (\lambda_2, \lambda_3, \dots, \lambda_d, \lambda_1, \lambda_2, \lambda_3, \dots, \lambda_d, \lambda_1, \lambda_2, \dots), \dots$$

$$\underline{\lambda}^{(d)} = (\lambda_d, \lambda_1, \dots, \lambda_{d-2}, \lambda_{d-1}, \lambda_d, \lambda_1, \dots, \lambda_{d-2}, \lambda_{d-1}, \lambda_d, \lambda_1, \dots),$$

Further, let the sequences $\underline{\lambda}^{(\nu)}$ describe the boundaries $\underline{u}^{(\nu)}$, $\nu = 1, \dots, d$ as in (A.9), i.e., the lattice points $(n, u_n)_{n=0,1,\dots}$ are not allowed to be touched by paths enumerated in the arrays $\gamma^{(\nu)}(n, k)$, $\nu = 1, \dots, d$

Conjecture Whenever $n > \lambda_1^{(\nu)} + \cdots + \lambda_k^{(\nu)}$ for all $\nu = 1, \dots, d$

$$\gamma^{(1)}(n, k) + \gamma^{(2)}(n, k) + \cdots + \gamma^{(d)}(n, k) = \beta(n, k)$$

where

$$\beta(n, 0) = d, \qquad \beta(-1, k) = -(\lambda_1 + \cdots + \lambda_d), \qquad \beta(n, k) = \beta(n-1, k) + \beta(n, k-1)$$

This conjecture would also imply the following generalization of Proposition 6.5(i).

Let μ_0 and $\theta_1, \dots, \theta_d$ be nonnegative integers with $\theta_1 + \cdots + \theta_d = c$. Further, let $f^{(j,0)}$ denote the function $f^{(0)}$ as in (59) for the choice of parameters as in (58) $(\mu_0^{(j)}, \mu_1^{(j)}, \dots, \mu_{d-1}^{(j)}) = (\mu_0, \theta_1, \dots, \theta_{j-1}, \theta_{j+1}, \dots, \theta_d)$ for $j = 1, \dots, d$. Then

$$f^{(1,0)}(t^d)+f^{(2,0)}(t^d)+\cdots+f^{(d,0)}(t^d)=q(t)^{-\mu_0}+q(\omega t)^{-\mu_0}+\cdots+q(\omega^{d-1}t)^{-\mu_0}$$

$$=\sum_{n=0}^{\infty}\frac{d\mu_0}{(c+d)n+\mu_0}\binom{(c+d)n+\mu_0}{dn}t^{dn}$$

Besides the period lengths $d=1$ and $d=2$, we could prove the conjecture for the following array η

	0	1	2	3	4	5	6	7	8 ...
−1	3	−2	−2	−2	−2	−2	−2	−2	−2 ...
0	3	1	−1	−3	−5	−7	−9	−11	−13 ...
1	3	4	3	0	−5	−12	−21	−32	−45 ...
2	3	7	10	10	5	−7	−28	−60	−105 ...
3	3	10	20	30	35	28	0	−60	−165 ...
4	3	13	33	63	98	126	126	66	−99 ...
5	3	16	49	112	210	336	462	528	429 ...
⋮	⋮	⋮	⋮	⋮	⋮	⋮	⋮	⋮	⋮

with $\eta(n,0)=3$, $\eta(-1,k)=-2$, and $\eta(n,k)=\eta(n,k-1)+\eta(n-1,k)$.

Proposition 6.9 *The positive entries $\eta(n,k)>0$ are the sum*

$$\eta(n,k)=\eta^{(1)}(n,k)+\eta^{(2)}(n,k)+\eta^{(3)}(n,k)$$

where $\eta^{(\nu)}(n,k)$ enumerates the number of paths from the origin to (n,k) not touching or crossing the boundaries $(m,u_m^{(\nu)})_{m=0,1,\ldots}$ with sequences $u_m^{(\nu)}$ being periodic of period length 2 defined for $\nu=1,2,3$ by

$$u_{2i}^{(1)}=1+3i,\ u_{2i+1}^{(1)}=2+3i,\ u_{2i}^{(2)}=1+3i,\ u_{2i+1}^{(2)}=3+3i,\ u_{2i}^{(3)}=2+3i,\ u_{2i+1}^{(3)}=3+3i,$$

Proof Observe that the boundaries via $\underline{u}$ arise for the choices $(s=1,\mu=1)$ for $\nu=1$, $(s=1,\mu=2)$ for $\nu=2$, and $(s=2,\mu=1)$ for $\nu=3$, respectively, which we studied intensively in Corollary 6.7.

The proposition is easily verified, when for all k some n is found where $\eta(n,k)=\eta^{(1)}(n,k)+\eta^{(2)}(n,k)+\eta^{(3)}(n,k)$. In order to do so, observe that application of Corollary 6.7 yields

$$\eta(2j,3j+1)=\eta^{(3)}(2j,3j+1)=\frac{1}{5j+2}\binom{5j+2}{2j+1}$$

the jth coefficient in $g^{(2,3,1)}(x)$ and

$$\eta(2j-1,3j-1)=\eta^{(2)}(2j-1,3j-1)+\eta^{(3)}(2j-1,3j-1)=\frac{1}{5j-1}\binom{5j-1}{2j}$$

the sum of the jth coefficients in $h^{(1,3,2)}$ and $h^{(2,3,1)}$.

Further, for all j it must be $\eta(2j-1,3j)=0$, since for all $\nu=1,2,3$ it is $\eta^{(\nu)}(2j,3j-1)=\eta^{(\nu)}(2j,3j)$ (all paths to $(2j,3j)$ must pass through $(2j,3j-1)$). □

Unfortunately, this is the only array with $d>2$ for which we could prove the conjecture. Actually, the analysis here was possible since the dual sequences $\underline{\upsilon}^{(\nu)}$ as in (65) are periodic with period length 2 and this case was considered before. The parameter d here is the period length of the corresponding sequences $\underline{\lambda}^{(\nu)}$, which for $(d=3,c=2)$ are $\underline{\lambda}^{(1)}=(1,1,0,1,1,0,\dots)$, $\underline{\lambda}^{(2)}=(1,0,1,1,0,1,\dots)$, $\underline{\lambda}^{(3)}=(0,1,1,0,1,1,\dots)$.

References

1. K.A.S. Abdel Ghaffar, H.C. Ferreira, On the maximum number of systematically – encoded information bits in the Varshamov – Tenengolts codes and the Constantin – Rao codes, in *Proceedings of 1997 IEEE Symposium on Information Theory, Ulm* (1997), p. 455
2. M. Aigner, Catalan-like numbers and determinants. J. Combin. Theory Ser. A **87**, 33–51 (1999)
3. M. Aigner, A characterization of the Bell numbers. Discret. Math. **205**(1–3), 207–210 (1999)
4. G.E. Andrews, Plane partitions (III): the weak Macdonald conjecture. Invent. Math. **53**, 193–225 (1979)
5. G.E. Andrews, Pfaff's method. I. The Mills-Robbins-Rumsey determinant. Discret. Math. **193**(1–3), 43–60 (1998)
6. G.E. Andrews, D. Stanton, Determinants in plane partition enumeration. Eur. J. Combin. **19**(3), 273–282 (1998)
7. E.E. Belitskaja, V.R. Sidorenko, P. Stenström, Testing of memory with defects of fixed configurations, in *Proceedings of 2nd International Workshop on Algebraic and Combinatorial Coding Theory, Leningrad* (1990), pp. 24–28
8. E.A. Bender, D.E. Knuth, Enumeration of plane partitions. J. Combin. Theory Ser. A **13**, 40–54 (1972)
9. E.R. Berlekamp, A class of convolutional codes. Information and Control **6**, 1–13 (1963)
10. E.R. Berlekamp, *Algebraic Coding Theory* (McGraw-Hill, New York, 1968)
11. E.R. Berlekamp, Goppa codes. IEEE Trans. Inf. Theory **19**, 590–592 (1973)
12. R.E. Blahut, *Theory and Practice of Error Control Codes* (Addison-Wesley, Reading, 1984)
13. R.E. Blahut, *Fast Algorithms for Digital Signal Processing* (Addison-Wesley, Reading, 1985)
14. D.L. Boley, T.J. Lee, F.T. Luk, The Lanczos algorithm and Hankel matrix factoriztion. Linear Algebr. Appl. **172**, 109–133 (1992)
15. D.L. Boley, F.T. Luk, D. Vandevoorde, A fast method to diagonalize a Hankel matrix. Linear Algebr. Appl. **284**, 41–52 (1998)
16. M. Bousquet Mélou, L. Habsieger, Sur les matrices à signes alternants. Discret. Math. **139**, 57–72 (1995)
17. D.M. Bressoud, *Proofs and Confirmations* (Cambridge University Press, Cambridge, 1999)
18. L. Carlitz, D.P. Rosselle, R.A. Scoville, Some remarks on ballot - type sequences. J. Combin. Theory **11**, 258–271 (1971)
19. L. Carroll, *Alice's Adventures in Wonderland* (1865)
20. P.L. Chebyshev, Sur l'interpolation par la méthode des moindres carrés. Mém. Acad. Impér. Sci. St. Pétersbourg (7) 1 (15), 1–24; also: Oeuvres I, 473–489 (1859)
21. U. Cheng, On the continued fraction and Berlekamp's algorithm. IEEE Trans. Inf. Theory **30**, 541–544 (1984)

22. S.H. Choi, D. Gouyou-Beauchamps, Enumeration of generalized Young tableaux with bounded height. Theor. Comput. Sci. **117**, 137–51 (1993)
23. W. Chu, Binomial convolutions and determinant identities. Discret. Math. **204**, 129–153 (1999)
24. G. Cohen, I. Honkala, S. Litsyn, A. Lobstein, *Covering Codes* (Elsevier, Amsterdam, 1997)
25. G.D. Cohen, S. Litsyn, A. Vardy, G. Zemor, Tilings of binary spaces. SIAM J. Discret. Math. **9**, 393–412 (1996)
26. S.D. Constantin, T.R.N. Rao, On the theory of binary asymmetric error correcting codes. Inf. Control **40**, 20–26 (1979)
27. N.G. de Bruijn, On the factorization of finite abelian groups. Indag. Math. Kon. Ned. Akad. Wet. Amst. **15**, 258–264 (1953)
28. N.G. de Bruijn, On the factorization of cyclic groups. Indag. Math. Kon. Ned. Akad. Wet. Amst. **15**, 370–377 (1953)
29. G. de Prony, Essai expérimental et analytique sur les lois de la dilatabilité de fluides élastiques et sur les celles de la force expansive de la vapeur de l' alcool, à différentes températures. J. de l'École Polytechnique 1, cahier **22**, 24–76 (1795)
30. P. Delsarte, Nombres de Bell et polynômes de Charlier. C. R. Acad. Sci. Paris (Ser. A) **287**, 271–273 (1978)
31. M. Desainte-Catherine, X.G. Viennot, Enumeration of certain Young tableaux with bounded height. *Combinatoire Énumérative (Montreal 1985)*. Lecture Notes in Mathematics, vol. 1234 (Springer, Berlin, 1986), pp. 58–67
32. C.L. Dodgson, Condensation of determinants. Proc. R. Soc. Lond. **15**, 150–155 (1866)
33. T. Etzion, A. Vardy, Perfect codes: constructions, properties, and enumeration. IEEE Trans. Inf. Theory **40**(3), 754–763 (1994)
34. T. Etzion, A. Vardy, On perfect codes and tilings: problems and solutions, in *Proceedings of 1997 IEEE Symposium on Information Theory, Ulm* (1997), p. 450
35. H. Everett, D. Hickerson, Packing and covering by translates of certain starbodies. Proc. Am. Math. Soc. **75**(1), 87–91 (1979)
36. P. Flajolet, Combinatorial aspects of continued fractions. Discret. Math. **32**, 125–161 (1980)
37. P. Flajolet, On congruences and continued fractions for some classical combinatorial quantities. Discret. Math. **41**, 145–153 (1982)
38. D. Foata, Combinatoire des identités sur les polynômes orthogonaux, in *Proceedings of the International Congress of Mathematicians, Warsaw* (1983), pp. 1541–1553
39. L. Fuchs, *Abelian Groups* (Pergamon Press, New York, 1960)
40. S. Galovich, S. Stein, Splittings of Abelian groups by integers. Aequationes Math. **22**, 249–267 (1981)
41. I. Gessel, A probabilistic method for lattice path enumeration. J. Stat. Plan. Inference **14**, 49–58 (1986)
42. I. Gessel, R. Stanley, Algebraic enumeration, *Handbook of Combinatorics*, vol. 2, ed. by R.L. Graham, M. Grötschel, L. Lovasz (Wiley, New York, 1996), pp. 1021–1069
43. I. Gessel, X.G. Viennot, Binomial determinants, paths and hook length formulae. Adv. Math. **58**, 300–321 (1985)
44. I. Gessel, X G. Viennot, Determinants, Paths, and Plane Partitions, Preprint (1989)
45. S. Golomb, A general formulation of error metrics. IEEE Trans. Inf. Theory **15**, 425–426 (1969)
46. S. Golomb, *Polyominoes*, 2nd edn. (Princeton University Press, Princeton, 1994)
47. S.W. Golomb, L.R. Welch, Algebraic coding and the Lee metric, in *Error Correcting Codes*, ed. by H.B. Mann (Wiley, New York, 1968), pp. 175–194
48. S.W. Golomb, L.R. Welch, Perfect codes in the Lee metric and the packing of polyominoes. SIAM J. Appl. Math. **18**, 302–317 (1970)
49. V.D. Goppa, A new class of linear correcting codes. Probl. Peredachi Informatsii **6**(3), 24–30 (1970) (in Russian)
50. V.D. Goppa, Rational representation of codes and (L,g) codes, Probl. Peredachi Informatsii **7**(3), 41–49 (1971) (in Russian)

51. V.D. Goppa, Decoding and diophantine approximations. Probl. Control Inf. Theory **5**(3), 195–206 (1975)
52. B. Gordon, A proof of the Bender - Knuth conjecture. Pac. J. Math. **108**, 99–113 (1983)
53. D.C. Gorenstein, N. Zierler, A class of error-correcting codes in p^m symbols. J. Soc. Indus. Appl. Math. **9**, 207–214 (1961)
54. R.L. Graham, D.E. Knuth, O. Patashnik, *Concrete Mathematics* (Addison Wesley, Reading, 1988)
55. S. Gravier, M. Mollard, On domination numbers of Cartesian products of paths. Discret. Appl. Math. **80**, 247–250 (1997)
56. A.J. Guttmann, A.L. Owczarek, X.G. Viennot, Vicious walkers and Young tableaux I: without walls. J. Phys. A: Math. General **31**, 8123–8135 (1998)
57. R.K. Guy, Catwalks, sandsteps and Pascal pyramids. J. Integer Seq. **3**, Article 00.1.6 (2000)
58. W. Hamaker, S. Stein, Combinatorial packing of R^3 by certain error spheres. IEEE Trans. Inf. Theory **30**(2), 364–368 (1984)
59. A.J. Han Vinck, H. Morita, Codes over the ring of integers modulo m. IEICE Trans. Fundam. Electron. Commun. Comput. Sci. **E81-A**(10), 1564–1571 (1998)
60. G. Hajós, Über einfache und mehrfache Bedeckungen des n-dimensionalen Raumes mit einem Würfelgitter. Math. Zeit. **47**, 427–467 (1942)
61. D. Hickerson, Splittings of finite groups. Pac. J. Math. **107**, 141–171 (1983)
62. D. Hickerson, S. Stein, Abelian groups and packing by semicrosses. Pac. J. Math. **122**(1), 95–109 (1986)
63. P. Hilton, J. Pedersen, Catalan numbers, their generalization, and their uses. Math. Intell. **13**(2), 64–75 (1991)
64. K. Imamura, W. Yoshida, A simple derivation of the Berlekamp - Massey algorithm and some applications. IEEE Trans. Inf. Theory **33**, 146–150 (1987)
65. E. Jonckheere, C. Ma, A simple Hankel interpretation of the Berlekamp - Massey algorithm. Linear Algebr. Appl. **125**, 65–76 (1989)
66. S. Klavžar, N. Seifter, Dominating Cartesian products of cycles. Discret. Appl. Math. **59**, 129–136 (1995)
67. V.I. Levenshtein, Binary codes with correction for deletions and insertions of the symbol 1. Probl. Peredachi Informacii **1**, 12–25 (1965). (in Russian)
68. V.I. Levenshtein, A.J. Han, Vinck, Perfect (d, k)-codes capable of correcting single peak shifts. IEEE Trans. Inf. Theory **39**(2), 656–662 (1993)
69. S. Lin, D.J. Costello, *Error - Control Coding* (Prentice-Hall, Englewood Cliffs, 1983)
70. B. Lindström, On the vector representation of induced matroids. Bull. Lond. Math. Soc. **5**, 85–90 (1973)
71. S.S. Martirossian, Single – error correcting close packed and perfect codes, in *Proceedings of 1st INTAS International Seminar on Coding Theory and Combinatorics, Thahkadzor, Armenia* (1996), pp. 90 – 115
72. M.E. Mays, J. Wojciechowski, A determinant property of Catalan numbers. Discret. Math. **211**, 125–133 (2000)
73. J.L. Massey, Shift register synthesis and BCH decoding. IEEE Trans. Inf. Theory **15**, 122–127 (1969)
74. W.H. Mills, Continued fractions and linear recurrences. Math. Comput. **29**(129), 173–180 (1975)
75. W.H. Mills, D.P. Robbins, H. Rumsey Jr., Enumeration of a symmetry class of plane partitions. Discret. Math. **67**, 43–55 (1987)
76. H. Minkowski, *Diophantische Approximationen* (Teubner, Leipzig, 1907)
77. S.G. Mohanty, *Lattice Path Counting and Applications* (Academic Press, New York, 1979)
78. T. Muir, *Theory of Determinants* (Dover, New York, 1960)
79. A. Munemasa, On perfect t-shift codes in Abelian groups. Des. Codes Cryptography **5**, 253–259 (1995)
80. T.V. Narayana, *Lattice Path Combinatorics* (University of Toronto Press, Toronto, 1979)

81. P. Peart, W.-J. Woan, Generating functions via Hankel and Stieltjes matrices. J. Integer Sequences **3**, Article 00.2.1 (2000)
82. O. Perron, *Die Lehre von den Kettenbrüchen* (Chelsea Publishing Company, New York, 1929)
83. W.W. Peterson, Encoding and error-correction procedures for the Bose-Chaudhuri codes. Trans. IRE **6**, 459–470 (1960)
84. M. Petkovšek, H.S. Wilf, A high-tech proof of the Mills-Robbins-Rumsey determinant formula. Electron. J. Comb. **3**(2), 19 (1996)
85. J.L. Phillips, The triangular decomposition of Hankel matrices. Math. Comput. **25**(115), 599–602 (1971)
86. G. Polya, G. Szegö, *Aufgaben und Lehrsätze aus der Analysis*, vol. II, 3rd edn. (Springer, Berlin, 1964)
87. C. Radoux, Déterminants de Hankel et théorème de Sylvester, in *Proceedings of the 28th Séminaire Lotharingien* (1992), pp. 115 – 122
88. C. Radoux, Addition formulas for polynomials built on classical combinatorial sequences. J. Comput. Appl. Math. **115**, 471–477 (2000)
89. L. Rédei, Die neue Theorie der endlichen abelschen Gruppen und eine Verallgemeinerung des Hauptsatzes von Hajós. Acta Math. Acad. Sci Hung. **16**, 329–373 (1965)
90. J. Riordan, *An Introduction to Combinatorial Analysis* (Wiley, New York, 1958)
91. H. Rutishauser, *Der Quotienten-Differenzen-Algorithmus* (Birkhäuser, Basel, 1957)
92. S. Saidi, Codes for perfectly correcting errors of limited size. Discret. Math. **118**, 207–223 (1993)
93. S. Saidi, Semicrosses and quadratic forms. Eur. J. Comb. **16**, 191–196 (1995)
94. A.D. Sands, On the factorization of finite abelian groups. Acta Math. **8**, 65–86 (1957)
95. A.D. Sands, On the factorization of finite abelian groups II. Acta Math. **13**, 45–54 (1962)
96. L.W. Shapiro, A Catalan triangle. Discret. Math. **14**, 83–90 (1976)
97. V. Sidorenko, Tilings of the plane and codes for translational metrics, in *Proceedings of 1994 IEEE Symposium on Information Theory, Trondheim* (1994), p. 107
98. F. Solove'eva, Switchings and perfect codes, in *Numbers, Information and Complexity, Special Volume in Honour of Rudolf Ahlswede*, ed. by I. Althöfer, N. Cai, G. Dueck, L. Khachatrian, M. Pinsker, A. Sárközy, I. Wegener, Z. Zhang (Kluwer Publishers, Boston, 2000), pp. 311–324
99. R.P. Stanley, Theory and application of plane partitions. Stud. Appl. Math. **50** Part 1, 167–189; Part 2, 259–279 (1971)
100. R.P. Stanley, A baker's dozen of conjectures concerning plane partitions, in *Combinatoire Énumérative (Montreal 1985)*.Lecture Notes in Mathematics, vol. 1234 (Springer, Berlin, 1986), pp. 285–293
101. R.P. Stanley, *Enumerative Combinatorics*, vol. 2 (Cambridge University Press, Cambridge, 1999)
102. S. Stein, Factoring by subsets. Pac. J. Math. **22**(3), 523–541 (1967)
103. S. Stein, Algebraic tiling. Am. Math. Mon. **81**, 445–462 (1974)
104. S. Stein, Packing of R^n by certain error spheres. IEEE Trans. Inf. Theory **30**(2), 356–363 (1984)
105. S. Stein, Tiling, packing, and covering by clusters. Rocky Mt. J. Math. **16**, 277–321 (1986)
106. S. Stein, Splitting groups of prime order. Aequationes Math. **33**, 62–71 (1987)
107. S. Stein, Packing tripods. Math. Intell. **17**(2), 37–39 (1995)
108. S. Stein, S. Szabó, *Algebra and Tiling*. The Carus Mathematical Monographs, vol. 25 (The Mathematical Association of America, Washington, 1994)
109. T.J. Stieltjes, Recherches sur les fractions continue. Ann. Fac. Sci. Toulouse **8**, J.1– 22 (1895); A.1–47 (1894)
110. T.J. Stieltjes, *Oeuvres Complètes* (Springer, Berlin, 1993)
111. V. Strehl, Contributions to the combinatorics of some families of orthogonal polynomials, mémoire, Erlangen (1982)
112. Y. Sugiyama, M. Kasahara, S. Hirawawa, T. Namekawa, A method for solving key equation for decoding Goppa code. Inf. Control **27**, 87–99 (1975)

113. R.A. Sulanke, A recurrence restricted by a diagonal condition: generalized Catalan arrays. Fibonacci Q. **27**, 33–46 (1989)
114. S. Szabó, Lattice coverings by semicrosse of arm length 2. Eur. J. Comb. **12**, 263–266 (1991)
115. U. Tamm, Communication complexity of sum-type functions, Ph.D. thesis, Bielefeld, 1991, also Preprint 91–016, SFB 343, University of Bielefeld (1991)
116. U. Tamm, Still another rank determination of set intersection matrices with an application in communication complexity. Appl. Math. Lett. **7**, 39–44 (1994)
117. U. Tamm, Communication complexity of sum - type functions invariant under translation. Inf. Comput. **116**(2), 162–173 (1995)
118. U. Tamm, Deterministic communication complexity of set intersection. Discret. Appl. Math. **61**, 271–283 (1995)
119. U. Tamm, On perfect 3–shift N–designs, in *Proceedings of 1997 IEEE Symposium on Information Theory, Ulm* (1997), p. 454
120. U. Tamm, Splittings of cyclic groups, tilings of Euclidean space, and perfect shift codes, *Proceedings of 1998 IEEE Symposium on Information Theory* (MIT, Cambridge, 1998), p. 245
121. U. Tamm, Splittings of cyclic groups and perfect shift codes. IEEE Trans. Inf. Theory **44**(5), 2003–2009 (1998)
122. U. Tamm, Communication complexity of functions on direct sums, in *Numbers, Information and Complexity, Special Volume in Honour of Rudolf Ahlswede*, ed. by I. Althöfer, N. Cai, G. Dueck, L. Khachatrian, M. Pinsker, A. Sárközy, I. Wegener, Z. Zhang (Kluwer Publishers, Boston, 2000), pp. 589–602
123. U. Tamm, Communication complexity and orthogonal polynomials, in *Proceedings of the Workshop Codes and Association Schemes*. DIMACS Series, Discrete Mathematics and Computer Science, vol. 56 (2001), pp. 277–285
124. U. Tamm, Some aspects of Hankel matrices in coding theory and combinatorics. Electron. J. Comb. **8**(A1), 31 (2001)
125. U. Tamm, Lattice paths not touching a given boundary. J. Stat. Plan. Interf. **2**(2), 433–448 (2002)
126. W. Ulrich, Non-binary error correction codes. Bell Syst. Tech. J. **36**(6), 1341–1388 (1957)
127. R.R. Varshamov, G.M. Tenengolts, One asymmetric error correcting codes (in Russian). Avtomatika i Telemechanika **26**(2), 288–292 (1965)
128. X.G. Viennot, *A combinatorial theory for general orthogonal polynomials with extensions and applications, in Polynômes Orthogonaux et Applications, Proceedings, Bar-le-Duc* (Springer, Berlin, 1984), pp. 139–157
129. X.G. Viennot, A combinatorial interpretation of the quotient – difference algorithm, Preprint (1986)
130. H.S. Wall, *Analytic Theory of Continued Fractions* (Chelsea Publishing Company, New York, 1948)
131. H. Weber, Beweis des Satzes, daß jede eigentlich primitive quadratische Form unendlich viele prime Zahlen darzustellen fähig ist. Math. Ann. **20**, 301–329 (1882)
132. L.R. Welch, R.A. Scholtz, Continued fractions and Berlekamp's algorithm. IEEE Trans. Inf. Theory **25**, 19–27 (1979)
133. D. Zeilberger, Proof of the alternating sign matrix conjecture. Electronic J. Comb. **3**(2), R13, 1–84 (1996)
134. D. Zeilberger, Proof of the refined alternating sign matrix conjecture. N. Y. J. Math. **2**, 59–68 (1996)
135. D. Zeilberger, Dodgson's determinant-evaluation rule proved by TWO-TIMING MEN and WOMEN. Electron. J. Comb. **4**(2), 22 (1997)

Further Readings

136. R. Ahlswede, N. Cai, U. Tamm, Communication complexity in lattices. Appl. Math. Lett. **6**, 53–58 (1993)
137. M. Aigner, Motzkin numbers. Eur. J. Comb. **19**, 663–675 (1998)
138. R. Askey, M. Ismail, Recurrence relations, continued fractions and orthogonal polynomials. Mem. Am. Math. Soc. **49**(300), 108 (1984)
139. C. Brezinski, *Padé-Type Approximation and General Orthogonal Polynomials* (Birkhäuser, Basel, 1980)
140. D.C. Gorenstein, W.W. Peterson, N. Zierler, Two-error correcting Bose-Chaudhuri codes are quasi-perfect. Inf. Control **3**, 291–294 (1960)
141. V.I. Levenshtein, On perfect codes in the metric of deletions and insertions (in Russian), Diskret. Mat. **3**(1), 3–20; English translation. Discret. Math. Appl. **2**(3), 1992 (1991)
142. H. Morita, A. van Wijngaarden, A.J. Han Vinck, Prefix synchronized codes capable of correcting single insertion/deletion errors, in *Proceedings of 1997 IEEE Symposium on Information Theory, Ulm* (1997), p. 409
143. J. Riordan, *Combinatorial Identities* (Wiley, New York, 1968)
144. S. Szabó, Some problems on splittings of groups. Aequationes Math. **30**, 70–79 (1986)
145. S. Szabó, Some problems on splittings of groups II. Proc. Am. Math. Soc. **101**(4), 585–591 (1987)

Appendix A
Supplement

Gedenkworte für Rudolf Ahlswede[1]

Rudi Ahlswede bin ich zum letzten Mal begegnet, als er am 30. Januar 2009 in Erlangen einen Festvortrag zur Nachfeier meines 80. Geburtstages hielt. Mathematiker wie Nichtmathematiker erlebten da einen Fürsten seines Fachs, der sein gewaltiges Souveränitätsgebiet begeistert und begeisternd durchstürmte und Ideen zu dessen fernerer Durchdringung und Ausweitung in großen Horizonten entwarf. Ich möchte noch kurz ein wenig über die "Anfangsbedingungen" berichten, die Rudi bei seinem 1966 mit der Promotion endenden Stochastik-Studium in Göttingen vorfand. Das Fach Stochastik, damals Mathematische Statistik genannt, war nach Kriegsende in West-Deutschland m.W. nur durch die Göttinger Dozentur von Hans Münzner (1906–1997) vertreten und mußte somit praktisch neu aufgebaut werden. Das begann mit der Übernahme neugeschaffener Lehrstühle durch Leopold Schmetterer (1919–2004) in Hamburg und Hans Richter (1912–1978) in München, die beide ursprünglich Zahlentheoretiker waren und sich in ihr neues Fach einarbeiteten. Dieser "1. Welle" folgte eine zweite, in der Jungmathematiker, wie Klaus Krickeberg (* 1929) und ich (* 1928), die in ihrem ursprünglichen Arbeitsgebiet bereits eine gewisse Nachbarschaft zur Stochastik vorweisen konnten. Bei mir war das durch Arbeiten zur Ergoden- und Markov-Theorie gegeben. Als ich 1958 in Göttingen das Münznersche Kleininstitut im Keller des großen Mathematischen Instituts an der Bunsenstraße übernahm, war ich für meine neue Aufgabe eigentlich zu jung und unerfahren. Ein Student, der damals zu meiner kleinen Gruppe stieß, konnte nicht erwarten, von einem souveränen, erfahrenen Ordinarius umfassenden Rat zu erhalten: ich hatte ihm damals nur einen Schritt der Einarbeitung in neue Themengebiete voraus. Meinen Zugang zur Shannon'schen Informationstheorie, auf die ich Rudi und andere "anzusetzen" versuchte, hatte ich über die Ergodentheorie gefunden, die mit der Einführung der Entropie-Invarianten (1959) durch A.N. Kolmogorov

[1] This obituary was hold during the conference at the ZiF in Bielefeld by Konrad Jacobs who died July 26th, 2015.

A. Ahlswede et al. (eds.), *Combinatorial Methods and Models*,
Foundations in Signal Processing, Communications and Networking 13,
DOI 10.1007/978-3-319-53139-7

(1903–1987) und Y. Sinai (* 1937) einen mich unmittelbar betreffenden Bezug zur Informationstheorie erhalten hatte, der in einem Uspehi-Artikel (1956) von A.Y. Chintchine (1894–1995) schon vorher systematisch ausgebreitet worden war; da diese Arbeit in Ostdeutschland sogleich ins Deutsche übersetzt worden war, hatten wir hier sprachlich sofort Zugang. Wesentlichere Impulse für uns ergaben sich allerdings aus dem Ergebnisbericht Coding Theorems of Information Theory (1961) von Jacob Wolfowitz (1910–1981). Nach Rudis Promotion kam es zu intensiven Kontakten mit J. Wolfowitz, mit dem er später mehere Arbeiten gemeinsam verfaßte, und dem er schließlich einen großartigen Nachruf widmete. Da ich Studenten wie R. Ahlswede und V. Strassen nur geringfügig "voraus" war, hatte ich später das beglückendste Erlebnis, das einem akademischen Lehrer zuteil werden kann: von seinen "Schülern" überholt zu werden und von ihnen lernen zu können. Auch nach der Erlanger Begegnung Anfang 2009 kam es immer wieder zu Telefonkontakten zwischen Rudi und mir. Bei einem der letzten (wohl 2010) schilderte ich ihm meine Erwägungen über die Frage, wie man sich als Mathematiker zu dem unvermeidlichen fachlichen Leistungsabfall - wie allmählich auch immer - nach der Emeritierung stellen solle. Ich hatte mich dafür entschieden, dann (bei mir nach 1993) nicht mehr forschungsaktiv zu sein, sondern mich anderen Interessengebieten zuzuwenden, wenn auch naturgemäß auf nunmehr amateurhaftem Niveau. Als ich ihn um seine Meinung hierzu fragte, kam die Antwort sogleich und in aller Entschiedenheit: seine Devise sei

Stirb in den Stiefeln!
(Die in your boots!).

Bei seinem Naturell kam nur in Frage, weiterzuarbeiten, so intensiv und so lange es nur angehen mochte. Rudi hatte noch eine Überfülle von Ideen und Problemen. In den Stiefeln, die ihm angewachsen waren, wäre er noch sehr lange weitermarschiert. So einen wie ihn vergißt man nie.

Commemorating Rudolf Ahlswede[2]

The last time I met with Rudi Ahlswede was in Erlangen on January 30, 2009, when he gave a lecture in honor of my 80th birthday. Mathematicians as well as non-mathematicians experienced a ruler in his field, one who stormed through his tremendous sovereign territory, inspired and inspiring, creating ideas, to which he penetrated and expanded upon to great horizons. I would like to say a little about the initial conditions that Rudi found himself in when he finished his Ph.D. program in stochastic studies in Göttingen in 1966. At the end of the war, the field of Stochastics in West Germany (at that time called Mathematical Stochastics) was, to my knowledge, represented only by one lecture position that was held by Hans Münzner (1906–1997) in Göttingen, and therefore had to be rebuilt practically from

[2]This obituary is the translation of the German obituary by Konrad Jacobs.

new. This began with the acquisition of two newly created institutes; in Hamburg by Leopold Schmetterer (1919–2004) and in Munich by Hans Richter (1912–1978), both of whom were originally number theorists and trained themselves in their new field. This first wave was followed by a second, in the form of the young mathematician Klaus Krickeberg (* 1929) and myself (* 1928); both of us originally came from areas of study that were in close proximity to the neighboring field of Stochastics. In my case, this was established through my work on Ergodic- and Markov Theory. In 1958, when I took over the Münzners' Klein Institute in Göttingen in the basement of the large mathematical institute in Bunsen Street, I was really too young and inexperienced for my new duties. A student, who at that time fell into my small group, could not expect a confident, experienced professor to give him comprehensive advice; compared to him, I was only a small step ahead in being familiar with the new topics. My approach to Shannon's Information Theory, to which I tried to push Rudi and others into researching, was made via Ergodic Theory. This, along with the introduction of entropy invariants (1959) through A.N. Kolmogorov and Y. Sinai, had, for me, a directly relevant connection to Information Theory that had already been widespread by an Uspehi (Advances in Physical Sciences) article (1956) by A.Y. Chintchine (1894–1995). This work, have been done in East Germany and translated into German, was immediately accessible because of the language. A crucial impulse for us however, came from the report, "Coding Theorems of Information Theory" (1961) by Jacob Wolfowitz (1910–1981). After Rudi finished his Ph.D., there was much contact with J. Wolfowitz and together they wrote many papers. Later, Rudi wrote a wonderful commemorative tribute to him. Because I had students like R. Ahlswede and V. Strassen, who I was only marginally ahead of in terms of research, I had the most exhilarating experience that a teacher can have: to be surpassed by their students and to be able to learn from them. After the meeting in Erlangen at the beginning of 2009, Rudi and I continued to have contact via telephone. During one of the last conversations (around 2010), I described to him my deliberations on the question of what a mathematician should do about the inevitable decline in performance – no matter how gradual it might be – and how one should position himself as a retired professor. I had decided then, starting around 1993, not to actively pursue research, but to turn to other areas of interest, which would naturally be on an amateur basis. When I asked his opinion of the matter, the answer came back immediately, with total resolve. His motto was

Die in your boots!

With his personality and temperament, it was only a matter of continuing to work as intensively and so long as one could. Rudi had a profusion of ideas and problems to solve. In the boots that he had grown into, he could have walked many more miles. You never forget a person like Rudi.

Comments by Alon Orlitsky

Rudi Ahlswede was truly a great information theorist. Not only did he make fundamental contributions to classical information theory, but he was also one of the first to explore the close connection between information theory and combinatorics. In addition, so many of his papers propose new problems, introduce new techniques, describe new results, and provide new insights.

To check how much I appreciated Rudi's research I resorted to a low-tech approach. Back in the old days there was an easy way to decide how much you liked someone's work: you went to your file cabinet and saw how many of their papers you had. When did that, I found a folder going from C to E, one from F to H, and then I to K, and so on - but when I looked at the A's there was one folder devoted to just "Ah" This folder had one paper by Al Aho, but the rest were by Rudi.

Of these papers, one of those I like most is "Coloring Hypergraphs - A new Approach to Multi-User Source Coding", which, I know, Ahlswede was very proud of. When you look at it, it's not exactly summer reading, unless you plan to spend the whole summer reading it. Rudi actually said that he wanted to write an elaborate paper but decided to keep it "short". In spite of the "brevity" of the paper - there are a lot of interesting and very useful results, and some of them I subsequently used.

Rudi himself used to joke (or not) that he thought that all results on combinatorial information theory were in this paper - just, that people didn't have the patience to find them. So, I wish that Rudi stayed longer with us and I wish that more of us had had the patience to read more of this and his other papers.

Author Index

A. Ahlswede et al. (eds.), *Combinatorial Methods and Models*,
Foundations in Signal Processing, Communications and Networking 13,
DOI 10.1007/978-3-319-53139-7

Subject Index

A. Ahlswede et al. (eds.), *Combinatorial Methods and Models*,
Foundations in Signal Processing, Communications and Networking 13,
DOI 10.1007/978-3-319-53139-7

MIX
Papier aus verantwortungsvollen Quellen
Paper from responsible sources
FSC® C105338

If you have any concerns about our products,
you can contact us on
ProductSafety@springernature.com

In case Publisher is established outside the EU,
the EU authorized representative is:
Springer Nature Customer Service Center GmbH
Europaplatz 3, 69115 Heidelberg, Germany

Printed by Libri Plureos GmbH
in Hamburg, Germany